STATISTICS FOR BUSINESS

STATISTICS FOR BUSINESS

A Course Unit Approach

PAUL TAYLOR

Principal Lecturer in Operational Research
The Polytechnic of the South Bank

DON DUNNING

Principal Lecturer in Statistics
Middlesex Polytechnic

POLYTECH PUBLISHERS LTD STOCKPORT

First Published 1977

ISBN 0 85505 019 5

Printed by:
Ashworths Print Services, Peel Mills, Chamber Hall Street, Bury, Lancs.

PREFACE

Who is this book for? It is designed for students preparing for examinations in statistics on courses in management, accountancy, business studies and economics. The majority of the exercises are *actual* examination questions from HNC, HND, BA and DMS courses as well as from the professional bodies ACCA, ICMA and IPM. We have borne in mind the needs of full-time students with little or no practical experience of statistics by incorporating practical exercises and orienting the text towards practical applications; and the needs of part-time students with limited class-contact by giving a fuller treatment of fundamental concepts and procedures than most other texts. We would like to think that we have written a genuine "student text": certainly we are indebted to the large number of students from a whole variety of courses who contributed to its development by their suggestions and constructive criticism.

Special features of the book. In place of chapters the book is made up of Course Units (CU) and Seminar Units (SU). These are more self-contained than conventional chapters and this should allow the student to tailor-make his own statistics course according to the availability of time and the particular syllabus to which he or she is working. In particular, starred (*) Course Units, sections and exercises are more advanced and may be omitted without seriously impairing understanding of the subject as a whole. This inevitably means that there is a certain amount of re-presentation throughout the course but it is hoped that the reader will find this no bad thing. The Seminar Units are devoted to the more practical aspects of the subject and are well suited to group discussion, particularly when supported by relevant newspaper articles.

An important feature of the course is its use of practical (P) exercises which can be carried out (at home, if necessary) using readily available materials such as playing cards, scissors, string, cloakroom tickets etc. These exercises are an integral part of the course – it is well established that practical work is the most effective way of appreciating statistical phenomena. Other exercises are labelled U, for those designed primarily to test understanding; D, those suited to discussion and E, examination questions.

Many text books give answers to some of their exercises: we have given *fully worked solutions* to many so that students studying alone can check their working step by step.

Guidance to students. We recommend the use of a pocket calculator for most statistical calculations (a model with memory and square root facilities is ideal). However, since calculators are not always permitted in examinations, we have also included hand calculation methods in the text.

Appendix 1 comprises Standard Data Sheets which are used in a variety of contexts throughout the course. Appendices 2–7 contain a short set of statistical tables presented in the most convenient form for the text. It is important, nonetheless, for the student to be familiar with the tables available for use in the examination – a set of these should be purchased and used throughout the course. Given a choice the authors recommend:

J. Murdoch & J. A. Barnes, *Statistical Tables*, Macmillan.

The solutions given to examination questions are for internally assessed courses. Students on ICMA and ACCA courses may obtain model answers to past examination questions from the professional bodies. Incidentally, the text meets the new (1976) ICMA regulations covering the statistics content of both Section B (Mathematics and Statistics) and Part I Quantitative Techniques. Students on IPM courses will find we have included a considerable number of examination questions set by the Institute under the New Scheme.

The reader will find that the Contents list is short but the Index is comprehensive, and hence is referred to the latter for chasing up any particular statistical topic.

Acknowledgements. We are grateful to the following professional bodies for their permission to reproduce past examination questions:

The Association of Certified Accountants
The Institute of Cost and Management Accountants
The Institute of Personnel Management

Appendices 5, 6 and 7 are printed by kind permission of the Biometrika Trustees.

Inevitably many of the ideas and some of the exercises in this book have come from our colleagues. We are pleased to acknowledge their help and assistance.

We are also indebted to the following people who have assisted us in a variety of ways: Anne Driver and Kay Iddiols (copy preparation), Penny Lester (editing) and Liz Willson (design of the cover). A major debt is due to our ever patient wives who have acted as *factota* throughout.

CONTENTS

COURSE UNITS

COURSE UNIT 1 – BASIC CONCEPTS AND DEFINITIONS

1.1 INTRODUCTION

Statistics is concerned with the collection, analysis, presentation and interpretation of numerical data. As with any scientific discipline a firm grasp of the subject can only be achieved by careful definition of concepts and terminology. In some other sciences the process of definition is made easier by coining new words (e.g. "entropy" in physics, "socialisation" in sociology) to take on a specialised and technical meaning. Statistics – in common with the other mathematical sciences – makes use of everyday words (e.g. "population", "probability", even the word "statistics" itself) and imbues them with a new, precise and technical meaning. So we have to be all the more careful with our definition of statistical terms and more particularly to be aware when we are using rather familiar words in their technically defined sense. This Course Unit introduces some of these terms and illustrates the contexts in which they are used.

1.2 SOME FUNDAMENTAL DEFINITIONS

To assist in the process of definition we next examine three situations in which a statistician might be involved.

Case A. A newly established off-the-peg tailoring firm manufactures, among other items, men's trousers. The company needs some numerical information on which to base their decisions on the "inside leg" sizes of their range. Of course, if all men had the same length legs there would be no problem: one would only need to take the inside leg measurement of one man and manufacture all trousers to that length. But men do vary in their leg lengths and so one single measurement would be of little use. What is needed is some data on leg lengths taken from a number of men and analysed in such a way as to reveal the pattern of variation from one individual to another. This analysis could then be used to assist the decision on leg sizing.

Case B. The government is continually reviewing its policy on Family Allowances (the scheme whereby families receive direct financial assistance depending upon the number of children in the family). In order to predict the cost to the nation of any alternative policy proposed, information is needed on the sizes of families throughout the country.

Case C. A company has just perfected a new form of gaming machine which externally looks much the same as any other "fruit" machine. What is new about the machine is the mechanism used to stop the revolving cylinders. The company is anxious to discover the properties of this machine – in particular the sequence of wins and losses – in order to check that the legal requirements of such machines are adhered to and that the machine proves attractive to a potential player.

In each of the above cases a statistician might be employed to collect, analyse and interpret relevant data. Let us now consider what is termed an **item of data** (or an **observation**) in each case. For comparison purposes we set this down in Exhibit 1.

EXHIBIT 1

Case	*Item of Data (Observation)*
A	An inside leg measurement of a man who buys off-the-peg trousers, e.g. "80 cm."
B	The number of children in a family, e.g. "3"
C	The result of "pulling the handle" once, e.g. "Win (cherry, cherry, apple)".

Of course, to perform any useful analysis, the statistician will have to collect many items of data, not just a single observation. In fact, the more the observations vary the greater the number of observations needed to gain an overall picture of the variability.

Now in each of the three cases the observations differ fundamentally in type. Put loosely, we say that the data in Case A is **continuous** and in Case B it is **discrete**: observations in Case C are known as **attribute** data. Such distinctions are important in that they often dictate which form of analysis is appropriate and, as such, the terms continuous, discrete and attribute need careful definition. We take them in turn.

In Case A the observations are measurements of length, and length itself is what is known as a **continuous variable**. That is to say, in principle, and given complete measurement accuracy, an observation could lie anywhere on a continuous scale of length and is not confined to certain fixed numerical values. Of course we cannot measure – nor would we normally wish to do so – with complete accuracy, so in practice observations of this type always involve some degree of approximation. We could, for instance, measure the leg lengths "to the nearest centimetre". As a result of this deliberate and to some extent unavoidable approximation such observations will indeed only have fixed values (in our example, only whole numbers of centimetres). So when we refer loosely to "continuous data" it should be taken to mean that the variable under study is continuous, not the data itself.

With a **discrete variable** the observations, by convention or by nature, *can only* have certain fixed numerical values. In Case B the "number of children in a family" is a discrete variable because the observations, obtained by counting (rather than by measurement, as in Case A) can only be whole numbers 0, 1, 2, 3, etc.[(1)] Notice that when recording such observations no approximations are involved, although we always have to admit the possibility of inaccuracy as the result of miscounting or incorrect recording.

Finally, with an **attribute variable** the observations do not, of themselves, have a numerical value at all, as in Case C. It is possible, however, to convert such observations into a numerical form. We could, for example, write a "1" for a win and a "0" for a loss in Case C – a convenient convention for some forms of analysis.

In each of the Cases A, B and C, we have been concerned with observations which reflect our interest in one aspect only of the situations outlined e.g. leg length for the Case A. An analysis of these observations is called a **univariate** ("one variable") analysis. Frequently in practice, however, we are forced to examine *jointly* two or more aspects of the same problem. As men differ in waist size as well as leg length, our tailoring concern will have to make joint decisions on the leg lengths and waist sizes of their range. To assist in these decisions **bivariate** data could be collected – each observation consisting of a leg measurement and a waist measurement of a man – and analysed. More complex problems still, demand **multivariate** data and analysis.

Note that the existence of data in the form of bivariate observations does not mean that all forms of analysis on this data will necessarily be correspondingly bivariate. Taking the example of bivariate observations of leg and waist measurement we could for example analyse the leg measurements (ignoring the waist measurements) and then, separately, analyse the waist measurements (ignoring the leg measurements). Here we would have performed two separate univariate analyses: a bivariate analysis would imply that the waist and leg measurements are analysed *jointly*. Most of this course is concerned with univariate statistical analysis although we shall be looking briefly at some bivariate analysis under the topics "Two Random Variables", "Correlation" and "Linear Regression", later in the course.

The set of all possible (or all conceivable) observations of a defined type is called the **population** of observations. Thus we can speak of the

---

(1)Examples of discrete variables which do not involve "counting" are shoe sizes (. . . 4, $4\frac{1}{2}$, 5, $5\frac{1}{2}$ . . .) and the prices of packaged goods in £p.

population of inside leg measurements of men who buy trousers off the peg – meaning a definitive list of leg measurements of all such men[(1)]. Similarly, for Case B, we might refer to the population of the number of children in families in the Country. Finally, for Case C, we have a population of observations for the results of pulling the fruit machine handle.

There is an important difference between the populations referred to in Cases A and B on the one hand and Case C on the other. For Cases A and B the populations are **finite**, that is to say there is a limit to the number of observations in the population – say 20m. for Case A and Case B. For Case C there is no limit (in principle) to the number of times the fruit machine handle can be pulled. That many of the observations will be the same does not concern us here – the fact is that these are separate observations and as such form part of the population. So for Case C the population is **infinite.**

Populations provide the main area of interest for statisticians. The tailoring company is interested in the leg sizes of *all* males (in a certain category), the Government in *all* families with children, the fruit machine manufacturer in the performance of their machines "in the long run". Unfortunately, it is normally the case that it is prohibitive, in terms of cost and time involved, to collect population data – for infinite populations it is clearly impossible – and so statisticians are forced to make do with less information than is provided by the population. The usual process is to take a **sample** of observations. A sample is simply a limited number of observations (technically, a *subset* of observations) taken from the population. (The manner in which these observations are *chosen* is a field of study in its own right but this does not concern us at the present time.) Thus the trouser manufacturer might measure the leg lengths of 1,000 "carefully chosen" males; the Government department might take a sample of family sizes from 5,000 "carefully chosen" families.

The relationship between the properties of a sample and the properties of the population from which that sample is drawn is a most important topic in the study of Statistics. It is important simply because information about the population is what we would ideally like to know, but sample data is normally all we are able to collect.

[(1)]Here we are using the word "population" in its strict sense to refer to the observations themselves. But it is also common-place to use the word in a looser *physical* sense, that is to refer to the items from which the observations are taken. The same is true of the word "sample".

1.3 POPULATION AND SAMPLE

To illustrate the relationship between the population and sample further we take a simple example. A company employs 100 representatives to sell its products to 8,000 (retail) customers. Each representative has a certain number of customers on whom he calls, as shown in Exhibit 2.

EXHIBIT 2: The Number of Customers Served by Each of 100 Representatives

90	77	112	82	78	94	88	60	67	85
53	100	62	86	83	78	88	19	89	109
99	99	83	104	87	81	97	86	83	83
107	61	70	98	56	105	70	92	90	95
95	67	86	79	99	41	63	65	104	87
51	74	120	39	75	85	112	78	109	82
65	62	46	119	78	72	88	107	80	55
105	86	68	67	71	77	49	72	91	78
46	97	55	102	69	92	66	99	103	94
71	59	83	91	60	74	62	75	52	67

N.B. The sum of these 100 figures is equal to the total number of customers, 8,000.

Bear in mind that although the population has been set down in full, this is purely for illustrative purposes and, as was stated earlier, a complete population of observations is rarely known. We have said that despite the problems of obtaining population data, it is a knowledge of that population that is desired. But this is perhaps too strong a statement. The management of the company employing the representatives will not in general be concerned with the *details* of such a population but would require the properties summarised in some convenient manner. There are of course a large number of summary figures that could be calculated with varying degrees of usefulness. We consider three such summary figures.

(a) Average number of customers per representative (μ) = 80.0 (i.e. the 8,000 customers divided by the 100 representatives).
(b) Proportion of representatives with 100 or more customers (i.e. 15 out of 100) (π) = 0.15
(c) "Range" of the observations (i.e. the largest observation, 120, minus the smallest, 19) (Ω) = 101

Any such summary figure relating to a population (and to a population, alone) is termed a **population parameter** or more shortly, simply a

parameter. It is customary to use Greek[1] symbols for parameters – the ones used in the above example are fairly standard for the summary figures given. Indeed, if we wish to refer to a parameter in general, without specifying which summary figure is being considered, we use the Greek letter θ.

There are two matters of fundamental importance relating to population parameters. The first is that when population data is not obtainable the population parameters cannot be evaluated directly: the best we can do is to *estimate*, not calculate, their values. Secondly, for a given population, parameters are fixed quantities, i.e. constants.

Having stated that it is normal practice to select a sample in order to find out about a population, how is this carried out? Of the many different *procedures* available for sampling, the study of elementary Statistics centres around one particular procedure called **simple random sampling**. This is a method where the element of human choice is eliminated, as far as possible, from the selection process itself. Without going into the technicalities of how this is achieved in practice – this will be covered in CU15 – the idea will be illustrated by the familiar "names out of a hat" procedure outlined below. Suppose, using our example, we write the name of each representative on a separate slip of paper, fold the slips, place them in a suitable container (the "hat") and thoroughly mix them up. To take a simple random sample of ten items from the population we would first take out ten slips, one at a time, making sure to mix up the slips after each selection. The actual sample (i.e. the number of customers served by each of these representatives) can then be determined by the records held by the company.

This description of a crude random sampling scheme has omitted one important feature – whether or not each slip is replaced in the "hat" before the next one is drawn. In the case where the slips are replaced we call this **sampling with replacement**; if the slips are not replaced we are **sampling without replacement**. The distinction between these modes of random sampling is only important when (as in the case under discussion) the number of items in the sample is a sizeable proportion of the number of items in the population.

Suppose, then, a random sample of 10 items is taken without replacement from the population of customers per representative and the result is shown by the ringed items in Exhibit 3. Remember that the items not ringed would not be known by a statistician following the above sampling procedure.

---

(1) The Greek letters used in this Course Unit are: μ, pronounced "mew"; π, pronounced "pie"; Ω, pronounced "omega", with the stress on the "o"; θ, pronounced "theeta".

EXHIBIT 3

90	(77)	(112)	82	78	94	88	(60)	67	85
53	100	62	86	83	(78)	88	(19)	89	109
99	99	83	104	87	81	97	86	(83)	83
107	61	70	98	56	105	70	92	90	(95)
95	67	(86)	(79)	99	41	63	65	104	87
51	74	120	39	75	85	112	78	109	82
65	62	46	119	78	72	88	107	80	55
105	86	68	67	71	77	49	72	91	78
46	97	55	102	69	92	66	99	103	94
71	59	83	(91)	60	74	62	75	52	67

N.B. The sum of the ringed figures is 780.

Now, how can this *sample* data be processed in order to give information about the *population*? We can summarise the sample data in exactly the same manner as for the population, that is by treating the sample as if *it were* the the population. The summary figures we then obtain are called **sample statistics** (or simply **statistics**) to underline the fact that they have been obtained from the sample rather than from the population.

(a) Average no. of customers per representative (i.e. 780/10), ($\bar{X}$) = 78.0
(b) Proportion of representatives with 100 or more customers (i.e. 1 out of 10) (P) = 0.10
(c) "Range" of the observations (i.e. largest observation, 112, minus smallest observation, 19) (W) = 93

These sample statistics can be used as *estimates* of their respective population parameters. You can see by comparing them with the population parameters if they are good as estimates in this particular case.

Using this terminology, we can say that the sample statistic[1] $\bar{X}$ estimates the population parameter, μ. Note that it is the convention to use Roman letters to denote sample statistics, just as Greek letters are reserved for parameters. It is important to maintain this convention; otherwise, if we refer to "the average number of customers per representative", it is not clear whether this is a sample statistic or a population parameter. The algebraic convention makes the meaning unambiguous.

---

[1]The notation X (pronounced X-bar) is used to represent the sample average rather than simply X, in order to distinguish between the particular use and other uses to which the symbol X may be put.

An important distinction between a parameter and a statistic is that although the former is fixed (for a given population), the latter is variable (i.e. depends upon the sample taken). This is illustrated for the "average number of customers per representative", in Exhibit 4 below, where the horizontal line represents a measurement scale.

EXHIBIT 4: (Relationship between μ and $\bar{X}$)

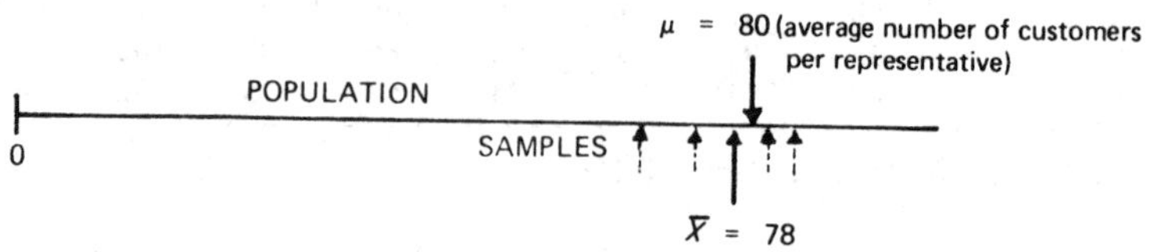

The solid "sample" arrow is the result we obtained ($\bar{X} = 78$) from our sample; the dotted arrows are other possible results obtained from different samples taken from the same population.

1.4 IDEAS OF ESTIMATION

As the main use to which sample statistics are put is in estimating the values of corresponding population parameters, it is natural to enquire how accurate these estimates are. This topic will be examined in greater depth in CU15, but for the present it is instructive to introduce a few terms used in the context of the accuracy of estimation. Our purpose is to illustrate that different types of statistics (averages, proportions, ranges or whatever), corresponding to different ways of summarising sample data, may behave differently in their estimation properties.

Consider, quite generally, a population parameter θ which is estimated by a sample statistic t. Suppose we were to take a large number of equal sized simple random samples from the same population and calculate a value of t for each. If it is true that t shows no overall tendency either to over- or underestimate θ, then we say that the statistic t is an **unbiased estimator** of θ. In such a case we might obtain results rather like those shown in Exhibit 5. If, on the other hand, t tends persistently to over- or underestimate θ we say that t is a **biased estimator** (Exhibit 6).

Finally, in Exhibit 7, we show some typical results from an unbiased statistic, but one where the variation in the value of t from one sample to another is (in relative terms) much greater than in Exhibit 5. We say that the statistic in Exhibit 5 is more **efficient** than in Exhibit 7.

EXHIBIT 5: Unbiased Estimation

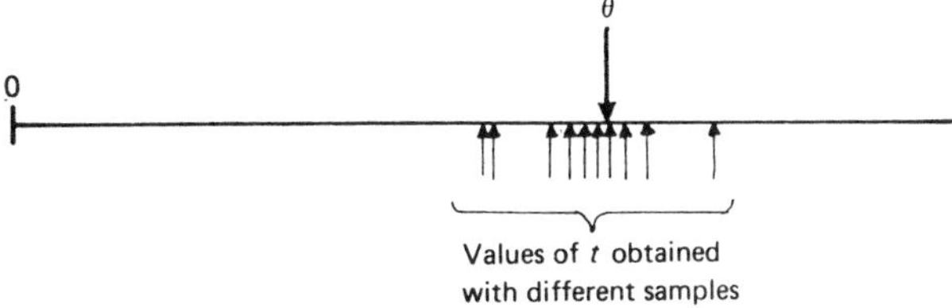

EXHIBIT 6: Biased Estimation

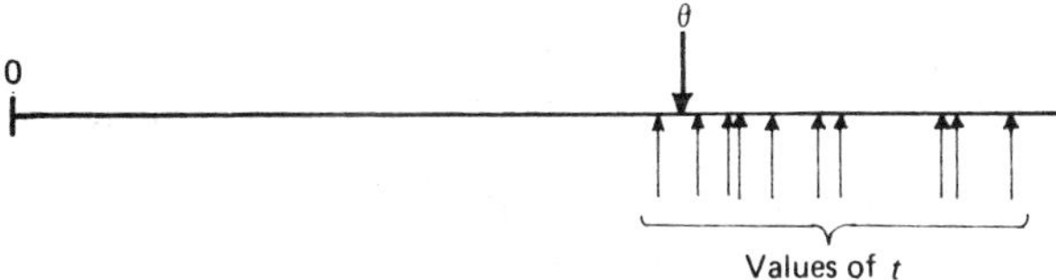

EXHIBIT 7: Inefficient Estimation

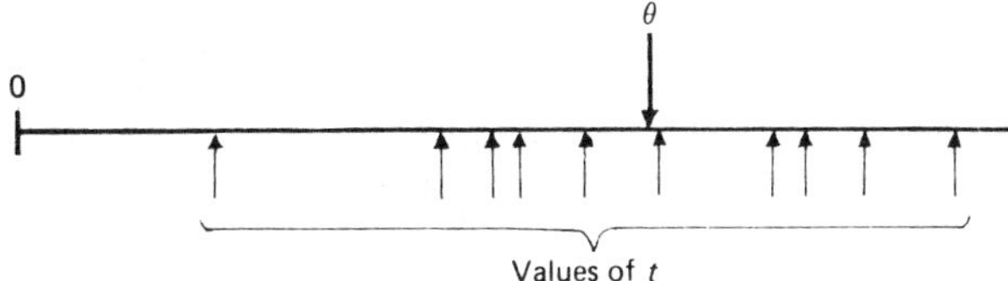

The above treatment is necessarily, at this stage, rather superficial. We must take care not to jump to the conclusion that a statistic is biased if, for example, in seven samples out of ten its value is greater than the corresponding population parameter (just as we ought not to conclude that a coin is biased if it shows seven heads in ten tosses). Our suspicions may be aroused that the statistic is biased in such a case, but they can only be confirmed if the effect persists over a much larger number of samples.

1.5 STATISTICAL REASONING

The main line of argument in this Course Unit can be summarised diagrammatically as in Exhibit 8.

EXHIBIT 8

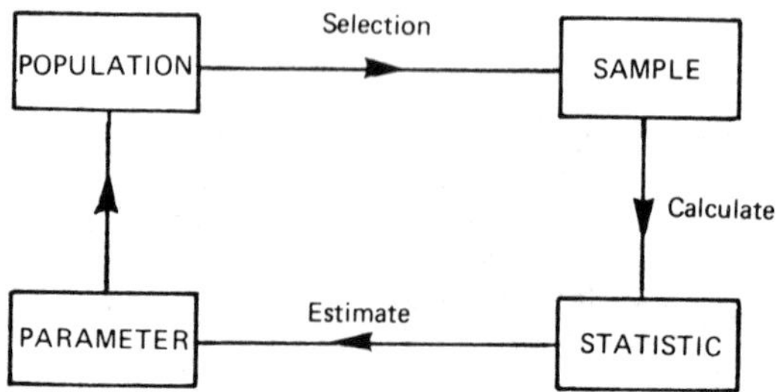

Having defined a population of interest, a sample is selected. From this sample various statistics can be calculated. These in turn enable us to estimate the values of the corresponding population parameters which summarise the properties of the population. This whole process is an example of **inductive** reasoning (arguing from the specific to the general).

But this is not the only line of argument used by statisticians. Sometimes we will know the parameters of a population with certainty (on theoretical grounds provided certain assumptions hold). From these we can predict the properties of samples taken from the population. This is a **deductive** reasoning process (arguing from the general to the specific).

Statistics as it is practised makes use of both inductive and deductive lines of reasoning, often in the context of one single problem.

READING

A very readable chapter covering the same ground as this CU may be found in:
R. Goodman, *Statistics* (Teach Yourself Series) EUP pp 11–17

EXERCISES

1(P) Throughout this Course Unit a number of terms with a specialised meaning in Statistics have been introduced. On their first mention they have been written in **bold type**. Begin to compile a glossary of statistical terms by writing down a definition of each. Leave room for improved or more formal definitions of the same terms as the course progresses.

2(P) Take 100 near identical slips of paper and write consecutive integers on each one (1, 2, 3, . . ., 100)[1]. Place them in a container and use this for taking a simple random sample without replacement of 10 items from the population tabulated in Section 1.3 of this Course Unit[2]. From this sample calculate the statistics $\bar{X}$, P and W as defined in the text. Repeat the sampling and calculation 12 times (i.e. a total of 12 samples, each of 10 items, remembering to replace the slips of paper after each complete sample has been taken). Tabulate your results as follows.

	Sample Data	$\bar{X}$	P	W
1				
2				
3				
.				
.				
.				
12				

Define a new statistic which might be of interest to the company employing the representative and calculate its value for each of your samples. Also calculate the corresponding population parameter from the population. For each of your four statistics draw up a diagram similar to Exhibit 4 in the text. Comment in general terms on how accurately the statistics estimate their respective parameters.

3(U) *Statistician, observe thyself!*
Write down a list of characteristics (at least 20) which apply to a person's physical or personal condition. Give the observation in your own case and also state whether the variable associated with each characteristic is discrete, continuous or attribute. For each continuous variable state the extent of the approximation used for the observation.

---

[1]Or use cloakroom tickets.

[2]Proceed in the following manner. First label the population data from 1 to 100 so that you can rapidly identify, say, the 67th item of data in the population. As each slip is drawn from the container read the number on it and then read off the value from the population table with that label. This process is entirely equivalent to, but more economical than, writing each population observation onto a separate slip of paper. (Our slips, for instance, can be used again with another population.)

Example:

Characteristic	*Observation*	*Type of Variable*
Number of brothers	2	Discrete
Weight	140 lbs (to nearest 5 lbs)	Continuous
Colour of hair	Brown	Attribute

4(U) Each of the following statements describes a sample. Describe the population as fully as possible in each case and also state the *source(s)* of variability in the observations comprising the sample.

(i) The electrical properties of a random sample of four electric motors taken from a consignment of 100.
(ii) The weights of five tins of cat food taken from a conveyor belt before labelling.
(iii) A sequence of numbers obtained by spinning a roulette wheel ten times.
(iv) A scientist's ten measurements of the speed of light using the same experimental technique.
(v) The weights of the first ten cabbages picked on a particular farm at harvest time.
(vi) The five directors' estimates of the sales revenue for a company in the next financial year.

SOLUTIONS TO EVEN NUMBERED EXERCISES

We shall not usually give solutions to the practical (P) exercises because there may be no unique solution and each person carrying out such an exercise may obtain different results. (With sampling exercises there is considerable benefit in comparing the results obtained by different individuals.) An exception is made in this case in order to give an indication of a suitable approach and layout of the answer.

2. The results are tabulated on facing page.

As our new statistic we define P' as the proportion of items of value 50 or less, while π', the corresponding population parameter, has the value 0.06 (i.e. 6 representatives out the 100 each served 50 or fewer customers).

Sample Number	Sample Data (10 Items)										$\bar{X}$	P	W	P' (see text)
1	72	91	71	55	78	100	66	102	83	71	78.9	0.20	47	0.00
2	75	82	109	56	78	94	51	85	86	99	81.5	0.10	58	0.00
3	79	109	74	105	62	61	49	99	92	82	81.2	0.20	60	0.10
4	62	73	103	86	95	105	51	66	62	86	78.9	0.20	54	0.00
5	81	112	72	91	89	99	55	41	51	88	77.9	0.10	71	0.10
6	86	55	41	99	49	86	104	119	78	41	75.8	0.20	78	0.30
7	83	94	65	62	62	78	104	95	92	65	80.0	0.10	42	0.00
8	49	46	104	60	82	81	97	112	70	90	79.1	0.20	66	0.20
9	80	105	59	62	86	103	60	82	78	99	81.4	0.20	46	0.00
10	82	95	94	51	70	112	86	98	94	87	86.9	0.10	61	0.00
11	66	86	107	82	104	119	95	78	105	88	93.0	0.40	53	0.00
12	99	98	109	56	78	41	51	88	86	52	75.8	0.10	68	0.10

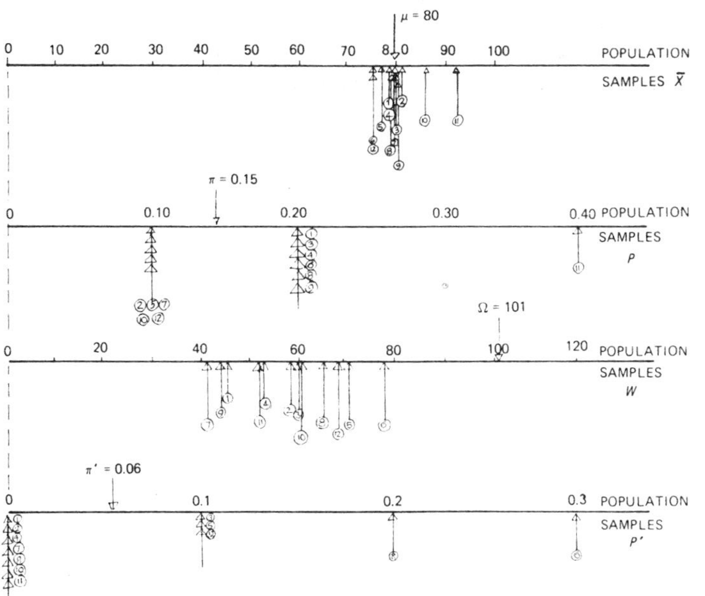

4. A description of a sample does not define uniquely the population. There are a number of differently defined populations which could all lead to the same sample – it all depends on how the items are selected. So the following are not necessarily the only solutions.

(i) Population: the electrical properties of all 100 motors.
Source of variability: manufacturing variations (human errors and material variations).

(ii) Population: the weights of all tins of cat food produced by the same filling machine under "identical" conditions.
Source of variability: imprecision in the filling process.

(iii) Population: the sequence of numbers obtained by spinning the roulette wheel an infinite number of times.
Source of variability: "chance".

(iv) Population: all the measurements the same scientist *could* obtain with the same apparatus; or, all the measurements *all* scientists could obtain using the same experimental technique.
Source of variability: experimental error.

(v) Population: the weights of all cabbages of the same type on the farm when picked.
Source of variability: variation in the seeds, fertility of the land and treatment; also the reasons *why* those ten cabbages were picked first (marketable size, convenience?).

(vi) Population: estimates by all similarly informed people at the same time.
Source of variability: variations in the detailed knowledge of the market, temperament (degree of optimism).

COURSE UNIT 2 – FREQUENCY DISTRIBUTIONS, HISTOGRAMS AND OGIVES

2.1 INTRODUCTION

In this Course Unit we consider how univariate data can be summarised in both tabular and pictorial form. Although we shall be thinking primarily of sample data, because that is what is usually collected, the principles apply to population data too.

There are a number of reasons why it is desirable to summarise data. The most important of these are listed below.

To assist in the interpretation of the data. Data is often included in reports to support particular arguments. "Raw" data is not very useful for this purpose as it is too bulky and requires more time and concentration to extract the essential features than most report readers are prepared to give. If proof of this statement is required, examine the data in Series A and B (of the Standard Data Sheets(1)) and attempt to describe the essential differences between these two sets of data.

To reduce computational effort. Sample data is normally used to calculate sample statistics of one kind or another. With large samples this places a considerable computational burden on the analyst – with the attendant likelihood of error – unless a computer is to hand. With the data in a summarised form the process of calculating sample statistics is made easier.

To "smooth out" sampling fluctuations. The basic reason for collecting sample data is of course to obtain partial information about the population. Sometimes a sample, despite the care with which it may have been selected, displays properties which are not truly representative of the population from which it was drawn. (We should however be very unlucky to find that our sample exhibited radically different properties.) Summarising the data will normally help to draw attention away from such irregularities. But in doing so we must be alive to the possibility of a real and important property of the population being obscured by the summarising procedure.

---

(1)See Appendix 1.

2.2 FREQUENCY DISTRIBUTIONS FOR DISCRETE DATA

One of the most useful ways of summarising univariate data is by the construction of a **frequency distribution** – this is simply a table expressing how many of the observations have particular values or lie within certain ranges of values. The construction of frequency distributions in the case of simple[(1)] discrete data is straightforward: unfortunately there are a number of additional problems which arise with continuous data, and so in order not to complicate the matter initially, our first example below deals with a simple discrete case.

A company has a large fleet of 10 wheeled trucks. In one week 50 of these trucks returned to the servicing bay and were subject to on-the-spot tyre checks. The number of tyres on each truck having serious faults was recorded and the results are given in Exhibit 1.

EXHIBIT 1: Number of Faulty Tyres per Truck

0	0	1	1	3
0	0	0	1	1
0	2	4	0	2
2	0	1	1	1
0	0	3	0	0
3	0	1	2	0
2	1	1	1	1
0	1	0	0	0
0	0	0	1	4
1	1	4	0	3

A quick glance at Exhibit 1 shows that the smallest observation is 0 and the largest 4. Thus our frequency distribution will have 5 rows (corresponding to the values 0, 1, 2, 3, 4). There are two ways in which the **frequencies** of each of these values can be determined. The first is simply to count the number of "noughts" in the data, the number of "ones" and so on. The second method, and this is to be preferred for reasons which will become apparent later, is to work systematically through the data, reading each observation in turn and placing a "tally" mark in the corresponding row in a prepared table. For convenience in final counting, tally marks are usually grouped in "fives". Carrying this out on the above data leads to the frequency distribution as shown in Exhibit 2.

---

(1)By "simple" we mean where the largest observation is a reasonably small integer (say 10). Where this is not the case, we have to resort to classifying in much the same way as for continuous data.

EXHIBIT 2

Number of faulty tyres per truck	*Tally marks*	*Frequency*
0	卌 卌 卌 卌 II	22
1	卌 卌 卌 I	16
2	卌	5
3	IIII	4
4	III	3
	Total	50

If this frequency distribution were to be used in a report the tally marks, having served their purpose, would be omitted and the technical term "frequency" would perhaps be replaced by some phrase such as "Number of trucks with stated number of faulty tyres".

In place of the frequencies, other related measures of the "distribution" are sometimes useful. The most common of these are illustrated in Exhibit 3.

EXHIBIT 3

Number of faulty tyres per truck	*Frequency*	*Relative frequency*	*Cumulative frequency*
0	22	0.44	22 (= 22)
1	16	0.32	38 (= 22 + 16)
2	5	0.10	43 (= 22 + 16 + 5)
3	4	0.08	47 (= 22 + 16 + 5 + 4)
4	3	0.06	50 (= 22 + 16 + 5 + 4 + 5)
	50	1.00	

The relative frequencies are simply the frequencies divided by the **sample size**[1]. Sometimes relative frequencies are expressed in percentage terms, e.g. 32% of the trucks had one faulty tyre.

The cumulative frequencies show the number of observations (i.e. sum of the frequencies) that are less than or equal to specified values. For example the cumulative frequency 43 in Exhibit 3 corresponding to 2 faulty tyres indicates there were 43 observations (22 + 16 + 5) of 2 faulty tyres or less. Sometimes cumulative frequencies are percentaged, too.

(1)The sample size, the total number of observations, is itself equal to the sum of the frequencies.

2.3 FREQUENCY DISTRIBUTIONS FOR CONTINUOUS DATA

To develop our ideas let us take a specific example – that of a company manufacturing gas central heating systems. The company wanted some information on which to base advertising claims on the running costs of a model which had been on the market for a year. The company took a random sample of gas consumption values in a particular quarter from 100 households each using this model (and not using gas for any other purpose). As gas consumption is a continuous variable, the company had to decide in advance to what degree of accuracy the data should be collected: they decided that the consumption values should be recorded *to the nearest therm*. The results of the survey are given in Exhibit 4 (Series A of the Standard Data Sheets).

EXHIBIT 4

55	82	83	109	78	87	95	94	85	67
80	109	83	89	91	104	90	103	67	52
107	78	86	19	72	66	92	99	60	75
88	112	97	88	49	62	70	66	88	62
72	85	81	78	77	41	105	92	94	74
78	75	87	83	71	99	56	69	78	60
119	39	104	86	67	79	98	102	82	91
46	120	73	62	68	86	70	55	112	83
62	74	99	100	86	67	61	97	77	59
65	51	99	53	105	95	107	46	90	71

We note that the lowest observation is 19 (therms) and the highest is 120 (therms).

Although we *could* form a frequency distribution by counting the number of observations at each consumption level (there are, for example, 4 observations of 83 therms), this would result in a frequency distribution with over a hundred rows – hardly a summary of the data! Consequently, it is normal practice to count the number of observations which fall within chosen *ranges* of consumption values. Each distinct range of values is called a **class** of values (sometimes a **group**). So we have the following problems: How many classes should be used? How should the classes be arranged? How do we make allowance for the fact that the observations are inevitably approximations?

It is considered reasonable to have between 5 and 15 classes – the former figure being more appropriate for small sets of data (say 25 observations) and the latter for large sets (say 1,000 observations). The point is that with too few classes, valuable detail may be obscured; but with too many, sample irregularities are likely to become undesirably prominent, giving a misleading impression of the population. In the example we use classes 10–19, 20–29, etc., up to 120–129 giving 12 classes, although the bulk of the data will fit into 8 classes. Constructing the frequency distribution on this basis we have the result shown in Exhibit 5.

EXHIBIT 5: Construction of a Frequency Distribution for Continuous Data

Nominal Class Limits			*Frequency*
10– 19	I		1
20– 29			0
30– 39	I		1
40– 49	IIII		4
50– 59	𝍸 II		7
60– 69	𝍸 𝍸 𝍸 I		16
70– 79	𝍸 𝍸 𝍸 IIII		19
80– 89	𝍸 𝍸 𝍸 𝍸		20
90– 99	𝍸 𝍸 𝍸 II		17
100–109	𝍸 𝍸 I		11
110–119	III		3
120–129	I		1
		Total	100

The classes in Exhibit 5 have been specified by **nominal class limits**. The word "nominal" used in this context means that these class limits only serve to determine in which class each observation belongs – they may have to be modified for other purposes, such as for the calculation of sample statistics. Note that the nominal class limits are specified to the same number of significant figures as the data itself, and that the classes are non-overlapping. This ensures a simple and unambiguous classifying procedure. (If the data had been recorded to the nearest 0.1 therms then the classes would read 10.0–19.9, 20.0–29.9 etc.).

There are two other features of the classification scheme used above which are of interest. Firstly all the classes have the same width and there are no "open-ended" classes (e.g. "below 40" or "above 120"). This is normally to be preferred as it simplifies any later calculations, although it has to be admitted that there are occasions when it is desirable to use unequal classes and/or open-ended classes. (See Exercise 4.) The second point is that we have used classes 10–19, 20–29 etc. in preference to, say, 12–21, 22–31 etc. We have chosen to do this solely on grounds of convenience – it is easier to count the observations within classes specified in the former way than in the latter way. Sometimes, however, there are other considerations. For the purpose of calculating sample statistics we shall later make the assumption that the class "mid-points" (e.g. 64.5 for a class of 60–69) are representative of the observations which fall within each class. We often make this assumption with impunity; but there are cases where, as the result of artificial "clustering" of observations (e.g. in the prices of packaged food), we must take care in the setting up of class limits to ensure that this assumption is as reasonable as possible.

If in our example of gas consumption we had been dealing with a discrete variable, then the frequency distribution obtained would be adequate for any further analysis. But because we are dealing with a continuous variable where the observations are recorded *to the nearest therm* we have to make an allowance for the fact that the observations are (rounded) approximations. Consider, for example, one of the classes, 60–69, specified by the nominal class limits. What gas consumption values, taken to a very high level of precision (i.e. readings to several decimal places), would fit into such a class? 59.499 would *not*, as it would be *recorded* as 59, whereas 59.501 would be included within the class. Similarly 69.499 would be in the class, 69.501 would not. Clearly all values within the range 59.5–69.5 would be recorded (to the nearest therm) as being within the nominal class 60–69. When dealing with continuous data it is conventional to present a frequency distribution using these **actual class limits** (e.g. 59.5–69.5) rather than the nominal class limits. (With discrete data the distinction between these two ways of specifying classes does not exist as discrete variables are normally recorded without approximation.)

There are two terms in common use related to frequency distributions specified in this way:

- **class width**[(1)] = Upper Actual Class Limit − Lower Actual Class Limit.
  e.g. = 69.5−59.5 = 10, for the class 59.5 to 69.5.

In our example a class width of 10 therms is used throughout.

- **class mid-point** = $\frac{1}{2}$(Upper Actual Class Limit + Lower Actual Class Limit).
  e.g. = $\frac{1}{2}(59.5 + 69.5) = 64.5$, for the class 59.5 to 69.5.

[(1)]This definition is appropriate for continuous data only. For discrete data (of the "counting" type) a suitable definition of class width is the "number of integer values in the class". Consider the following classification of discrete data:

Number of Errors in an Invoice	*Frequency*	*Class Width*
0	150	1
1	20	1
2–4	10	3
5–8	4	4

Finally, in Exhibit 6, we present our gas consumption data in the conventional manner (i.e. using *actual* class limits) and, for later reference, calculate the relative frequencies and cumulative frequencies. Note that the cumulative frequency of a class states how many observations are less than the *upper* actual class limit.

EXHIBIT 6

Class Limits (actual)	*Class Mid-Point*	*Frequency*	*Relative Frequency*	*Cumulative Frequency*
9.5– 19.5	14.5	1	0.01	1
19.5– 29.5	24.5	0	0.00	1
29.5– 39.5	34.5	1	0.01	2
39.5– 49.5	44.5	4	0.04	6
49.5– 59.5	54.5	7	0.07	13
59.5– 69.5	64.5	16	0.16	29
69.5– 79.5	74.5	19	0.19	48
79.5– 89.5	84.5	20	0.20	68
89.5– 99.5	94.5	17	0.17	85
99.5–109.5	104.5	11	0.11	96
109.5–119.5	114.5	3	0.03	99
119.5–129.5	124.5	1	0.01	100
	Totals	100	1.00	

Note that presented in this fashion there are no "gaps" in the consumption values (compare with nominal class limits). Some people, on being shown a frequency distribution such as this, ask the question, "What class would a consumption value of, say, exactly 39.50 fit into?". The answer depends on how the reading of 39.50 was recorded – it would either be recorded as 39 (rounding down) or 40 (rounding up). In the former case it would have gone into 29.5–39.5, in the latter case 39.5–49.5.

2.4 HISTOGRAMS

It is often helpful in the interpretation of data to depict tabular results in the form of a graph or diagram. The pictorial representation of a frequency distribution is called a **histogram** (see Exhibits 8 and 9). For simple discrete data the histogram is formed by displaying vertical lines[1], with lengths proportional to the frequencies, on a graph (see Exhibit 8).

For continuous data the histogram takes the form of bars (reminding us that the data has been put into classes) having widths proportional to the class widths. An important feature of such histograms is that the *area* of the bars should be proportional to the frequencies depicted. Where the data has been put into classes of equal width throughout this is easily achieved: the bars are drawn with *heights* proportional to the frequencies, as shown for our example in Exhibit 9. But where the class widths are not all equal we have to be more careful. Consider the following simple frequency distributions constructed from the *same* set of data.

(a) *Class Limits*	*Frequency*	(b) *Class Limits*	*Frequency*
0–10	20	0–10	20
10–20	6	10–30	10
20–30	4		30
	30		

Note that three equal width classes are used in (a), but only two classes of unequal width in (b). One would expect the histograms drawn from these two frequency distributions to look *roughly* the same.
Exhibit 7 shows that this is only the case when the *area* method is used (where the frequency of the *double*-width class, 10, has been divided by 2 before drawing). For another example see Exercise 4.

EXHIBIT 7: Procedure for Drawing Histograms with Unequal Class Widths

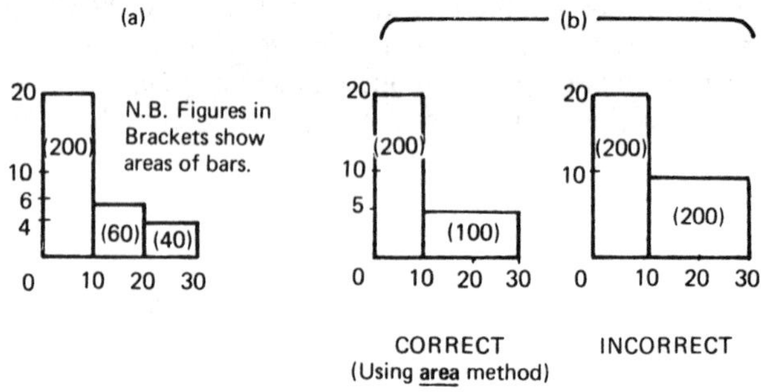

(1)In some texts (and reports) histograms for simple discrete data are drawn with bars rather than lines, similar to the manner in which we have drawn the histogram (Exhibit 9) for *continuous* data. We prefer to accentuate the important difference between discrete and continuous data by drawing their histograms differently.

EXHIBIT 8: Histogram for Discrete Data

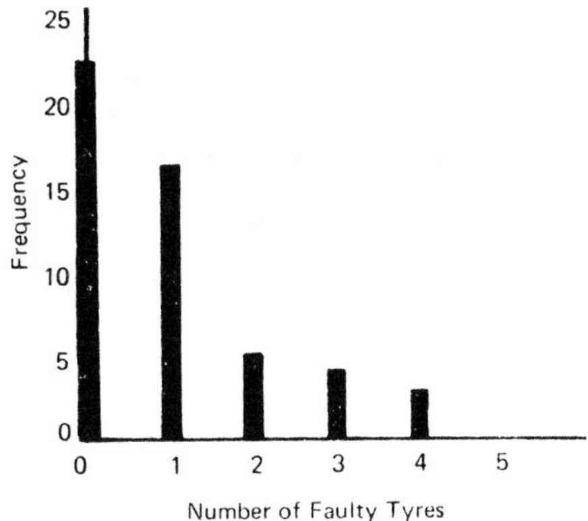

EXHIBIT 9: Histogram for Continuous Data

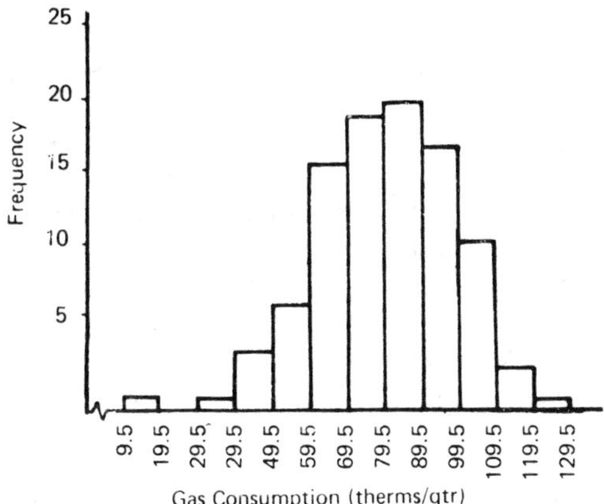

The shape of the histogram itself is of some interest. The histogram of our gas consumption data, shown in Exhibit 9 might be said to be (roughly) **symmetrical** and what is termed **bell-shaped**. By symmetrical we mean that the histogram does not tend to lean one way or the other. When histograms have a longer righthand "tail" the distribution is called **positively skewed** (see Exhibit 10); if the left-hand "tail" is longer the distribution is **negatively skewed** (Exhibit 11). In practice most economic data (e.g. incomes of employees in a particular industry) is positively skewed, some age-related data (e.g. age of false-teeth wearers) is negatively skewed.

EXHIBIT 10 **EXHIBIT 11**

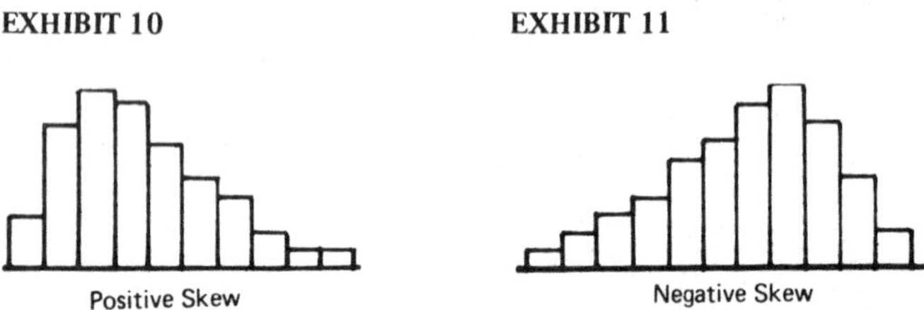

Positive Skew

Negative Skew

As an example of an extremely positively skewed distribution, look at Exhibit 8 from our faulty tyre data. In this case the skewness is so pronounced that the distribution is given the special name of a **reversed J-shape** distribution.

Another pictorial representation is the **frequency polygon**. This is simply another way of drawing the frequency distribution. Points are placed on a graph at the mid-points of the top of where the histogram bars would appear and these points are joined up with straight lines. (Exhibit 12 shows a frequency polygon superimposed on a histogram to illustrate their relationship.) If, instead of joining these points up with straight lines, we attempt to draw a smooth curve through them (see Exhibit 13) we are in fact estimating the form of the *population* distribution (of continuous data) instead of simply presenting the factual sample results. When in later course units we represent a distribution by a smooth curve, it is to indicate that the population is being considered. The justification for this is a belief that if a frequency polygon were constructed from population data it would have a smooth shape. Indeed, we shall find later that there exist a number of distributions derived from theoretical considerations – and thus which refer to populations rather than samples – all of which plot as smooth curves.

EXHIBIT 12 **EXHIBIT 13**

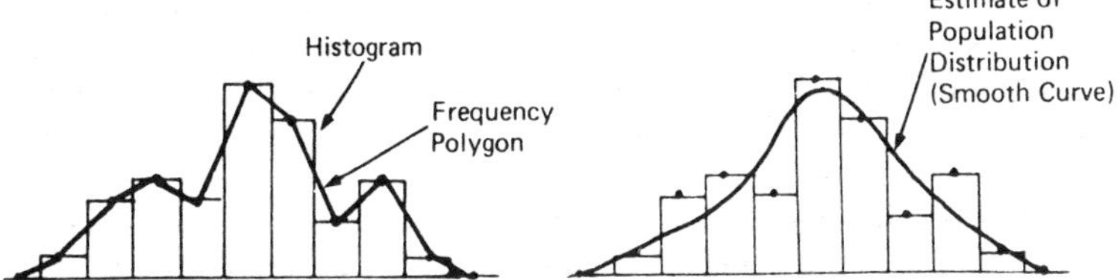

2.5 OGIVES

Another frequently used pictorial representation of a frequency distribution is the **ogive**. This is essentially a plot of the *cumulative* frequency (and for this reason is sometimes called a cumulative frequency graph).

Ogives for simple discrete distributions are formed by plotting each observation value against its cumulative frequency and joining the points up in a stepwise manner (to emphasize the discreteness of the distribution). Exhibit 14 shows the ogive for the faulty tyre data. Look carefully to see how the ogive is drawn. As a test to see if the construction is correct we might ask, in this case, the slightly absurd question, "How many observations are less than 1.2?" for instance. Reading off the graph we see this gives a cumulative frequency of 38, which is correct.

EXHIBIT 14: Ogive for Discrete Data

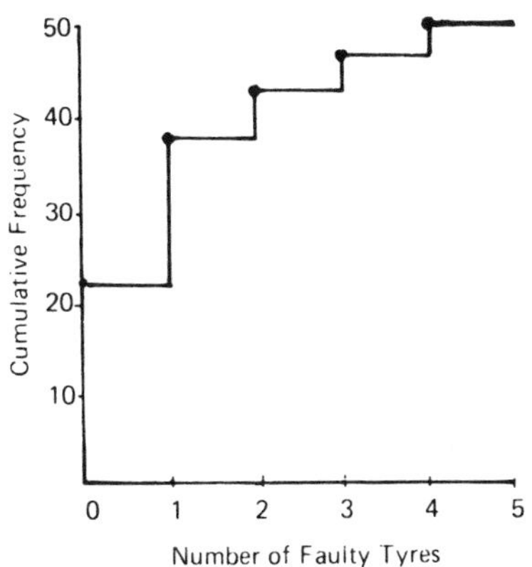

With a continuous distribution we plot the *upper actual class limit* against the cumulative frequency, for each class, and join the points so formed with straight lines. Exhibit 15 shows the result for the gas consumption data. Note that we have had to "invent" a low-valued class with upper actual class limit of 9.5 and cumulative frequency 0 to start off the construction.

The ogive shows graphically for the sample the number of observations having values *less than* a specified amount. By using an ogive we can readily estimate the number of observations lying within any chosen range of values as Exhibit 16 demonstrates.

If instead of using cumulative frequencies we use cumulative percentages, and if instead of joining the points on the ogive with straight lines we attempt to draw a smooth curve through the points, we can use the ogive so formed to estimate percentages of the *population* less than certain values or between certain values.

EXHIBIT 15: Ogive for Continuous Data

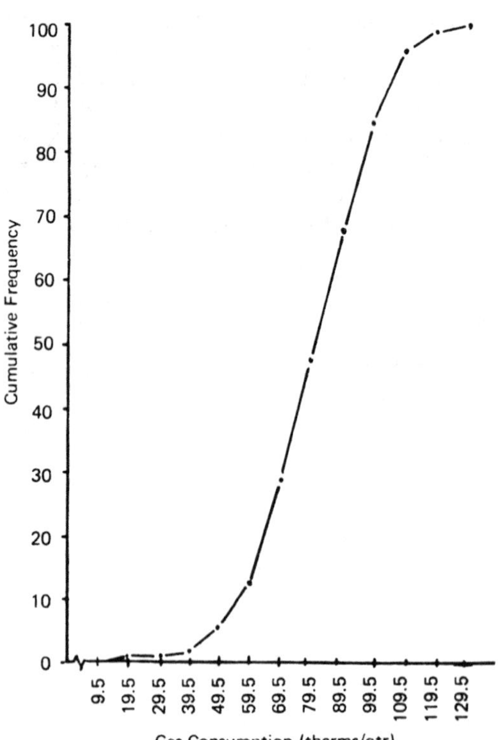

EXHIBIT 16: Use of the Ogive for Estimation

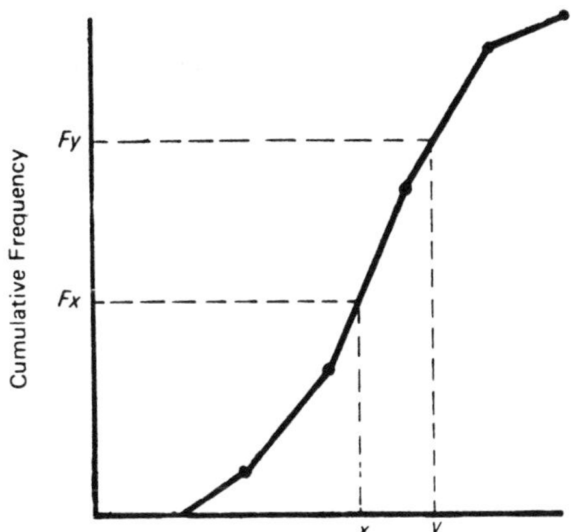

F_x estimates the number of observations less than x
F_y estimates the number of observations less than y
Therefore $(F_y - F_x)$ estimates the number of observations lying between x and y.

READING

K.A. Yeomans, *Introducing Statistics*, Statistics for the Social Scientist; Volume One, Penguin. Chapter 2.

EXERCISES

1(P) In this Course Unit we summarised the data of Series A (in the Standard Data sheets). The interpretation put on this data was that it was a sample of quarterly gas consumption (to the nearest therm) of certain households. Suppose Series B is similarly a sample of quarterly gas consumption of other households employing a different kind of gas central heating system. Once again assume that the readings are taken to the nearest therm. Construct a frequency distribution for Series B and draw up its histogram and ogive. Present the results in such a way as to facilitate comparison between Series A and Series B. Explain to a layman, as it were, the differences between the consumption patterns as revealed by your analysis.

2(U) In a survey of the ages of all 5,000 employees in a particular industry, the data could be recorded (in years) in one or other of the following ways:

A. In the normal manner – that is, recording whole years of age only. Thus employees of age 21 years 1 month and 48 years

10 months would be recorded as 21 and 48 respectively.

B. To the nearest whole year of age. Our 21 year 1 month and 48 year 10 months employees would in this case be recorded as 21 and *49*.

Draw up specimen frequency distributions showing nominal class limits, actual class limits and class mid-points if the data was recorded as in A. Repeat the exercise for the case of the data being recorded as in B.

3(U) Give further examples of data you would expect to have a distribution which is

(i) Symmetrical
(ii) Positively Skewed
(iii) Negatively Skewed

Can you suggest any physical reasons why the data you have chosen has the skewness characteristics you attribute to it?

4(U) The following frequency distribution, taken from the 1963 Census of Production, gives an analysis of firms by size (number of employees) in the general mechanical engineering industry:

Average number employed	Number of firms
25– 49	390
50– 99	285
100– 199	187
200– 299	59
300– 399	42
400– 499	27
500– 749	21
750– 999	15
1,000–1,499	24
1,500–1,999	15
2,000–2,999	19
3,000 and over	40
Total	1,144

(i) Why have the compilers not used equal class widths?
(ii) Why have the compilers used an open-ended class "3,000 and over"?
(iii) Attempt to draw a histogram for this data, commenting on any difficulties you encounter.

5(U) Some texts suggest that class limits should be chosen so that the class mid-point is a convenient whole number. Show how this can be achieved by suggesting nominal and actual class limits suitable for the gas consumption data.

6(E) Three identical "cold forming" machines produce rods to the same nominal specifications. The rods produced by these machines are funnelled down a single chute to the next production process. From time to time a sample of rods emerging from the chute is subjected to quality control checks. One of these is a measurement check where the difference between a rod's actual length and its design specified length is measured on a "clock gauge". On one occasion a sample of 50 rods yielded the following results (in units of, and to the nearest 0.001").

−2	3	2	−2	4	−1	−3	−1	4	1
−1	5	3	−1	5	0	4	1	0	−2
−3	−1	3	0	−1	4	−2	0	−1	0
3	1	−1	5	0	0	−3	4	0	4
1	4	1	−1	2	−2	−1	4	−2	−1

(i) Analyse this data in the form of a frequency distribution and draw its histogram.

(ii) What deductions can be made from the shape of histogram about the production process? What action would you propose taking, if any, and why?

HND (Business Studies)

7(E) (i) *Explain* and *justify* the procedure you would adopt in classifying the overtime earnings, in a particular month, of 200 employees. You may assume that exact figures are available.

(ii) Describe how you would construct an ogive from (a) the frequency distribution of overtime earnings, and (b) the raw data (i.e. unclassified). Discuss briefly the relative advantages and disadvantages of methods (a) and (b).

HNC (Business Studies)

8(E) A company manufacturing off-the-peg men's trousers has recently commissioned a market research survey into the socio-economic characteristics of men who buy trousers of various styles and prices. For one particular (conservative) style the survey report included a frequency distribution of a random sample of 100 inside leg measurements from the group of men who favoured this style and who were "high spenders" on clothes. This frequency distribution is shown on the following page:

Inside Leg Measurement (to nearest $\frac{1}{2}$")	No. of Men with Stated I.L.M.
28	2
$28\frac{1}{2}$	0
29	4
$29\frac{1}{2}$	7
30	12
$30\frac{1}{2}$	16
31	20
$31\frac{1}{2}$	14
32	10
$32\frac{1}{2}$	9
33	3
$33\frac{1}{2}$	3
Sample size	100

This company intends manufacturing this style of trouser in three sizes, small ($29\frac{1}{2}$" inside leg), medium (31" inside leg) and large ($32\frac{1}{2}$" inside leg). The company knows from experience that a customer will only accept trousers without alteration provided that the inside leg measurement is within $\frac{1}{2}$" (either way) of his size. Trousers which are too long for a customer can be shortened (and it is assumed that the customer will accept this), but because of the style no lengthening alterations are possible.

Using the data given construct an approximation to the ogive of the *population* of leg lengths of potential customers and use this to produce estimates of the following quantities:

(i) % of potential customers who will not find a leg size to fit them with or without alterations.

(ii) proportions of the company's production (of this style) which should be planned in short, medium and long fittings.

(iii) the proportions of each size which will require subsequent shortening alterations.

BA (Business Studies)

SOLUTIONS TO EVEN NUMBERED EXERCISES

2.

Nominal Class Limits (A and B)	*Actual Class Limits* (A)	*Class Mid Points* (A)	*Actual Class Limits* (B)	*Class Mid Points* (B)
16–20	16–21	18.5	15.5–20.5	18
21–25	21–26	23.5	20.5–25.5	23
26–30	26–31	28.5	25.5–30.5	28
31–35	31–36	33.5	30.5–35.5	33
36–40	36–41	38.5	35.5–40.5	38
41–45	41–46	43.5	40.5–45.5	43
46–50	46–51	48.5	45.5–50.5	48
51–55	51–56	53.5	50.5–55.5	53
56–60	56–61	58.5	55.5–60.5	58
61–65	61–66	63.5	60.5–65.5	63
66–70	66–71	68.5	65.5–70.5	68

4. (i) Equal class widths would have resulted in either (a) an excessively large number of narrow classes (e.g. of width 25) needed to pick out the detail at the "low" end of the distribution, or (b) a reasonable number of wide classes (e.g. of width 500) which would have obscured most of the interesting detail. The problem arises because of the extreme positive skew of the distribution.

(ii) There may be one or two firms with a very large number of employees (e.g. 20,000). Rather than having a number of classes with zero frequencies towards the tail of the distribution, an open ended class has been used.

(iii) The lengths of the histogram bars are first calculated by "modifying" the frequencies, in order to ensure that the areas under the bars are proportional to the frequencies, e.g.

Class Limits	Frequency	Class Width†	Length of Bar (any convenient units)
25– 49	390	25	390 = 390
50– 99	285	50	285/2 = $142\frac{1}{2}$
100–199	187	100	187/4 = $46\frac{3}{4}$
200–299 etc.	59	100	59/4 = $14\frac{3}{4}$

†Note that the data is *discrete*.

It can be seen that the length of the bars decrease so rapidly that the histogram drawn from them would convey little useful information.

6. (i)

Nominal Class Value	*Actual Class Limits*		Frequency
−3	−3.5 to −2.5	|||	3
−2	−2.5 to −1.5	𝍸 |	6
−1	−1.5 to −0.5	𝍸 𝍸 |	11
0	−0.5 to 0.5	𝍸 |||	8
1	0.5 to 1.5	𝍸	5
2	1.5 to 2.5	||	2
3	2.5 to 3.5	||||	4
4	3.5 to 4.4	𝍸 |||	8
5	4.5 to 5.5	|||	3

The histogram is shown below

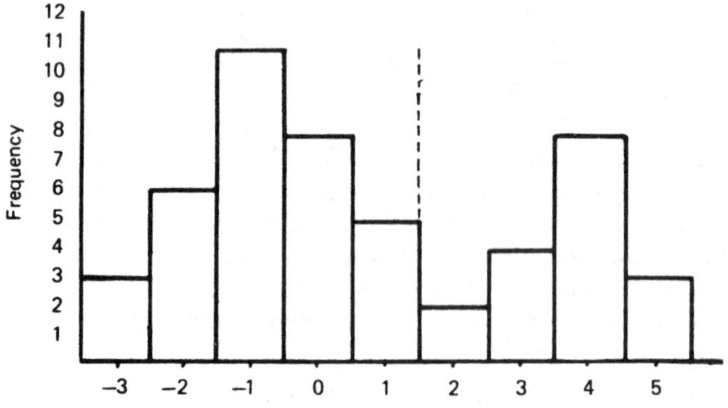

(ii) The distribution is two-peaked (bimodal) indicating that there are possibly two separate sources of variability. Arbitrarily dividing the observations into those less than 2 and those greater than 1 (see dotted line on histogram), there are 33 observations in the former category, 17 in the latter (approximately in the ratio 2:1). This could indicate that two (of the three) production machines have given rise to the former observations – producing rods tending to be shorter than the specification length – and the other machine producing rods which are rather greater than the specification length.

In deciding what course of action to take we must first determine if a variation of −0.003" to +0.005" from the specification length is acceptable at any one time: if it is, no action need be taken. If it is not and yet a variation from −0.003" to +0.002" (say, corresponding to the two machines) is acceptable, we need to determine which machine (of the three) is tending to produce oversize rods. To do this, samples, say of 50 rods each, should be taken from each machine separately. Having isolated the machine making the oversize rods it can then be reset (adjusted) and the output checked before continuing production.

8.

Nominal Class Value (")	*Actual Class Limits*	*Frequency*	*Cumulative Frequency*
28	27.75–28.25	2	2
$28\frac{1}{2}$	28.25–28.75	0	2
29	28.75–29.25	4	6
$29\frac{1}{2}$	·	7	13
30	·	12	25
$30\frac{1}{2}$	·	16	41
31	·	20	61
$31\frac{1}{2}$	·	14	75
32	·	10	85
$32\frac{1}{2}$	·	9	94
33	·	3	97
$33\frac{1}{2}$	33.25–33.75	3	100

The cumulative frequency is plotted against the upper actual class limits and the resultant points "fitted" by a *smooth curve* to approximate to the population ogive (see graph). As the sample size was 100 we may use the cumulative frequency axis as a cumulative percentage scale for the population.

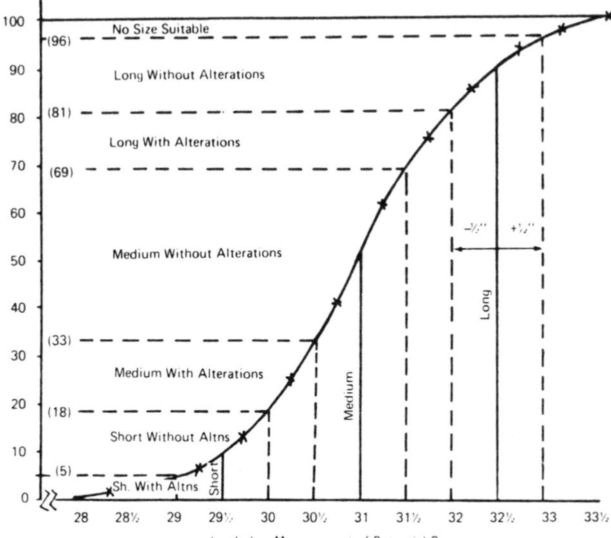

Having inserted the information about sizing on the graph we can immediately obtain the following estimates as *%'s of total potential demand.*

No size available	4%	
Long without alterations	15%	27%
Long with alterations	12%	
Medium without alterations	36%	51%
Medium with alterations	15%	
Short without alterations	13%	18%
Short with alterations	5%	
	100%	96%

Therefore

(i) 4% of potential customers will not find a leg size to fit them *with or without alteration*

(ii) $27 \times \frac{100}{96}\% \simeq 28\%$ of total production should be long

$51 \times \frac{100}{96}\% \simeq 53\%$ of total production should be medium

$18 \times \frac{100}{96}\% \simeq 19\%$ of total production should be short

(iii) $\frac{12}{27} \times 100\% \simeq 44\%$ of long trousers will require alterations

$\frac{15}{51} \times 100\% \simeq 29\%$ of medium trousers will require alterations

$\frac{3}{18} \times 100\% \simeq 17\%$ of short trousers will require alterations

COURSE UNIT 3 – USEFUL NOTATIONS

3.1 A NOTATION FOR OBSERVATIONS

Statistical calculations are often carried out on large volumes of data in the form of univariate or multivariate sets of observations. In order to express concisely how such calculations may be performed in general, it is convenient to employ some mathematical symbolism.

First, a notation for observations themselves. Consider the familiar 100 observations of Series A, reproduced as Exhibit 1.

EXHIBIT 1: Series A

55	82	83	109	78	87	95	94	85	67
80	109	83	89	91	104	90	103	67	52
107	78	86	19	72	66	92	99	60	75
88	112	97	88	49	62	70	66	88	62
72	85	81	78	77	41	105	92	94	74
78	75	87	83	71	99	56	69	78	60
119	39	104	86	67	79	98	102	82	91
46	120	73	62	68	86	70	55	112	83
62	74	99	100	86	67	61	97	77	59
65	51	99	53	105	95	107	46	90	71

We can refer to this data by making use of an algebraic notation, using the symbol X for example. In Statistics, just as in Algebra, we normally reserve the letters in the latter part of the alphabet for variables, but X is not a variable in the normal algebraic sense (i.e. unknown): in fact in order to distinguish between the algebraic and the statistical usage we refer to X as a **random variable**[(1)] (or **variate**). So we could describe the above data as observed values of a random variable X.

When we wish to refer to a *particular* observation we use a subscript notation. Thus, adopting the convention that we read from left to right along a row, and then from one row to the next, we can say $X_1 = 55$ (being the *first* observation in the set of data), $X_2 = 82$, and so on. X_{100}, the last observation, has the value 71. If we wish to refer to an individual observation in general, without immediately specifying which one, we use the notation X_i meaning the *i*th observation.

For some purposes it is convenient to perform calculations on an

---

[(1)]It is sufficient for the present to define random variable as a numerical quantity, the properties of which can be described by a frequency distribution.

ordered set of data, that is to say the observations arranged in ascending order of magnitude (see Exhibit 2).

EXHIBIT 2: Series A in Ascending Order

19	39	41	46	46	49	51	52	53	55
55	56	59	60	60	61	62	62	62	62
65	66	66	67	67	67	67	68	69	70
70	71	71	72	72	73	74	74	75	75
77	77	78	78	78	78	78	79	80	81
82	82	83	83	83	83	85	85	86	86
86	86	87	87	88	88	88	89	90	90
91	91	92	92	94	94	95	95	97	97
98	99	99	99	99	100	102	103	104	104
105	105	107	107	109	109	112	112	119	120

Thus using this ordered set of data, $X_1 = 19, X_2 = 39$, etc., and $X_{100} = 120$.

When dealing with bivariate or multivariate data we need a separate notation for each type of measurement. If, quoting the example given in CU1, we have collected bivariate data on the leg and waist measurements of men, then we could use X_i to refer to the ith leg measurement and Y_i to the ith waist measurement. Note that, as an important feature of the notation, X_{10} and Y_{10} for instance refer to the leg and waist measurements of the *same* man (i.e. the 10th in a certain list).

3.2 A NOTATION FOR FREQUENCY DISTRIBUTIONS

Exhibit 3 shows the frequency distribution of the number of faulty tyres on fifty trucks, taken from CU2.

EXHIBIT 3

No. of Faulty Tyres, z_i	*No. of Trucks with Stated No. of Faulty Tyres,* f_i
0	22
1	16
2	5
3	4
4	3
Total	50

Let us call the "number of faulty tyres" the random variable Z. The left-hand column of figures in Exhibit 3, it should be realised, is not a set of observations as such, but a set of class values (compare with "class limits") which has been used in summarising the original data: we shall use the notation z_i to denote these class values[1] (giving $z_1 = 0, z_2 = 1, z_3 = 2, z_4 = 3, z_5 = 4$). Note that we have created a distinction between the observations themselves Z_i (which are not given above) and the class values z_i (which are) by using upper-case letters for the former and lower-case for the latter. The frequencies (the right hand column) are conveniently referred to by the notation f_i (so $f_1 = 22, f_2 = 16$ etc.) – it is usual to reserve the letters f,g, and h to denote frequencies.

3.3 THE SUMMATION NOTATION

Suppose we wished to denote the process of adding up the first four observations in a set of data. Using the notation we have developed, this would be written as:

$$X_1 + X_2 + X_3 + X_4. \tag{1}$$

If we wished to denote adding up 100 items of data we could avoid writer's cramp by using dots to indicate obvious intermediate steps as follows:

$$X_1 + X_2 + X_3 + \ldots + X_{100}. \tag{2}$$

But the process could be simplified by making use of the **summation notation**. In place of (1) we could write[2]

$$\sum_{i=1}^{4} X_i$$

and in place of (2):

$$\sum_{i=1}^{100} X_i.$$

Note that the summation starts with the item of data specified by the value given to i *below* the Σ sign and stops at the item of data specified by the value given to i *above* the Σ sign.

In general we define:

$$\sum_{i=k}^{m} X_i = X_k + X_{k+1} + X_{k+2} + \ldots + X_{m-2} + X_{m-1} + X_m.$$

---

(1) We shall see in CU4 that with frequency distributions having finite class widths the class mid-points are used as "representative" class values.

(2) The symbol Σ is the upper case Greek letter "sigma" and corresponds to the instruction "add up".

So, to take a specific example, using the data in Exhibit 1 we have:

$$\sum_{i=11}^{20} X_i = 80 + 109 + 83 + 89 + 91 + 104 + 90 + 103 + 67 + 52$$
$$= 868$$

It is conventional to use the letter n to denote the number of observations in a sample. So in order to denote, quite generally, the sum of *all* the observations in a sample we could write:

$$\text{Total} = \sum_{i=1}^{n} X_i,$$

but where there is no ambiguity it is more usual to write simply ΣX, implying that *all* the values are summed. For the data in Exhibit 1, we obtain,

$$\sum X = 55 + 82 + 83 + 109 + \ldots + 107 + 46 + 90 + 71 = 8{,}000.$$

The summation notation can be generalised in a number of ways of which the following are examples:

ΣX^2 denotes "square each of the values, then add". Care must be taken to distinguish ΣX^2 from $(\Sigma X)^2$, which simply means "square ΣX".
E.g. for the data in Exhibit 1 we have:

$$\sum X^2 = 55^2 + 82^2 + 83^2 + 109^2 + \ldots + 107^2 + 46^2 + 90^2 + 71^2$$
$$= 676{,}184.$$

Some summations involve the product of observations, e.g.

$$\sum_{i=1}^{n-1} X_i X_{i+1} = X_1 X_2 + X_2 X_3 + \ldots + X_{n-2} X_{n-1} + X_{n-1} X_n,$$

or the product of class values and frequencies,

$$\sum_{i=1}^{m} z_i f_i = z_1 f_1 + z_2 f_2 + \ldots + z_{m-1} f_{m-1} + z_m f_m.$$

Using the data in Exhibit 3 (where m, the number of classes, equals 5) we obtain,

$$\sum_{i=1}^{5} z_i f_i = 0 \times 22 + 1 \times 16 + 2 \times 5 + 3 \times 4 + 4 \times 3 = 50.$$

3.4 PROPERTIES OF THE SUMMATION NOTATION

There are three rules related to the summation notation which are useful for later work. The first is:

Rule 1[(1)] $\sum (U_i \pm V_i) = \sum U_i \pm \sum V_i$

The proof is simple. As no summation limits are given it is assumed that the summations run from 1 to n. So the left-hand side of rule 1 can be expanded as follows:

$$\begin{aligned} \text{LHS} = \quad & U_1 \pm V_1 \\ + & U_2 \pm V_2 \\ & \cdot \quad \cdot \\ + & \cdot \quad \cdot \\ & \cdot \quad \cdot \\ + & \cdot \quad \cdot \\ & \cdot \quad \cdot \\ + & U_n \pm V_n \\ = \quad & U_1 + U_2 + U_3 + \ldots + U_n \\ \pm & (V_1 + V_2 + V_3 + \ldots + V_n), \text{ by adding up the columns,} \\ = \quad & \sum U_i \pm \sum V_i, \text{ which is the required right-hand side.} \end{aligned}$$

As an example, $(3 \pm 2) + (4 \pm 1) + (6 \pm 4) = 13 \pm 7$.

Rule 2 $\sum kU_i = k\sum U_i$, where k is a constant.

That is, constants may be taken outside Σ signs as they may from brackets in algebra.

E.g. $(3 \times 2) + (3 \times 4) + (3 \times 6) = 3(2 + 4 + 6) = 3 \times 12$.

Rule 3 $\sum_{i=1}^{n} k = nk$, where k is a constant.

That is, if we add a constant k, n times, the answer is nk,

---

(1) The "±" sign, read as "plus or minus" is to be interpreted as follows. The rule holds true with the upper sign (+) throughout and also with the lower sign (−) throughout. It is simply a more compact way of writing what would otherwise be *two* rules:

$$\Sigma (U_i + V_i) = \Sigma U_i + \Sigma V_i$$

and

$$\Sigma (U_i - V_i) = \Sigma U_i - \Sigma V_i.$$

E.g. $(4 + 4 + 4 + 4 + 4) = 5 \times 4$.

Rules 2 and 3 are important in many statistical proofs, bearing in mind the statements in CU1 that parameters are constants for a population, while statistics can also be considered fixed values for the sample data on which they are based. The above rules will be required for CU's 4 and 5.

EXERCISES:

Use the following data for Exercises 1–5

Able Discount Stores	*Jan*	*Feb*	*Mar*	*Apr*	*May*	*June*
Sales of Washing Machines (X)	2	0	4	6	2	3
Sales of Dishwashers (Y)	1	0	0	2	0	1

1. Identify X_1, X_3, X_n and $X_{(n/2+1)}$ where n has its usual meaning.
2. Evaluate from first principles (i.e. *without* using Rules 1, 2 or 3):

(i) $\sum_{i=1}^{4} X_i$ (ii) $\sum_{i=3}^{6} X_i$ (iii) $\sum X$ (iv) $\sum X^2$

(v) $\sum Y$ (vi) $\sum XY$ (vii) $\sum(X + Y)$ (viii) $\sum(X - 2)$

(ix) $\sum 3X$ (x) $\sum_{i=1}^{5} X_i X_{i+1}$ (xi) $\sum_{i=1}^{10} 3$

3. For what values of i does $Y_i = 0$.
4. If the X's refer to the *ordered set* of washing machine sales, repeat Exercise 1.
5. (a) Use the values you obtained in Exercise 2 to demonstrate the truth of Rules 1, 2 and 3 given in Section 3.4.
   (b) Prove, from first principles, Rules 2 and 3.
6. Use the rules of summation to prove that:

$$\sum_{i=1}^{n} (X_i - k) = \sum_{i=1}^{n} X_i - nk$$

7. Similarly prove that:

$$\sum_{i=1}^{n} (X_i - k)^2 = \sum_{i=1}^{n} X_i^2 - 2k \sum_{i=1}^{n} X_i + nk^2$$

Use calculating machines, where appropriate, for the following exercises

8. The table shown compares the durations of absence (in days) from a company as a result of two types of illness, A and B.

Duration (days) (x)	Number of Absences	
	from illness A (f)	from illness B (g)
1	1	–
2	2	–
3	4	2
4	3	0
5	7	8
6	10	15
7	8	18
8	4	10
9	3	12
10	1	5
11	–	3
12	–	2

Calculate:

(i) $\sum_{i=1}^{12} f_i$ (ii) $\sum_{i=1}^{12} x_i f_i$ (iii) $\sum_{i=1}^{12} x_i^2 f_i$

(iv) $\sum_{i=1}^{12} g_i$ (v) $\sum_{i=1}^{12} x_i g_i$ (vi) $\sum_{i=1}^{12} x_i^2 g_i$

Give a *physical* interpretation of your values for (i), (ii), (iv) and (v).

9. The table shown compares average salaries in £1,000's of U.K. executives for 1967 and 1970:

JOB	1967	1970
Chief Accountant	3.3	3.7
Works Manager	3.5	3.8
Chief Buyer	2.2	2.6
Personnel Manager	2.4	3.2
Head of R & D	3.3	3.9
Home Sales Manager	3.5	3.8
Company Secretary	3.2	4.2
Export Sales Manager	3.1	3.5
Head of Work Study	1.9	2.5
Marketing Manager	3.2	3.8

Source: Associated Industrial Consultants Ltd.

Denoting 1967 average salaries by the variable X and 1970 average salaries by the variable Y, calculate:

(i) $\sum X^2$ (ii) $\sum Y^2$ (iii) $\sum XY$

Where possible calculate the following quantities directly from the values obtained for (i), (ii), and (iii).

(iv) $\sum (X+Y)^2$ (v) $\sum X^2 Y^2$ (vi) $\sum (X-Y)^2$ (vii) $\sum (X/Y)^2$
(viii) $\sum (X+0.8Y)^2$

If it is not possible to calculate these quantities in the manner suggested, explain why.

SOLUTIONS TO EVEN NUMBERED EXERCISES

2. (i) 12, (ii) 15, (iii) 17, (iv) 69, (v) 4, (vi) 17, (vii) 21, (viii) 5, (ix) 51, (x) 42, (xi) 30.

4. 0; 2; 6; 3. (N.B. as $n = 6$, $n/2 + 1 = 4$; hence $X_4 = 3$).

6. $$\sum_{i=1}^{n} (X_i - k) = \sum_{i=1}^{n} X_i - \sum_{i=1}^{n} k \quad \text{(from Rule 1)}$$

$$= \sum_{i=1}^{n} X_i - nk \quad \text{(from Rule 2).}$$

8. (i) 43, (ii) 249, (iii) 1619, (iv) 75, (v) 557, (vi) 4403
The physical interpretations are as follows:

(i) In total there were 43 absences resulting from illness A.
(ii) A total of 249 days of absence resulted from illness A.
(iv) In total there were 75 absences resulting from illness B.
(v) A total of 557 days of absence resulted from illness B.

COURSE UNIT 4 – MEASURES OF AVERAGE

4.1 THE SEARCH FOR A MEASURE OF AVERAGE

"Everyone knows" that the average height of an Englishman is 175 cm. (5′9″), but not everyone knows precisely the meaning of such a statement.

One interpretation would be that although men's heights do vary from individual to individual the value of 175 cm. is somehow a representative value for an Englishman's height – we certainly would not be surprised if we measured the height of an Englishman and found it to be that value (at least, to the nearest cm.). But such an interpretation is rather imprecise for the purpose, say, of comparing the heights of Englishmen and Frenchmen.

In the quest for a more precise meaning for the word "average" compare the three histograms shown in Exhibits 1, 2 and 3.

EXHIBIT 1 **EXHIBIT 2** **EXHIBIT 3**

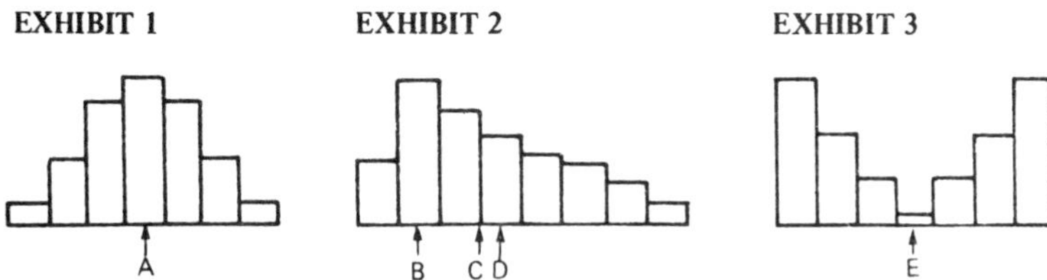

In Exhibit 1, which is a symmetrical bell-shaped distribution, few people would disagree that the value A is the value which best represents the "average". By contrast, in Exhibit 2, which is positively skewed, argument could rage as to whether B, C or D (or indeed, other values) gives the most representative value.

In Exhibit 3, which is symmetrical but rather curiously U-shaped, it would be a brave man indeed who could maintain that the value E was at all a representative value, although it is very definitely in the middle of the distribution.

In view of this it is not surprising that there is no single **measure of average** (sometimes called a **measure of location** or a **measure of central tendency**) which has found universal acclaim for the purpose of singling out a representative value. Instead, statisticians make a choice from a number of different measures of average, depending on the particular circumstances. The following list gives the names of various measures of average which are in common use: each has its own advantages and disadvantages as a way of measuring "representativeness":

- Mode
- Median
- Arithmetic Mean (or simply "Mean")
- Geometric Mean
- Harmonic Mean
- Weighted Mean (especially in the form of an Index Number – see SU2)

In fact, the matter is such that statisticians rarely use the word "average" at all; they refer to the particular measure of average they have chosen. In this Course Unit we examine the Mode, the Median and the Mean as measures of average. Details of the others can be found elsewhere (see reading recommendations).

Details of the calculations involved will be illustrated with reference to two sets of data. The first, given in Exhibit 4, is a set of discrete data and the second is the sample of gas consumption (therms/quarter) taken from Series A of the Standard Data Sheets. It is convenient, for the purpose of calculating *some* of the statistics, to use an ordered set of data. For this reason Series A, reorganised as an ordered set, is reproduced as Exhibit 5.

EXHIBIT 4: Washing Machine Sales in First Six Months

Month	*Jan*	*Feb*	*Mar*	*Apr*	*May*	*Jun*
Sales of washing machines	2	0	4	6	2	3

EXHIBIT 5: Sample of Gas Consumption (Therms/Qtr.) Arranged in Ascending Order

19	39	41	46	46	49	51	52	53	55
55	56	59	60	60	61	62	62	62	62
65	66	66	67	67	67	67	68	69	70
70	71	71	72	72	73	74	74	75	75
77	77	78	78	78	78	78	79	80	81
82	82	83	83	83	83	85	85	86	86
86	86	87	87	88	88	88	89	90	90
91	91	92	92	94	94	95	95	97	97
98	99	99	99	99	100	102	103	104	104
105	105	107	107	109	109	112	112	119	120

4.2 THE MODE

The **mode** is the most commonly occurring value (the phrase *à la mode* means fashionable in French).

With discrete data the concept of a mode (or "modal value") is fairly clear-cut. In Exhibit 4 the most commonly occurring value is 2 (washing machines), so this is the mode. But suppose in July there were 4 sales and this value were added to the table. We would then have a case where the value 2 occurred twice (Jan. and May) and also where the value 4 occurred twice (March and July). In this case the mode is not unique and the data is then sometimes referred to as **bimodal**. Suppose instead that the data ran only from February to June inclusive; then the data has no mode.

With discrete data formed into a frequency distribution the determination of the mode is simple: it is that value with the largest frequency – visually identifiable by the longest histogram line. Examine the example given in CU2 of the number of faulty tyres on 10-wheeled trucks. The modal number of faulty tyres is zero.

There is a fundamental problem of mode determination in the case of continuous data. If the data is recorded to a high level of accuracy then the chances are that there will be no two identical observations and hence no mode. If the observations are recorded to an approximation in some convenient units (e.g. "to the nearest therm" in the case of the gas consumption data) the mode can sometimes be determined directly. Examining Exhibit 5 we see that the value *78* occurs most frequently (5 times); but with only a slight change in the data (say, with two of the *78* values missing) we could have identified the mode as *62*, *86* or even *99*, each of which occurs 4 times. In fact the difficulties are such that an alternative approach is often used. One technique is illustrated in Exhibit 6 (which is itself a detail taken from the gas consumption histogram we derived in CU2).

The construction lines shown join the top right-hand corner of the modal class bar (i.e. the *longest* histogram bar) to the top right-hand corner of the pre-modal class bar, and the top left-hand corner of the

EXHIBIT 6: Mode Estimation by Construction from Histogram of Continuous Data

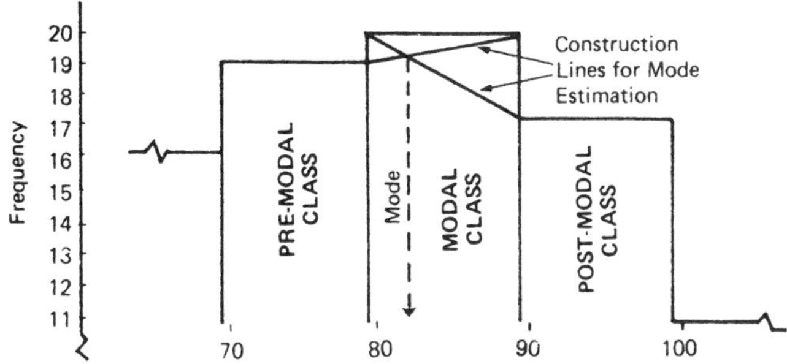

modal class bar to the top left-hand corner of the post-modal class bar. Their point of intersection estimates the mode (82). This value can also be obtained by direct calculation – details are given in Exercise 2. It should be noted, however, that whichever method is used the value obtained for the mode depends, sometimes critically, on the way in which the data is classified. It should perhaps be emphasised that these techniques for mode estimation are only valid where the modal class, the pre-modal class and the post-modal class have the same width.

Somewhat ironically, the mode is not a very commonly used measure of average, as a result of the difficulties we have mentioned in determining its value. However, the mode does have its uses. It is easy for the layman to understand as being the most "typical", "popular" or "common" value. Conceptually it is easier to imagine the average family having 2 children, rather than 2.213 children. We have seen that it is sometimes easy to obtain – a useful feature when quick answers are required – and a further advantage is that it is unaffected by extra-ordinarily large or small observations in the data, which might arise, for example, as the result of inaccurate measurement or recording.

The mode is particularly useful in multi-variate situations. What, for instance, do we mean by an "average" car? Do we measure its engine capacity, top speed, petrol consumption, number of passenger seats? Similar questions may be asked about such items as computers, aero-planes and schools. In these situations we may quote as "typical" an actual physical example – that is, the most popular model of car on the road, or the most widely used computer installation. In market research, modes (e.g. modal preferences for brands of soup) may be obtained for different segments of the population – according to age, social class or part of the country.

4.3 THE MEDIAN

The **median** is defined as the value of the middle item in an ordered set of data. It is sometimes referred to as the "50-percentile", expressing the fact that 50% of the observations are below (and 50% above) this value. It should be emphasised that the median is *not* equal to ½ (largest observation + smallest observation) – this is yet another measure of average called the mid-range.

If we have an *ordered set* of data (X) with an *odd* number of observations, n, then the median, by definition, is the value[(1)] of $X_{(n+1)/2}$. With an even number of observations the convention is to

---

(1)For the purposes of satisfying the definition given the value $X_{(n+1)/2}$ is conceptually regarded as being allocated equally above and below the median.

take a simple average of the middle *two* observations of the ordered set; that is, the median is given by $\frac{1}{2}(X_{n/2} + X_{(n/2+1)})$. Thus, ordering the data in Exhibit 4 we have:

0 2 2 3 4 6

In this case we have 6 observations, so $n/2 = 3$ and $(n/2 + 1) = 4$ and therefore the median is given by:

$$\tfrac{1}{2}(X_3 + X_4) = \tfrac{1}{2}(2 + 3) = 2.5 \text{ (washing machines)}$$

It can be seen that there are three observations smaller than this value and three observations larger. Using the data in Exhibit 5 (100 observations) the median is

$$\tfrac{1}{2}(X_{50} + X_{51}) = \tfrac{1}{2}(81 + 82) = 81.5 \text{ (therms).}$$

So far, we can see that the median does not suffer from the main disadvantage that afflicts the mode: the median can always be calculated without difficulty, although the necessity of ordering the data beforehand can be an onerous task with a large number of observations. The median shares with the mode the desirable property of not being affected by outlying observations.

Just as we could estimate the population mode from the *histogram*, the population median may be estimated from the *ogive*. The technique is illustrated in Exhibit 8 which is a detail from the ogive of the gas consumption data of Series A reproduced here as Exhibit 7.

EXHIBIT 7: Ogive of Series A

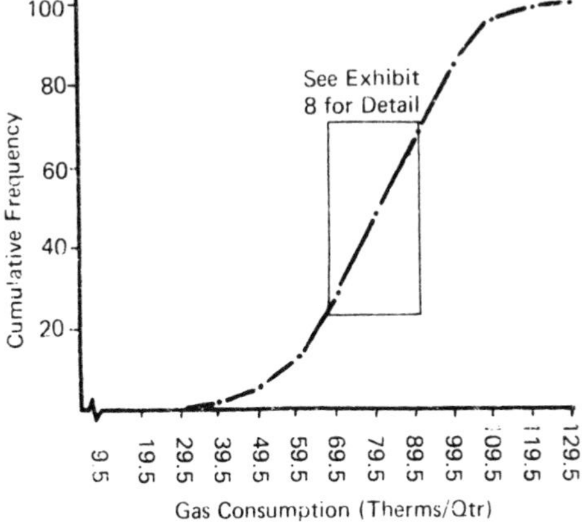

EXHIBIT 8: Median Estimation from Ogive of Continuous Data

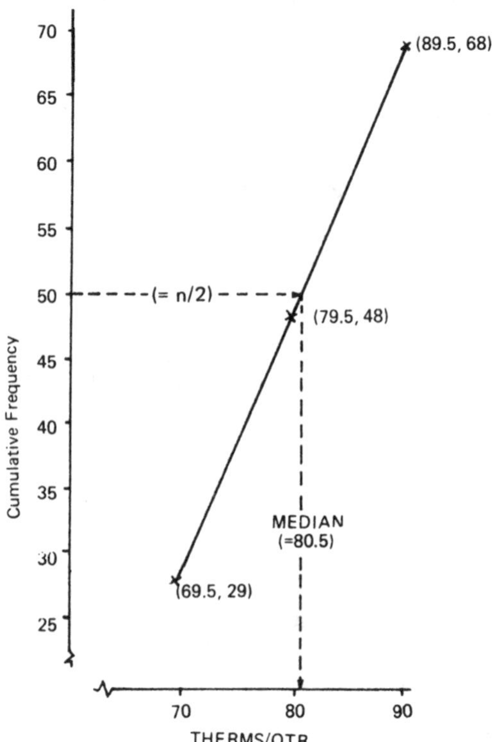

The construction illustrated can be described in general for a set of n observations as follows. A horizontal line is drawn corresponding to the cumulative frequency[1] of $n/2$. The measurement value at which this line cuts the ogive is an estimate of the median. Note that the value obtained by construction (80.5) differs from that obtained by the direct application of the definition (81.5) due to the assumption implicit in the construction that the observations are evenly spread within the median class (and this, in fact, is not quite true).

The median may be used in several types of situation in preference to the mode or the mean (Section 4.4): the fact that the data has to be

---

[1] With grouped continuous data $n/2$ is always used irrespective of whether n is odd or even, since the median value obtained in this way cuts the ogive exactly in half and this would also be the value that divides the *histogram* into two equal *areas.*

ordered for the calculation of the median is now turned to advantage. The median may be obtained in those situations where we cannot give a numerical "score", but we can nonetheless "rank" order (i.e. place in order of preference) members of the set. Rank orders are commonly used in judging competitions (beauty, baby, gardening etc.) and other "preference" situations such as flavours of soups, patterns of carpets and the design of buildings.

The median may also be calculated when the extreme ends of the distribution are missing. Two examples: a pair of scales cannot cope with very light weights or very heavy weights (see Exhibit 9); in surveys it is frequently the case that the very rich and the very poor are difficult to locate and interview. In such cases, provided we know the tail of the distribution from which the observations are missing, we may still obtain an average.

EXHIBIT 9: Calculation of the Median with Missing Data

$\underbrace{X_1, X_2, X_3}$, 1.0, 2.1, 3.2, 4.4, 4.5, 4.8, 5.1, $\underbrace{X_{11}, X_{12}}$

Unmeasurable (below 1 mgm)	Unmeasurable (more than 10 mgms)

We obtain the median as $\frac{1}{2}(3.2 + 4.4) = \frac{1}{2}(7.6) = 3.8$ mgms.

Further applications of the median are found in "life-testing" in industrial quality control (the "life" of electric light bulbs, batteries, tyres) and in manpower planning (length of service completed with a company before leaving). In these examples we wish to determine an average value before all the data becomes available – after all, we may have to wait 40 years for a long-service employee to leave! By calculating the median as soon as 50% of the bulbs have burned out, or 50% of employees have left, we have a measure of average which may be used to tackle urgent problems.

4.4 DEFINITION OF THE MEAN

Statisticians use the word **mean** in exactly the same way as the layman would use the word "average"; that is, the sum of the values divided by the number of values. Thus for a sample of size n we define the mean, $\bar{X}$, as:

$$\bar{X} = \frac{1}{n} \sum_{i=1}^{n} X_i \quad \left(= \frac{1}{n} \sum X \text{ in our simplified notation} \right) \qquad (1)$$

(Of course if we were dealing with a finite population of size N, instead of a sample, we can similarly define the population mean as $\mu = \frac{1}{N}\sum_{i=1}^{n} X_i$).

Taking the sample in Exhibit 4 we have:

$$\bar{X} = \frac{1}{6}(2+0+4+6+2+3) = \frac{17}{6} \simeq 2.83$$

Using the sample of gas consumption (Exhibit 5) we should need a calculating machine[1] to work out the value of the mean from the definition. ΣX in this case is 8,000 and so the mean of the 100 observations is given by $\bar{X} = 8{,}000/100 = 80.0$.

4.5 CALCULATION OF THE MEAN BY HAND FOR UNCLASSIFIED DATA

Obviously it is only practicable to calculate the mean of unclassified data by hand when there are a small number of observations. Normally one would use the definition of the mean given in Section 4.4. But there are occasions when because the observations are (large) "awkward" values it is somewhat easier to subtract a suitable constant quantity from each observation before finding the mean. The principle of this approach (though hardly a suitable application!) is illustrated in Exhibit 10 on the washing machine sales data.

EXHIBIT 10:

	X	$(X-2)$
Jan	2	0
Feb	0	−2
Mar	4	2
Apr	6	4
May	2	0
Jun	3	1
	Total	7−2 = 5

[1] Note that when evaluating ΣX for a large set of data on a calculating machine without a tally-roll (i.e. where one cannot check each entry keyed in) it is a wise precaution to perform the summation in stages (column by column or row by row) working out a sub-total for each. By this means errors can be tracked down more easily and corrected without reworking all the data, simply by rechecking each subtotal.

Here we have subtracted the constant quantity 2 from each observation and found the sum $\Sigma(X - 2)$. (Try it for yourself by subtracting 3.)
It can be shown[1] that for *any* constant a,

$$\bar{X} = a + \frac{1}{n}\sum (X - a) \qquad (2)$$

In our case $a = 2$, so we have

$$\bar{X} = 2 + 5/6 \simeq 2.83, \text{ as before.}$$

The constant a is known as the "assumed average". Whatever value we choose for a the result for $\bar{X}$ will always be the same; however, the closer our assumed average is to $\bar{X}$ the simpler will be the arithmetic. The assumed average will be used again in Section 4.6 and yet again in CU5.

4.6 CALCULATION OF THE MEAN FOR CLASSIFIED DATA

It will be recalled (from CU3) that with classified data we use the notation x_i (*not* X_i) to denote the class value and f_i the corresponding frequency (different i's correspond to the different classes). If we can assume, as we usually do, that each class mid-point is a representative value for all items in the class (or at least, that the inaccuracies resulting from this assumption throughout the frequency distribution cancel each other out) then we can calculate the mean value from a frequency distribution using the following formula (which we shall later justify):

$$\bar{X} = \frac{1}{n}\sum_{i=1}^{m} x_i f_i \;=\; \sum_{i=1}^{m} x_i f_i \Big/ \sum_{i=1}^{m} f_i, \qquad (3)$$

where m denotes the *number of classes*

Note that n, the number of observations and Σf_i, the sum of the frequencies are the same quantity. For this reason some texts use

[1] *Proof* $\Sigma(X - a) = \Sigma X - na$ (using the rules of summation in CU3)

or $\Sigma X = na + \Sigma(X - a)$

and dividing throughout by n to obtain $\bar{X}$, we obtain

$$\bar{X} = \frac{1}{n}\sum X = a + \frac{1}{n}\sum (X - a).$$

formulae with n, others with Σf_i – they are completely interchangeable. We illustrate the use of this formula on the frequency distribution for Series A (derived in CU2) in Exhibit 11.

EXHIBIT 11: Calculation of the Mean with Classified Data

x_i *(mid-point)*	f_i	$x_i f_i$
14.5	1	14.5
24.5	0	0
34.5	1	34.5
44.5	4	178.0
54.5	7	381.5
64.5	16	1032.0
74.5	19	1415.5
84.5	20	1690.0
94.5	17	1606.5
104.5	11	1149.5
114.5	3	343.5
124.5	1	124.5
	$\Sigma f_i = 100$	$\Sigma x_i f_i = 7970.0$

Therefore $\bar{X} = 7970.0/100 = 79.7$

Each of the $x_i f_i$ quantities serves as an approximation to the *subtotal* of all observations in its respective class. Take the class 49.5 – 59.5 (with mid-point 54.5), for instance. The total of the 7 observations in this class is in fact (51 + 52 + 53 + 55 + 55 + 56 + 59) = 381: the $x_i f_i$ value for this class as calculated in Exhibit 11, is 381.5, a very close approximation. Consequently the grand total, $\Sigma x_i f_i$, serves as an approximation to the sum of all observations in the sample, ΣX. This provides the justification for formula (3).

Note that the value of $\bar{X}$ obtained by this method, 79.7, is *not* exactly the same as the value obtained by applying the definition to the unclassified data (the result then was 80.0). The reason is that the assumption on which the formula $\bar{X} = \frac{1}{n} \sum x_i f_i$ was based does not hold exactly, i.e. the mid-points are *not* truly representative of the items in the classes. In particular the class with mid-point 14.5 contained only one item of value 19, the class with mid-point 124.5 also contained just one value, 120. Despite such effects the overall result is only in error by some 0.4%, sufficient for most purposes.

4.7 CALCULATION OF THE MEAN BY HAND FOR CLASSIFIED DATA

The formula used in Section 4.6 provides the easiest way of calculating the mean of classified data if a calculating machine is available, or even by hand if the numbers involved are relatively manageable. It is sometimes the case, however, that we need to calculate the mean when the numbers are difficult and yet no machine is available. In such a case we can simplify the arithmetic by **coding**[1] the class mid-points before performing the calculations.

The technique of coding can be *described* as follows. Firstly, from each class mid-point, x_i, we subtract a convient constant quantity a (the "assumed average"). We then divide all the differences obtained in this way by another convenient constant, c. The results, denoted by d_i, are the coded values of the class mid-points. (As a matter of fact, provided that all the class widths are the same, there is no need actually to carry out these calculations, as we shall see.)

A formula for the mean based on coded mid-points is as follows:

$$\bar{X} = a + \frac{c}{n} \sum d_i f_i \qquad (4)$$

Of course this formula is only useful if it is easier to evaluate $\Sigma d_i f_i$ than to evaluate $\Sigma x_i f_i$. We can help to ensure this by choosing a as one of the class mid-points close to the mean and c as the (smallest) class width: this procedure leads to simple values for the d_i's.

We rework the example in Exhibit 11 in this manner (Exhibit 12), on the following page.

Knowing that the mean is somewhere in the middle of the distribution, the assumed average, a, may be selected from any of the mid-points 54.5, 64.5, 74.5, 84.5. We have chosen $a = 74.5$ and $c = 10$. This has led to a simple progression for the d_i's which are most easily filled in by starting with "0" at the mid-point chosen as a. Note that this

---

[1]The word "coding" comes from "code", a device used to disguise or simplify information. Coded data is sometimes used in published reports where the original data on which it is based is confidential, as well as for the purpose (as here) of simplifying calculations. As an illustration of what is meant by coding, consider the following. An employee earns £2,750 a year which is 3 increments up (each of £250) on a salary scale starting at £2,000. The coded value of his salary is simply 3 (increments).

EXHIBIT 12: Hand Calculation using Classified Data

x_i *(mid-points)*	f_i	d_i *(coded values of x_i)*	$d_i f_i$
14.5	1	−6	−6
24.5	0	−5	0
34.5	1	−4	−4
44.5	4	−3	−12
54.5	7	−2	−14
64.5	16	−1	−16
74.5	19	0	0 −52
84.5	20	1	20
94.5	17	2	34
104.5	11	3	33
114.5	3	4	12
124.5	1	5	5
	Σf_i = 100		104 positive subtotal
			−52 negative subtotal
			$\Sigma d_i f_i$ = 52

N.B. $a = 74.5$, $c = 10$.

simple progression is the result of having classified data with equal class widths(1).

Returning to the calculation of the mean we have,

$$\bar{X} = a + \frac{c}{n}\sum d_i f_i = 74.5 + \frac{10}{100} \cdot 52 = 79.7,$$

a result identical to that obtained in Section 4.6, but easier (with practice) to calculate.

4.8* PROPERTIES OF THE MEAN

The mean, as a measure of average, has one serious disadvantage – it is affected by outlying observations and errors in recording to a greater extent than the mode or the median. Consider the lowest observation *19* in Series A. Had this value arisen as an inaccurate recording of the true value of *49* then such an error would not have led to any inaccur-

(1)With a distribution where the class widths are *not* all equal it is wiser to code each x_i individually by substituting in $d_i = (x_i - a)/c$.

acy in the calculation of the mode or the median, but the mean would have been in error by 0.3.

Despite this disadvantage the mean is the most widely used measure of average. The main reason is that the mean has many more convenient analytical properties than the other measures, the most important of which are now discussed briefly.

The sample mean, $\bar{X}$, is an unbiased estimator of the population mean and is furthermore a reasonably efficient estimator. We can demonstrate its unbiased nature in the following way – the matter of its efficiency is examined in the exercises. Consider a very simple population consisting of just three observations 1, 2 and 3. The population mean, μ, is clearly 2. Now consider taking all possible samples, with replacement, of just two observations (Exhibit 13).

EXHIBIT 13: All Possible Samples from a Population of 3 Items

Sample		*Sample Mean $(\bar{X})$*
1	1	1
1	2	1.5
1	3	2
2	1	1.5
2	2	2
2	3	2.5
3	1	2
3	2	2.5
3	3	3
	Total	18

The average value of the sample means themselves is 18/9 = 2, a value exactly equal to the population mean. This is precisely what is meant when we say that the sample mean is an unbiased estimator – *on average* (over all samples) the sample mean is equal to the population mean. Symbolically we write $\hat{\mu} = \bar{X}$, the hat (ˆ) over μ indicating that the "best estimate" of μ is the value of $\bar{X}$.

Two final analytical properties of $\bar{X}$. Firstly, the sum of the differences of observations from the mean, $\Sigma(X - \bar{X})$, is zero – a result easily proved by using the rules of summation. So conversely, if we were to find $\Sigma(X - a) = 0$ when evaluating the mean using Formula (2), we would then have $\bar{X} = a$, directly. Secondly, a related property, important for later work, is that the sum of the *squared* differences of the observations from the mean is smaller than the sum of the squared differences of the observations from any other value (e.g. the mode or median). Formally we could express this property as:

$$\text{Min.}\left\{\sum (X_i - a)^2\right\} \text{ is obtained when } a = \bar{X}.$$

4.9 COMPARISON OF THE MODE, MEDIAN AND MEAN

Summarising the results we have obtained for Series A, we have:

Mode = 82.0 (by construction)
Median = 80.5 (by construction)
Mean = 80.0 (using the unclassified data).

These three values are closely similar. This is a property of a symmetrical distribution – in fact if the population distribution were exactly symmetrical and we were able to estimate the population parameters precisely we would find them exactly equal. With a noticeably positively skewed distribution we should find the mean was greater than the median which in turn was greater than the mode. With a noticeably negatively skewed distribution the opposite condition obtains: the mean is less than the median which is less than the mode. In fact we may use the differences between these measures of average to obtain a measure of skewness (CU5). The various relationships between these measures of average are illustrated in Exhibit 14. It is a useful aid in memorising these relationships to note that for the skewed distribution the mean, median and mode are always in alphabetical order from the direction of the "long tail" of the distribution.

EXHIBIT 14: Symmetrical and Skewed Distributions

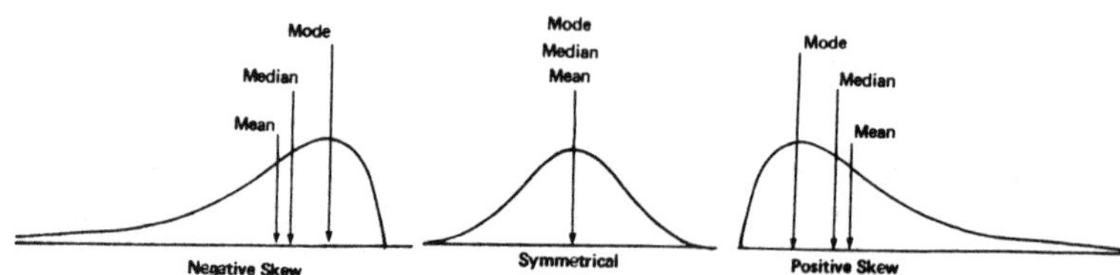

EXHIBIT 15: Physical Interpretation of Measures of Average

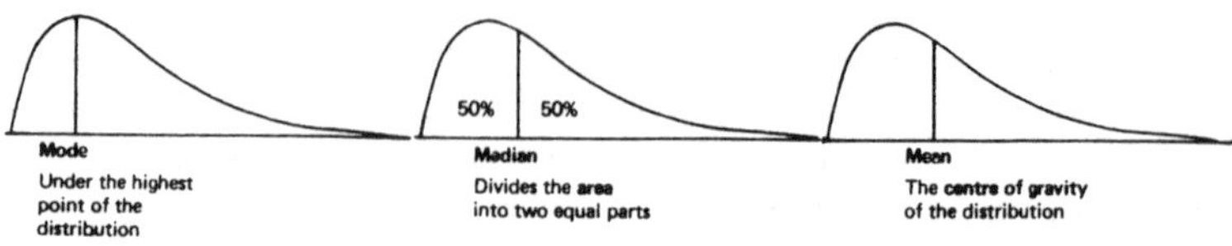

A physical interpretation of these three measures is illustrated in Exhibit 15.

Hence the mean *could* be obtained by cutting out (in wood or cardboard) the distribution and finding its point of balance on a knife edge – but this is even more tedious and less accurate than the method of calculation! However, it does explain why the mean is susceptible to change by the odd abnormally high or low value since it is the same principle as a see-saw where a light person seated some distance from the pivot (centre of gravity) balances a heavier person seated nearer to the pivot. By analogy it would need several normal observations to "balance" a single abnormal observation in the mean.

READING

Details of some other measures of average may be found in: M.J. Moroney, *Facts from Figures*, Penguin, Chapter 4 and some parts of Chapter 6.

EXERCISES

1(P) Use Series B (together with the frequency distribution you have already worked out for this data) to perform the following calculations.

(i) Calculate the mode by construction from the histogram.
(ii) Calculate the median by construction from the ogive.
(iii) Calculate the mean using the unclassified data.
(iv) Calculate the mean from the frequency distribution using a calculating machine.
(v) Calculate the mean from the frequency distribution by hand.
(vi)*Cut out a large histogram in cardboard and use the physical method (Section 4.9) to estimate the mean, commenting on any difficulties you encounter.

Comment on any discrepancies you find between the results for (iii), (iv), (v) and (vi). Indicate the values of the mode, median and mean on your histogram for Series B.
In some older texts a relationship which is said holds for moderately skewed distributions is given by:

$$\text{Mean} - \text{Mode} \simeq 3(\text{Mean} - \text{Median})$$

Test this relationship with the values you have obtained. If you get a nonsense result, try to explain why.

2(U) (i)* An estimate of the mode for grouped data can be obtained by calculation in the following manner:

$$\text{Mode} = L_z + \frac{f_z - f_{z-1}}{2f_z - f_{z-1} - f_{z+1}} \cdot C_z$$

Where, L_z is the lower actual class limit of the modal class
f_z is the frequency of the modal class
f_{z-1} is the frequency of the premodal class
f_{z+1} is the frequency of the post-modal class
C_z is the class width of the modal class.

Show by using similar triangles that this formula is identical to the construction used in the text (Exhibit 6). Demonstrate its use on Series A.

(ii) It is often stated that the mode cannot be calculated for grouped data when unequal class widths are employed. Sketch an example where this is (a) true, (b) untrue.

3(U) (i)*An estimate of the population median can be obtained by calculation in the following manner:

$$\text{Median} = L_m + \frac{(n/2 - F_{m-1})}{f_m} \cdot C_m$$

Where, L_m is the lower actual class limit of the median class
n is the number of observations in the sample
f_m is the frequency of the median class ($f_m = F_m - F_{m-1}$)
F_{m-1} is the *cumulative* frequency of the premedian class
C_m is the class width of the median class.

Show by using similar triangles that this formula is identical to the construction used in the text (Exhibit 8).

(ii) Demonstrate the use of this formula on Series A.

4(U) Rework Exhibit 12 using 64.5 as the assumed average and verify that $\bar{X}$ still equals 79.7.

5(E) "The most common housing unit has three bedrooms" (Mode)
"The median salary of Work Study managers in 1967 was £1,900" (Median)
"The average family has 2.3 children" (Mean)
Do you agree with the measures of average the writers of these three quotations have chosen? Explain why.

(HND, part question)

6(U) Prove the statement in the text that for a given set of X_i the smallest value of $\Sigma(X_i - a)^2$ occurs when $a = \bar{X}$, either by

(a) using the calculus
or (b) evaluating $\Sigma(X_i - a)^2$ for the washing machine sales data with values of a equal to 2, $2\frac{1}{2}$, $2\frac{5}{6}$, 3.

7(U) For each of the samples you selected in CU1 calculate the median. Present your results in the form of a diagram similar to those used in CU1 and in such a way as to facilitate comparison between the efficiency of the sample median and sample mean as estimators of their respective parameters. Comment on your results.

8(E) (a) Compute from the information shown below, the median monthly salary of accountants employed by Regional Consultants Ltd.

REGIONAL CONSULTANTS LTD.,

Monthly Salaries – Accountants

£	Number of Accountants
118–126.9	3
127–135.9	5
136–144.9	9
145–153.9	12
154–162.9	5
163–171.9	4
172–180.9	2

(b) What does knowledge of the median salary tell you about the salaries of accountants employed by this company, and why would you prefer to use the median rather than the arithmetic mean when considering the above data?

(ACCA, June 1971)

9(E) From the following frequency distribution of the hourly earnings of the skilled employees of Electronics Ltd. (Workshop 1).

(a) Calculate the mean hourly wage, and modal hourly wage, and
(b) explain the characteristics of the mean and mode as representative values.

ELECTRONICS LTD. (Workshop 1)

Hourly Wage (p)	Number of Employees
50– 59.9	8
60– 69.9	10
70– 79.9	16
80– 89.9	14
90– 99.9	10
100–109.9	5
110–119.9	2

(ACCA, December 1971)

SOLUTIONS TO EVEN NUMBERED EXERCISES

2(i)

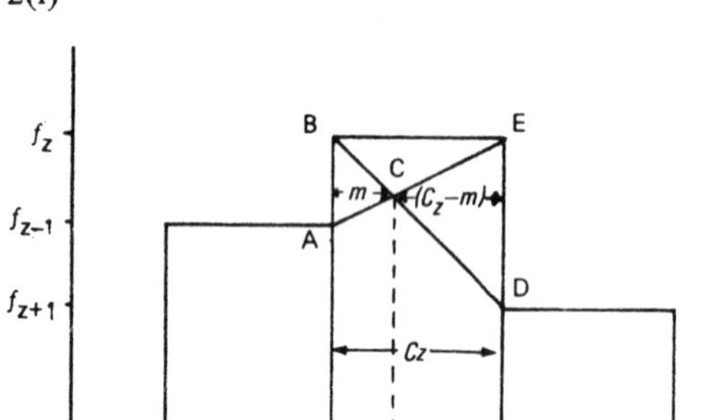

As drawn, we have Mode = $L_z + m$ (1)

Now the Δs ABC and EDC are similar; hence linear dimensions will be in proportion. Calling AB and ED the "base" of the Δs, we have

$$\frac{\text{Height}}{\text{Base}} = \frac{m}{\text{AB}} = \frac{C_z - m}{\text{ED}}.$$

So solving for m, $m = \dfrac{C_z \cdot \text{AB}}{\text{AB} + \text{ED}}$.

Now AB $= f_z - f_{z-1}$ and ED $= f_z - f_{z+1}$.

So $m = \dfrac{f_z - f_{z-1}}{2f_z - f_{z-1} - f_{z+1}} \cdot C_z$, which by substitution in (1) gives the result.
Referring to Exhibit 6, $f_z = 20$, $f_{z-1} = 19$, $f_{z+1} = 17$, $C_z = 10$ and $L_z = 79.5$.

$$\text{So Mode} = 79.5 + \frac{(20 - 19)}{(40 - 19 - 17)} \cdot 10$$

$$= 79.5 + \tfrac{1}{4} \cdot 10 = 79.5 + 2.5 = 82,$$

the same result as by the construction.

2(ii)

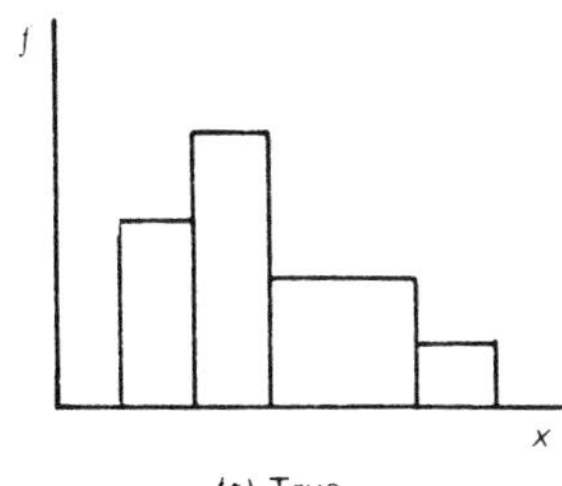

(a) True

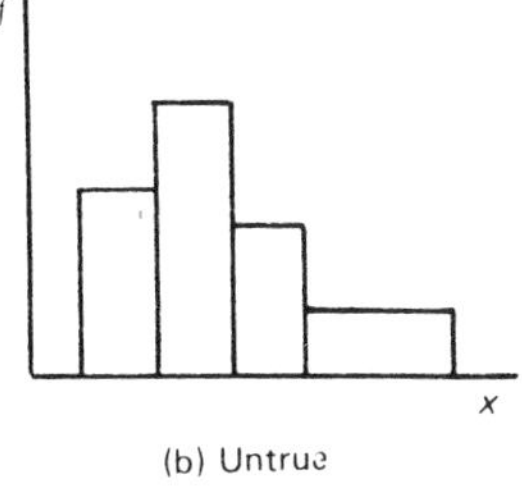

(b) Untrue

4.

x_i	f_i	d_i	d_if_i	
14.5	1	−5		−5
24.5	0	−4		0
34.5	1	−3		−3
44.5	4	−2		−8
54.6	7	−1	0	−7
64.5	16	0	0	−23
74.5	19	+1	19	
84.5	20	+2	40	
94.5	17	+3	51	
104.5	11	+4	44	
114.5	3	+5	15	
124.5	1	+6	6	
	100		+175	
			− 23	
		$\Sigma d_if_i =$	+152	

$$\bar{X} = 64.5 + \frac{10}{100} \cdot 152 = 64.5 + 15.2 = 79.7$$

6(b) *Washing Machines data (Exhibit 4)*

X	$(X-2)$	$(X-2)^2$	$(X-2\frac{1}{2})$	$(X-2\frac{1}{2})^2$	$(X-2\frac{5}{6})$	$(X-2\frac{5}{6})^2$	$(X-3)$	$(X-3)^2$
2	0	0	$-\frac{1}{2}$	.25	$-\frac{5}{6}$	25/36	−1	1
0	−2	4	$-2\frac{1}{2}$	6.25	$-2\frac{5}{6}$	289/36	−3	9
4	2	4	$1\frac{1}{2}$	2.25	$+1\frac{1}{6}$	49/36	1	1
6	4	16	$3\frac{1}{2}$	12.25	$+3\frac{1}{6}$	361/36	3	9
2	0	0	$-\frac{1}{2}$	.25	$-\frac{5}{6}$	25/36	−1	1
3	1	1	$\frac{1}{2}$	.25	$\frac{1}{6}$	1/36	0	0
		25		21.50		750/36 (= $20\frac{5}{6}$)		21

N.B. $\bar{X} = \dfrac{17}{6} = 2\frac{5}{6}$.

COURSE UNIT 5 – MEASURES OF VARIATION

5.1 THE NEED FOR A MEASURE OF VARIATION

In CU4 we developed ways of measuring average values of univariate sets of data. In effect what we did was to "replace" a complete frequency distribution by one single point on the histogram axis. Although the *average* provides an important and easily understandable summary figure (a sample statistic), it is insufficient by itself to summarise many of the differences between sets of data encountered in practice. Consider, for example, the following three histograms shown in Exhibit 1.

EXHIBIT 1: Three Histograms with the Same Mean Value

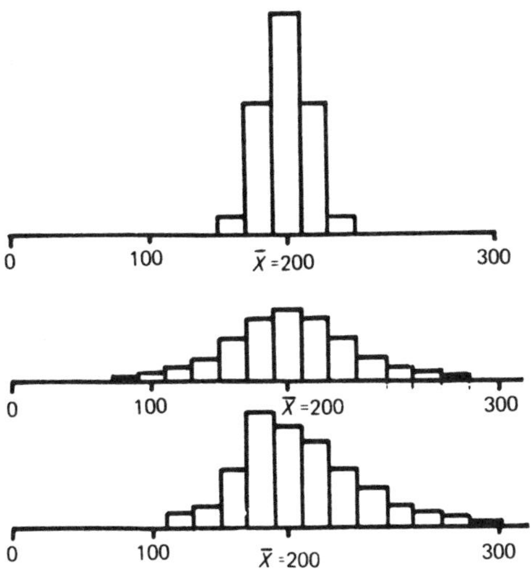

The three sets of data from which the histograms were drawn all have the same mean value (200), although the pattern of variation is clearly quite different in each case. It would thus be quite inadequate to describe each by its mean value and leave it at that. The main difference between (a) and (b) is that histogram (b) is more widely "spread", reflecting the greater degree of variability in the observations from which it was drawn. Similarly (c) is more widely spread than (a), although the comparison in this case is made more difficult by the fact that the histograms differ considerably in shape – (a) is symmetrical and bell-shaped whereas (c) is positively skewed – but despite these difficulties the notion of "spread" or "variation" is still a useful one in effecting a comparison. Such comparisons may be made quantitatively using summary figures called **measures of variation** (or alternatively, **measures of dispersion**). Just as with the various measures of average dealt with in CU4 there is no single measure of variation in universal use. The following measures of variation are used to a greater or lesser extent.

- Range
- Interquartile range (IQR)
- Mean Absolute Deviation (or MAD)
- Variance (and its square root the *standard deviation*).

As in CU4 we illustrate these measures of variation using the washing machine sales data used in that Course Unit and the data of Series A (both in a classified and unclassified form).

It is perhaps unfair to inflict upon the reader a study of such statistics without first giving some indication of why it is important to measure variation at all. If a population consists of observations all having the same value (i.e. displaying no variability) a sample of just one observation would be sufficient to find out all there is to know about the population. If, on the other hand, the population is very variable, a large sample would be needed to gain a representative picture of the population. To put it another way, the extent to which the observations vary in a given situation determines the reliablity of results taken from a sample. So if we *measure* the variability we can assess the reliability of sample results. Later (CU15) we shall find that, having measured the variability, we may be able to *design* a sample so as to provide a desired degree of reliability of precision.

5.2 RANGE

The **range**, originally introduced in CU1, is simply the largest observation minus the smallest observation in a set of data. We have already seen that the sample range is a biased and somewhat inefficient estimator of the population range. The reason for the bias is that unless the sample happens to contain the largest and smallest observation in the population (and that is most unlikely if the sample size is small in comparison with the population) the sample range must underestimate the population range. The sample range is an inefficient estimator of the population range simple because it only takes account of two outlying (and hence unreliable) values, the largest and the smallest, in a set of data.[1]

Despite these drawbacks the range is sometimes used as a measure of variation on account of its being easy to calculate (at least with small samples) and easy to understand. A particular application is in the field of Statistical Quality Control. Here a sample of manufactured components (usually between 5 and 10) is taken from a production line at regular intervals (say, hourly) and a critical dimension of each item measured. The range of these observations is then calculated and indicated on a "Range Chart" – the idea being to try and identify when the variability of the critical dimension has exceeded some predetermined norm. The reason why it is satisfactory to use the range as a measure of variability in this way is that we are not relying on one single sample but rather on a series of samples taken at regular intervals to produce evidence of the manufacturing process changing its variability: in particular we are *comparing* ranges based on the same (small) sample size. This topic is considered in more detail in SU4.

5.3 INTERQUARTILE RANGE

Just as the median represents that value above which 50% of the observations lie, the **upper quartile**, Q_3, is that value above which 25% of the observations lie, and, similarly, the **lower quartile**, Q_1, is that value *below* which 25% of observations lie[2]. Note the rather curious notation here; by comparison we could denote the median by Q_2, the "two-quartile". We define the **inter-quartile range** as $(Q_3 - Q_1)$ i.e. the range of values in which the central 50% of the observations lie. The calculation of Q_3 and Q_1 follows a closely similar process to the calculation of the median – either by using the definition on the ordered data, or by construction from the ogive, or by calculation from the frequency distribution. Using the definition on the ordered set of data we have:

$Q_3 = 94$, $Q_1 = 67$, Interquartile range $= 94 - 67 = 27$

---

[1]But the smaller the sample the more efficient is the range (relative to other measures of variation). For a sample consisting of just two observations the range makes use of all available sample information and is thus as efficient as any other measure of variation.

[2]This idea of dividing up an ordered set of data into a number of equal parts can be extended as follows. **Percentiles** divide the ordered set of data into 100 equal parts (so the 35-percentile is that value below which 35% of the observations lie), **deciles** divide the data into 10 equal parts, and **fractiles** into an unspecified number of parts. Note that the median = 50-percentile = 5th decile, and is the most widely used fractile!

and the so-called **semi-interquartile range** (or **quartile deviation**) is half this value, $13\frac{1}{2}$.

The relationship between the range and the interquartile range is shown in Exhibit 2 with reference to a positively skewed histogram and its associated ogive.

EXHIBIT 2: Graphical Illustration of Range and IQR

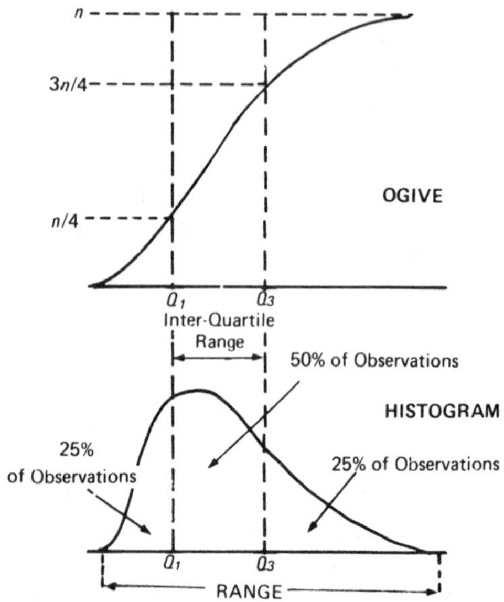

5.4 THE MEAN ABSOLUTE DEVIATION (MAD)

So far the measures of variation presented have attempted to measure the *overall* spread in the observations, rather than the extent to which the observations differ from a fixed reference value. The approach of measuring spread relative to a fixed value forms the basis for the Mean Absolute Deviation (sometimes called simply, the Mean Deviation) and the Variance as measures of variation. A natural reference value from which to measure the differences (or "deviations") of individual observations is the mean value. (One could argue that the median might be a better value since it splits the data into two groups with equal numbers of observations in each – those larger than the median, and those smaller. But the median as a convenient reference value for these purposes is rarely used on account of its unattractive analytical properties.) Let us develop this idea further, making use of the washing machine sales data, which has a mean value $\bar{X}$ of $2\frac{5}{6}$.

EXHIBIT 3: Deviations from the Mean and their Squares for the Washing Machines Sales Data

	X_i	$(X_i - \bar{X})$		$(X_i - \bar{X})^2$
Jan	2		$-\frac{5}{6}$	0.69 (i.e. 25/36)
Feb	0		$-2\frac{5}{6}$	8.01
Mar	4	$1\frac{1}{6}$		1.37
Apr	6	$3\frac{1}{6}$		10.05
May	2		$-\frac{5}{6}$	0.69
June	3	$\frac{1}{6}$		0.03
	Total	$4\frac{1}{2}$	$-4\frac{1}{2} = 0$	20.84

N.B. $\bar{X} = 2\frac{5}{6}$.

It seems an attractive proposition to derive a measure of variation from the sum of the "deviations from the mean" of all observations, i.e. $\Sigma(X_i - \bar{X})$, but as Exhibit 3 shows this value is zero. This is a direct consequence of using the mean as a reference value as can be seen from:

$$\sum(X_i - \bar{X}) = \sum X_i - \sum \bar{X} = n\bar{X} - n\bar{X} = 0.$$

If we ignore the fact that some of the deviations are negative (corresponding to observations smaller than the mean) and form the sum of the *absolute* deviations, i.e.[(1)] $\Sigma|X_i - \bar{X}|$ we obtain a statistic which has intuitive appeal as a measure of variation. As it stands this statistic has the undesirable property of being dependent on the number of observations in the sample – larger samples giving higher values. This disadvantage can be overcome, quite simply, by division by the sample size. So as a formal definition of the **mean absolute deviation** we have:

$$\text{MAD} = \frac{1}{n} \sum |X_i - \bar{X}|$$

$$= \frac{1}{6}(\tfrac{5}{6} + 2\tfrac{5}{6} + 1\tfrac{1}{6} + 3\tfrac{1}{6} + \tfrac{5}{6} + \tfrac{1}{6}) = \frac{9}{6} = 1\tfrac{1}{2}, \text{ in our example.}$$

Notice the name of this statistic – it is literally the mean of the absolute deviations (from the mean value).

The MAD has not found wide application as a measure of variation for much the same reason as that which accounts for the unpopularity

(1)The vertical brackets, | . . . |, are used in mathematics to denote "without regard to sign" and are known as **modulus** brackets (**mod**, for short).

of the median as a measure of average – its analytical intractability.[1] This is perhaps a pity because not only is the MAD a readily understandable measure of variation but it is to some extent a much "fairer" measure than the statistic that is in much wider use – the standard deviation. On occasions where analytical tractability is of little importance the MAD is indeed sometimes used. One example is on stock control where it is important to monitor the variability in demand for stocked items from week to week. Computer packages for stock control incorporate routines for evaluating the MAD and updating its value. This information provides the basis for deciding when a particular stocked item should be re-ordered.

5.5 THE DEFINITION OF VARIANCE

We saw in the previous section how the fact that deviations of individual observations from the mean cancelled each other out when added gave rise to a problem in developing a statistic for measuring variation. The solution given there was to sum the *absolute* deviations (i.e. ignore any minus signs). An alternative approach is to square the deviations before summing, as all squared quantities are positive. In this way we could form the sum $\Sigma(X_i - \bar{X})^2$ (= 20.84 in our example of washing machine sales – see Exhibit 3.)

Note that the effect of squaring the deviations before adding them is to emphasise those observations which differ most from the mean value. Referring to Exhibit 3 it can be seen that the observation 6, the April figure, accounts for about 35% of the "sum of *absolute* deviations" but almost 50% of the "sum of *squared* deviations".

Dividing $\Sigma(X_i - \bar{X})^2$ by n to obtain a statistic which is not dependent on the sample size, we arrive at the definition of **variance** – the most important measure of variation for analytical purposes.

$$\text{Variance, } s^2 = \frac{1}{n} \sum (X_i - \bar{X})^2 \qquad (1)$$

Notice that the symbol chosen as the sample variance, s^2, is itself a squared quantity, reminding us that statistic has dimensions of (measurement)^2. The value of s, the square-root of the variance, is called the **standard deviation** (SD) which is more widely used in applied statistics. It is sometimes referred to as the "Root Mean Squared Deviation" – a direct translation of the mathematical formula:

[1] The expression for the MAD, for example, cannot be differentiated.

Standard Deviation, $s = \sqrt{\frac{1}{n}\sum (X_i - \bar{X})^2}$

Thus for the data in Exhibit 3 we have,

$$s^2 = \frac{1}{6} \cdot 20.84 = 3.47$$

and

$$s = \sqrt{3.47} = 1.86$$

Note that this standard deviation is somewhat greater than the MAD, (=1.50), although of the same order of magnitude. (See Exercise 2.)

5.6 CALCULATION OF THE VARIANCE FOR UNCLASSIFIED DATA

Formula (1), with which we defined the variance, is extremely cumbersome to apply in practice. The reason is that from each observation the mean has to be subtracted and then the result squared. As it is usual to find that the mean value is an awkward valued decimal the process involves squaring and adding awkward decimal numbers, however "neat and tidy" were the original observations. Consequently it is more convenient to calculate the variance using a different but totally equivalent formula[(1)]:

$$s^2 = \frac{1}{n}\sum X_i^2 - \bar{X}^2 \qquad (2)$$

This formula, which is probably the most useful of all formulae for the variance, can most easily be remembered by the phrase:

"The mean value of X^2 – (mean value of $X)^2$"

Its application to the washing machine sales data is shown in Exhibit 4.

---

(1) Proof of $s^2 = \frac{1}{n}\sum X_i^2 - \bar{X}^2$

The definition of variance is:

$$s^2 = \frac{1}{n}\sum (X_i - \bar{X})^2$$

$$= \frac{1}{n}\left\{\sum X_i^2 - 2\bar{X}\sum X_i + \sum \bar{X}^2\right\}$$, using the rules of summation,

$$= \frac{1}{n}\left\{\sum X_i^2 - 2n\bar{X}^2 + n\bar{X}^2\right\}$$, using the definition of $\bar{X}$,

$$= \frac{1}{n}\left\{\sum X_i^2 - n\bar{X}^2\right\} = \frac{1}{n}\sum X_i^2 - \bar{X}^2.$$

EXHIBIT 4: Use of Formula (2) in Calculating Variance

	X_i	X_i^2
Jan	2	4
Feb	0	0
Mar	4	16
Apr	6	36
May	2	4
Jun	3	9
	$\Sigma X_i = 17$	$\Sigma X_i^2 = 69$

$$s^2 = \frac{69}{6} - \left(\frac{17}{6}\right)^2 = \frac{414 - 289}{36} = \frac{125}{36} = 3.47, \text{ as before.}$$

On the more advanced calculating machines (and on a computer) we can accumulate the value of ΣX^2 and ΣX (the latter for the purpose of evaluating $\bar{X}$) at the same time.

For Series A we have:

$$\sum X = 8{,}000, \sum X^2 = 676{,}184,\ n = 100.$$

So

$$s^2 = \frac{676{,}184}{100} - \left(\frac{8{,}000}{100}\right)^2 = 6{,}761.84 - 6{,}400 = 361.84.$$

Therefore

$$s = \sqrt{361.84} = 19.02.$$

5.7 CALCULATION OF THE VARIANCE FOR CLASSIFIED DATA

Using the standard notation of x_i to denote the class mid-points and f_i the associated frequencies, and making the usual assumption that x_i's are representative values within the classes, the following is an equivalent formula for variance calculation:

$$s^2 = \frac{\sum x_i^2 f_i}{\sum f_i} - \bar{X}^2 \qquad (3)$$

where $\bar{X}^2 = (\sum x_i f_i / \sum f_i)^2$

Exhibit 5 illustrates its use on Series A.

EXHIBIT 5: Use of Formula (3) in Calculating Variance

x_i	f_i	$x_i f_i$	$x_i^2 f_i$
14.5	1	14.5	210.25
24.5	0	0	0
34.5	1	34.5	1,190.25
44.5	4	178.0	7,921.00
54.5	7	381.5	20,791.75
64.5	16	1,032.0	66,564.00
74.5	19	1,415.5	105,454.75
84.5	20	1,690.0	142,805.00
94.5	17	1,606.5	151,814.25
104.5	11	1,149.5	120,122.75
114.5	3	343.5	39,330.75
124.5	1	124.5	15,500.00
	$n = \Sigma f_i = 100$	$\Sigma x_i f_i = 7{,}970.0$	$\Sigma x_i^2 f_i = 671{,}705.00$

Thus $s^2 = \dfrac{671{,}705}{100} - \left(\dfrac{7{,}970}{100}\right)^2 = 6{,}717.05 - 6{,}352.09 = 364.96$

Therefore $s = \sqrt{364.96} = 19.104$.

A number of features of this calculation merit comment. Firstly the values of $x_i f_i$ have been calculated for the purpose of evaluating $\bar{X}$, which itself is required for the variance calculation. Secondly the values of $x_i^2 f_i$ are most easily calculated from x_i and $x_i f_i$ in the form of $x_i \cdot (x_i f_i)$. (This avoids squaring the x_i's as a separate operation.) Lastly the result is not quite the same as that obtained from the unclassified data (the result then was $s^2 = 361.84$) – the reason being that the x_i's are not exactly representative values within the classes.

5.8 CALCULATION OF THE VARIANCE FOR CLASSIFIED DATA BY HAND

It will have been seen that calculation of the variance in the manner shown in the last section involves some unwieldy arithmetic. This is acceptable only if the calculations are performed by calculating machine or computer or if the values of the x_i's and f_i's are relatively small. As was the case for the calculation of the mean, hand calculation can be simplified by coding the data. Using the notation developed in Section 4.7 of CU4, a formula directly equivalent to Formula (3) is given by[(1)]

[(1)]The quantity in the square brackets in Formula (4) is in fact the variance of the d's (coded values). The multiplier c^2 converts the variance of the coded values to the variance of the original values.

$$s^2 = c^2 \left[\frac{\sum d_i^2 f_i}{\sum f_i} - \left(\frac{\sum d_i f_i}{\sum f_i} \right)^2 \right] \quad (4)$$

where d_i is the coded value of the class mid-point x_i.

We rework the variance of the Series A data in this manner (Exhibit 6).

EXHIBIT 6: Use of Formula (4) in Calculating Variance

x_i	f_i	d_i	$d_i f_i$		$d_i^2 f_i$	
14.5	1	−6	−6		36	
24.5	0	−5	0		0	
34.5	1	−4	−4		16	
44.5	4	−3	−12		36	$a = 74.5$
54.5	7	−2	−14		28	$c = 10$
64.5	16	−1	−16		16	
74.5	19	0	−52	negative subtotal	0	
84.5	20	1	20		20	
94.5	17	2	34		68	
104.5	11	3	33		99	
114.5	3	4	12		48	
124.5	1	5	5		25	
	$n = f_i =$ 100		104	positive subtotal	$\Sigma d_i^2 f_i =$ 392	
			−52			
			$\Sigma d_i f_i =$ 52			

We have $s^2 = c^2 \left[\frac{\Sigma d_i^2 f_i}{\Sigma f_i} - \left(\frac{\Sigma d_i f_i}{\Sigma f_i} \right)^2 \right]$

$$= (10)^2 \left[\frac{392}{100} - \left(\frac{52}{100} \right)^2 \right]$$

$$= (10)^2 \times 3.6496$$

so $s = 10\sqrt{3.6496} = 19.104$, as in Exhibit 5.

5.9 PROPERTIES OF THE SAMPLE VARIANCE

Our definition of the sample variance was given (Section 5.5) as

$$s^2 = \frac{1}{n} \sum_{i=1}^{n} (X_i - \bar{X})^2$$

Similarly we could define the variance of a finite population of N observations as

$$\sigma^2 = \frac{1}{N} \sum_{i=1}^{N} (X_i - \mu)^2$$

Note the essentially similar forms of those two definitions (N replacing n and μ replacing $\bar{X}$) and the use of the Greek symbol σ (sigma)[1] for the population standard deviation.

It is now appropriate to consider, as we have done for other statistics, how good an estimator s^2 is of σ^2. We approach this problem in two separate ways – neither of them too rigorous. Suppose, for the sake of argument, we have a sample of n observations where we happen to know the population mean, μ, of the population from which that sample was selected. We would then expect that the quantity

$$\frac{1}{n} \sum_{i=1}^{n} (X_i - \mu)^2$$

would be a more accurate estimate of the population variance than that given by the direct application of the sample variance definition

$$\frac{1}{n} \sum_{i=1}^{n} (X_i - \bar{X})^2$$

But we have already seen that $\Sigma_{i=1}^{n} (X_i - \bar{X})^2$ is a smaller quantity than $\Sigma_{i=1}^{n} (X_i - \mu)^2$ – see CU4 Section 4.8 and Exercise 6 in that course unit. Consequently the sample variance must on average *underestimate* the value of the population variance, i.e. the sample variance is a "downward biased" estimator of the population variance.

Let us approach this problem another way. Consider, as we did in CU4, the simple population having just three observations, 1, 2 and 3. The population mean and variance are given by:

$$\mu = \frac{1 + 2 + 3}{3} = 2$$

$$\sigma^2 = \frac{1}{3} [(1-2)^2 + (2-2)^2 + (3-2)^2] = \frac{2}{3}$$

Let us now consider all possible samples, with replacement, of just two observations and calculate the sample variance for each (Exhibit 7).

---

[1] Don't confuse the lower-case sigma, σ, the population SD with the upper-case sigma, Σ, the summation operator.

EXHIBIT 7: All Possible Samples from the Population 1, 2, 3

Sample	*Sample Mean* ($\bar{X}$)	*Sample Variance* (s^2)
1 1	1	0
1 2	$1\frac{1}{2}$	$\frac{1}{4}$
1 3	2	1
2 1	$1\frac{1}{2}$	$\frac{1}{4}$
2 2	2	0
2 3	$2\frac{1}{2}$	$\frac{1}{4}$
3 1	2	1
3 2	$2\frac{1}{2}$	$\frac{1}{4}$
3 3	3	0
		Total 3

The average sample variance of all possible samples is $\frac{3}{9}$ $(=\frac{1}{3})$, a firgure considerably less than the population variance. Indeed, we would need to multiply the average sample variance by the "correction factor" 2 to obtain the population variance. In general, for a sample of n observations we need to multiply the sample variance by a factor $(\frac{n}{n-1})$ in order to obtain an unbiased estimate of the population variance[1]. This factor, called Bessel's correction factor, enables us to state:

$$\hat{\sigma}^2 = \left(\frac{n}{n-1}\right)s^2 = \frac{1}{n-1}\sum(X_i - \bar{X})^2.$$

That is to say, the "best" estimate of the population variance is obtained by replacing n in the definition of sample variance by $(n - 1)$. This explains why many text books *define* variance with $(n - 1)$ as the divisor. We have taken the view that it is more logical to have consistent definitions for the sample and population variance and to admit that the statistic is a biased estimator of the parameter. (It is interesting to note that no texts define the sample range to be an unbiased estimator of the population range.) Strictly speaking Bessel's correction factor should always be applied when an unbiased estimate is being sought: in practice, however, it is often ignored for large samples (where n is at least 30).

---

(1) $n = 2$ in our example, so $\left(\frac{n}{n-1}\right) = \frac{2}{1} = 2.$

5.10 RELATIONSHIP BETWEEN THE STANDARD DEVIATION AND THE RANGE

Providing one refers to distributions of a defined shape there are well-known relationships between the standard deviation and the range. The most important distributions in statistics are those which we have previoulsy described as symmetrical and bell shaped – in later Course Units we shall refer to these as "normal" distributions. For such distributions the relationship between the standard deviation and the average sample range is given in Exhibit 8.

EXHIBIT 8: Conversion Factors from Range to Standard Deviation

Sample Size (n)	*Conversion Factor*	*n*	*Conversion Factor*	*n*	*Conversion Factor*
2	1.13	20	3.73	200	5.49
3	1.69	30	4.09	300	5.75
4	2.06	40	4.32	400	5.93
5	2.33	50	4.50	500	6.07
6	2.53	60	4.64	600	6.18
7	2.70	70	4.75	700	6.28
8	2.85	80	4.85	800	6.35
9	2.97	90	4.94	900	6.42
10	3.08	100	5.01	1000	6.48

The conversion factors give the amount by which we must *divide* the range (or even better, the range averaged over a number of equal sized samples) in order to estimate the population standard deviation.

One use to which such a conversion table could be put is in producing a "quick and dirty" estimate of the standard deviation from the range. Take Series A as an example – we are justified in doing so because its histogram is symmetrical and bell-shaped. The range (of a sample of 100) is 101. Thus our estimate of the standard deviation, using the conversion factor of 5.01 in the table, is 101/5.01 = 20.2 (which is quite close to the result obtained by lengthy calculation, 19.0).

Another use of the table is in Statistical Quality Control, where, because we are taking samples on a regular basis we can average the ranges from the samples and thus arrive at a reasonably accurate value for the standard deviation (SU4).

Lastly we can use the notion of the range/SD relationship to enable us to sketch rough histograms (which in themselves help to guide our thinking) from a knowledge of the mean and standard deviation. It can be seen that for reasonably large samples (*n* greater than 100, say), the average range is very approximately 6 times the standard deviation. As we are considering a symmetrical bell-shaped distribution the mean will be centrally disposed in the distribution, and thus we might say that the largest observed item is about three standard deviations more than the mean, and the smallest observed item is about three standard deviations less than the mean. Using these crude guidelines we could sketch a distribution whose mean is 100 and standard deviation 20, say, in the manner shown in Exhibit 9.

EXHIBIT 9: Sketch of a Distribution from the Knowledge of the Mean and Standard Deviation

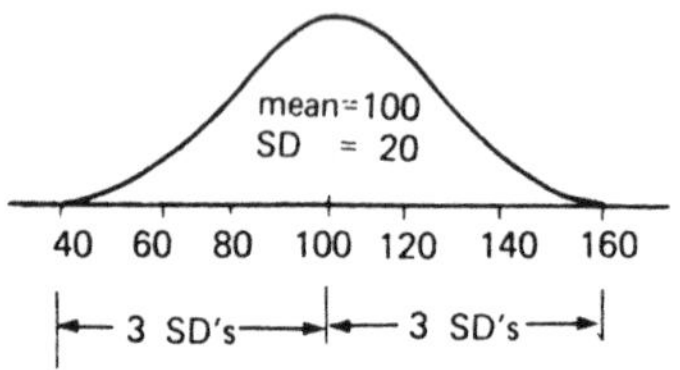

5.11 OTHER STATISTICS RELATED TO THE STANDARD DEVIATION

Consider two products C and D, the weekly demand for which can be summarised as follows.

Product	Mean Demand per Week	SD of Demand per Week
C	1000	40
D	50	40

In absolute terms these two products have an equally variable demand pattern, as expressed by the fact that their standard deviations are equal. Over a period of a year, for example, the demand for product C may vary between 900 and 1100 units in a week, and that for product D between 0 units and (exceptionally) 200 units in a week. But clearly the variability in the demand for D will be much more noticeable (and consequently more important, in terms of policies for stocking and ordering the product) than that of product C, which could be considered as being a product with a sensibly constant demand rate. To measure the *relative importance* of variation in a sample, a statistic called the **coefficient of variation** is often employed.

$$\text{Coefficient of Variation (CV)} = \frac{\text{SD}}{\text{Mean}} \tag{5}$$

Thus the CV's of products C and D are 40/1000 (=0.04) and 40/50 (= 0.8) respectively. Sometimes CV's are given as percentages, and so for products C and D they can be expressed as 4% and 80%.

The coefficient of variation may be used to *compare* the variability of two samples which differ in their mean values or where the samples relate to quite different types of measurement (see, for example, Exercise 7). It is worth noting, in this context, that the coefficient of variation is independent of the units of measurement. A group of adult males may have a mean height of 69″ and SD of 6″ when measured in inches (or 175.3 cm and 15.2 cm, respectively, in metric units). Although the SD values depend on the scale of measurement used (inches or cms) the CV (≃0.09) is the same for both.

The coefficient of variation is also *indirectly* related to the skewness of the distribution. For variables which cannot take negative values (such as the demand for products, heights and weights of components) symmetrical bell-shaped distributions have CV's of less than 0.3 (Series A data, for instance, has a CV of 0.24), skewed distributions tend to have a higher CV than 0.3, and very skewed distributions, such as reversed-J distributions, often have a CV of approaching (or possibly exceeding) 1.0.

To measure skewness *directly* a number of measures have been suggested. One of the simplest to evaluate is known as **Pearson's measure of skewness.**

$$\text{Skewness} = \frac{\text{Mean} - \text{Mode}}{\text{SD}} \tag{6a}$$

$$\simeq \frac{3\,(\text{Mean} - \text{Median})}{\text{SD}} \tag{6b}$$

Because the mode (and the median) is greater than the mean for a positively skewed distribution and less than the mean for a negatively skewed distribution the *sign* (+ or −) of this measure of skewness indicates the *direction* of skewness and its numerical value, the *strength* of the skewness.

Consider the following statistics taken from a survey of the "age at death" of men in a certain professional occupation:

Mean	Median	Mode	SD
72.2	74.1	78.0	8.1

We have, using (6a): Skewness = (72.2 – 78.0)/8.1 = –0.72
and using (6b): Skewness = 3 (72.2 – 74.1)/8.1 = – 0.70

We see that the two methods of calculation give similar, though not equal, figures. (Formula 6(b) is the preferred method of evaluation with *small* samples because of the difficulty of mode determination.) These figures indicate a *negative* skew and the numerical value of about 0.7 means that the skewness is quite pronounced – it is rare in practice to find numerical values of greater than one. Symmetrical distributions have a skewness value of zero because the mean, median and mode are equal.

This measure of skewness shares with the coefficient of variation the property of being independent of the units of measurement, by virtue of being a *ratio* of statistics in the same units.

READING

M.J. Moroney *Facts from Figures* Penguin, Chapter 5 and some parts of Chapter 6.

EXERCISES

1(P) Use Series B (together with the frequency distribution you have already worked out for this data) to perform the following calculations.

(i) Calculate the interquartile range in any way you choose
(ii) Calculate the variance and standard deviation using the unclassified data.
(iii) Calculate the variance and standard deviation from the frequency distribution using a calculating machine.
(iv) Calculate the variance and standard deviation from the frequency distribution by hand.

Comment on any discrepancies you find between the results for (ii), (iii) and (iv). Indicate 1, 2 and 3 standard deviations greater than (and less than) the mean on your histogram.

2(U) Evaluate the Mean Absolute Deviation for Series A (using the mean value of 80.0). Comment on the following relationship which holds for symmetrical bell-shaped distributions.

Standard deviation $\simeq$ 1.25 • MAD

3(U) A formula sometimes useful for calculating the variance of small samples by hand is:

$$s^2 = \frac{1}{n}\sum (X_i - a)^2 - (a - \bar{X})^2$$

where a is any convenient quantity (the assumed average). Use this formula to calculate the variance of the washing machine sales data. Also prove that the formula is correct.

4(E) A sample of 400 returns from a survey of the independent sector of the grocery trade gave the number of employees for each shop, in addition to the owner/manager and his family

Number of Employees	*Number of Shops*
0	129
1	92
2	76
3	49
4–6	30
7–9	16
10 or more	8

(a) Calculate the mean and standard deviation of the number of employees per shop by hand.
(b) Comment on the skewness of the distribution and suggest one measure of skewness that could be used.

(N.B. you will need to make some reasonable assumption about the open-ended class).

HNC (Business Studies)

5(E) (a) What does the information in the table below indicate about the earnings of full-time men and women workers in the retail drapery, outfitting and footwear trades at October 1969 and April 1970?

(b) What does the data indicate about the relative earning positions of the lowest paid men and women workers in these trades?

Weekly Earnings in the Retail Drapery, Outfitting and Footwear Trades

	Full-time Men Workers			Full-time Women Workers		
	Lowest Decile £	Lower Quartile £	Median £	Lowest Decile £	Lower Quartile £	Median £
October 1966	12.36	14.33	16.89	7.78	8.43	10.00
April 1970	15.20	17.60	21.10	9.20	10.30	12.00

(ACCA June 1972)

6(E) The distribution shown below is the output of the factories of Quality Clothing Ltd., for the month of May 1972. You are required to:

(a) Calculate the standard deviation from these figures, and
(b) contrast the mean deviation and the standard deviation as measures of dispersion and indicate briefly what the standard deviation calculated in (a) means for the monthly output of Quality Clothing Ltd.

QUALITY CLOTHING LTD.

Monthly output men's suits (000's)	Number of factories
23 and under 28	10
28 and under 33	20
33 and under 38	20
38 and under 43	24
43 and under 48	20
48 and under 53	16
53 and under 58	8
58 and under 63	2

(ACCA June 1972)

7(E) Machinery Suppliers Ltd. employ a force of 50 travelling salesmen. The sales manager has initiated a statistical analysis of their daily travel records over a six month period. The analysis disclosed the distances travelled, time spent on travel, and petrol used by each salesman, each day. A report showing mean and standard deviation of these figures was submitted to the sales manager.

MACHINERY SUPPLIES LTD.

Salesmen's Travel Records Analysis. January–July 1972.

	Distance (Miles)	Time (Minutes)	Petrol Consumption (Gallons)
Mean	100	120	4
Standard Deviation	20	30	0.5

(a) Calculate the variability in (i) the salesmen's mileage, (ii) the time spent on travel and (iii) petrol consumption.

(b) Explain why the sales manager might be interested in the calculations in (a) above and justify the method you have used.

(ACCA June 1973)

Hint: Interpret "variability" as the coefficient of variation.

8(E) The distribution of the lengths of service of random samples of 100 workers from each of two factories are shown below:

Period of service (months)	Factory A	Factory B
0–2	20	7
3–5	25	7
6–8	14	6
9–11	8	7
12–17	11	12
18–23	10	6
24–35	8	17
36–47	2	15
48 and over	2	23

Compare the two distributions,

(a) using an appropriate graphical method;

and (b) using appropriate measures of location (or central tendency) and of dispersion.

(IPM November 1972)

SOLUTIONS TO SOME EVEN NUMBERED QUESTIONS

2. $$\text{MAD} = \frac{\sum |X - \bar{X}|}{n} = \frac{1536}{100} = 15.36$$

$s \simeq 1.25 \cdot \text{MAD} = 19.20$ which is a close approximation to $s = 19.10$ (Section 5.7).

4. (a)

Number of Employees x_i	*Number of Shops* f_i	$x_i f_i$	$x_i^2 f_i$
0	129	0	0
1	92	92	92
2	76	152	304
3	49	147	441
5*	30	150	750
8*	16	128	1024
12*	8	96	1152
	400	765	3763

$$\bar{X} = \frac{765}{400} = 1.9125$$

$$s^2 = \frac{3763}{400} - \left(\frac{765}{400}\right)^2$$

$$= 9.4075 - (1.9125)^2$$

$$= 5.750$$

$$s = \sqrt{5.750} = 2.40$$

*Class mid-points (the last of which, 12, is an estimate).

(b) The distribution is very positively skewed, so much so that its mode is zero, i.e. a reversed J-shape distribution. The skewness could be calculated from:

$$\text{Skewness} = (\text{Mean} - \text{Mode})/\text{SD} = (1.9125 - 0)/2.40$$

$$= 0.80$$

COURSE UNIT 6 – INTRODUCTION TO PROBABILITY

6.1 INTRODUCTION

The course so far has dealt with various ways of summarising information taken from samples. In addition we have given a little thought to the relationship between the properties displayed by samples and the properties of the parent population from which the samples are drawn. We now turn our attention to an entirely different matter. By the introduction of the concept of probability we shall find that we are able, in certain circumstances, to predict the properties of samples *without physically taking any samples at all.*

The origins of the concept of probability lie in a simple mathematical theory of games of chance. During the 1650's a French gambler, de Méré, consulted a well-known mathematician, Pascal, who in turn corresponded with a colleague, Fermat; this correspondence forms the origin of modern probability theory. It is thus not surprising that most elementary treatments of probability refer to simple games of chance such as tossing coins and rolling dice. If games of chance seem far removed from any business activity, it is hoped that the reader will be reassured by the following thought. Businessmen make decisions in the face of uncertainty. When a company launches a new product, for example, the marketing manager will be by no means certain of the eventual sales potential of that product. When a gambler places his bet he will, to a greater or lesser extent, be assailed by similar doubts as to the final outcome. It is the uncertainty which is common to these two situations, and this is why a study of probability, based on games of chance, provides the necessary analytical tools for measuring and controlling the various forms of uncertainty with which the businessman is faced.

The word probability is a technical word but corresponds very closely to what is commonly understood by the word "chance". Consider the following three statements, each of which uses the phrase "50% chance":

Statement 1 There is a 50% chance that a tossed coin will show "heads".

Statement 2 There is a 50% chance that a television tube will last for more than 3 years.

Statement 3 There is a 50% chance that a certain company's turnover will exceed £1 m. in 5 years' time.

To rewrite the statements in statistical language one would simply need to replace the phrase "50% chance" by the technical phrase "probability of 0.5" – it is usual among statisticians to express probabilities in fractional or decimal form, rather than as a percentage. However, when conversing with non-statisticians the percentage method is commonly used.

But how do these statements differ in the use of the phrase "50% chance"? It might be agreed that the most significant difference is the extent to which the statements rely, for their truth, on historical information.

Statement 1, for example, could well rely on no historical information at all. We might examine a given coin and aver that Statement 1 were true, simply on the basis that the coin appears "true" and has one "head" and one "tail".

Statement 2, on the other hand, can only be a statement based on historical information. It must mean that, for similar television tubes used in similar circumstances, 50% of the tubes *in the past* lasted for more than three years – there is nothing intrinsic about a television tube as there was for the coin that enables us to make such a statement without relying on historical evidence.

Finally, Statement 3 is neither a statement about the intrinsic nature of the company's growth rate, nor directly related to historical evidence (because the next 5 years for this company will be a unique experience). Thus Statement 3 is *an opinion* expressing a measure of confidence in the firm's prospects over the next five years.

So the notion of chance, or probability, can arise and take meaning in any one of these three ways. In cases where probability statements are made "by the very nature of things", as in Statement 1 we say that we are using an *a priori*[(1)] notion of probability. Where the probability statement is essentially related to historical evidence, as in Statement 2, we are using an *empirical*[(2)] notion of probability. Where the probability is simply to formalise an opinion, as in Statement 3, we are using a *subjective* or *personalistic* notion[(3)].

Probability theory can be developed along either *a priori* or *empirical* lines, giving exactly the same results; so once we have established any

[(1)] Literally, "before the event". It is sometimes called *theoretical* or *symmetric* probability.

[(2)] By contrast with *a priori*, empirical probability is sometimes called *a posteriori* – "after the event".

[(3)] Whether a probability is *a priori*, empirical or subjective depends not on whether historical or theoretical information *can* be used to evaluate it, but whether, in a given situation, it *is* used.

basic probabilities we can manipulate them without regard to whether they are *a priori*, empirical or, indeed, subjective. (As a matter of interest, purists will insist that the *development* of probability theory must proceed with the use of none of these types of probability, but with the more abstract notion of *axiomatic* probability – the reasons for this will be seen later.)

6.2 SOME DEFINITIONS

As a prerequisite to defining probability in any of these three ways, we shall need to introduce a number of technical terms. These may, at the outset, appear somewhat abstract but it is hoped the illustrations which follow them will make the matter clear. The reader is advised to work quickly through the definitions and then turn to the illustrations, referring back to the definitions in order to check that these are well understood.

Trial A trial is a description of a process, the results of which are unknown until that trial takes place. Notice the connexion between this technical use of the word "trial" and the word used in its normal (legal) context. (As we shall see later, a trial may take the form of an experiment using a new machine in production, or a survey of consumers in marketing.)

Outcome An outcome is a *detailed* description of a possible result of a trial. By "detailed" we mean here "as detailed as necessary to separate out results which are, in principle, different". Thus if a trial consisted of tossing a coin twice, then one outcome would be "tails first, heads second". If on the other hand two coins were tossed simultaneously, we would need a notional labelling scheme (one coin being called "coin A", the other "coin B") in order to separate out the result "one head, one tail" into the two different outcomes, "coin A heads, coin B tails" and "coin A tails, coin B heads".

Event An event is a description of a possible result of a trial which is of interest to us – the main difference between an outcome and an event being the degree of detail used in the description. In fact, it is often the case that an event is a description of a set of outcomes which have certain features in common (never the other way around!).

Random Variables The term "random variable" is the name given to numerical values associated with events. In other words the notion of a random variable enables us to refer to events by their associated numerical values – a form of shorthand.

We now give three illustrations to clarify the meaning of these terms.

Illustration 1

Trial: Toss two coins.

Outcomes: There are only four alternatives if we ignore the possibility of a coin landing on its edge.

1. HH (meaning "first coin heads, second coin heads")
2. HT (meaning "first coin heads, second coin tails")
3. TH
4. TT

Events: The following are examples:

(a) Just one head (i.e. outcomes 2 and 3).
(b) At least one tail (i.e. outcomes 2, 3 and 4).
(c) "Both the same" (i.e. outcomes 1 and 4).

Random Variables: Again these are examples taken from a large number of possibilities.

(i) The number of heads shown (can take values 0, 1 or 2), corresponding to events of type (a) above.
(ii) £1 if both the same, nothing otherwise; corresponding to events of type (c) and appropriate to a simple game of chance.
(iii) The number of heads less the number of tails (can take values –2, 0, 2).

Illustration 2

Trial: Select an employee and determine the amount of overtime worked in the past month.

Outcomes: An infinite number of possibilities here, since time is a continuous variable. They will be of the type "43 hrs. 39 mins.", but the numerical value will be different for each outcome.

Events: The following are examples:

(i) Overtime worked, to the nearest half-hour, of $43\frac{1}{2}$

hrs. (comprising all outcomes from 43 hrs. 15 mins. to 43 hrs. 45 mins.).

(ii) Overtime worked of at least 40 hrs.

Random[1] Variables

(i) Overtime worked to the nearest minute.

(ii) Overtime worked to the nearest half-hour.

(iii) The value of wages paid, in £p, for the overtime worked.

Illustration 3

Trial: Take a sample of 100 quarterly gas consumption values, to the nearest therm, of consumers using a certain type of central heating installation.

Outcome: The data given in Series A, in the order originally given, is a single outcome. Different outcomes are given by changing the order of the observations and/or their values.

Events: Examples are:

(i) The sum of the values is 8000 (note that this event occurred in the trial which resulted in Series A).

(ii) 12% of the values are greater than 100.

(iii) A frequency distribution identical to the one given in CU2. Note that there are a large number of outcomes comprising this event.

Random Variables: Examples are:

(i) The sum of the consumption values.

(ii) The sample mean.

(iii) The sample range.

Note that sample statistics are themselves random variables (a fact that we noted in CU1 when investigating the relationship between sample statistics and population parameters). This important result will be used in later course units when we shall be asking the question "what is the probability that a given sample statistic will lie within a certain range of values?".

---

[1]If we actually carried out the trial referred to in Illustration 2 and recorded the value of one of these random variables, this would be a single "observation" – which illustrates the relationship between our current use of the term "random variable" and its previous use in CU3.

6.3 DEFINITIONS OF PROBABILITY

Making use of our definitions of trial, outcome, event and random variable we are now in a position to define probability in each of the three ways – *a priori*, empirical and subjective. In each case a definition will be given of an event. Suppose we consider tossing two coins and wish to determine the probability of just one head being shown. We can use one or other of the following notations:

(a) Pr (just one head),

or (b) Pr (E) if we have already defined the event E as "E: just one head"[(1)],

or (c) Pr ($R = 1$) if we have already defined R as the random variable "the number of heads shown",

or even (d) $p(1)$ if the context is clear. Note that in this case we use the notation $p(1)$ not[(2)] Pr(1).

We define ***a priori*** probability in the following manner:

$$\text{Pr(Event)} = \frac{\text{Number of equally likely outcomes which comprise the event}}{\text{Total number of equally likely outcomes}} \quad (1)$$

So taking the above example (of tossing two coins) and assuming that the coins are "true" (i.e. flat, uniform thickness) we might assert that each of the four possible outcomes, HH, HT, TH and TT, are equally likely. Two of these outcomes, HT and TH, comprise the event in which we are interested, so,

$$\text{Pr (exactly one head)} = \frac{2}{4} = \frac{1}{2}.$$

Similarly, if we roll a "fair" die[(3)] there are *six* equally likely outcomes, corresponding to the six faces. Suppose we wish to determine the probability of the event, A, defined by A: 1 or 6 is shown. Clearly there are two equally likely outcomes which comprise this event, so:

$$\text{Pr}(A) = \frac{2}{6} = \frac{1}{3}$$

The careful reader will have noticed the circularity in the above

(1)The colon (:) in the definition of E is read "is defined by".

(2)The change in notation occurs because with Pr(. . .) the item in the brackets is an *event*, whereas with $p(.)$ it is the numerical value of a random variable.

(3)"die" is the singular form of the word dice: one die, two dice.

definition. We should not, strictly, use the phrase "equally likely", a notion involving probability, to define probability itself. Notice too that the words "fair" and "true" also beg the question. But the concept of equal likelihood is so fundamental to a study of Statistics and is also so intuitively clear that it is hoped that the writers will be forgiven for its use in this way.

Even allowing for these shortcomings, the *a priori* definition has its limitations. If we have a bent coin or a loaded die we cannot invoke the concept of equal likelihood and so an *a priori* evaluation of probability is not possible. Another difficulty arises if the numerator and denominator in equation (1) are infinite: the probability is then indeterminate. Consider, for example, the probability of selecting an odd-valued whole number from all possible whole numbers. The main limitation, however, is that we cannot derive *a priori* probabilities for events which are not concerned with "simple" games of chance, such as for evaluating the probability of an employee staying with a company for at least two years. Fortunately we can overcome most of these difficulties by using the empirical notion of probability.

It is more difficult to *define* **empirical** probability than to calculate its (approximate) value in a particular case: we start off by attempting a definition and then move on to some evaluations. Suppose a defined trial is repeated n times and we observe the number of times, f_E, that an event E occurs. The *relative frequency* of the event E is given by f_E/n. The empirical probability that a *new trial*, carried out under "identical" conditions, will result in E is defined as the "long run" relative frequency, i.e. the value of f_E/n as n approaches infinity. To see what we mean, suppose a coin were tossed 10 times, then 100 times, then 1000 times and the number of heads shown counted in each of these three cases. The results might be as shown in Exhibit 1.

EXHIBIT 1: The Concept of Empirical Probability

n	(number of tosses)	10	100	1000
f_E	(number of heads observed)	6	47	510
$\frac{f_E}{n}$	(relative frequency of heads)	0.6	0.47	0.510

It does not require too much imagination to visualise that if the coin were tossed a million times the relative frequency of heads might be very close to 0.500. If that were the case we could maintain that the coin were "true" (the *assumption* that would be made for an *a priori* statement of probability.

The definition we have given does not enable us to *calculate* empirical probabilities, simply because we never have the experience of an infinite number of trials. So in practice we have to content ourselves with approximate statements of empirical probability, viz:

$$\Pr(E) \simeq \frac{f_E}{n}, \qquad (2)$$

where the event E occurred with a frequency f_E in a *large number, n,* of "identical" trials.

As an example, consider the frequency distribution of the "length of stay" of the last 100 female employees to leave a packaging organisation, as shown in Exhibit 2. It can be seen that 38 out of the 100 employees stayed with the company for at least two years. So we can say that the

EXHIBIT 2: Distribution of Length of Stay

Length of Stay (Years)	*No. of Employees*	
0–1	26	
1–2	36	
2–3	16	38 employees stayed for at least two years.
3–4	20	
4–5	2	
over 5	0	
Total n =	100	

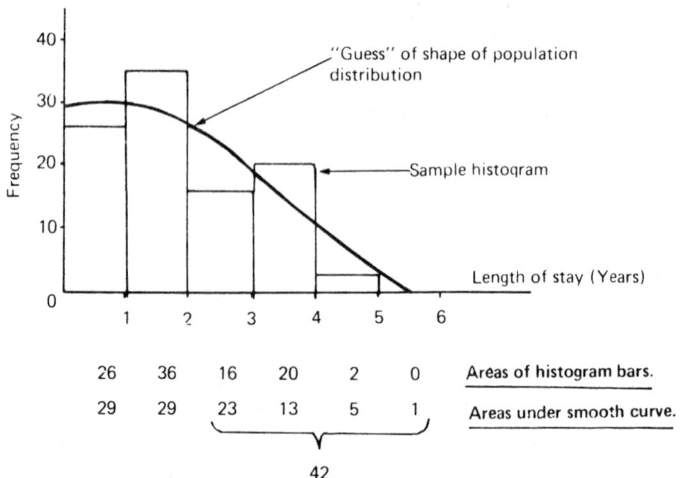

26	36	16	20	2	0	Areas of histogram bars.
29	29	23	13	5	1	Areas under smooth curve.

42

(empirical) probability of the next employee to join the company staying for at least two years is approximately 38/100 or 0.38[1].
This value rests on the assumption of "identical trials" – that is, that recruitment policy, working conditions and wage rates have remained the same (relative to alternative forms of employment) over the whole period of data collection and for the foreseeable future.

Where empirical probabilities are to be derived from a frequency distribution, more accurate estimates may sometimes be obtained by "smoothing the histogram". The smooth curve, drawn by eye and superimposed on the sample histogram in Exhibit 2 represents a belief in the "true" shape of the distribution had a much larger sample been available for analysis[2]. We are simply saying that the irregularities displayed by the sample histogram are purely local (sample) effects and should be disregarded for the purpose of probability estimation. The curve has been drawn so that the total area under it is the same as the total area of the histogram bars (=100): the areas under the curve within each class (obtained by "counting the squares" on graph paper) are shown beneath the histogram. We see that an improved estimate of the probability of the next employee staying for at least two years is given by 42/100 = 0.42, using this method.

Turning to **subjective** probability, an adequate definition is very difficult to provide. Our only attempt is as follows: the probability of an event E is a measure of confidence a particular individual has in the truth of the proposition that event E will occur if a trial were performed.

The assessment of subjective probabilities is fast approaching an art form. Here is one example of a commonly used technique. Suppose a Sales Manager wished to estimate the sales revenue of a newly introduced product in the coming year. He might start off by giving the single figure £10,000, in the belief that there is the same chance of the actual revenue exceeding this figure as there is of the actual revenue falling short of this figure. Then, working on the temporary assumption that the revenue will exceed £10,000, he might quote a higher figure £15,000, to express his view that the revenue is as likely to be between £10,000 and £15,000 as it is to be more than £15,000. Similarly, working on the opposite assumption that the revenue will be less than £10,000, he might give £6,000 as the revenue he believes will divide the

---

(1)Note that had we asked "what is the probability that a randomly chosen employee *from the original 100 employees of Exhibit 2* stayed for at least two years?" the answer would be *exactly* 0.38 *a priori*.

(2)In later course units we shall find that there are more sophisticated ways of fitting a "smooth curve" to a frequency distribution in particular cases.

possible revenues from 0 – £10,000 into two equally likely regions. In summary he will have expressed his opinion that the four events,

A: Revenue less than £6,000
B: Revenue between £6,000 and £10,000
C: Revenue between £10,000 and £15,000
D: Revenue more than £15,000

are all equally likely. So proceeding *as if* these events were equally likely *a priori*, the probability of each of these events is 1/4 = 0.25.

6.4 COMPARISON OF THE THREE TYPES OF PROBABILITY

A fundamental property of probability can be seen from an examination of the *a priori* and empirical definitions – that is, probabilities must lie between the values of 0 and 1 inclusive. The interpretations of the extreme values of Pr(E) = 0 and Pr(E) = 1 in *a priori*, empirical and subjective terms are given in Exhibit 3.

EXHIBIT 3: Interpretations of Zero and Unit Probability

Probability type	*Pr(E) = 0*	*Pr(E) = 1*
A priori	*E* is absolutely impossible	*E* is absolutely certain
Empirical	*E* has never occurred	*E* has always occurred
Subjective	The belief that *E* cannot occur	The belief that *E* is certain to occur

EXHIBIT 4: A Scale of Probability

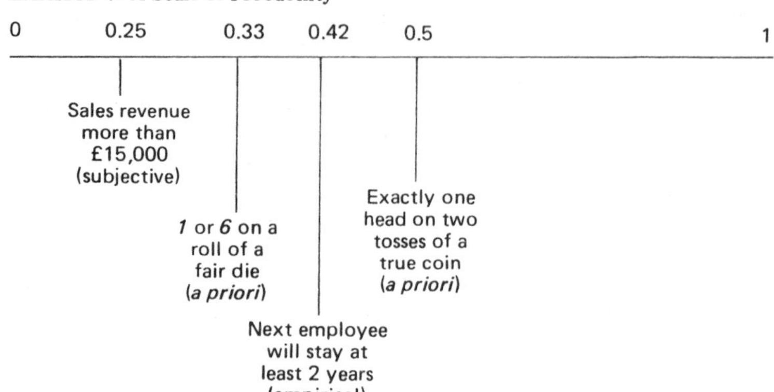

Bearing these results in mind it is convenient to think of probabilities on a scale extending from 0 to 1 – the higher up the scale a probability is, the more likely is the event to occur. This idea is illustrated in Exhibit 4 using the probabilities evaluated in this Course Unit.

We have seen that the process of defining probability is not a simple one. Underlying the three definitions of probability we have given there are assumptions (as in most applications of mathematics to the real world) which can be summarised briefly as follows:

A priori – the existence of equally likely outcomes. This assumption is usually made by appeal to physical considerations, especially that of symmetry. Thus we speak of a true coin, a fair die, a perfectly balanced roulette wheel, or a "well shuffled" pack of cards – all with the implicit assumption of equally likely outcomes.

Empirical – the assumption of trials being carried out under "identical" conditions, and that it is possible to carry out a large number of such trials.

Subjective – that the person making the assessment is rational, and that other persons with precisely the same experience and ability (if they could be found) would all arrive at the same probability values.

As a rough guide to the current usage of the three types of probability it may be said that *a priori* methods apply principally to games of chance, either in a direct sense (rolling dice) or where we deliberately set out to create a "gamelike" situation such as when *randomly* selecting an item from a production line. More particularly, in the context of this course, we use *a priori* concepts as a teaching aid in developing rules for manipulating probabilities which, when learned, can be applied equally to empirical and subjective probabilities.

The notion of empirical probability has been responsible for the bulk of applications of statistical methods in such fields as life assurance, market research, quality control and agricultural experiments.

The use of the subjective probability concept is much more recent and is rapidly gaining ground in areas of application previously dominated by the empirical methods. In particular it has become intimately associated with "Decision Theory" in which subjective probabilities and the use of costs of wrong decisions are combined to aid the decision maker in choosing between alternative courses of action. An example of such an application is given in CU8.

EXERCISES

1(P) Take a pack of 52 playing cards and shuffle well. Deal, face down, the first 12 cards and put these cards aside. The object of this exercise is to determine the probability of randomly selecting a black card from the remaining pack of 40 cards.

(i) Give an estimate of the *subjective probability* of drawing a black card at random, stating the reasons for your answer.

(ii) Cut the pack and record the colour of the card shown. Shuffle the pack. Repeat this process of cutting and shuffling fifty times, setting out your results in the manner shown in the table.

Trial Number n	Event Black(B) or Red(R)	Total Number of Black Cards so far f_B	Relative Frequency of Black Cards f_B/n
1	R	0	0/1 = 0
2	R	0	0/2 = 0
3	B	1	1/3 = 0.33
4	B	2	2/4 = 0.50
5	R	2	2/5 = 0.40
6	B	3	3/6 = 0.50
.	(Specimen results)	(Specimen results)	(Specimen results)
.	.	.	.
.	.	.	.
.	.	.	.
50	.	.	.

Complete the table (as illustrated) by working out the Total Number of Black Cards shown, f_B, and the Relative Frequency of Black Cards, f_B/n, at each stage. Draw a graph showing how f_B/n changes with n. Finally make an assessment of the empirical probability of drawing a black card (or cutting the pack at a black card) at random.

(iii) By examing the pack of 40 cards evaluate the *a priori* probability of randomly selecting a black card.

(iv) Comment on any discrepancies between the three probability value you obtained.

2(P) Choose a person in the public eye (a politician or a film star) whose age is unknown to you. Using the technique discussed in Section 6.3 for assessing *subjective* probabilities obtain some probability estimates that his/her age lies within certain ranges. (It is helpful to carry this out as a group exercise where a single person is agreed on and then each member of the group produces their own estimates, which are later compared.)

3(D) The World Heavyweight Boxing Champion has just retired. Two leading contenders are nominated to box for the championship. Contender Ernie Wallop has had 22 bouts in all, of which he won 18, drew 3 and lost 1. Contender Tom Thumper is four years younger and has won all 14 of his previous bouts. Wallop and Thumper have fought 6 opponents in common (but never each other) and they both won all 6.
Discuss if it is possible to apply (a) *a priori*, (b) empirical and (c) subjective probability concepts to the result of the Wallop-Thumper championship bout. Under the championship rules a draw is not allowed.

4(U) A bookmaker quotes "odds" of 3 to 2 against Wallop and 2 to 1 on Thumper. Convert those into (a) percentage (b) fractional probabilities.

5(E) A company in which the training period for apprentices is five years is considering its intake for the coming year. Information concerning apprentices recruited in previous years is given below:

Year of intake	1963	1964	1965	1966
Number of apprentices recruited	700	500	150	250
Number of apprentices leaving in				
First year	28	18	8	14
Second year	29	18	6	10
Third year	17	8	2	1
Fourth year	13	8	1	2
Fifth year	2	2	1	1

(a) What is the probability that an apprentice will qualify?
(b) What is the probability that an apprentice will stay for longer than two years?
(c) How many apprentices should be recruited in 1971 to provide the company with 300 qualified men in 1976?

(ICMA June 1971)

6(E) The table given below shows a frequency distribution of the lifetimes of 500 light bulbs made and tested by ABC Limited:

Lifetime (hours)	Number of light bulbs
400 and less than 500	10
500 and less than 600	16
600 and less than 700	38
700 and less than 800	56
800 and less than 900	63
900 and less than 1,000	67
1,000 and less than 1,100	92
1,100 and less than 1,200	68
1,200 and less than 1,300	57
1,300 and less than 1,400	33
	500

(a) Using the information in this table construct an **ogive**.

(b) Determine the percentage of light bulbs whose lifetimes are at least 700 hours but less than 1,200 hours.

(c) What risk is ABC Limited taking if it guarantees to replace any light bulb which lasts less than 1,000 hours?

(d) Instead of guaranteeing the life of the light bulb for 1,000 hours, ABC Limited suggests introducing a 100 day money back guarantee. What is the probability that refunds will be made, assuming the light bulb is in use:

(i) 5 hours per day;
(ii) 8 hours per day?

(ICMA May 1972)

COURSE UNIT 7 – RULES OF PROBABILITY

7.1 INTRODUCTION

In this Course Unit (Sections 7.2 to 7.4) we develop some fundamental rules of probability theory in a non-rigorous, though it is hoped intuitively appealing, manner. In Section 7.5 these same rules are proved more formally by means of set theory. The whole treatment is based on an *a priori* approach (because this is the simplest way) but this is of course subject to the same shortcomings as is the *a priori* definition itself (CU6).

There are basically only two rules of probability, an *addition* rule and a *multiplication* rule – so the arithmetic involved is quite straightforward. What is more difficult is the identification of those situations where the two rules can be applied. While reading the text and carrying out the exercises the reader should continually ask himself, "What is the justification for using this particular rule?"

7.2 EVALUATING PROBABILITIES IN THE FORM Pr (A or B)

Consider rolling a "fair" die once and defining two events A and B as follows:

A: Face 2 is shown (just one outcome),
B: 4 or more is shown (three outcomes; faces 4, 5 and 6).

The probabilities of these two events are readily evaluated using the *a priori* definition: $\Pr(A) = 1/6$, $\Pr(B) = 3/6 = 1/2$.

It can be seen that these two events are of the "either/or" type: if A occurs, B cannot occur; if B occurs then A cannot. Such events where the occurrence of one event precludes the occurrence of the other(s) are called **mutually exclusive** events, because they have *no outcomes in common.*

By $\Pr(A \text{ or } B)$ we mean the probability of "A occurring or B occurring" – that is, the probability of any of the faces 2,4,5 or 6 showing (4 outcomes). Hence, by definition, $\Pr(A \text{ or } B) = 4/6 = 2/3$. We note that $\Pr(A \text{ or } B) = \Pr(A) + \Pr(B)$, i.e. $2/3 = 1/6 + 1/2$, in this case, but this is only true because A and B have no outcomes in common. Formally:

If A and B are mutually exclusive events,
$\Pr(A \text{ or } B) = \Pr(A) + \Pr(B)$ (1)

This important result is called the **special rule for the addition of probabilities**. The rule can be extended and generalised.

As an extension, consider m mutually exclusive events, E_1, E_2, $E_3, \ldots, E_m$ relating to a given trial. Then:

$$\Pr(E_1 \text{ or } E_2 \text{ or } E_3 \text{ or} \ldots E_m) = \Pr(E_1) + \Pr(E_2) + \Pr(E_3) + \ldots + \Pr(E_m).$$

Now suppose $E_1, E_2, \ldots, E_m$, taken together, describe *all possible* results of a trial; then $\Pr(E_1 \text{ or } E_2 \text{ or} \ldots E_m) = 1$, because one or other of these events is certain to occur. In this case the mutually exclusive events $E_1, E_2 \ldots, E_m$ are called **collectively exhaustive**, or to use an equivalent phrase, form a **partition of the sample space.** (The reason for this rather curious wording should become clear in Section 7.5.) A very special case of such events arises when we deal with **complementary events.**

For any event, E, the complementary event[(1)] $\bar{E}$ is defined as follows:

$\bar{E}: E$ does not occur.

So $\Pr(E \text{ or } \bar{E}) = \Pr(E) + \Pr(\bar{E})$, because E and $\bar{E}$ are mutually exclusive. But $\Pr(E \text{ or } \bar{E}) = 1$, because E and $\bar{E}$ are collectively exhaustive. Consequently $\Pr(E) + \Pr(\bar{E}) = 1$, and so by rearrangement,

$$\Pr(E) = 1 - \Pr(\bar{E}). \qquad (2)$$

This is an equation which often proves useful in practice since frequently it is easier to compute $\Pr(\bar{E})$ rather than $\Pr(E)$. The formula gives a quick method for arriving at $\Pr(E)$ in such cases. As an example, suppose 10 items are to be selected from a large batch of manufactured items and we wish to evaluate the probability that "one or more of these items are substandard". Now the complementary event is that "*none* of the items is substandard" and this has a probability which is much easier to work out (as we shall see in CU9). So we would calculate the required probability in the form:

Pr(one or more substandard items) = 1 − Pr(no substandard items)

We next turn to the problem of evaluating Pr(A or B) in conditions where A and B are *not* mutually exclusive (i.e. a generalisation of Equation (1). In fact in such a case the event "A or B" is somewhat

[(1)]The "bar" over the E has quite a different meaning from that in the symbol $\bar{X}$, the sample mean. $\bar{E}$ is read "not E" and means everything that could happen if event E does not occur.

ambiguous. To be precise, what we are seeking is the probability that either the event A occurs or the event B occurs or (possibly) that both A and B occur. An example should make this clear.

Consider drawing a single card from a pack of well-shuffled playing cards. We define two events:

A: the card is a picture card (12 outcomes)
B: the card is a heart (13 outcomes)

In this case the event "A or B" means that the card is either a picture card or a heart or possibly both.

Now it can be seen that events A and B are *not* mutually exclusive because there are some three cards which are both picture cards and hearts. Hence we cannot use Equation (1) directly to evaluate Pr(A or B). One possible approach to this problem is to define a new set of events which *are* mutually exclusive and which, when taken together, mean exactly the same thing as "A or B", viz:

C: the card is a picture card but not a heart (9 outcomes)
D: the card is a picture card *and* a heart (3 outcomes)
E: the card is not a picture card but a heart (10 outcomes)

So, Pr(A or B) = Pr(C or D or E) = Pr(C) + Pr(D) + Pr(E)
(because C, D and E *are* mutually exclusive)
$= 9/52 + 3/52 + 10/52 = 22/52.$

Note that the answer is *not* 25/52, the result that would be obtained by blindly applying Equation (1) without first checking to see whether A and B were mutually exclusive or not.

This device of defining a new set of mutually exclusive events equivalent to a set which are not mutually exclusive sometimes requires considerable ingenuity in more complex problems – although when it can be done, it is often worthwhile. As an alternative it is worth developing a completely general formula for Pr(A or B). To do this we must return to the original definition of *a priori* probability.

How many (equally likely) outcomes comprise the event "A or B"? Taking the above example, the answer is clearly *not* (12 + 13) because of the "double counting" of the picture card hearts. To make allowances for this we must subtract the number of outcomes which are common to both A and B. Hence there are (12 + 13 – 3) distinct outcomes in the event ("A or B"). So

$$\Pr(A \text{ or } B) = \frac{12 + 13 - 3}{52} = \frac{12}{52} + \frac{13}{52} - \frac{3}{52}$$

$$= \frac{22}{52}, \text{ as before}$$

The fraction 12/52 can be identified with Pr(*A*), the fraction 13/52 with Pr(*B*) and the fraction 3/52 with the value of Pr(*A and B*). "*A* and *B*", of course, simply means a picture card heart.

We have arrived at the **general rule for the addition of probabilities:**

$$\Pr(A \text{ or } B) = \Pr(A) + \Pr(B) - \Pr(A \text{ and } B) \quad (3)$$

Comparing this rule with the special rule of Equation (1) we see that if the events *A* and *B* are mutually exclusive then Pr(*A* and *B*) must be zero. This is of course entirely reasonable as it implies that the events *A* and *B* can never occur together.

7.3 EVALUATING PROBABILITIES OF THE FORM Pr(*A* and *B*)

Events *A* and *B* are termed **independent** if the occurrence of event *A* does not affect the chances of occurrence of event *B* and vice versa. Consider rolling a "true" die *twice*. We define two events:

A: the *first* roll shows 2 or more, (5 outcomes)
B: the *second* roll shows 4 or more. (3 outcomes)

These events are clearly independent since what happens on the first roll does not affect in any way what happens on the second.

Let us evaluate Pr(*A and B*) from first principles. The number of (equally likely) outcomes which comprise the so-called **joint event**, "*A* and *B*", is obtained by combining each of the five ways by which *A* can occur with each of the three ways by which *B* can occur. The result is 5 × 3 (= 15) distinct outcomes. The *total* number of equally likely outcomes, by similar reasoning, is 6 × 6 = 36 (i.e. each of the six faces of the first roll combined with each of the six faces of the second roll).

$$\text{Hence, } \Pr(A \text{ and } B) = \frac{5 \times 3}{6 \times 6} = \frac{5}{6} \times \frac{3}{6} \,(= 5/12)$$

Now the fraction 5/6 can be identified with Pr(*A*) and the fraction 3/6 with Pr(*B*). This leads us to the **special rule for the multiplication of probabilities.**

If two events *A* and *B* are *independent*,
$\Pr(A \text{ and } B) = \Pr(A) \times \Pr(B)$. (4)

Another example. Suppose a dispatch department has eight large orders to deal with, five of which are for the home market and the remaining three for export. If two of the orders are randomly assigned to a particular group of packers, what is the probability that both these orders are for the home market? We define two events:

> A: the *first* order assigned to the packers is for the home market,
> B: the *second* order assigned to the packers is for the home market.

What we require is clearly Pr(A and B).

Now the probability that the *second* order is for the home market, Pr(B), will itself depend on whether the first order is for the home market or not. If the first order is, in fact, for the home market then the remaining choice for the second order is from seven orders, ***four*** of which are for the home market: if, on the other hand, the first order is for export then this leaves all *five* home orders to choose from. Consequently the events A and B are *not* independent[(1)] and so we are not able to use Equation (4) to evaluate Pr(A and B). Again, let us evaluate Pr(A and B) from first principles.

The number of (equally likely) outcomes comprising the event "A and B" is $5 \times 4 = 20$ (there are five home market choices for the first order and four remaining choices for the second order). By similar reasoning the *total* number of equally likely outcomes is $8 \times 7 = 56$.

$$\text{Hence Pr (both home market)} = \Pr(A \text{ and } B) = \frac{5 \times 4}{8 \times 7} = \frac{5}{8} \times \frac{4}{7} \left(= \frac{5}{14} \right)$$

The fraction 5/8 can simply be identified with Pr(A), but what of the fraction 4/7? Well, this is simply the probability of B occurring *on the assumption* that A has occurred, i.e. in conditions where the first order *is* for the home market (leaving four home market orders out of seven remaining for the second choice). Such a probability is called a **conditional probability**; it is written Pr($B|A$) and read "the probability of B given A." We could define Pr($B|A$) in an ***a priori*** fashion as follows:

$$\Pr(B|A) = \frac{\text{No. of equally likely outcomes which comprise } B \text{ in conditions where } A \text{ has occurred}}{\text{Total no. of equally likely outcomes in conditions where } A \text{ has occurred}}$$

So for *any* A and B,

$$\Pr(A \text{ and } B) = \Pr(A) \times \Pr(B|A). \qquad (5)$$

This result is known as the **general rule for the multiplication of probabilities.**

[(1)]Formally, this is because the "sampling" is taking place *without* replacement.

Comparing (4) and (5) we see that if events A and B are independent $\Pr(B) = \Pr(B|A)$, a result often used to define independence.
Simple manipulation of Equation (5) gives:

$$\Pr(B|A) = \frac{\Pr(A \text{ and } B)}{\Pr(A)} \tag{6}$$

This result is often used to define conditional probability, and is used to evaluate conditional probabilities in more complex problems (Section 7.4).

Proceeding further with the dispatch department problem, let us try to evaluate the probability that (i) both orders are for export and (ii) that one is for the home market, the other for export. We have,

$$\text{Pr (both export)} = \frac{3}{8} \times \frac{2}{7} = \frac{3}{28}, \text{ by direct application of (5).}$$

The simplest way of evaluating the probability that "one is for the home market, the other for export" is by recognising that the three events "both home", "both export", and "one home, one export" are mutually exclusive and collectively exhaustive. So:

$$\begin{aligned}\text{Pr("both home" or "both export")} &= \text{Pr(both home)} + \text{Pr(both export)}\\ &= 5/14 + 3/28 = 13/28\\ &\quad \text{(by using earlier results).}\end{aligned}$$

$$\begin{aligned}\text{So, Pr(one home, one export)} &= 1 - \text{Pr("both home" or "both export")}\\ &\quad \text{(by using Equation (2))}\\ &= 1 - 13/28 = 15/28\end{aligned}$$

Another way (less convenient in this case) of evaluating this probability is to "break down" the event "one home, one export" into the two mutually exclusive events:

C: first order home, second export,
D: first order export, second home.

Hence

$$\begin{aligned}\text{Pr(one home, one export)} &= \Pr(C \text{ or } D) = \Pr(C) + \Pr(D)\\ &= \left(\frac{5}{8} \times \frac{3}{7}\right) + \left(\frac{3}{8} \times \frac{5}{7}\right)\\ &= 15/28 \text{ as before.}\end{aligned}$$

The second procedure for evaluating Pr(one home, one export) is so important that it is worth stating it formally as a means of evaluating probabilities of complex events.

(a) Describe the complex event by writing down a set of mutually exclusive events $E_1, E_2, \ldots, E_L$ such that (E_1 or E_2 . . . or E_L) is equivalent to the complex event.
(b) Evaluate the probabilities $\Pr(E_i)$ for each i.
(c) Sum these probabilities to obtain the probability of the complex event.

7.4 TREE DIAGRAMS, MARGINAL PROBABILITY AND BAYES' THEOREM

Consider the case of a company that recruits its management trainees from two sources – graduates (G) and short service commissioned officers from the armed services (S, for service entry). On average 80% of new entrants are graduates and 20% are ex-servicemen. It has been found from past experience that 30% of the graduate entrants and 40% of the service entrants receive promotion within a period of two years. If we met a new trainee whose method of entry was unknown to us, what is the probability of his being promoted within the two year period?

Let us first of all summarise the statistical information contained within this statement.

(a) Pr (graduate entry) = $\Pr(G) = 0.8$
(b) Pr (service entry) = $\Pr(S) = 0.2$
(c) Pr (graduate entrant gets promotion) = $\Pr(P|G) = 0.3$
∴ (d) Pr (graduate entrant fails to get promotion) = $\Pr(\overline{P}|G) = 0.7$
(e) Pr (service entrant gets promotion) = $\Pr(P|S) = 0.4$
∴ (f) Pr (service entrant fails to get promotion) = $\Pr(\overline{P}|S) = 0.6$

It should be noted that (c) to (f) are conditional probabilities (using the obvious notation that P means "obtains promotion" and $\overline{P}$, "fails to obtain promotion"). All this information can be represented most conveniently in the form of the **tree diagram** shown in Exhibit 1.

The branching points I, II and III represent the "chance" effects of the type of entry and promotion prospects. More generally for tree diagrams, the branching points represent *decisions* – in our example we might say "taken by nature" – and for this reason such representations are often called **decision trees**. It is important to note that at each

EXHIBIT 1: Tree diagram

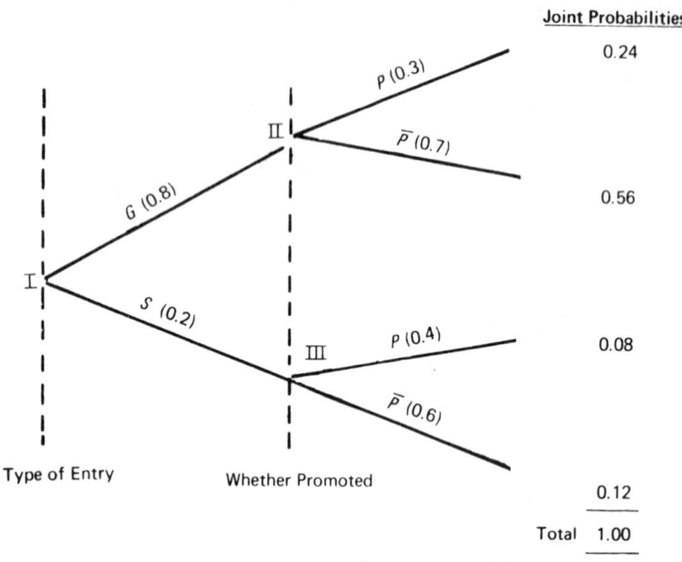

branching point the probabilities associated with the emerging branches add up to one.

Each *path* through the tree (from left to right) represents a distinct joint event. Consider the uppermost path, "graduate and promotion". The probability of this joint event is easily evaluated as:

$$\Pr(G \text{ and } P) = \Pr(G) \times \Pr(P|G) = 0.8 \times 0.3 = 0.24$$

In terms of the diagram we simply move along a path, from left to right, multiplying up the probabilities on the way. The (joint) probabilities of the other events are shown against the end-points of the paths in the diagram. It can be seen that these probabilities add up to one: this is because the various paths represent *all possible* mutually exclusive events.

Let us now return to the original question – what is the probability that our new entrant will gain promotion? As we do not know whether the trainee is a graduate or a service entrant we must consider both paths through the tree which result in promotion – "graduate and promotion" and "service entrant and promotion". Since these represent mutually exclusive events the probability of one or other of these occurring is simply the sum of their probabilities, i.e. 0.24 + 0.08 = 0.32. Symbolically, we have:

$$\Pr(P) = \Pr(\{G \text{ and } P\} \text{ or } \{S \text{ and } P\})$$
$$= \Pr(G \text{ and } P) + \Pr(S \text{ and } P)$$
$$= 0.24 + 0.08 = 0.32$$

Notice how easy it is to evaluate such probabilities using the decision tree. We simply identify those paths (which, by virtue of the construction of a decision tree, represent mutually exclusive events) which satisfy our requirement ("promotion", in this case) and add up their associated probabilities.

A probability, such as Pr(*P*), which is evaluated by a process of summing joint probabilities is called a **marginal probability**[(1)]: the reader should take care not to confuse this use of the word *marginal* with its other uses – the economic use, for instance. This terminology is used because marginal probabilities occur as sub-totals "in the margin" of a joint probability table as shown in Exhibit 2. Marginal probability can be thought of as being the probability of one type of classification feature (e.g. promotion) ignoring the other classification (e.g. entry type).

EXHIBIT 2: Joint Probability Table Showing Marginal Probabilities

	G	*S*	Sub-total	
P	0.24	0.08	0.32 = Pr(*P*)	marginal probabilities.
$\bar{P}$	0.56	0.12	0.68 = Pr($\bar{P}$)	
Sub-total	0.80 Pr(*G*)	0.20 Pr(*S*)	1.00	

Returning to our example, suppose we meet a trainee who has been promoted within two years of joining the company. What is the probability that he was a graduate entrant? The answer is *not* 0.8 because this does not take account of the additional information that he has been promoted. (If, for instance, it were the case that all graduates, but no service entrants, gained promotion within two years the answer

(1)We have $\Pr(P) = \Pr(G \text{ and } P) + \Pr(S \text{ and } P) = \Pr(G) . \Pr(P|G) + \Pr(S) . \Pr(P|S)$. More generally if an event A always occurs in conjunction with one or other of a set of mutually exclusive events $E_1, E_2, \ldots E_m$, then the marginal probability of A is given by:

$$\Pr(A) = \Pr(E_1) \cdot \Pr(A|E_1) + \Pr(E_2) \cdot \Pr(A|E_2) + \ldots + \Pr(E_m) \cdot \Pr(A|E_m).$$

would be 1.0). In this case we need to evaluate $\Pr(G|P)$. By using Equation (5), we have:

$$\Pr(G|P) = \frac{\Pr(G \text{ and } P)}{\Pr(P)} = \frac{0.24}{0.32} = 0.75$$

In layman's terms this simply means that 75% of such promotions are gained by graduate entrants – a figure somewhat lower than the proportion of all entrants who are graduates (80%) on account of the rather smaller promotion chances of the graduates.

This idea of "revising" probabilities in the light of additional information is central to a whole branch of Statistics called **Bayesian Statistics.** The idea can be generalised as follows. The most crude estimate of the probability of an event (the trainee being a graduate, in our case) is given the name **prior** probability (do not confuse this with *a priori*, although there is a clear link between these two phrases) because this probability can be estimated *prior* to any immediately relevant information becoming available. Having obtained the additional information (that the trainee has been promoted, in our case) we can revise our estimate of the probability of the event in the manner as shown. This revised probability is called a **posterior** probability. To sum up, using our example, we have:

Prior Probability $\Pr(G) = 0.8$,
Posterior Probability $\Pr(G|P) = 0.75$.

The idea of estimating posterior probabilities is so important that a further example is given from the field of auditing. Auditors know from experience the likely "error rate" in certain accounts, i.e. they know the prior probability of entries being in error in some way. Their normal procedure is to sample a number of entries and count the number of errors observed. This additional information enables them to evaluate the posterior probability that the whole account contains more than a certain "acceptable" number of errors.

In simple cases posterior probabilities can be evaluated in the manner shown, possibly making use of a decision tree to clarify the problem. An alternative way, which is completely equivalent, is to make use of **Bayes' theorem**[1] as follows:

---

[1] A proof of this theorem is as follows: Using the general rule for the multiplication of probabilities:

$\Pr(A \text{ and } B) = \Pr(A) \cdot \Pr(B|A)$ and $\Pr(B \text{ and } A) = \Pr(B) \cdot \Pr(A|B)$

But $\Pr(A \text{ and } B) = \Pr(B \text{ and } A)$; so by equating the right hand sides and by rearranging, Equation (7) is obtained immediately.

$$\Pr(A|B) = \Pr(B|A) \cdot \frac{\Pr(A)}{\Pr(B)} \tag{7}$$

In words, Bayes' theorem states that the *posterior* probability, $\Pr(A|B)$, is obtained by multiplying the *prior* probability, $\Pr(A)$, by the *conditional* probability, $\Pr(B|A)$, and dividing by the *marginal* probability, $\Pr(B)$. It should be noted that although Equation (7) is the *simplest* way of writing Bayes' theorem it does disguise the fact that the marginal probability, $\Pr(B)$, has to be evaluated as a separate exercise by summing the appropriate joint probabilities.

Using Bayes' theorem on our "promotion" example, we have (by writing G for A and P for B):

$$\Pr(G|P) = \Pr(P|G) \cdot \frac{\Pr(G)}{\Pr(P)}$$

$$= 0.30 \times \frac{0.80}{0.32} = 0.75, \text{ as before.}$$

7.5 A SET-THEORETIC TREATMENT OF PROBABILITY

This section gives a more formal presentation of the basic rules of probability introduced earlier. It may appeal to those who prefer to think in terms of diagrams. The treatment given here is again based on *a priori* probability with all the philosophical objections that this entails. Some other writers give a similar presentation, though based on empirical probability.

Consider the familiar trial of tossing two "true" coins. We can represent all possible (equally likely) outcomes by writing the convenient symbols HH, HT, TH, TT inside a rectangle which we shall refer to as a **Venn diagram** (Exhibit 3).

EXHIBIT 3: Venn Diagram for Outcomes of Tossing Two Coins

HH	TT
HT	TH

Alternatively we can refer to the **set**, S, of all possible outcomes, termed the **sample space**, by the notation:

$$S = \{\text{HH, HT, TH, TT}\}$$

Furthermore we can refer to a **sub-set** of outcomes defining particular events in a similar manner. For example,

A: Exactly 1 head = {HT, TH}
B: First coin "heads" = {HT, HH}

Such events can be represented on the Venn diagram by enclosing the relevant outcomes within a closed curve as shown in Exhibit 4.

EXHIBIT 4: Definition of Events A and B

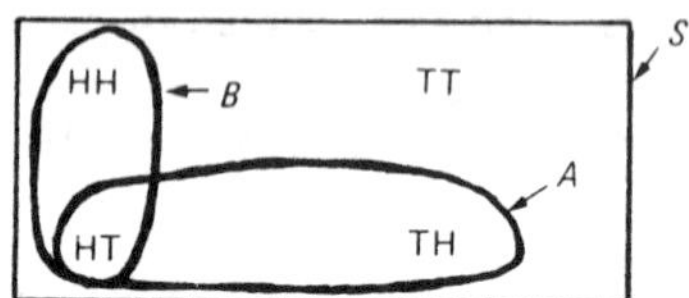

We use the notation $n(A)$, for instance, to denote the number of (equally likely) outcomes comprising event A (i.e. in sub-set A). Thus in our example we have: $n(A) = 2$, $n(B) = 2$, and $n(S) = 4$, by an obvious extension of the notation.

We define the composite event "$A \cup B$", read "A union B", as the subset which contains all the outcomes in event A and all the outcomes in event B. Thus the event "$A \cup B$" occurs if event A occurs or event B occurs or (possibly) both occur. It can be seen that the set-theoretic notation "$A \cup B$" is exactly equivalent to the phrase "A or B" used earlier in this course unit. In our example we have:

$A \cup B$ = {HH, HT, TH}, and so $n(A \cup B) = 3$.

We also define the composite event "$A \cap B$", read "A intersection B", as the subset which contains only those outcomes which are common to both A and B. So "$A \cap B$" is exactly equivalent to our earlier use of "A and B". In the example, we have:

$A \cap B$ = {HT}, so $n(A \cap B) = 1$.

EXHIBIT 5

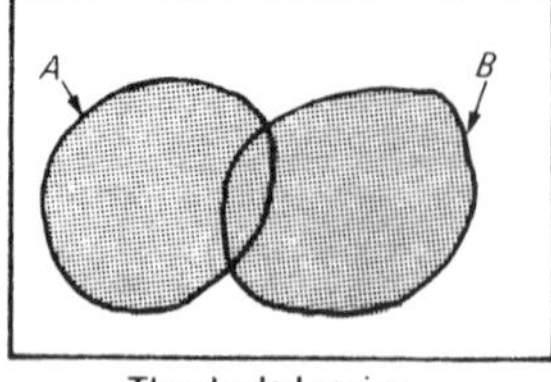

The shaded region represents $A \cup B$

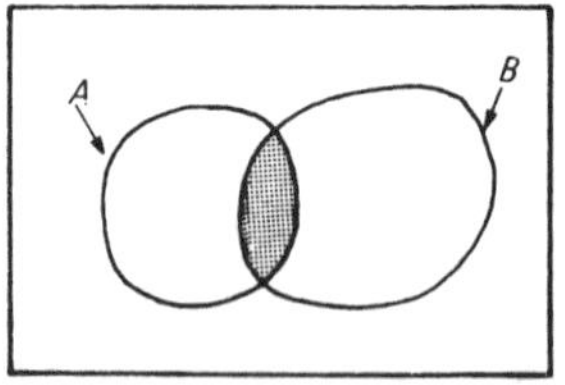

The shaded region represents $A \cap B$

The meaning of $A \cup B$ and $A \cap B$ in general is illustrated in Exhibit 5.

For *any* event, E, we *define* $\Pr(E) = n(E)/n(S)$. Applying this definition to our example, we find:

$$\Pr(A) = n(A)/n(S) = 2/4 = 1/2;\ \Pr(B) = n(B)/n(S) = 2/4 = 1/2, \text{ and}$$

$$\Pr(A \cup B) = n(A \cup B)/n(S) = 3/4;\ \Pr(A \cap B) = n(A \cap B)/n(S) = 1/4.$$

All the elementary rules of probability can be derived from these set-theoretic definitions. To make matters easier, and at the same time more general, consider the Venn diagram in Exhibit 6 where the lower-case letters denote the number of (equally likely) outcomes in each "region" of the diagram (drawn up for any trial for which events A and B have been defined).

EXHIBIT 6

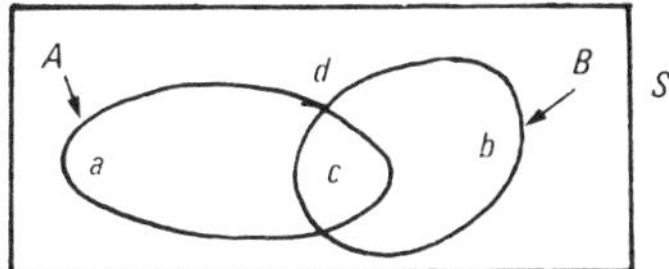

$$\text{Now } \Pr(A \cup B) = \frac{n(A \cup B)}{n(S)} = \frac{a+b+c}{a+b+c+d} = \frac{(a+c)+(b+c)-c}{a+b+c+d}$$

$$= \frac{a+c}{a+b+c+d} + \frac{b+c}{a+b+c+d} - \frac{c}{a+b+c+d}$$

$$= \Pr(A) + \Pr(B) - \Pr(A \cap B),$$

which proves the general rule for the addition of probabilities. In particular if $c = 0$, the events A and B are mutually exclusive (i.e. non-overlapping, in Venn diagrams) and hence we obtain the special rule for the addition of probabilities.

Consider the special case of the mutually exclusive events, E_1, $E_2, \ldots, E_m$ forming a partition of the sample space. This can be represented as a Venn diagram, as shown in Exhibit 7.

EXHIBIT 7

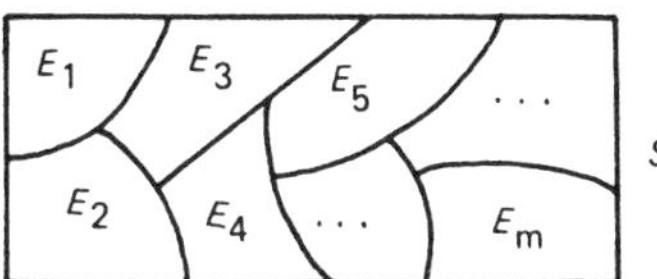

(The reason for the phrase "partition of the sample space" should now be apparent.)

Clearly,

$$n(E_1) + n(E_2) + n(E_3) + \ldots + n(E_m) = n(S)$$

So

$$\frac{n(E_1)}{n(S)} + \frac{n(E_2)}{n(S)} + \frac{n(E_3)}{n(S)} + \ldots + \frac{n(E_m)}{n(S)} = 1, \text{ or}$$

$$\Pr(E_1) + \Pr(E_2) + \Pr(E_3) + \ldots + \Pr(E_m) = 1.$$

The rule for complementary events is a special case of this result. Finally, we define

$$\Pr(B|A) = \frac{n(A \cap B)}{n(A)} = \frac{n(A \cap B)}{n(S)} \cdot \frac{n(S)}{n(A)} = \frac{\Pr(A \cap B)}{\Pr(A)}$$

Hence $\Pr(A \cap B) = \Pr(A).\Pr(B|A)$, the general rule for multiplication of probabilities.

READING

A treatment similar to ours, giving a more extended treatment of Bayes' rule, can be found in:
George J. Brabb, *Quantitative Management,* Holt, Rinehart and Winston, Ch.4, "Measuring Chance".

EXERCISES

There are no practical exercises for this Course Unit.

1(U) DELTA PACKING COMPANY LTD.: Number of Employees

	Males		*Females*		
Department	Under 18	18 & over	Under 18	18 & over	Total
Senior Staff	0	15	0	5	20
Junior Staff	6	4	8	12	30
Transport	2	34	2	2	40
Stores	1	20	2	17	40
Maintenance	1	3	4	12	20
Production	0	4	24	72	100
Total	10	80	40	120	250

(i) All employees are included in a weekly draw for a £10 prize; what is the probability that in one particular week the winner is (a) male (b) a member of staff (c) a female member of staff (d) from the Maintenance or Production Departments (e) is male or a transport worker?

(ii) Two employees are to be selected at random to represent the company at the annual Packaging Exhibition in London. What is the probability that (a) both are female (b) both are production workers (c) one is a male under 18, the other a female over 18?

(iii) If three young persons under 18 are to be chosen at random for the opportunity to go on Outward Bound courses, what is the probability that (a) all are male (b) all are female (c) only one is male (d) only one is female?

(iv) If two persons report sick what is the probability (a) one is from Stores and one is from Maintenance (b) both are from Stores (c) if both are from Stores that one is male and one is female? What assumption do you have to make in your answers?

In each of the above questions state where appropriate

(a) whether you are using the addition of multiplication rule;
(b) whether the probabilities are marginal, independent or conditional.

2(U) Three events, related to the Sales Volume in the coming year for a newly introduced product, are as follows:

A: Sales will be *more than* 5,000
B: Sales will be *more than* 10,000
C: Sales will be *less than* 2,000

The probabilities of these events have been estimated as $\Pr(A) = 0.3$, $\Pr(B) = 0.1$, $\Pr(C) = 0.2$.

(i) Can the terms *mutually exclusive* and *independent* be used in the context of these events? Give as many illustrations as you can.

(ii) Evaluate the probabilities of the two following events,

D: Sales will be between 5,000 and 10,000
E: Sales will be between 2,000 and 5,000

(iii) Explain the meaning in words, and evaluate the associated probabilities of "A or B", "A and B", "A or C", and "A and C".

3(U) A shipment of 15 automatic weighing machines contains 3 defectives. If two of the machines are selected randomly, what is the probability that (a) both machines are defective, (b) at least one of the machines is defective.

4(U) A six-sided die is so biased that it is twice as likely to show an even number as an odd number when thrown. It is thrown twice. What is the probability of the sum of the two numbers thrown being even?

5(U) A certain toothpaste company – the one which manufactures toothpaste with stripes – finds that on average each batch of 1,000 tubes contains ten with no stripes (though possibly some toothpaste) and three with no toothpaste (though perhaps a full set of stripes). What is the probability that a tube chosen at random will prove faulty if it is known that one tube in a thousand contains nothing at all?

6(E) In a television service organisation, sets arriving for repair are first passed to an apprentice serviceman. If an apprentice fails to diagnose and correct the fault in a set it is then passed on to a fully trained serviceman. If he in turn is unable to correct the fault, the set is finally passed over to the service manager for attention.
It is known that on average an apprentice, a fully trained serviceman and the service manager are able to correct the faults of 60%, 50% and 40% respectively of sets *with which they deal*. Find the probability that:

(i) a fault in a set will not be rectified by the organisation,
(ii) the service manager will have to deal with a set which he is able to correct,
(iii) the service manager will *not* have to examine a set.

(HND, part question).

7(E) (a) Explain briefly the value of conditional probability calculations to the businessman.
(b)

	Defective Electron Tubes per box of 100 units			
Firm	0	1	2	3 or more
Supplier A	500	200	200	100
Supplier B	320	160	80	40
Supplier C	600	100	50	50

Table: Number of Boxes

From the data given in the above table, calculate the conditional probabilities for the following questions:

(i) If one box had been selected at random from this universe,[1] what are the probabilities that the box would have come from Supplier A; from Supplier B; from Supplier C?
(ii) If a box had been selected at random, what is the probability that it would contain two defective tubes?
(iii) If a box had been selected at random, what is the probability that it would have no defectives and would have come from Supplier A?
(iv) Given that a box selected at random came from Supplier B, what is the probability that it contained one or two defective tubes?
(v) If a box came from Supplier A, what is the probability that the box would have two or less defectives?
(vi) It is known that a box selected at random has two defective tubes. What is the probability that it came from Supplier A; from Supplier B; from Supplier C?

(ACCA, June 1972)

8(E) Automatic Document-Readers are a recent development in Data Processing equipment. At present they are highly expensive and rather unreliable. The Computacom XYZ 4 is a document-reader capable of reading with perfect accuracy all figures from 0 to 9 except 3 and 8. Exhaustive research has shown that:

3 is read as 8 with probability 0.2
8 is read as 3 with probability 0.1

However, these probabilities do not apply when identical figures are read consecutively. In such cases misreading the first doubles the probability that the second is misread, while if the first is read correctly, the probability of misreading the second is halved. For example, if 33 appears and the first 3 is interpreted as 8, there is a 40% chance that the second 3 is also misread.
The XYZ 4 can also perform arithmetic calculations. The numbers 2, 8 and 8 are read and multiplied in order.

(i) Draw a tree diagram to represent the situation and work out the probability that the result is 48.
(ii) On a particular occasion 48 was actually obtained. What is the probability that the first 8 was misread?

[1] "Universe" is another word for "population".

(iii) An appointment card is read and duplicated by the XYZ 4. The appointment is timed at 3.30 p.m. 3.8.74. With what probability is the appointment wrongly made if figures are held to be consecutive even when separated by punctuation?

BA (Business Studies)

9*(U) Prove by using Venn diagrams, or otherwise, that

(i) $\Pr(A \cup B \cup C) = \Pr(A) + \Pr(B) + \Pr(C) - \Pr(A \cap B) - \Pr(A \cap C) - \Pr(B \cap C) + \Pr(A \cap B \cap C)$

(ii) $\Pr(A \cap B \cap C) = \Pr(A) \cdot \Pr(B|A) \cdot \Pr(C|A \cap B)$

Demonstrate these rules using an example of randomly cutting a pack of cards where *A:* Red card, *B:* Value of card greater than 7 (Ace low), and *C:* Even number of spots.

10*(E) Three towns X, Y and Z are in telecommunications contact by means of three telephone lines A, B and C as shown below.

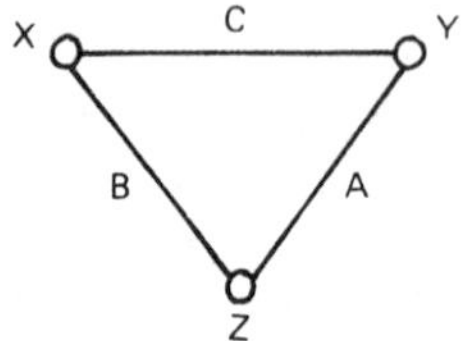

As an illustration of how the system operates, towns X and Y remain in contact only if line C is operative or if both lines A and B are operative. Lines A, B and C are subject to failure in any one day with probabilities 0.1, 0.2 and 0.3 respectively. Such failures occur independently.

(i) Evaluate the probability that X and Y will not be able to make contact on a particular day.

(ii) If on a particular day X and Y are not able to make contact, what is the probability that line A has failed.

(iii) What is the reliability of the whole system? (Reliability is defined as the probability that all three towns are able to contact each other.)

SOLUTIONS TO EVEN NUMBERED EXERCISES

2.

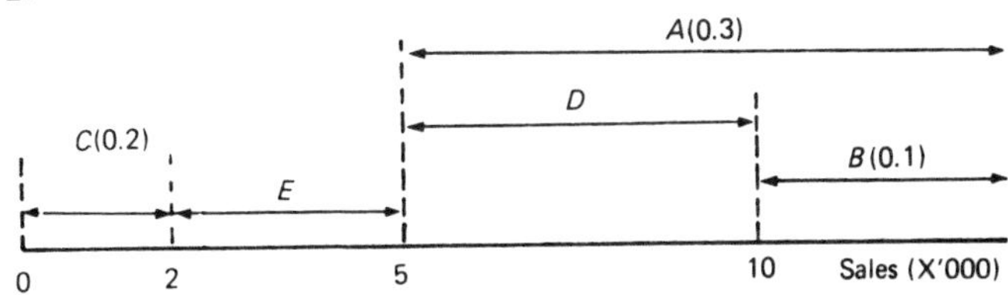

(i) A and C are mutually exclusive, B and C are mutually exclusive, but A and B are *not* mutually exclusive. The term *independent* cannot be applied in this context.

(ii) $\Pr(A) = \Pr(D \text{ or } B) = \Pr(D) + \Pr(B)$, because D and B are mutually exclusive.
So $\Pr(D) = \Pr(A) - \Pr(B) = 0.3 - 0.1 = 0.2$
Now $\Pr(C) + \Pr(E) + \Pr(A) = 1$, because C, D and A are mutually exclusive and collectively exhaustive.
So $\Pr(E) = 1 - \Pr(C) - \Pr(A) = 1 - 0.2 - 0.03 = 0.5$.

(iii) "A or B" means that the sales will be greater than 5,000 *or* greater than 10,000. Effectively this means that the sales will be greater than 5,000.
So $\Pr(A \text{ or } B) = \Pr(A) = 0.3$.
"A and B" means that the sales will be greater than 5,000 *and* greater than 10,000, the net effect of which is that the sales will be greater than 10,000.
So $\Pr(A \text{ and } B) = \Pr(B) = 0.1$
"A or C" means that the sales will either be less than 2,000 or more than 5,000.
So $\Pr(A \text{ or } C) = \Pr(A) + P(C) = 0.3 + 0.2 = 0.5$, as A and C are mutually exclusive.
"A and C" means that the sales will be both less than 2,000 and more than 10,000, which is a clear impossibility.
Hence $\Pr(A \text{ and } C) = 0$.

4. A die can show either an even number (E) or an odd number (O) when thrown. E and O are mutually exclusive, collectively exhaustive events. So $\Pr(E) + \Pr(O) = 1$, but $\Pr(E) = 2\Pr(O)$, and so $\Pr(E) = 2/3$ and $\Pr(O) = 1/3$. On two throws only "OO" or "EE" result in an even sum.
$\Pr(\text{even sum}) = \Pr(OO \text{ or } EE) = \Pr(OO) + \Pr(EE)$

$$= \frac{1}{3} \cdot \frac{1}{3} + \frac{2}{3} \cdot \frac{2}{3} = \frac{5}{9}.$$

6.

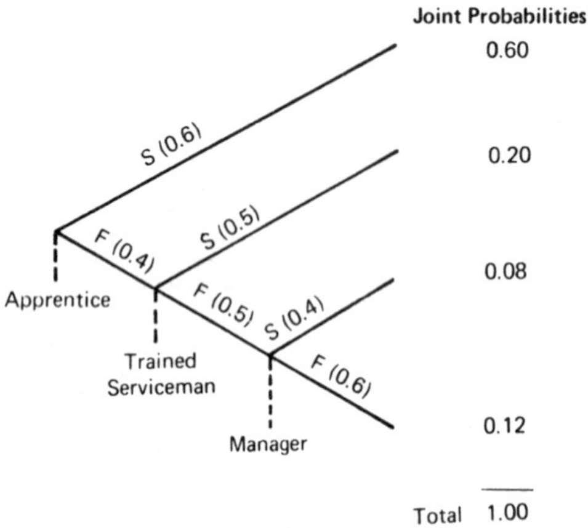

N.B. "S" denotes success, "F" denotes failure.

Hence (i) 0.12, (ii) 0.08, (iii) 0.60 + 0.20 = 0.80.

8.

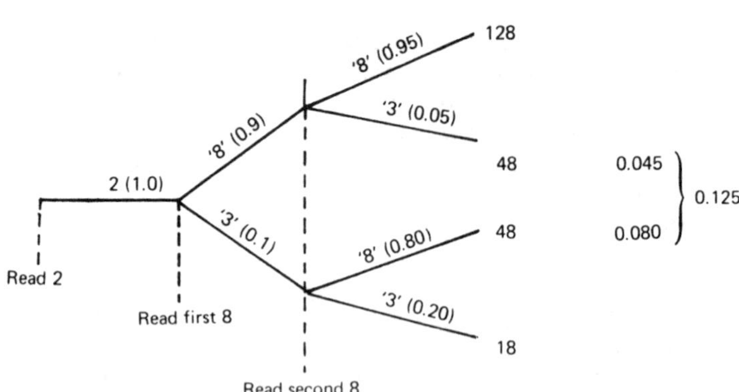

(i) Pr(48 obtained) = 0.045 + 0.080 = 0.125 (see tree).

(ii) Pr(first 8 misread|48) = Pr(first 8 misread *and* 48)/Pr(48)
= 0.080/0.125 = 0.64

(iii) Pr(appointment *correctly* made) = (0.8)(0.9)(0.8)(0.9)
= 0.5184

Number read → 3 3 3 8

Therefore Pr(appointment *incorrectly* made)

$= 1 - 0.5184 = 0.4816.$

10. Let A represent event that line A is operative: $\bar{A}$ that line A has failed.
Let F represent event that X and Y are not in contact. F occurs only when ($\bar{C}$ and $\bar{B}$) *or* ($\bar{C}$ and $\bar{A}$) occur.

(i) $\Pr(F) = \Pr([\bar{C} \text{ and } \bar{B}] \text{ or } [\bar{C} \text{ and } \bar{A}])$
$= \Pr(\bar{C} \text{ and } \bar{B}) + \Pr(\bar{C} \text{ and } \bar{A}) - \Pr(\bar{A} \text{ and } \bar{B} \text{ and } \bar{C})$
$= (0.3 \times 0.2) + (0.3 \times 0.1) - (0.1 \times 0.2 \times 0.3)$
$= 0.084$ (or by enumeration, as in table).

	(0.1)	(0.2)	(0.3)	
	A	B	C	Jt. Prob.
F	X	X	X	0.006
	X	X		
F	X		X	0.024
	X			
F		X	X	0.054
		X		
			X	
			Total	0.084

"X" denotes line failure

(ii) $\Pr(\bar{A} \mid F) = \dfrac{\Pr(\bar{A}) \Pr(F \mid \bar{A})}{\Pr(F)}$, Bayes' Rule

$$= \frac{\Pr(\bar{A}) \cdot \Pr(\bar{C})}{\Pr(F)} = \frac{0.1 \times 0.3}{0.084} = 0.357$$

(or from definition of $\Pr(\bar{A} \mid F) = \Pr(\bar{A} \text{ and } F) | \Pr(F)$.

(iii) For all 3 towns to remain in contact, *at most* one line can fail.
Pr(no lines fail) = Pr(A and B and C) = 0.9 × 0.8 × 0.7 = 0.504
Pr(A only fails) = Pr(A and B and C) = 0.1 × 0.8 × 0.7 = 0.056
Pr(B only fails) = Pr(A and B and C) = 0.9 × 0.2 × 0.7 = 0.126
Pr(C only fails) = Pr(A and B and C) = 0.9 × 0.8 × 0.3 = 0.216

0.902

As the left hand side events are mutually exclusive, the probabilities add. Reliability is 90.2%.

COURSE UNIT 8 – PROBABILITY DISTRIBUTIONS AND EXPECTED VALUES

8.1 INTRODUCTION

Consider the following situation. A contracting company owns four earth-moving machines. From time to time a machine breaks down and is out of commission, while being repaired, until the following week. Knowledge of the probability that two (say) of these machines will break down and so will be out of commission in a given week would be useful to the company in planning its work schedules. But more useful still would be the knowledge of the probabilities associated with all possible numbers of machines (0,1,2,3 or 4) being out of commission. Such a set of probabilities is called a **probability distribution** – it gives a comprehensive "picture" of a particular form of uncertainty with which the company is faced.

Stated in formal terms, if we have a trial (in our example, a given week's activity) which results in the realisation of a random variable R, (in our case, the number of machines breaking down) then the set of probabilities corresponding to each possible value R can take (0,1,2,3 or 4 in the example) is termed the probability distribution of that random variable. A probability distribution may be thought of as a prediction of what will happen when the trial occurs – as compared with a *frequency distribution* which is a statement of what actually happened over a number of trials. The individual probabilities of a distribution must, of course, obey the rules of probability: in particular they must each lie in the range 0 to 1, but more significantly *their sum must be unity*. The reason for this latter property is that different values of the random variable correspond to mutually exclusive events and the probability distribution gives the probabilities of *all possible* values of the random variable. (Formally, the set of all possible values of the random variable forms a partition of the sample space.)

There are a number of different ways in which probability distributions can be expressed – depending on how specific or general we wish to be. Returning to the example, suppose that the contractor can evaluate the probabilities on the basis of certain theoretical considerations and/or a process of sieving past experience. (The precise manner in which this can be achieved does not concern us here – we are assuming that this can be done, in order to illustrate certain matters of principle.) These probabilities might be expressed in one or other of the following forms:

(a) **A purely numerical statement.** For example we might have:

$$\begin{aligned}
\Pr(R=0) &= p(0) = 0.7017 \\
\Pr(R=1) &= p(1) = 0.2105 \\
\Pr(R=2) &= p(2) = 0.0632 \\
\Pr(R=3) &= p(3) = 0.0189 \\
\Pr(R=4) &= p(4) = 0.0057 \\
\text{and } \Pr(R=5 \text{ or more}) &= 0.0000 \\
\text{Total} &= 1.0000
\end{aligned}$$

The statement $\Pr(R = 5 \text{ or more}) = 0$ is added to demonstrate that only values or R of 0,1,2,3 and 4 are physically realisable in this situation and to underline the fact that the probabilities sum to unity

(b) **A probability function.** Instead of listing all the probabilities numerically, we might write down a general expression involving r (a particular value that R can take) as follows:

$$\Pr(R=r) = p(r) = 0.7017 \cdot (0.3)^r, \quad r=0,1,2,3,4$$
$$= 0, \text{ other } r.$$

This compact statement "generates" the probability values given in (a) by substituting each value, in turn, for r.

(c) **A parametric probability function.** More generally we might write:

$$\Pr(R=r) = p(r) = \left(\frac{1-\gamma}{1-\gamma^5}\right) \cdot \gamma^r, \quad r=0,1,2,3,4$$
$$= 0, \text{ other } r.$$

The quantity γ is a parameter which specifies the particular conditions to which the probability distribution can be applied (its value must be in the range 0 to 1). When $\gamma = 0.3$ the expression is identical to that given in (b). Other values of γ give *different* probabilities. The contractor could *choose* a value of γ appropriate to the type of terrain on which the earth-movers would be working – the more difficult the terrain the larger the chosen value of γ, corresponding to a greater likelihood of breakdowns.[1]

---

[1] It is left as an exercise for the reader familiar with geometric progressions to demonstrate that *whatever* the value of γ (provided that $0 \leqslant \gamma < 1$) the individual probabilities will sum to unity.

It should be emphasised that the example given has no general validity: it was used solely to illustrate alternative ways in which probability distributions can be expressed. In later Course Units we shall study particular probability distributions, expressed as parametric functions (some with one parameter, others with two parameters) which are appropriate for describing the behaviour of random variables in certain commonly occurring situations.

The example we used to illustrate a probability distribution is *discrete* and *finite*. It is a "discrete" distribution because it describes the probabilities of a discrete random variable (the number of machines broken down); it is "finite" because the random variable only has a finite number of values (0,1,2,3 and 4). Continuous distributions (and to a lesser extent, infinite distributions) involve slight further complications and are considered later (Section 8.3). We next look at another discrete finite distribution, but one which can be derived from elementary probability considerations.

8.2 AN EXAMPLE OF A DISCRETE PROBABILITY DISTRIBUTION

Back to coin tossing! Consider the familiar example of the trial of tossing two "fair" coins. We define the random variable R as being the number of heads shown. Analysing this situation from first principles along the lines developed in CU7, we obtain the results shown in Exhibit 1.

EXHIBIT 1: Tossing Two Coins

Outcome	*Event*	*Value of R (= r)*	*Pr(R=r) = p(r)*
TT	No heads	0	0.25
TH, HT	One head	1	0.50
HH	Two heads	2	0.25
			1.00

Thus as a complete statement of the probability distribution[1] of R we have:

$$p(0) = 0.25, p(1) = 0.50, p(2) = 0.25, \text{ and } p(r) = 0, \text{ other } r.$$

We can represent this probability distribution in the form of a (probability) histogram. As R is discrete, the preferred form of the histogram

[1] As will be shown in CU9 this is a special case of the binomial distribution.

EXHIBIT 2: Histograms of a Discrete Probability Distribution

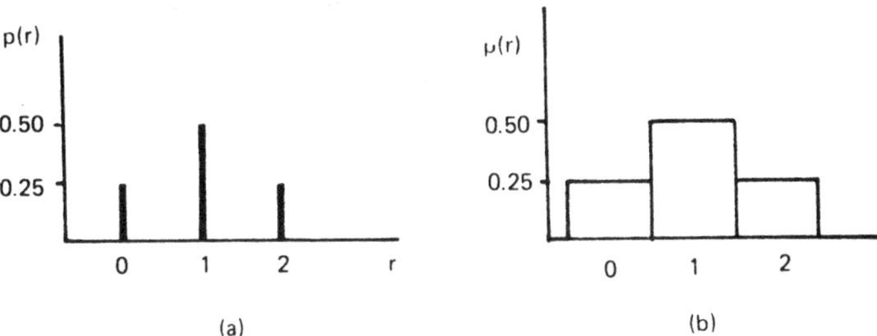

is shown in Exhibit 2a, although the representation shown in Exhibit 2b is often adopted.

Resulting from the fact that the random variable R can only take integer values, the histogram "bars" in Exhibit 2(b) are of unit width. This being so, the indvidual probabilities are represented by the *areas* of the bars, and in particular the *total area* under the histogram bars is unity – a result which will be useful, by analogy, when considering continuous probability distributions.

8.3 SPECIFYING CONTINUOUS PROBABILITY DISTRIBUTIONS

When considering the meaning of a continuous probability distribution an immediate problem arises. As a continuous random variable (X) can in principle take *any* value (possibly within certain limits) then the probability that it has a particular value (x), $\Pr(X = x)$, must be vanishingly small. (Consider for example, the probability of a particular casting being of length 5.21634109 . . . cm, when such castings vary in length from 5.20 cm to 5.30 cm.) The best we can do with a continuous random variable is to specify the probability that its value lies within a certain range of values, say from x_L (a lower value) to x_U (an upper value) – that is, $\Pr(x_L < X < x_U)$. The manner in which this is achieved is as follows.

EXHIBIT 3: Concept of a Probability Density Function, $f(x)$

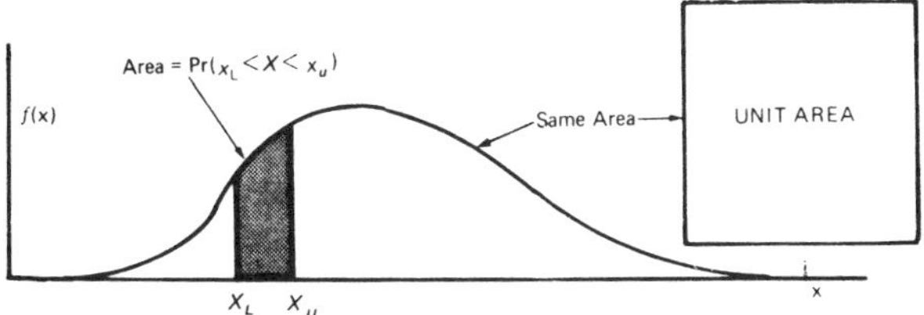

Exhibit 3 shows a graph of a function of x, $f(x)$, which is called a **probability density function** (p.d.f.) of a continuous random variable X.

The precise mathematical form of $f(x)$ does not concern us here – it is sufficient to note that, firstly, $f(x)$ is never negative (just as probabilities are never negative) and secondly that the total area under the curve is unity (as in the case of the histogram of a discrete probability distribution). Expressing these properties formally, using the calculus[1] notation we have:

$$f(x) \geqslant 0, \text{ all } x \tag{1}$$

$$\int_{-\infty}^{+\infty} f(x)\,dx = 1 \tag{2}$$

Requirements for a p.d.f. $f(x)$

Now the value of $\Pr(x_L < X < x_U)$ is obtained from the *area under* the p.d.f. curve, $f(x)$, between x_L and x_U, i.e.

$$\Pr(x_L < X < x_U) = \int_{x_L}^{x_U} f(x)\,dx \tag{3}$$

Finally it can be visualised from Exhibit 3 that if x_L and x_U are close together then the area representing this probability will be approximately rectangular in shape with base length $(x_U - x_L)$ and height equal to the value of $f(x)$ half way between x_U and x_L.

$$\Pr(x_L < X < x_U) \simeq f\left(\frac{x_L + x_U}{2}\right) \cdot (x_U - x_L) \tag{4}$$

if x_U and x_L are close together

We can also use this result to *estimate* probability densities from empirical data. For example, we found that in Series A there were 16 (out of 100) observations in the class 49.5–59.5. So our (empirical) estimate of the probability that a random variable from the distribution will fall into this class is 16/100 of 0.16. Now as we have:

$$x_L = 49.5,\ x_U = 59.5, \text{ so } \frac{x_L + x_U}{2} = 54.5 \text{ and } (x_U - x_L) = 10,$$

[1]A knowledge of the calculus is not necessary for an understanding of this Course Unit. The formal statements may be ignored by non-calculus readers.

Equation (4) gives $0.16 \simeq f(54.5).10$ or $f(54.5) \simeq 0.016$. Stated quite simply, we can obtain an estimate of the probability density at the mid-point of a class by dividing the relative frequency (our estimate of the probability) by the class width – a result we shall use in CU11.

In summary, it is not possible to express continuous probability distributions in purely numerical form as there are an infinite number of values of the random variable. Continuous distributions are expressed by means of their probability density functions, areas under which have a simple probability interpretation.

8.4 AN EXAMPLE OF A CONTINUOUS PROBABILITY DISTRIBUTION

Consider the following p.d.f.

$$f(x) = 2(1-x),\ 0 \leqslant x \leqslant 1$$
$$= 0, \text{ otherwise.}$$

This has the graph shown in Exhibit 4.

EXHIBIT 4: The Probability Density Function 2(1 – x)

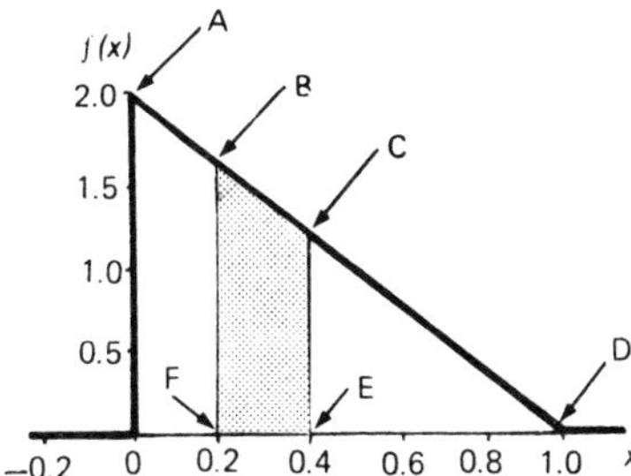

We see immediately that this is a "valid" p.d.f. because:

- $f(x)$ is greater than or equal to zero for all values of x
- the total area under the "curve" is unity ($= \frac{1}{2}$ base $\times$ height of triangle OAD $= \frac{1}{2} \times 1 \times 2 = 1$)

Note especially that the value of $f(x)$ at $x = 0$ is 2 – there is no requirement that probability densities, unlike probabilities, must be less than one.

Let us evaluate $\Pr(0.2 < X < 0.4)$, the shaded area. There are a number of possible approaches in this case.

- We note that the shaded area is a trapezium. Hence the area is

  $\frac{1}{2}(\text{BF} + \text{CE}) \times (\text{FE}) = \frac{1}{2}[f(0.2) + f(0.4)] \times 0.2$

  $= \frac{1}{2}[1.6 + 1.2] \times 0.2 = 0.28$

- The calculus method.

  $$\Pr(0.2 < X < 0.4) = \int_{0.2}^{0.4} 2(1 - x)dx = 2\,[x - x^2/2]_{0.2}^{0.4} = 0.28$$

- Using the rectangle approximation (Equation (4)), on the assumption that $x_L = 0.2$ and $x_U = 0.4$ can be considered "close together". The value of $f(x)$ at the mid-point of 0.2–0.4 (0.3) is 1.4. Hence the probability,

  $\Pr(0.2 < X < 0.4) = 1.4 \times 0.2 = 0.28$

  (This method is generally an approximation – it is exact in this case because the p.d.f. is a straight line.)
- We note that the required area = Areas of Δ BFD – Area of Δ CED

  $= \frac{1}{2} \times 0.8 \times 1.6 - \frac{1}{2} \times 0.6 \times 1.2$

  $= 0.64 - 0.36 = 0.28.$

  This method, which amounts to subtracting two different "areas in the tail of the distribution", is particularly convenient in cases where the "areas in the tail" are given in standard statistical tables (e.g. as for the normal distribution, CU11).

8.5 EXPECTED FREQUENCY

We return to the example of tossing two "fair" coins and counting the number of heads (Section 8.2). Now visualise repeating this trial 100 times. In how many of these trials would we expect to observe two heads (i.e. where the random variable $R = 2$)? The answer, of course, is 25 because we can interpret the probability of this event (= 0.25) as the overall proportion of trials in which $R = 2$. We are saying that the **expected frequency** of $R = 2$ is 25. This expected frequency is denoted by $\hat{f}_2$ – the "hat" indicating that this value is an *estimate* of the *observed* frequency, f_2, which would result from physically tossing a pair of coins 100 times and counting the number of times two heads were observed.

Similarly, taking the example in Section 8.4, visualise repeating the trial which results in the value of X (whose p.d.f. is $2(1 - x)$) 30 times. The expected frequency of X lying between 0.2 and 0.4 is given by (number of trials) × (probability) = 30 × 0.28 = 8.4. Note that expected frequencies, unlike observed frequencies, can be non-integral – so the technical word "expected" corresponds only loosely with the meaning of the word in normal usage. Quite generally then:

$$\text{Expected frequency of an event } E = \begin{pmatrix}\text{number of trials}\\ \text{visualised}\end{pmatrix} \times \begin{pmatrix}\text{probability}\\ \text{of event } E\end{pmatrix}$$

$$\text{or } \hat{f}_E = n \times \Pr(E), \qquad (5)$$

where E is of the form $(R{=}r)$ for a discrete distribution and $(x_L < X < x_U)$ for a continuous distribution.

Expected frequency can be interpreted in two ways:

- It is the frequency that would be observed with a "perfect" sample of n items (trials). We mean by "perfect", a sample exhibiting in strict proportion all the characteristics of the population. Of course we are not able, in general, to select such a perfect sample – to do so would require perfect knowledge of the population and that would defeat the whole purpose of sampling.
- If a series of n trials were repeated very many times and the frequency of the event observed in each, then the *average* of these observed frequencies would equal the expected frequency. (Proof of this statement is given in a footnote in CU9.)

8.6 COMPARISON OF OBSERVED AND EXPECTED FREQUENCIES

Exhibit 5 shows the expected frequencies resulting from 100 "visualised" trials of our coin tossing example.

EXHIBIT 5: Expected Frequencies from 100 Trials

r	$Pr(R{=}r) = p(r)$	$100\,p(r) = \hat{f}_r$
0	0.25	25
1	0.50	50
2	0.25	25
Totals	1.00	100

Of course it is somewhat unlikely that if we actually carried out this experiment (of tossing a pair of coins 100 times) our observed frequencies would exactly equal these expected frequencies. We should expect some discrepancy between the observed and expected frequencies simply because the expected frequencies are truly representative of the (infinite) population, whereas the observed frequencies result from a sample of just 100 observations. But suppose that the observed frequencies in a particular case were markedly different from the expected frequencies. What would we conclude? There are essentially three possibilities: we could conclude that any one or combination of these is true.

- The assumptions on which the probability distribution was based may be in error. In our example the coins may not be "fair" – they may be bent or non-uniform.
- The trials may have not been performed randomly, and so the observed frequencies are not the result of a simple random sample. In our case a poor "tossing" technique could give this result.
- We have simply encountered a *rare* sample result by chance.

The comparison of observed and expected frequencies plays an important part in statistical methodology. It enables us to check theoretical assumptions and (sometimes) to check on the validity of a sampling procedure.

8.7 EXPECTED VALUES

The ideas in this section will be developed using a discrete probability distribution. Consider the following formulae for the *sample* mean and variance of grouped data.

$$\text{Mean} = \frac{1}{n}\sum x_i f_i \qquad \text{(see CU4)}$$

$$\text{Variance} = \frac{1}{n}\sum x_i^2 f_i - \left[\frac{1}{n}\sum x_i f_i\right]^2 \qquad \text{(see CU5)}$$

Now if the observations are values of a discrete random variable (0,1,2 etc.), then we can write $x_i = r$ and $f_i = f_r$, and so

$$\text{Mean} = \frac{1}{n}\sum r f_r$$

$$\text{Variance} = \frac{1}{n}\sum r^2 f_r - \left[\frac{1}{n}\sum r f_r\right]^2$$

Now suppose that *instead* of physically taking a sample of n observations and using the observed frequencies f_r, we "visualise" a sample of n observations and use the *expected* frequencies $\hat{f}_r$ (= $np(r)$) in these formulae. We obtain:

$$\text{Mean} = \frac{1}{n}\sum r \cdot np(r) = \sum rp(r)$$

$$\text{Variance} = \frac{1}{n}\sum r^2\, np(r) - \left[\sum rp(r)\right]^2$$

$$= \sum r^2\, p(r) - \left[\sum rp(r)\right]^2$$

Note that these results do not contain n; that is, the values are independent of the sample size "visualised". In fact these results do not give the *sample* mean and variance at all; they give the values of the *population* mean (μ) and variance (σ^2). This can be seen by remembering that the expected frequencies are those frequencies that would occur with a "perfect" sample, or alternatively by considering the case where n represents the *population* size.

$$\mu = \sum_{\text{all } r} rp(r) \qquad \text{or}^{(1)} \qquad \mu = \int_{\text{all } x} xf(x)dx \tag{6}$$

$$\sigma^2 = \sum_{\text{all } r} r^2\, p(r) - \mu^2 \qquad \qquad \sigma^2 = \int_{\text{all } x} x^2 f(x)dx - \mu^2 \tag{7}$$

The importance of these results is that in situations where we know the probability distribution of a random variable, we can *calculate* its population mean and variance; this should be contrasted with the process of taking a sample, calculating the sample mean and variance and using these to *estimate* the population values.

These results are often expressed in **expected value** notation. The "expected value of R", denoted by $\text{E}(R)$, is *defined by*:

$$\text{E}(R) = \sum_{\text{all } r} rp(r) \qquad \text{E}(X) = \int_{\text{all } x} xf(x)dx \tag{8}$$

From (6) we see that E(R) is equal to the population mean. In fact the expressions "expected value" and (population) "mean" are used interchangeably in statistical work. Extending this notation we can write:

$$\text{E}(R^2) = \sum r^2 p(r) \qquad \text{E}(X^2) = \int x^2 f(x)dx \tag{9}$$

or quite generally

$$\text{E}[g(R)] = \sum g(r)p(r) \qquad \text{E}[g(X)] = \int g(x)f(x)dx \tag{10}$$

where $g(r)$ is *any* function of r.

(1)The results for a continuous distribution are quoted alongside those for a discrete distribution.

Consequently the results for the mean and variance are often written as follows:

$\mu = \mathrm{E}(R)$	$\mu = \mathrm{E}(X)$	(11)
$\sigma^2 = \mathrm{E}(R^2) - [\mathrm{E}(R)]^2$	$\sigma^2 = \mathrm{E}(X^2) - [\mathrm{E}(X)]^2$	(12)

The result that the population variance is "the expected value of R^2 – (expected value of $R)^2$" is directly equivalent to the sample result that the sample variance is "the mean value of R^2 – (mean value of $R)^2$".

8.8 EXAMPLES OF EXPECTED VALUES

Taking the discrete probability distribution of Section 8.2 we have:

$$\mu = \mathrm{E}(R) = \sum rp(r) = (0 \times \tfrac{1}{4}) + (1 \times \tfrac{1}{2}) + (2 \times \tfrac{1}{4}) = 1,$$

a result that should not surprise us, as we should intuitively expect to average one head on tossing two coins.

Similarly,

$$\mathrm{E}(R^2) = \sum r^2 p(r) = 0 \times \tfrac{1}{4} + 1 \times \tfrac{1}{2} + 4 \times \tfrac{1}{4} = 1\tfrac{1}{2}$$

Hence

$$\sigma^2 = \mathrm{E}(R^2) - [\mathrm{E}(R)]^2 = 1\tfrac{1}{2} - 1^2 = \tfrac{1}{2}.$$

As an example of a continuous distribution, consider the p.d.f. $2(1 - x)$ used in Section 8.4. We can calculate the mean and variance of this distribution either by dividing up the distribution into "classes" – which gives an approximate result by treating the continuous distribution as though it were discrete as shown in Exhibit 6 – or alternatively (and exactly) by using the calculus as follows:

$$\mu = \mathrm{E}(X) = \int_0^1 x \cdot 2(1 - x)dx = 2[x^2/2 - x^3/3]_0^1 = 1/3$$

Now

$$\mathrm{E}(X^2) = \int_0^1 x^2 \cdot 2(1 - x)dx = 2[x^3/3 - x^4/4]_0^1 = 1/6$$

Hence

$$\sigma^2 = \mathrm{E}(X^2) - [\mathrm{E}(X)]^2 = 1/6 - (1/3)^2 = 1/18$$

EXHIBIT 6: Mean and Variance of X with p.d.f. $f(x) = 2(1 - x)$. Approximate (Non-calculus) Method, Treating the Continuous Distribution as if it were Discrete

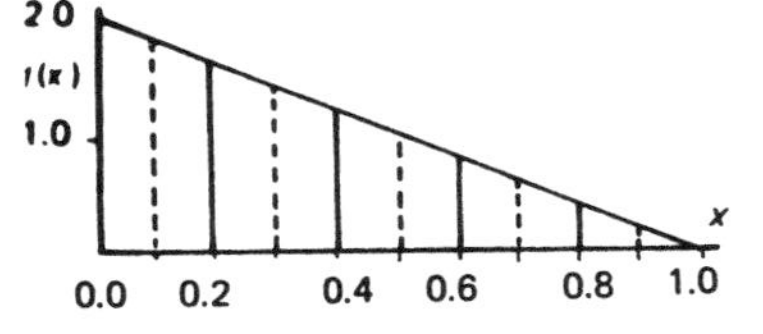

Class x_L	*Limits* x_U	*Mid-point* $x_m = (x_U + x_L)/2$	*Value of p.d.f. at mid-point* $= f(x_m)$		*Probability Within Each Class* $\simeq f(x_m) \cdot (x_U - x_L)$ $(= p(x_m)$ *say*$)$	$x_m \cdot p(x_m)$	$x_m^2 p(x_m)$
0.00	0.20	0.10	1.80		0.36	0.036	0.0036
0.20	0.40	0.30	1.40		0.28	0.084	0.0252
0.40	0.60	0.50	1.00		0.20	0.100	0.0500
0.60	0.80	0.70	0.60		0.12	0.084	0.0588
0.80	1.00	0.90	0.20		0.04	0.036	0.0324
				Totals	1.00	0.340	0.1700

$\mu = E(X) \simeq \Sigma x_m p(x_m) \simeq 0.34$

$E(X^2) \simeq \Sigma x_m^2 p(x_m) \simeq 0.17$

So

$\sigma^2 = E(X^2) - [E(X)]^2 \simeq 0.17 - (0.34)^2 = 0.0544$

8.9 PAYOFF TABLES AND EXPECTED PAYOFF

Business risks arise as the result of uncertainties about the future. A particularly useful way of expressing these uncertainties and relating them to courses of action available at present is by means of a **payoff table**. Such a table lists (vertically) possible courses of action and (horizontally) the "states of nature", values of a random variable which represents the inherent uncertainty. The entries in the table give the monetary "payoffs" resulting from all possible combinations of courses of action and states of nature.

Consider the case of the owner of an independent furniture store. Each year he is confronted with the problem of ordering new suites of furniture, six months in advance, from a leading manufacturer's catalogue. Of course, he need not order any – but he is in business to sell furniture. The more he sells, the more profit he could make, but if he over-orders he is left with goods which have to be sold at a loss.

The suites cost £100 each wholesale and are to be sold at £140 (which is £40 below the recommended retail price). Unsold suites are sold off in the winter sales for £80. Exhibit 7 shows the payoff table for this problem.

EXHIBIT 7: Payoff Table for Furniture Store (£'s Profit)

Course of Action	*"States of Nature" No. of suites that could be sold at £140 each*					
(order)	*0*	*1*	*2*	*3*	*4*	*5*
1	–20	40	40	40	40	40
2	–40	20	80	80	80	80
3	–60	0	60	120	120	120
4	–80	–20	40	100	160	160
5	–100	–40	20	80	140	200

The table is constructed by finding the profit (or loss) associated with each action and state of nature, where £40 profit is made on each suite sold and £20 loss (expressed as a negative profit) on each suite not sold.

Hence, if he orders 3 suites but sells only 2, the payoff is 2 × 40 (on the two he sold) – 1 × 20 (on the one not sold) = £60. Similarly, if he orders 5 but sells only one, the payoff = 1 × 40 – 4 × 20 = –£40. It can be seen that he can make the most profit if he orders 5 suites and sells all of them to give a payoff of £200. But he also stands to make the largest loss of £100 if he buys 5 but cannot sell any of them.

Note that if he orders 2 but could have sold 3, 4 or 5, his payoff remains £80, the same as ordering and selling 2 – but he will have lost some customer "good-will" through not ordering sufficient suites.

The problem facing the store owner is how many to order – 1, 2, 3, 4 or 5? The "classical"(1) statistical approach would be to use his sales statistics for past years, or carry out market research. However, furniture suites are to some extent fashion goods and there may be no history of sales of this new style suite. The cost of commissioning a survey amongst potential customers is perhaps ten times that of the largest possible profit. In short, the classical methods are of no use to our decision maker.

In this situation, we may have to use the owner's subjective assessment of the probabilities of selling 0–5 suites. We assume he is rational and that intuitively he takes into account all his business background of the numbers and type of customer visiting his store and previous sales of similar types of suites. Suppose he assesses the probabilities as follows:

Number of suites that could be sold, r	0	1	2	3	4	5
Probability, $p(r)$	0.0	0.1	0.4	0.2	0.2	0.1

Note that the probability of selling 2 suites is considerably higher than for any other number: consequently it is intuitively appealing for the owner to order 2 suites. However if we approach this problem using "expected payoffs" a different conclusion is reached.

The expected payoff of a given course of action is determined by using Equation (10), repeated below:

$$E[g(R)] = \sum g(r)p(r),$$

where $g(r)$, in this context, is the payoff associated with r suites that could be sold. So if the owner orders 2 suites his expected payoff is given by:

$$\begin{aligned} &(-40 \times 0) \\ &+ (20 \times 0.1) \\ &+ (80 \times 0.4) \\ &+ (80 \times 0.2) \\ &+ (80 \times 0.2) \\ &+ (80 \times 0.1) = 74(£) \end{aligned}$$

This expected payoff of £74 corresponds to an estimate of the *average* profit the owner would make if he were able to repeat this decision in

(1)In recent literature many authors refer to empirical probability based methods as "classical" in order to distinguish them from the newer Bayesian based method.

identical circumstances a large number of times. Exhibit 8 gives the expected payoffs, calculated in a similar manner, for all courses of action.

EXHIBIT 8: Expected Payoff for Furniture Store

Course of Action (Order)	*Expected Payoff (£)*
1	40
2	74
3	84
4	82
5	68

One possible approach to deciding on the "best" course of action (there are other approaches) is to choose the one giving the largest expected payoff. On this basis our owner should order 3 suites.

Another area of application of expected payoffs is in connexion with decision problems which can be represented most conveniently in the form of a *decision tree*. Exercise 6 is an example which the reader should try for himself and then work carefully through the solution given.

READING

J.E. Freund and F.J. Williams, *Elementary Business Statistics*, 2nd Ed., Prentice-Hall.

EXERCISES

1(P) A continuous random variable X has the following p.d.f.

$$f(x) = 3(1-x)^2, \quad 0 \leqslant x \leqslant 1.$$
$$= 0, \text{ otherwise.}$$

(i) Draw the graph of $f(x)$.
(ii) Verify that $f(x)$ is a "proper" probability density function.
(iii) Evaluate the mean and variance of this distribution.
(iv) Attempt to define "median" and "mode" of a continuous random variable. Using your definitions, calculate (approximately if necessary) their value, in this example.

Readers familiar with the calculus should use exact methods where possible. If non-calculus methods are used (as in Exhibit 6 in the text) the answers given should include a statement explaining whether the values obtained are greater than or less than the exact values.

2(U) State which of the following are valid probability distributions and why the others are not:

(i) $\Pr(R < 1) = 0$, $\Pr(R=1) = 1$, $\Pr(R > 1) = 0$
where R is a discrete random variable which can only take integer values.
(ii) $p(0) = 0, p(1) = \frac{1}{2}, p(2) = \frac{1}{4}, p(3) = 1/8 \ldots p(r) = 1/2^r$
where the random variable R can only be a positive integer.
(iii) $p(0) = 1/3, p(1) = 1/3, p(2) = 1/3$
where the random variable can only be a positive integer.
(iv) $\Pr(R < -1) = 0$, $\Pr(R = -1) = 1/8$, $\Pr(R = 0) = 3/4$, $\Pr(R = 1) = 1/8$ and $\Pr(R > 1) = 0$,
where R is an integer random variable.
(v) $p(1) = \frac{1}{2}, p(2) = \frac{1}{2}, p(3) = \frac{1}{2}$ and $p(r) = 0$, other r.
(vi) $p(0) = \frac{1}{4}$, $p(1) = -\frac{1}{4}$, $p(2) = \frac{1}{2}, p(3) = \frac{1}{4}, p(4) = \frac{1}{4}$ and $p(r) = 0$, other r.

3(U) A certain trial can only result in a random variable R taking the values 1,2 or 3. The probability distribution of R is given by:
$p(1) = 0.4, p(2) = 0.5, p(3) = 0.1$.

(i) Evaluate the expected value of R, $E(R)$.
(ii) Evaluate the expected value of R^2, $E(R^2)$.
(iii) Evaluate the variance of R.
(iv) If the trial were repeated four times, what is the expected frequency that $R = 1$.
(v) If the trial were repeated very many times, what would be the mean value of R.

4(E) Marketing executives, about to launch a new product, were asked to estimate the probability of sales volumes of 1,000, 2,000 and 3,000 in the first year. A "Delphi" technique was used to obtain agreement on the following figures.

Sales Volume in the First Year	Probability
1,000	0.3
2,000	0.5
3,000	0.2

Probabilities of various sales volumes were similarly estimated at the same time for the second year. Ready agreement was reached on the assumption that the sales in the second year would be dependent on the sales in the first year, and the final estimates were:

Sales Volume in the Second Year	Probability
1,000 more than sales volume in the first year	0.6
2,000 more than sales volume in the first year	0.3
3,000 more than sales volume in the first year	0.1

By drawing up a probability "tree" for this problem obtain the probability distribution for the total sales in the first **two** years. Calculate the mean and SD of the two-year sales figure. Assume that the sales figures in the first and second years will be whole numbers of thousands.

BA (Business Studies)

5(E) An advertising agency is given first option for five sixty-second "breaks" on commercial television for a Saturday evening six months ahead. These breaks may be shared among the various advertising accounts held by the agency. On each break sold to one of its clients the agency will make a profit of £150. However, for each break not taken up by its clients the agency will have to sell it to another agency and will incur a loss of £200. The media director responsible for the forward booking of T.V. time makes the following assessment of the number of breaks that will be taken up by his clients.

No. Breaks Taken	Probability
0	0.1
1	0.1
2	0.2
3	0.3
4	0.2
5	0.1

(i) Draw up a "payoff" table for the above problem, showing the profits and losses associated with each possible act by the agency, and the demands by its clients.

(ii) On how many breaks should the agency take up its option? What are the assumptions underlying this approach to solving the problem?

(iii) Suppose the agency could, at a cost, determine exactly how many of the "breaks" would be taken upon each occasion. Assume that over a long period the given probabilities still represent the relative frequencies of the various numbers of breaks taken. How high would the cost of this "perfect" information have to be before it would be worth buying?

BA (Business Studies)

6(E) An oil company has purchased drilling rights on a particular site. They have the choice of either drilling directly or conducting a survey to indicate the likelihood of striking oil. From previous experience of similar sites they know that if a survey proves positive there is an 80% chance of subsequently striking oil in workable quantities – even if the survey results are negative there will still be a 5% chance. It is also known that 10% of such surveys prove positive in that area.

(i) Draw up a decision tree for this problem.
(ii) Evaluate the probability of there being oil in workable quantities on the site.
(iii) What should the company's policy be? Should they conduct a survey? What if the survey proves negative? The following economic information applies:
- a survey costs £100,000
- drilling costs £2m.
- if workable quantities of oil are found the net revenue from the oil is expected to be £20m.

BA (Business Studies)

7(E)* A computer bureau running customers' programs has found that on average a proportion π of the programs fail to execute for one reason or another. Derive an expression for $p(k)$, the probability that in a batch of programs the kth program will be the first program which fails to execute. Calculate the mean and variance of k.
Hints:

$$1 + 2x + 3x^2 + 4x^3 + \ldots = (1 - x)^{-2}$$

and

$$1 + 4x + 9x^2 + 16x^3 + \ldots = (1 + x)(1 - x)^{-3}$$

for $|x| < 1$.

SOLUTIONS TO EVEN NUMBERED EXERCISES

2. (i) Yes, this is a probability distribution, but of a very elementary type. The random variable R is *certain* to take the value 1.
   (ii) Yes, an *infinite* discrete distribution because:

$$\frac{1}{2}+\frac{1}{4}+\frac{1}{8}+\frac{1}{16}+\ldots=1 \text{ (Sum of an infinite geometric progression)}$$

   (iii) This is only a probability distribution if it is also true that $p(r) = 0$, other r.
   (iv) Yes, as $\frac{1}{8}+\frac{3}{4}+\frac{1}{8}=1$. Note that it is possible to have a discrete random variable which can take negative values.
   (v) No, because the probabilities add up to more than one.
   (vi) No. The negative probability value is nonsense (even though the probabilities add up to one).

4.

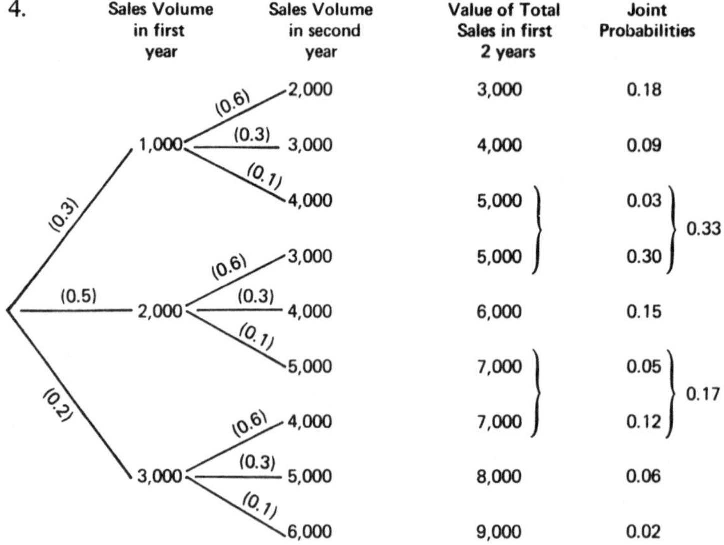

(Note that there are two possible "routes" to achieving a 2-year sales of 5,000 and 7,000.)

Working in units of thousands, we have:

2-year Sales (x)	*Prob. of x* $p(x)$	$xp(x)$	$x^2p(x)$
3	0.18	0.54	1.62
4	0.09	0.36	1.44
5	0.33	1.65	8.25
6	0.15	0.90	5.40
7	0.17	1.19	8.33
8	0.06	0.48	3.84
9	0.02	0.18	1.62
	1.00	5.30	30.50

Mean 2 Year Sales $= \Sigma xp(x) = 5.3$ (thousands).

Variance of 2 Year Sales $= \Sigma x^2 p(x) - [\Sigma xp(x)]^2 = 30.50 - (5.30)^2 = 2.41$

SD of 2 Year Sales $= \sqrt{2.41} = 1.55$ (thousands).

6. (i)

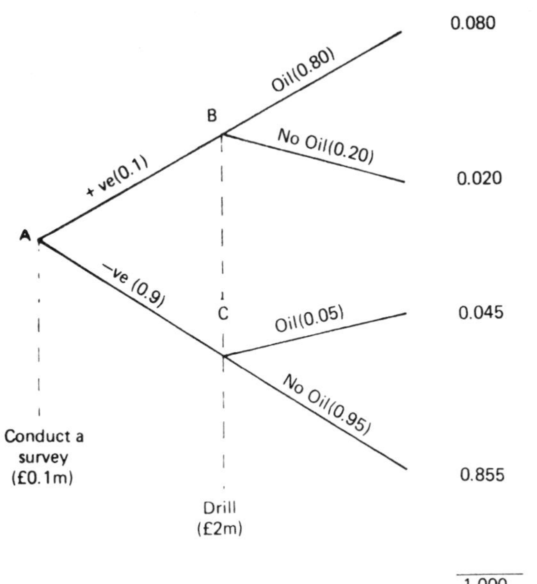

(ii) Pr(Oil) = 0.080 + 0.045 = 0.125.

(iii) Working *backwards* through the decision tree, and by using monetary units of £ millions, we have

expected payoff of *taking decision to drill* at B
$= (0.80 \times 20 + 0.20 \times 0) - 2 = 14,$

expected payoff of *taking decision to drill* at C
= (0.05 × 20 + 0.95 × 0) – 2 = –1 (i.e. a loss)

Hence we would decide not to drill if the survey proved negative. We can now simplify the decision tree to the following form:

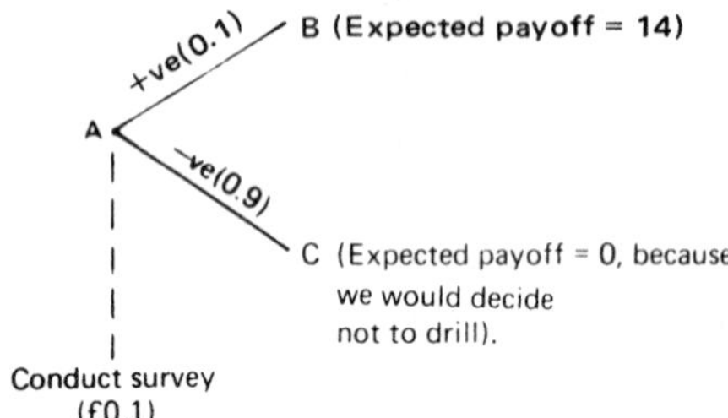

Expected payoff when conducting survey = (0.1 × 14 + 0.9 × 0) – 0.1 = 1.3. We finally compare this with the expected payoff of drilling *without conducting a survey*. Making use of the result in (i), we obtain the simple decision tree:

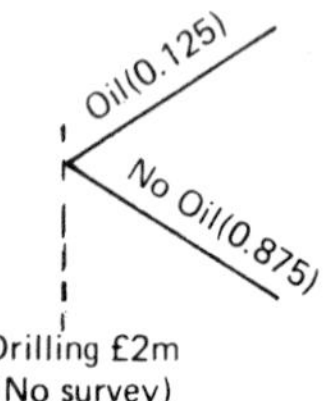

Expected payoff when *not* conducting a survey = (0.125 × 20 + 0.875 × 0) – 2 = 0.50.
Consequently it is worth conducting a survey because it increases the expected payoff from £0.5m to £1.3m.

COURSE UNIT 9 – THE BINOMIAL DISTRIBUTION

9.1 INTRODUCTION

Many statistical enquiries are conducted by asking the same question of each member of a sample. The questions may either be put *directly*, as when asking employees about a new bonus scheme; or they may be *implied*, as when examining manufactured components for defects. In this Course Unit we are concerned only with very simple questions – those that can be framed in such a way that the answer is either "yes" or "no". But it is surprising how many situations to which this applies.

To take just one example, suppose an industrial process is periodically checked by selecting a sample of 10 items from a production line and counting the number of these which have faults. We could, if we so wished, visualise this situation in the following way. As an item is selected we ask the question, "Is this item faulty?" The sampling technique amounts to repeating this process for each of the 10 items and counting the number of times we answered "yes".

The statistical analysis of situations of this type is the province of the *binomial distribution*.

9.2 A BERNOULLI TRIAL

First let us tie up this idea of an enquiry conducted by asking "yes/no" questions with the terminology developed in earlier Course Units. The question asked may be expressed generally as, "Has a certain event, E, occurred?" The answer will be "Yes" if E has occurred or "No" if the complementary event $\bar{E}$ has occurred. But as it is more convenient to work with numerical quantities we "code" the answers by defining a random variable S which takes the value *1* if E has occurred and *0* if $\bar{E}$ has occurred (*1* is thus the code for "yes" and *0* for "no")[(1)]. Finally we use the symbol[(2)] π to denote the probability of E occurring, the value of which may be known (or estimated) in a particular situation.

---

(1)The introduction of the random variable S is purely a device to associate a numerical value with essentially non-numerical events. Similar devices are used outside the field of statistics, e.g. when football results are announced the number "3" means a "scoring draw", the number "1" means a "home win".

(2)The symbol π is used as a population parameter corresponding to the proportion of all conceptually identical trials in which E would occur. The symbol should not be confused with the mathematical constant π(= 3.14159. . .).

So as not to become overwhelmed with this formalism, we quote a few concrete examples in Exhibit 1.

EXHIBIT 1: Bernoulli Trials

Trial	*Event, E*	π	*Values of S*
(i) Toss an unbiased coin	Heads	1/2	$S = 1$ if heads (E occurs) $S = 0$ if tails ($\bar{E}$ occurs)
(ii) Roll a true die	Face 6 is shown	1/6	$S = 1$ if 6 shows $S = 0$ if any other face shows
(iii) Measure the length of a standard casting	Length is greater than upper quartile, Q_3, of such castings	1/4	$S = 1$ if length $\geqslant Q_3$ $S = 0$ if length $< Q_3$
(iv) Select a screw at random from a bin of 1,000 screws, 100 of which have no head slots	Screw has no head slot	1/10	$S = 1$ if screw has no head slot $S = 0$ if screw has a head slot
(v) Ask a housewife to say which of three sandwiches, A, B & C, each containing the *same* meat paste, she prefers	Housewife prefers B	1/3	$S = 1$ if she prefers B $S = 0$ if she prefers A or C

It is common practice to refer to event E (however defined) as being a "success", irrespective of whether E is a desirable event or not. Similarly $\bar{E}$ can be referred to as a "failure".

Such trials (with E, π and S defined as indicated) are called **Bernoulli** trials after the mathematician who investigated their properties.

The probability distribution of S has the delightfully simple form of

$$\Pr(S = 1) = \pi$$
$$\Pr(S = 0) = (1 - \pi)$$
$$\Pr(S = s) = 0, \text{ other } s$$

or more compactly,

$$\begin{aligned}\Pr(S = s) &= \pi^s (1 - \pi)^{1-s}, \; s = 0,1\\ &= 0, \text{ otherwise.}\end{aligned}$$

We can use this probability distribution to evaluate the expected value (mean) and variance of S.

We have

$$\mu = E(S) = 1 \cdot \pi + 0 \cdot (1 - \pi) = \pi$$

and

$$\sigma^2 = E(S^2) - [E(S)]^2 = 1 \cdot \pi + 0 \cdot (1 - \pi) - \pi^2 = \pi(1 - \pi).$$

Referring to our Example (i) in Exhibit 1 consider tossing a coin a very large number of times, writing down *1* each time a head occurs and *0* each time a tail occurs. The mean value of these *0*'s and *1*'s is given by $\mu = \pi = 0.5$. A not very surprising result! What is perhaps not so intuitively obvious is that the variance of these values is given by $\sigma^2 = \pi(1 - \pi) = 0.25$.

9.3 THE BINOMIAL DISTRIBUTION

So far we have considered the Bernoulli trial and the Bernoulli probability distribution. Such a distribution is not very important in its own right: its importance lies in its relationship to the binomial (i.e. "two names") distribution, which is a fundamental distribution of sampling theory.

We define a **binomial experiment** as a sequence of n independent Bernoulli trials having identical characteristics. By "identical characteristics" we mean that the trials are carried out under identical conditions and that we are interested in the same event, E, in each case. By "independent" we mean that π, the probability of E occurring in each Bernoulli trial, is the same throughout and does not depend on whether, at some point, the last Bernoulli trial resulted in E or not.

To make these distinctions clear, consider Example (iv) in Exhibit 1 concerning a bin of 1,000 screws, 100 of which are slotless. Suppose we were to take a random sample of 10 screws. Then we could either:

(a) sample *with* replacement, or
(b) sample *without* replacement.

If we adopt method (a) then we can consider the process as 10 independent Bernoulli trials of identical characteristics (i.e. a binomial experiment with $n = 10$ and $\pi = 0.1$). But if we adopt method (b), then it is clear that π will change throughout the process. For example the first screw selected will have a probability of 0.1 of being slotless. If it actually is slotless then the probability of the second screw being headless as well is not 0.1 but 99/999 ($\simeq 0.099$). So we cannot strictly regard sampling *without* replacement as a binomial experiment. The exception to this rule is where the sample size is a very small fraction (say less than 5%) of the population size and so, for all practical purposes, we can ignore the minute changes in π that occur. If, in such a case, we consider the situation as a binomial experiment we say that we are regarding it as a "model" of the situation – an imperfect but probably adequate representation of the truth. In most practical cases, in fact, we make use of the notion of a binomial experiment as a model because the conditions required for a true binomial experiment are difficult to achieve or justify.

Suppose, in a binomial experiment, we define the random variable R as the number of Bernoulli trials in which the event E occurs – the number of "successes". R is clearly the sum of the individual Bernoulli random variables[(1)] and can only take values $0,1,2,\ldots, n$. The probability distribution of R, $\Pr(R = r)$, is called the **binomial distribution** and expresses the probability of obtaining r "successes" in n independent Bernoulli trials. In its most general form $\Pr(R = r)$ will be expressed in terms of n and π. We next consider a particular example.

Three production lines each produce different plastic soldiers (types A, B and C) at the same rate. The outputs from these production lines are thoroughly mixed and boxed in "fives".[(2)] What is the probability that a box selected at random will contain R soldiers ($R = 0,1,2,3,4,5$) of type A?

We can consider the process of examining the contents of the box as a binomial experiment with $n = 5$. Each soldier in the box is either of type A (with probability $\pi = \frac{1}{3}$) – a "success" – or of some other type (with probability $1 - \pi = \frac{2}{3}$).

Consider the particular event that the box contains just 2 soldiers of type A ($R = 2$). We can evaluate the probability of this event from first principles as follows, using "A" to denote that a soldier is of type A and "O" to denote that it is not (Exhibit 2).

We see that each (or the 10) mutually exclusive outcomes is equally likely. Hence the probability of the event ($R = 2$) is:

$$\Pr(R = 2) = p(2) = 10(\tfrac{1}{3})^2(\tfrac{2}{3})^3 = 0.3292.$$

---

(1) $R = \Sigma_{i=1}^{n} S_i$. The reason why the *Bernoulli* random variable was defined as it was should now be clear – it is used as a device to count the number of successes in a binomial experiment. Readers familiar with counting procedures in computer program loops will see a close parallel here.

(2) We are thinking here of the familiar promotional scheme of including a number of "novelties" in a packet of cereal. The variation in the type of novelties from one packet to another, it is claimed, adds to the appeal.

EXHIBIT 2: Outcomes and Probabilities Associated with Two Soldiers of Type A

	Outcomes (Soldier 1 2 3 4 5)	*Joint Probability*
10 mutually exclusive outcomes	A A O O O	$(\frac{1}{3})(\frac{1}{3})(\frac{2}{3})(\frac{2}{3})(\frac{2}{3}) = (\frac{1}{3})^2(\frac{2}{3})^3$
	A O A O O	$(\frac{1}{3})(\frac{2}{3})(\frac{1}{3})(\frac{2}{3})(\frac{2}{3}) = (\frac{1}{3})^2(\frac{2}{3})^3$
	A O O A O	$(\frac{1}{3})(\frac{2}{3})(\frac{1}{3})(\frac{2}{3})(\frac{2}{3}) = (\frac{1}{3})^2(\frac{2}{3})^3$
	A O O O A	$(\frac{1}{3})(\frac{2}{3})(\frac{1}{3})(\frac{2}{3})(\frac{2}{3}) = (\frac{1}{3})^2(\frac{2}{3})^3$
	O A A O O	$(\frac{1}{3})(\frac{2}{3})(\frac{1}{3})(\frac{2}{3})(\frac{2}{3}) = (\frac{1}{3})^2(\frac{2}{3})^3$
	O A O A O	$(\frac{1}{3})(\frac{2}{3})(\frac{1}{3})(\frac{2}{3})(\frac{2}{3}) = (\frac{1}{3})^2(\frac{2}{3})^3$
	O A O O A	$(\frac{1}{3})(\frac{2}{3})(\frac{1}{3})(\frac{2}{3})(\frac{2}{3}) = (\frac{1}{3})^2(\frac{2}{3})^3$
	O O A A O	$(\frac{1}{3})(\frac{2}{3})(\frac{1}{3})(\frac{2}{3})(\frac{2}{3}) = (\frac{1}{3})^2(\frac{2}{3})^3$
	O O A O A	$(\frac{1}{3})(\frac{2}{3})(\frac{1}{3})(\frac{2}{3})(\frac{2}{3}) = (\frac{1}{3})^2(\frac{2}{3})^3$
	O O O A A	$(\frac{2}{3})(\frac{2}{3})(\frac{2}{3})(\frac{1}{3})(\frac{1}{3}) = (\frac{1}{3})^2(\frac{2}{3})^3$

Similarly we can evaluate $Pr(R = r)$ for other values of r. The results are tabulated in Exhibit 3.

EXHIBIT 3: The Binomial Distribution with $n = 5$, $\pi = \frac{1}{3}$

r	$Pr(R = r) = p(r)$	$rp(r)$	$r^2p(r)$
0	$(\frac{2}{3})^5 = 0.1317$	0	0
1	$5(\frac{1}{3})(\frac{2}{3})^4 = 0.3292$	0.3292	0.3292
2	$10(\frac{1}{3})^2(\frac{2}{3})^3 = 0.3292$	0.6584	1.3168
3	$10(\frac{1}{3})^3(\frac{2}{3})^2 = 0.1646$	0.4938	1.4814
4	$5(\frac{1}{3})^4(\frac{2}{3}) = 0.0412$	0.1648	0.6592
5	$(\frac{1}{3})^5 = 0.0041$	0.0205	0.1025
Totals	1.0000	1.6667	3.8891

We see that the probabilities corresponding to each possible value of R sum to unity – a result we would expect for a "proper" probability distribution. Calculating the mean and variance of R by using expected values we obtain:

$$\mu = E(R) = \sum_{r=0}^{n} r\,p(r) = 1.667 \text{ (to 3 decimal places)}$$

and

$$\sigma^2 = E(R^2) - [E(R)]^2 = 3.8891 - (1.667)^2 = 1.111$$

(to 3 decimal places)

We shall see later that there is a much easier way of determining the mean and variance of a random variable which has a binomial distribution.

We next make use of this example to deduce some *general* results about the binomial distribution. Firstly, looking at the probabilities we calculated, we see that they form the following pattern:

$$\Pr(R = r) = \begin{pmatrix} \text{number of mutually} \\ \text{exclusive outcomes} \\ \text{which give rise to } r \\ \text{"successes"} \end{pmatrix} \cdot (\tfrac{1}{3})^r (\tfrac{2}{3})^{5-r}$$

It is not hard to visualise that for a general binomial experiment, specified by n and π we would have:

$$\Pr(R = r) = \begin{pmatrix} \text{number of mutually} \\ \text{exclusive outcomes} \\ \text{which give rise to } r \\ \text{successes} \end{pmatrix} \cdot \pi^r (1 - \pi)^{n-r}$$

Now the "number of mutually exclusive outcomes which give rise to r successes" is the number of different ways in which n items can be arranged in a straight line if r of them are of one type (successes) and $(n - r)$ are of another type (failures). It is a simple result of the theory of combinations that this number is given by the expression:

$$\frac{n!}{r!(n-r)!}$$

where $n!$ (pronounced n-factorial) means

$n \times (n - 1) \times (n - 2) \times \ldots 2 \times 1$

(and similarly for $r!$). Taking our particular example of $n = 5$, $r = 2$, we obtain

$$\frac{5!}{2!3!} = \frac{5 \cdot 4 \cdot 3 \cdot 2 \cdot 1}{2 \cdot 1 \cdot 3 \cdot 2 \cdot 1} = 10,$$

a result we obtained earlier by writing down each possible outcome and finding that there were 10 distinct outcomes.

Finally then, the general expression for the terms of the binomial distribution is given by:[(1)]

$$\Pr(R = r) = p(r) = \frac{n!}{r!\,(n-r)!}\,\pi^r (1 - \pi)^{n-r}, \quad \text{for } r = 0,1,2, \ldots n$$
$$= 0, \text{ otherwise} \qquad (1)$$

[(1)]The reader might recognise this expression as the general term (rth) in the expansion of $(\pi + [1 - \pi])^n$. Incidentally this leads to a very simple proof that the binomial probabilities sum to unity since $(\pi + [1 - \pi])^n = 1^n = 1$.

At this point it is worth restating that in order to evaluate a binomial distribution we need to know the values of the two parameters n and π. (Some texts refer to n as an index rather than a parameter.) Consequently the binomial distribution is a two-parameter distribution, and for this reason the notation $b(r; n, \pi)$ is sometimes used for $\Pr(R = r)$ – the "b" stands for *b*inomial.

Proceeding generally we could derive an expression for the mean value and variance of the binomial distribution in terms of n and π by algebraic manipulation on the expected value formulae – a process completely anologous to our arithmetic calculation of the mean and variance in our earlier example.

Details of this manipulation are not given here but the results, which are of fundamental importance, are quoted below:

Mean Value of binomial distribution (μ) = $n\pi$ (2)
Variance of binomial distribution (σ^2) = $n\pi(1 - \pi)$ (3)

Checking these results with our earlier example ($n = 5$, $\pi = \frac{1}{3}$), we obtain:

$$\mu = 5 \cdot \tfrac{1}{3} = 1.667 \text{ (to 3 decimal places)}$$

and

$$\sigma^2 = 5 \cdot \tfrac{1}{3} \cdot \tfrac{2}{3} = 1.111 \text{ (to 3 decimal places),}$$

precisely the same values as we obtained earlier by working from numerical probabilities. It is clearly easier to make use of the general results for the mean and variance rather than to evaluate these quantities using the numerical probabilities.

The following comments may help the reader "feel at home" with (and remember) the general formulae for the mean and variance.

- The formula for the mean ($n\pi$) has an easy intuitive interpretation. If the proportion of "successes" in a population is π then in a sample of size n we would expect, on average, a proportion of π of those n to be successes, that is $\pi \times n$ or $n\pi$.[(1)] For example, if 30% ($\pi = 0.30$) of homes have a telephone, then in a sample of n = 1,000 homes we would expect to find $n\pi$ = 300 homes with telephones.
- The random variable R can only take values 0,1,2, . . . n. Hence the

---

[(1)]In CU8 we stated without proof that the "expected" frequency $\hat{f}_r = np(r)$, could be interpreted as the average frequency over a large number of samples each of size n. If we interpret a "success" as an observation "r" (with probability $p(r) = \pi$), then the binomial result for the mean provides an immediate proof.

variability that R can exhibit will be greater if n is large than if n is small. This "explains" why the formula for the variance has n as a multiplier. If $n = 1$ then we are referring to a single Bernoulli trial whose variance was shown to be $\pi(1 - \pi)$ in Section 9.2 (i.e. the same as the binomial variance, $n\pi(1 - \pi)$ when n is put equal to 1).

- If $\pi = 0$ then a "success" is impossible. Consequently the binomial random variable R will *always* take the value 0. It will consequently exhibit no variability – a result confirmed by putting $\pi = 0$ into the binomial variance, $n\pi(1 - \pi)$. If $\pi = 1$ then R will always be equal to n – another condition of zero variance.

9.4 THE RECURSIVE FORM OF THE BINOMIAL DISTRIBUTION

The individual binomial probabilities are difficult to compute using Equation (1), especially if n is large. In practice the probabilities are obtained in one of three ways: using published tables or using a suitable approximation (this process will be dealt with in CU's 11 and 12), or using the **recursive** form of the binomial distribution.

The recursive form of calculation works as follows:

Firstly $p(0) = (1 - \pi)^n$ is calculated; $p(1)$ is then calculated from $p(0)$, $p(2)$ is calculated from $p(1)$, and so on. The formula for recursive calculation, which is totally equivalent to the use of Equation (1)[(1)], is shown in Equations (4).

$$p(0) = (1 - \pi)^n$$

$$p(r) = \frac{n - r + 1}{r} \cdot \frac{\pi}{1 - \pi}\, p(r - 1), \quad r = 1, 2, \ldots, n \tag{4}$$

The recursive form of the binomial distribution

Taking our toy soldier example ($n = 5$, $\pi = \frac{1}{3}$), we have:

$$p(0) = (1 - \tfrac{1}{3})^5 = (\tfrac{2}{3})^5 = 0.1317$$

$$p(1) = \frac{5 - 1 + 1}{1} \cdot \frac{1}{3} \cdot \frac{3}{2} \cdot p(0) = \frac{5}{2} p(0) = 0.3292$$

$$p(2) = \frac{5 - 2 + 1}{2} \cdot \frac{1}{3} \cdot \frac{3}{2} \cdot p(1) = \frac{4}{4} p(1) = 0.3292$$

$$p(3) = \frac{5 - 3 + 1}{2} \cdot \frac{1}{3} \cdot \frac{3}{2} \cdot p(2) = \frac{1}{2} p(2) = 0.1646$$

etc.

[(1)] Proof of Equations (4) is left to the reader. Hint: write down the expression for $p(r)$, write down an expression for $p(r - 1)$, and divide the former by the latter.

9.5 TABLES OF THE BINOMIAL DISTRIBUTION

Tables of the binomial distribution are published in two forms – individual probabilities,[1] $\Pr(R = r)$, for given values of n and π, and so-called cumulative probabilities,[2] the probabilities of r *or more* successes. The use of tables giving individual probabilities is straightforward: cumulative probability tables, on the other hand, can sometimes lead to confusion, although they are more convenient for many applications. In this section we illustrate the use of cumulative probability tables with a simple example.

Suppose a production line produces, on average, 5% defective items. Every hour 10 items are selected at random from the output. What is the probability that in the sample (a) *at least* 2 items are defective, (b) *exactly* 2 items are defective, and (c) *at most* 2 items are defective?

For all practical purposes we can consider the sampling process as being 10 independent Bernoulli trials of identical characteristics. Hence the number of defectives in a sample of 10 can be described by the binomial distribution with $n = 10$ and $\pi = 0.05$. Referring to cumulative binomial tables we find the values shown in Exhibit 4.

EXHIBIT 4: Cumulative Binomial Probabilities ($n = 10$, $\pi = 0.05$)

$r = 0$	1.0000
$r = 1$	0.4013
$r = 2$	0.0861
$r = 3$	0.0115
$r = 4$	0.0010
$r = 5$	0.0001
$r = 6$	0.0000

The value alongside $r = 2$, 0.0861, gives the probability of *2 or more* defectives (the required probability for (a)), that is

$$p(2) + p(3) + p(4) + \ldots + p(10) = 0.0861.$$

Similarly alongside $r = 3$, we have

$$p(3) + p(4) + \ldots + p(10) = 0.0115.$$

By subtraction, we obtain the answer to (b): $p(2) = 0.0746$.

Now we also require the probability of "at most 2 defectives" (c). This

---

[1] See Appendix 2.

[2] For example, J. Murdoch and J.A. Barnes *Statistical Tables for Science Engineering and Management*, Macmillan.

is given by $p(0) + p(1) + p(2)$. From the tables we have:

$$p(3) + p(4) + \ldots + p(10) = 0.0115,$$

but we also know

$$p(0) + p(1) + p(2) + p(3) + p(4) + \ldots + p(10) = 1.000.$$

Hence by subtraction we obtain an answer to (c), 0.9885.

Examining a set of binomial tables it will be seen that the probabilities are only tabulated for selected combinations of n and π in order to save space. There is no simple way of interpolating between the tables if the particular combination of n and π for which the probabilities are required is not tabulated. Consequently for many applications approximations to the binomial distribution are used: in some cases the so-called Poisson distribution (CU10) and in others, the normal distribution (CU11) is used.

It will also be seen that tables are only published for values of π up to (and including) 0.5. The probabilities for higher values of π (up to 1.0) can be obtained as follows. Suppose we require $\Pr(R = 14)$ from a binomial distribution with $n = 20$, $\pi = 0.7$. We are seeking the probability of 14 "successes" in 20 Bernoulli trials each with a probability of success at 0.7. This is equivalent to the probability of 6 (i.e. 20–14) failures in 20 trials each with a probability of failure of 0.3 (i.e. 1–0.7). So:

$$\begin{aligned} &\Pr(R = 14) \text{ with } n = 20, \pi = 0.7 \\ &= \Pr(R = \;\;6) \text{ with } n = 20, \pi = 0.3 \\ &= 0.1916 \text{ (from tables using } n = 20, \pi = 0.3). \end{aligned}$$

Finally Exhibit 5 shows the histograms of a number of binomial distributions. Notice how the shape becomes more symmetrical as n increases (with π constant) and as π increases (with n constant).

9.6 AREAS OF APPLICATION OF THE BINOMIAL DISTRIBUTION

The main application of the binomial distribution is to "attribute" sampling. By "attribute" sampling we mean that each item in the sample can be described simply by whether or not it possesses a predefined characteristic or property. The type of attribute will of course depend on the area of study: in quality control we may be interested in whether an item is "defective" (i.e. outside design specifications); in a social survey, whether a person is a pensioner or alternatively whether a household has an income in excess of £3,000 p.a; in market research, whether a housewife prefers a new formulation of detergent.

EXHIBIT 5: Shapes of the Binomial Distribution

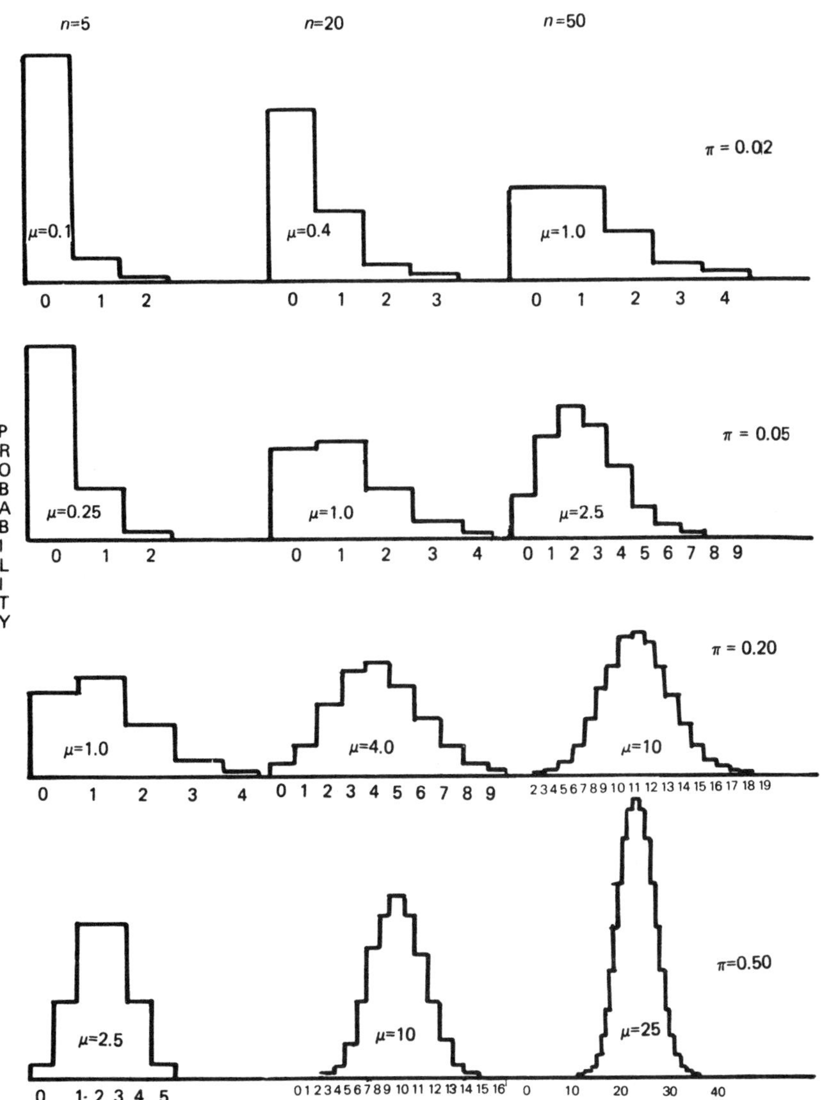

N.B. The histograms have been drawn using "bars" rather than lines and the probability scales adjusted to arrange for the same area under each, in order to achieve maximum comparability.

Strictly, the sampling should be random (to meet the conditions for a binomial experiment) and should be conducted either "with replacement" or if not, the sample size should be a very small fraction of the population size.

In such cases the binomial distribution, with n being the sample size and π the probability of each item possessing the particular attribute, describes the probability of a sample containing r (= 0,1,2, . . ., n) items possessing that attribute.

There are other situations where, although at first sight no physical sampling takes place, it is possible to think about the problems in terms of "samples". Consider for example the examination questions given at the end of this Course Unit and the material in the following section.

9.7 "FITTING" A BINOMIAL DISTRIBUTION TO DATA

In CU2 we formed 50 observations (of the number of faulty tyres on ten wheeled trucks) into a frequency distribution as shown in Exhibit 6.

EXHIBIT 6: Frequency Distribution of Faulty Tyre Data

Number of faulty tyres per truck, x_i	*Observed frequency*, f_i	$x_i f_i$	$x_i^2 f_i$
0	22	0	0
1	16	16	16
2	5	10	20
3	4	12	36
4	3	12	48
Totals	$\Sigma f_i = 50$	$\Sigma x_i f_i = 50$	$\Sigma x_i^2 f_i = 120$

We have $\bar{X} = \dfrac{\Sigma x_i f_i}{\Sigma f_i} = \dfrac{50}{50} = 1.0$

and $s^2 = \dfrac{\Sigma x_i^2 f_i}{\Sigma f_i} - \bar{X}^2 = \dfrac{120}{50} - (1.0)^2 = 1.4$

Does it seem reasonable to describe such data in terms of the binomial distribution? Certainly each truck can be considered as a "sample" of 10 tyres, and we are interested in the number of these which are faulty. The situation appears to have the properties to which the binomial distribution can be applied. But can we say (as is required for the binomial distribution) that the probability of a tyre being faulty, π, is the same for each wheel, and the same from one truck to another? We can investigate this point by "fitting" a binomial distribution to the data and comparing the observed frequencies (from the data) with the expected frequencies from the binomial distribution.

Of course we cannot fit a *general* binomial distribution to the data: we have to fit a *particular* distribution with chosen values of n and π. It is reasonable to use $n = 10$ (as 10 is our "sample" size). The value of π cannot be specified *a priori* and so in such circumstances we choose π so that the *mean of the sample data is equal to the mean of the binomial distribution*. In our case, we have:

$$\bar{X} = 1.0 \quad \text{(sample mean)}$$
$$= n\pi = 10\pi \quad \text{(binomial mean)},$$

from which we obtain[(1)] $\pi = 0.1$.

Now the binomial distribution with $n = 10$, $\pi = 0.1$ has a variance of $10 \times 0.1 \times 0.9$ $[= n\pi(1 - \pi)] = 0.9$. This is considerably lower than our sample variance of 1.4 and gives us the first clue that the binomial distribution may not be an appropriate theoretical distribution with which to describe the data. Nonetheless, proceeding with evaluating individual binomial probabilities (using tables for convenience) and scaling these to obtain expected frequencies, we obtain the results shown in Exhibit 7.

EXHIBIT 7: Comparison of Observed and Expected Frequencies

$r = x_i$	*Binomial Probability* $Pr(R = r) = p(r)$	*Expected Frequency* $\hat{f}_r = 50p(r)$ *(to 1 decimal place)*	*Observed Frequencies* f_i
0	0.3487	17.4	22
1	0.3874	19.4	16
2	0.1937	9.7	5
3	0.0574	2.9	4
4	0.0112	0.6	3
5	0.0015	0.0	0
6	0.0001	0.0	0
Totals	1.0000	50.0	50

Note that in performing these calculations we must take care to distinguish between n in the binomial formula and the number of observations (50) by which the probabilities are scaled to obtain the

(1) A more intuitively appealing, but equivalent, argument for choosing π is as follows. In our sample the total number of defective tyres $= \Sigma X = \Sigma x_i f_i = 50$. Now the total number of tyres examined equals (the number of trucks examined) $\times$ (number of tyres per truck) $= 50 \times 10 = 500$. Hence the *proportion* of tyres which were faulty is $50/500 = 0.1$. This is the value chosen for π.

expected frequencies – it was for this reason we used the notation Σf_i in place of the number of observations n in the formulae for calculating the mean and variance of the data.

We can see that there are certain discrepancies between the observed and expected frequencies. In particular we observe greater frequencies for low (0) and for high (3 and 4) values of r than are given by the expected frequencies of the binomial distribution. What are the reasons for this? We could argue that these discrepancies are simply due to the chance effects of sampling and if we took another sample (of another 50 trucks) the results would be quite different. Now there are formal tests one can employ to check whether this is so (some of these tests are explored later in the course). For the moment let us assume that an appropriate test has been carried out and the conclusion is that chance alone is inadequate to explain the difference between the observed and expected frequencies (or, for that matter, between the sample variance and the binomial variance). What alternative explanation can we offer for these discrepancies? We might suggest that we are more likely to find trucks with no faulty tyres (e.g. new trucks or good drivers) or with several faulty tyres (e.g. old trucks or bad drivers) than would be predicted on the basis of a constant probability (π) of each tyre being faulty.

Although we have concluded that the binomial distribution does not "fit" the data very well, and hence it is not reasonable, without further examination, to use the binomial distribution in other similar cases, the process of fitting a distribution to the data has forced us to look more critically at the data than we might otherwise have done.

More important, we have established the general principles for fitting a probability distribution to observed data. The steps involved can be summarised as follows:

(a) Obtain the values of the required parameters (by estimation or otherwise) from the sample (e.g. n and π).

(b) Generate the probabilities from the probability distribution using the parameters calculated.

(c) Multiply the probabilities by the total number of observations in the sample (Σf_i) to obtain the expected frequencies.

We attempt to interpret the results obtained by considering whether or not the process conforms to our probability model. By using this procedure we can often learn a lot about the nature of the process which gave rise to the data.

EXERCISES

1(P) "Simulate" the example on toy soldiers in the text as follows. Mark 2 sides of a six-sided pencil (or ball-point pen). Consider a binomial experiment of rolling the pencil 5 times and counting the number of times a marked side comes uppermost (r). This represents the number of type "A" soldiers in a box of 5, as the probability of getting a marked side uppermost in a single roll is 2/6 or 1/3.

(i) Perform 50 such experiments, noting the value of r in each case.

(ii) Arrange the values of r in a frequency distribution and calculate the mean $\bar{r}$ and variance s_r^2 of r. Compare with μ and σ_r^2 respectively.

(iii) Calculate the expected frequencies from 50 experiments and compare with the observed frequencies. Comment on any differences between the observed and expected frequencies

(iv) Fit a binomial distribution *with the same mean* to the frequency distribution you have observed. Is the "fit" better or worse than that obtained in part (iii), where the data was compared with the expected frequencies of the binomial distribution with $n = 5$, $\pi = \frac{1}{3}$? Explain why.

2(U) Show that (a) $6! = 720$, (b) $8! = 56 \times 6!$, (c) $\frac{10!}{8!} = 90$, (d) $\frac{8!}{6!2!} = 28$.
Also demonstrate (by listing all possibilities) that the result in (d) gives the number of distinct outcomes in 8 trials of which 6 are "successes".

3(E) A consignment of castings contains 10 castings with major cracks, another 10 with only hairline cracks and 30 with no cracks at all. Six inspectors on separate occasions select a casting at random, examine it, and replace it. What is the probability that just three inspectors find a major crack in the casting they examined? Would this hold if the castings are not replaced? Give your reasons.

Part question HND (Business Studies)

4(E) In a large company 40% of the employees are men and 60% are women. Each employee automatically enters for the Xmas "draw" in which there are 8 prizes. What is the probability that the prizes will be awarded to 3 men and 5 women. Is the answer you obtain the same as the probability of finding 3 men and 5 women on a particular table seating 8 in the company canteen? Give your reasons.

Part question HND (Business Studies)

5(E) A British coastal resort claims that in June there are on average only 6 rainy days. On the assumption that the binomial distribution applies, what is the probability that there will be 10 rainy days next June. Precisely what assumptions have been made in your answer. If these assumptions are *not* justified, in what way will the distribution of the number of rainy days in June differ from the binomial distribution. (Give rough sketches.)

Part question HND (Business Studies)

6(U) (i) On average 90% of a company's orders are for the home market. On a randomly chosen day four orders are dispatched. Evaluate the probabilities that 0,1,2,3, or 4 of these are for the home market. Plot the histogram of this binomial distribution.

(ii) Calculate the terms of the binomial distribution with $n = 6$, $\pi = 0.40$ and also for $n = 4$, $\pi = 0.05$. (Do not use tables.)

7(E) Suppose a sample of 10 is taken from a day's output of a machine that normally produces 5% defective parts. If the day's production is inspected 100% whenever the sample of 10 gives 2 or more defectives. What is the probability that a day's production will be inspected 100% if the machine has been working normally? What would the answer be if on a particular day the machine was incorrectly set and it produced 10% defective parts?

Part question HND (Business Studies)

8(E) A consignment of 10 spectrometers has been received by a research laboratory. The consignment appears to have suffered from superficial transit damage. As a check the laboratory intends to subject 2 randomly chosen spectrometers to an exhaustive series of tests. If, in fact, only 7 out of the 10 spectrometers would pass such tests,

(i) obtain the probability distribution for the number of spectrometers (of the 2 to be tested) that *fail*;

(ii) calculate the mean and variance of this distribution;

(iii) compare your results in (i) and (ii) with those for the (unrealistic) situation of selecting the two spectrometers for testing by sampling *with replacement*.

What conclusions can you draw from this comparison on the difference between sampling with or without replacement?

BA (Business Studies)

N.B. The solution to part (i) is not a binomial distribution.

9(E) (i) Explain carefully the meaning of the symbol $E(R^2)$ for a discrete probability distribution. Why is it useful to evaluate $E(R^2)$?

(ii) A firm produces goods with an average proportion "defective" (goods which fail before the guaranteed life) of 5%. A buyer offers to pay £5 over the usual price for a batch of 40 providing the firm will accept the following penalty for defectives.

For no defective	no penalty
For 1 defective	£1 penalty
For 2 defectives	£4 penalty
For 3 defectives	£9 penalty
For 4 defectives	£16 penalty
etc.	etc.

If the firm is interested in maximising profits "in the long run" should they accept the offer?

BA (Business Studies)

10(E) A company uses a written test as part of the selection procedure for all job applicants to its Data Processing Division. The test consists of twenty questions and for each question four answers are provided. Candidates are required to indicate which of the four answers is correct in each case. Ample time is given to the candidates to complete the test.

Assume that a candidate's knowledge can be represented either by complete ignorance of the question, in which case he will answer the question at random, or else complete knowledge, in which case he gives the correct answer. If this assumption is correct where should the pass level be set in order to give the maximum number of passes subject to less than 1% of totally ignorant candidates being passed? What proportion of candidates who know half the answers to the questions will pass this test? What proportion of candidates who know the answers to three-quarters of the questions will pass the test?

BA (Business Studies)

SOLUTIONS TO EVEN NUMBERED EXERCISES

2. (d) $\dfrac{8!}{6!2!} = \dfrac{8.7.6.5.4.3.2.1}{6.5.4.3.2.1.2.1} = 28$

OUTCOMES (S = Success, F = Failure)

1	2	3	4	5	6	7	8	9	10	11	12	13	14	15	16	17	18	19	20	21	22	23	24	25	26	27	28
S	S	S	S	S	S	F	S	S	S	S	S	F	S	S	S	S	F	S	S	S	F	S	S	F	S	F	F
S	S	S	S	S	F	S	S	S	S	S	F	S	S	S	S	F	S	S	S	F	S	S	F	S	F	S	F
S	S	S	S	F	S	S	S	S	S	F	S	S	S	S	F	S	S	S	F	S	S	F	S	S	F	F	S
S	S	S	F	S	S	S	S	S	F	S	S	S	S	F	S	S	S	F	S	S	S	F	F	F	S	S	S
S	S	F	S	S	S	S	S	F	S	S	S	S	F	S	S	S	S	F	F	F	F	S	S	S	S	S	S
S	F	S	S	S	S	S	F	S	S	S	S	S	F	F	F	F	F	S	S	S	S	S	S	S	S	S	S
F	S	S	S	S	S	S	F	F	F	F	F	F	S	S	S	S	S	S	S	S	S	S	S	S	S	S	S
F	F	F	F	F	F	F	S	S	S	S	S	S	S	S	S	S	S	S	S	S	S	S	S	S	S	S	S

4. Consider a man to be a "success" – then $n = 8$, $\pi = 0.40$, and we require:

$$\Pr(R = 3) = \frac{8!}{3!5!}(0.4)^3(0.6)^5 = 56 \times 0.064 \times 0.0778 = 0.279$$

The probability of finding 3 men and 5 women on a table for 8 is *not* the same because:

- employees will not choose tables randomly – they may wish to sit with their friends (i.e. the Bernoulli trials are not independent).
- we do not know that 40% of the people *using the canteen* will be men.

6. (i) $n = 4$, $\pi = 0.9$, and so using the recursive form

$$\frac{n - r + 1}{r} \cdot \frac{\pi}{1 - \pi} p(r - 1)$$

we obtain

$$p(0) = (1 - \pi)^n = (0.1)^4 = 0.0001 \text{ and } p(r) = \frac{5 - r}{r} \cdot 9p(r - 1).$$

r	0	1	2	3	4
$p(r)$	0.0001	0.0036	0.0486	0.2916	0.6561

(ii) $n = 6$, $\pi = 0.40$

r	0	1	2	3	4	5	6
$p(r)$	0.0467	0.1866	0.3110	0.2765	0.1382	0.0369	0.0041

$n = 4$, $\pi = 0.05$

r	0	1	2	3	4
$p(r)$	0.8145	0.1715	0.0135	0.0005	0.0000

8. (i)

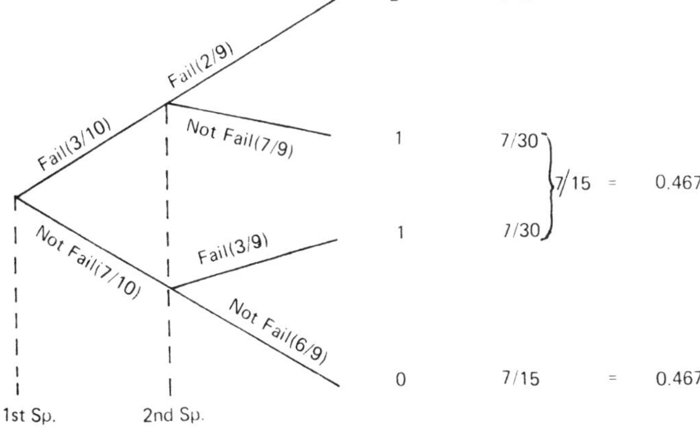

Total 15/15 = 1

(ii) $E(R) = 0(7/15) + 1(7/15) + 2(1/15) = 3/5 = 0.6$. (Mean)
$E(R^2) = 0(7/15) + 1(7/15) + 4(1/15) = 11/15 \simeq 0.73$
So $\mathrm{Var}(R) = 0.73 - (0.6)^2 = 0.37$. (Variance)

(iii) When sampling with replacement the distribution of failures will be binomial with $n = 2$, $\pi = 0.3$.
This has a probability distribution: $p(0) = (0.7)^2$, $p(1) = 2(0.7)(0.3)$ and $p(2) = (0.3)^2$. Comparing with (i), we have

	r	0	1	2	Mean	Variance
Without replacement	$p(r)$	0.467	0.467	0.067	0.6	0.37
With replacement	$p(r)$	0.490	0.420	0.090	$n\pi = 0.6$	$n\pi(1 - \pi) = 0.42$

It can be seen that by sampling without replacement a smaller variance is obtained, although the mean value is the same. (This will lead to a more accurate estimate of population proportions.)

10. The probability that a candidate ignorant of a particular question obtains the correct answer = 0.25. For a *totally* ignorant candidate the probability of scoring 0,1,2,3 . . . 20 is given by the binomial distribution with $n = 20$, $\pi = 0.25$.
Using binomial tables, we have for a totally ignorant candidate:

Pr (scoring 10 or more) = 0.0139,
Pr (scoring 11 or more) = 0.0039,
Pr (scoring 12 or more) = 0.0009.

Hence to ensure that less than 1% of totally ignorant candidates pass, the pass mark must be set at 11 or higher.
But to pass the *maximum* number of candidates, subject to this proviso, we need to set the pass mark at 11.
A candidate who knows half the answers will need to gain one more mark by chance from the remaining 10 in order to pass. Using tables ($n = 10$, $\pi = 0.25$) we have Pr(one or more) = 0.9437. Thus 94% of such candidates are expected to pass.
All candidates knowing the answers to three-quarters of the questions (i.e. 15) will pass.

COURSE UNIT 10 – THE POISSON DISTRIBUTION

10.1 INTRODUCTION

In CU9 we found that *binomial* probabilities are difficult to calculate and tabulate. In this Course Unit we consider another discrete probability distribution, the Poisson[(1)] distribution, (having probabilities which are easier to calculate and tabulate) which can serve as a convenient approximation to the binomial distribution under certain conditions on n and π. We shall introduce the Poisson distribution in this role as an approximation, and then, having familiarised ourselves with its properties, turn to its other quite separate applications arising in connexion with what is called a Poisson process.

10.2 THE POISSON DISTRIBUTION AS AN APPROXIMATION TO THE BINOMIAL DISTRIBUTION

In CU9 it was stated that:

Mean value of the binomial distribution = $n\pi$,

Variance of the binomial distribution = $n\pi(1 - \pi)$.

Now consider a binomial distribution where n is quite large (say 100) and π is quite small (say 0.02). We then have:

Mean value (μ) = $100 \times 0.02 = 2.0$, and

Variance (σ^2) = $100 \times 0.02 \times 0.98 = 1.96$.

It can be seen that $\mu = 2.0$ is very nearly equal to $\sigma^2 = 1.96$; and if we could discover a distribution where μ was *exactly equal* to σ^2, then we should be introducing only a negligible error by using this new-found distribution as an approximation to the binomial distribution.

We can "discover" this new distribution by making certain simplifications to the following recursive form of the binomial distribution (Formula (4) in CU9):

$$p(r) = \frac{n - r + 1}{r} \cdot \frac{\pi}{1 - \pi} \cdot p(r - 1) \qquad (1)$$

---

(1) Named after the mathematician who first published details of the distribution in 1837. This explains why the word Poisson, unlike binomial or normal, starts with a capital letter.

Under the conditions mentioned earlier (large n, small π), we can make the following drastic approximations:

$$1 - \pi \simeq 1, \text{ and}$$

$$n - r + 1 \simeq n.$$

This second approximation is reasonable because, with large n and small π, only values of r *much smaller* than n will have probabilities which are large enough to be worth considering (e.g. with $n = 100$ and $\pi = 0.02$, $p(10)$ is less than 0.00001). Substituting these approximations into (1) gives:

$$p(r) \simeq \frac{n\pi}{r} \cdot p(r-1). \qquad (2)$$

As it stands this formula can be used to obtain approximate binomial probabilities under conditions of large n and small π. But if we replace the $\simeq$ sign in (2) by an = sign and substitute the symbol λ for the quantity $n\pi$ (in order to anticipate applications which are not necessarily binomial in nature), we arrive at:

$$p(r) = \frac{\lambda}{r} p(r-1). \qquad (3)$$

This is the recursive form of the **Poisson** distribution which, while serving as an approximation to the binomial distribution under the stated conditions on n and π (and by putting $\lambda = n\pi$), is a valid distribution in its own right with its own properties and applications.

We can explore some of these properties by putting $r = 1,2,3 \ldots$ into (3), as follows;

$p(1) = \frac{\lambda}{1} \cdot p(0)$, (i) by putting $r = 1$ into (2),

$p(2) = \frac{\lambda}{2} p(1) = \frac{\lambda^2}{2 \cdot 1} p(0)$, (ii) by putting $r = 2$ into (2) and substituting from (i),

$p(3) = \frac{\lambda}{3} p(2) = \frac{\lambda^2}{3 \cdot 2 \cdot 1} p(0)$, (iii) by putting $r = 3$ into (2), and substituting from (ii).

It is not difficult to see that the general term $p(r) = \frac{\lambda^r}{r!} p(0)$, but unfortunately this does not enable us to evaluate the probabilities because we do not yet know an expression for $p(0)$. What is $p(0)$? We can find out by remembering that the sum of *all* the probabilities must be unity for a proper probability distribution. So,

$$p(0) + p(1) + p(2) + p(3) + \ldots = 1,$$

or by substitution of the individual terms,

$$p(0) + \frac{\lambda}{1!}p(0) + \frac{\lambda^2}{2!}p(0) + \frac{\lambda^3}{3!}p(0) + \ldots = 1$$

or

$$p(0)\left\{1 + \frac{\lambda}{1!} + \frac{\lambda^2}{2!} + \frac{\lambda^3}{3!} + \ldots\right\} = 1$$

The quantity in the{ }brackets is in fact equal to the mathematical constant e (= 2.71828 . . .) raised to the power λ, that is[1] e^{λ}. So

$$p(0)e^{\lambda} = 1$$

or

$$p(0) = 1/e^{\lambda} = e^{-\lambda} \qquad (4)$$

Substituting this value for $p(0)$ into the general term for $p(r)$ we have:

$$\Pr(R = r) = p(r) = \frac{\lambda^r}{r!} \cdot e^{-\lambda}, \quad r = 0,1,2,3,\ldots \qquad (5)$$

$$= 0, \text{ otherwise}$$

where λ is the parameter of the Poisson distribution, and e is the constant 2.71828 . . .

We note two major differences between the Poisson distribution and the binomial distribution. Firstly, the Poisson distribution has *one* parameter only, λ, in comparison with the two parameters n and π of the binomial distribution: it is this fact which makes it an easier distribution to tabulate. Secondly, there is no upper limit to the value of r (i.e. the Poisson distribution has an infinite number of terms), unlike the binomial, where r can only take the values $0,1,2,\ldots, n$. It may seem strange in view of this that we can use the Poisson distribution as an

---

(1) If we take the special case of $\lambda = 1$ we find:

$$e^{\lambda} = e^1 = e = 1 + \frac{1}{1!} + \frac{1}{2!} + \frac{1}{3!} + \ldots$$

$$= 1 + 1 + 0.50 + 0.17 + \ldots = 2.67 + \ldots$$

With more terms we would obtain a result close to 2.71828 . . .

approximation to the binomial; but there is, in fact, no real difficulty since the Poisson probabilities rapidly approach zero for large values of r as a consequence of the $r!$ term in the denominator of (5).

It is easy to find the mean and variance of the Poisson distribution by regarding it as a binomial approximation. For large n and small π, we have:

Mean $= n\pi = \lambda$, in Poisson notation.

Variance $= n\pi(1 - \pi) \simeq n\pi = \lambda$, by making precisely the same approximation $(1 - \pi \simeq 1)$ as was made in deriving the distribution itself.

So formally we have[1]:

For a Poisson distribution (parameter λ)

Mean, $\mu = \lambda$

Variance, $\sigma^2 = \lambda$ (6)

The fact that the mean and variance of the Poisson distribution are equal provides a quick check on whether it is reasonable to describe empirical data by the Poisson distribution (see Section 10.7).

10.3 INDIVIDUAL TERMS OF THE BINOMIAL AND POISSON DISTRIBUTIONS

If the Poisson distribution is to be of use in providing an easier method of obtaining binomial probabilities we must establish the relationship between the individual terms of the distributions and be confident about the size of the error introduced by the approximation. In Exhibit 1 we compare the individual terms for probabilities of events $r = 0,1,2,3$ etc. and the value of these terms for the binomial distribution with $n = 100$, $\pi = 0.02$ and the corresponding Poisson distribution with $\lambda = (100 \times 0.02) = 2.0$. We see that there is in general close agreement between the values. The reason, clearly, is that our requirements of large n and small π are amply satisfied in this case.

It should be appreciated that since the Poisson distribution has only a single parameter, λ, the same approximation will apply for any combination of the two binomial parameters n and π, having the same value

---

[1] We have not *proved* that the mean and variance are equal, but this can be done by algebraic manipulation of expected values.

of $n\pi$ (= λ). For example, the Poisson distribution with $\lambda = 2$ can serve as an approximation to all the following binomial distributions: $n = 100$, $\pi = 0.02$; $n = 200$, $\pi = 0.01$; $n = 400$, $\pi = 0.005$ etc. Note, however, that while the binomial distribution with $n = 4$, $\pi = 0.5$ also has a value $n\pi = 2$, we cannot use the Poisson distribution with $\lambda = 2$ as an approximation because n is not large and π is not small. As a useful rule-of-thumb the Poisson distribution should only be used as an approximation when n is *at least* 10 and π is *at most* 0.1.

EXHIBIT 1: Comparison of Binomial and Poisson Terms

	Pr(R = r)		*Values of Probabilities*	
r	*Binomial*	*Poisson*	*Binomial* *(n = 100)* *(π = 0.02)*	*Poisson* *(λ = 2.0)*
0	$(1-\pi)^n$	$e^{-\lambda}$	0.1326	0.1353
1	$n\pi(1-\pi)^{n-1}$	$\lambda e^{-\lambda}$	0.2707	0.2707
2	$\frac{n(n-1)}{2!}\pi^2(1-\pi)^{n-2}$	$\frac{\lambda^2}{2!}e^{-\lambda}$	0.2734	0.2707
3	$\frac{n(n-1)(n-2)}{3!}\pi^3(1-\pi)^{n-3}$	$\frac{\lambda^3}{3!}e^{-\lambda}$	0.1823	0.1804
4	$\frac{n(n-1)(n-2)(n-3)}{4!}\pi^4(1-\pi)^{n-4}$	$\frac{\lambda^4}{4!}e^{-\lambda}$	0.0902	0.0902
5†	$\frac{n!}{5!(n-5)!}\pi^5(1-\pi)^{n-5}$	$\frac{\lambda^5}{5!}e^{-\lambda}$	0.0353	0.0361
6	$\frac{n!}{6!(n-6)!}\pi^6(1-\pi)^{n-6}$	$\frac{\lambda^6}{6!}e^{-\lambda}$	0.0114	0.0121
7	$\frac{n!}{7!(n-7)!}\pi^7(1-\pi)^{n-7}$	$\frac{\lambda^7}{7!}e^{-\lambda}$	0.0032	0.0034
8	$\frac{n!}{8!(n-8)!}\pi^8(1-\pi)^{n-8}$	$\frac{\lambda^8}{8!}e^{-\lambda}$	0.0007	0.0009
9	$\frac{n!}{9!(n-9)!}\pi^9(1-\pi)^{n-9}$	$\frac{\lambda^9}{9!}e^{-\lambda}$	0.0002	0.0002
	etc	etc	–	–
		Totals	1.0000	1.0000

†Note change of notation for the binomial term at this point.

10.4 THE RECURSIVE FORM OF THE POISSON DISTRIBUTION

We have already stated the recursive form of the Poisson distribution in Equation (3). Combining this with the expression for $p(0)$, Equation

(4), we have the following convenient computational form:

$$p(0) = e^{-\lambda}$$

$$p(r) = \frac{\lambda}{r} \cdot p(r-1), \quad r = 1,2,3,\ldots \tag{7}$$

Since values of the exponential function, e^{-x}, are given in most sets of mathematical tables (and in some statistical tables) we may readily obtain the value of $p(0)$. From $p(0)$ we obtain $p(1)$ in the form $p(1) = \lambda p(0)$; and from $p(1)$ we obtain $p(2)$ as $p(2) = \frac{\lambda}{2} p(1)$, and so on – an example of the recursive calculation of probabilities is given in Section 10. 6. Alternatively, for certain values of λ, the probabilities may be obtained from tables of the Poisson distribution. As with the binomial distribution it is common to find tables arranged in the form of *cumulative* probabilities.[(1)] The use of such tables follows the same principles as outlined for cumulative binomial tables (CU9, Section 9.5).

10.5 THE SHAPE OF THE POISSON DISTRIBUTION

Exhibit 2 shows the histograms of Poisson distributions with $\lambda = 0.1$, 1.0 and 10.0.

In general, for values of λ close to zero, the distribution is very positively skewed, the mode of the Poisson distribution being zero for all values of λ up to 1.0. For higher values of λ the distribution becomes less skewed and it is generally regarded as being symmetrical (for all practical purposes) for values of λ greater than 10. This point is taken up again when considering the *normal* distribution (CU11).

10.6 USING THE POISSON APPROXIMATION

In CU9 we considered the situation of taking a random sample of 10 screws from a bin of 1,000 screws of which 100 were headless. This is illustrative on a small scale of a wide range of situations arising in industry. Now screws (and many similar low cost items) are bought and used in large quantities. It is doubtless inconvenient when a headless screw turns up, but since a visual check on quality can be carried out quickly we may take a large sample without incurring too much

[(1)] For example, see J. Murdoch and J.A. Barnes, *Statistical Tables*, but *individual* probabilities are given in our Appendix 3.

EXHIBIT 2: Three Poisson Distributions

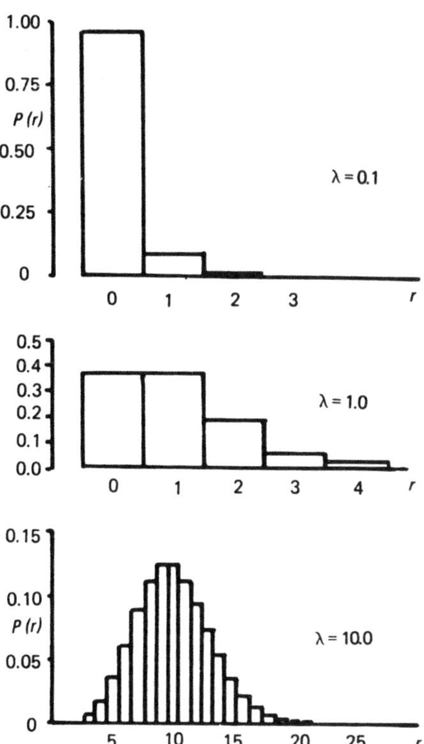

(The probability distributions have been drawn with bars rather than lines to facilitate comparison)

expense. So given a delivery of 10,000 screws (a realistic order quantity) we may quickly check a random sample of 300,[(1)] for example, in order to verify that only an acceptably low percentage of them is defective. On the basis of this sample we either accept the delivery into our warehouse stores, or, if the percentage of defectives is unacceptably high, we "reject" the delivery – either for sorting or for returning to the supplier. This procedure is essentially that of **acceptance sampling** in Statistical Quality Control.

[(1)] This could be achieved quickly by filling a tray known to hold (about) 300 screws. There is no need to *count* out the screws – minor fluctuations in the sample size are of no practical significance.

Suppose that there are usually 1% of defective screws in a delivery and that this is considered a "satisfactory" standard of quality. We can design an Acceptance Sampling rule from a consideration of the probability distribution of the number of defective screws in our sample of 300. Firstly, using the Poisson approximation, we calculate the value of $\lambda = n\pi = 300(0.01) = 3.0$. For the purposes of illustration we use the recursive form of calculation, as follows:

$$\Pr(0 \text{ defectives}) = p(0) = e^{-3.0}(\text{from tables}) = 0.0498,$$

$$\Pr(1 \text{ defective}) = p(1) = 3.0 \times 0.0498 = 0.1494,$$

$$\Pr(2 \text{ defectives}) = p(2) = \frac{3.0}{2} \times 0.1494 = 0.2241,$$

$$\Pr(3 \text{ defectives}) = p(3) = \frac{3.0}{3} \times 0.2241 = 0.2241,$$

$$\Pr(4 \text{ defectives}) = p(4) = \frac{3.0}{4} \times 0.2241 = 0.1681,$$

$$\Pr(5 \text{ defectives}) = p(5) = \frac{3.0}{5} \times 0.1681 = 0.1008,$$

$$\Pr(6 \text{ defectives}) = p(6) = \frac{3.0}{6} \times 0.1008 = 0.0504,$$

and so on.

By adding up the probabilities we have calculated, $\Sigma_{r=0}^{6}\, p(r)$, we see that the probability of finding *6 or fewer* defectives in our sample is equal to 0.9667. Consequently the probability of finding *7 or more* defectives is 1 − 0.9667 = 0.0333. So, if we were to use the rule, "Reject the delivery if 7 or more defectives are found in the sample", we would reject about 3% of deliveries which are considered satisfactory (1% defectives).

But suppose a sub-standard delivery arrived which actually had 5% defective screws and we applied the same Acceptance Sampling procedure to it. Clearly, in this case, $\lambda = 300(0.05) = 15$. Working out the probabilities in a similar manner (or more conveniently, by using Poisson tables) we now find that the probability of *7 or more* defectives in the sample is 0.9924 (compared with 0.0333, for "satisfactory" deliveries). So we would reject over 99% of deliveries having 5% defectives.

It can be seen that Acceptance Sampling enables us to reject the majority of unsatisfactory deliveries and yet accept most satisfactory deliveries. Although there is a small chance of accepting an unsatisfactory delivery (and also of rejecting a satisfactory delivery), this chance is often well worth taking as it cuts the inspection cost of deliveries dramatically in comparison with a *complete* quality check on all deliveries.

10.7 FITTING THE POISSON DISTRIBUTION TO "NON-BINOMIAL" DATA

In this section we fit a Poisson distribution to some data (the weekly sales of peramulators from a small shop). In the example chosen we are not using the Poisson distribution as an approximation to the binomial distribution (it would be impossible, for instance, to say what the value was of the binomial parameter n) but as a probability distribution in its own right. In such cases, where it is found that the "fit" is a good one, we may speculate that we are dealing with a fundamental statistical process called a **Poisson process**, the general conditions for which are to be discussed in Section 10.8. But first the example.

A shop selling toys, bicycles and prams in a country town keeps a record of the weekly sales of prams, in the form of a frequency distribution, as shown in the first two columns of Exhibit 3.

EXHIBIT 3: The Weekly Sales of Prams

No. of Prams Sold (x_i)	*No. of Weeks* (f_i)	*Poisson Distribution,* $\lambda = 1.5$	
		Probabilities	*Expected Frequencies,* $\hat{f}_i$
0	22	0.2231	22.3
1	34	0.3347	33.5
2	24	0.2510	25.1
3	12	0.1255	12.6
4 or more(1)	8	0.0657(1)	6.5(1)
Totals	100	1.0000	100.0

To fit a Poisson distribution to this data we first find the mean number of prams sold per week, $\bar{X}$, and equate this to the Poisson mean, λ. So $\bar{X} = \Sigma x_i f_i / \Sigma f_i = 150/100 = 1.5 = \lambda$. Incidentally, we could also calculate the variance $s^2 = 1.41$, the closeness of the mean and variance of the data indicating by itself the suitability of a Poisson model because, for a Poisson distribution, the mean and the variance are exactly equal.

(1)Note that since the Poisson distribution has an infinite number of terms we combine the probabilities and expected frequencies for the terms 5 to ∞ (where the observed frequencies are zero) with those for 4 prams by expressing them as "4 or more".

From tables (or recursive calculations) we then obtain the corresponding probabilities of the Poisson distribution with $\lambda = 1.5$; finally we multiply the probabilities by the total frequency (100) to obtain the expected frequencies. (Compare this example with that given for the binomial distribution in CU9, Section 9.6.)

Pending a formal examination of "Goodness of Fit" in CU19, a visual inspection shows a close agreement between the observed and expected frequencies in Exhibit 3.

10.8 THE GENERAL CONDITIONS FOR A POISSON PROCESS

The line in Exhibit 4 represents a continuum and the crosses represent "events" which occur along it. (It is perhaps easier to think of this in terms of an example: the line might represent a time scale running from left to right and the crosses the particular times at which a telephone call is received by a switchboard.)

EXHIBIT 4: Illustration of a Poisson Process

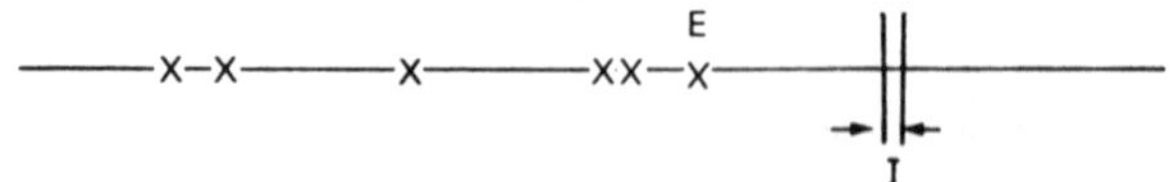

The sequence of events is said to form a Poisson process if the following conditions are satisfied:

(a) the probability of an event occurring in a short interval, I, is dependent *only* on the width of the interval. In particular, the probability is *not* dependent on how close the interval is to the last event (*E*) that occurred, nor is it dependent on where precisely, along the continuum, I is placed.
(b) the probability of an event occurring in a short interval, I, is *proportional* to the width of the interval.
(c) the probability of two or more events occurring in a short interval, I, is approximately zero.

If these conditions are fulfilled then the number of events occurring in *any* fixed interval has a Poisson distribution.

Applying these conditions to the pram sales example (Section 10.7), (a) implies that a buyer between 2.00 p.m. and 3.00 p.m. is not influenced (directly or otherwise) by when the last pram was sold (friendly competition between mothers, a sudden spate of births, or a short promotional campaign by the shop would mean that this is not true): it also implies that the chances of a sale between 10.00 a.m. and 11.00 a.m. are the same as those between 2.00 p.m. and 3.00 p.m. Condition (b) means that, for example, the chances of a sale in an hour are twice those in half an hour. Finally condition (c) states that the chances of two or more sales in an hour are negligible. In particular, note that if two mothers went *together* to buy their prams, this would invalidate both conditions (a) and (c).

In the frequency distribution of Exhibit 3 we have used a time interval (based on tradition) of the Sales Week – in effect, by "lumping together" a number of hourly intervals. But the analysis is valid for "intervals" of other types – intervals of length, area and volume. What we are in fact saying is that the Poisson distribution describes the number of *random* events per unit time, length, area or volume, where the word *random* is to be interpreted, in this context, by the adherence to conditions (a), (b) and (c).

In all such cases the width of the interval convenient for analysis can be considered as the grouping together of very large number of very short intervals (e.g. the sales week may be considered as 2,400 minutes, a page of print in a book as a near infinite number of squares each of a microcentimetre in width). But there is no simple way of equating this number of short intervals with the sample size, n, in the binomial distribution because, by considering smaller and smaller intervals, this number could become infinite. In short, then, the Poisson distribution may be compared with the binomial distribution where $n = \infty$ and $\pi \simeq 0$. This fact explains the connexion between the Poisson distribution as applied to a Poisson process and its use as an approximation to the binomial distribution.

10.9 THE IMPORTANCE OF THE POISSON DISTRIBUTION

The importance of the distribution is that, like the binomial distribution, the Poisson distribution provides a suitable model for many situations and processes met with in industry and business. In Quality Control it is used as a basis for Acceptance Sampling Schemes. In Operational Research it is used as a suitable model for withdrawals from stock, the pattern of arrivals in a queue and the number of telephone calls. It also has applications in fields as diverse as medicine, biology, accident studies and bird migration.

Why is the Poisson distribution so widely applicable? A consideration of the way in which telephone calls arise provides a clue. If you have a telephone at home then there are potentially a very large number of persons who could call you at a given time. Not only your neighbours and friends but potentially anyone in the rest of the world could (if only be incorrect dialling!) phone you. The probability of their doing so is, of course, quite small for a small unit of time, while the probability of two or more doing so at the same instant of time is practically zero: consequently a Poisson distribution of calls per day is quite possible. Obviously, if you break one of the conditions given in Section 10.8 (for instance, by having large numbers of friends phone you on your birthday) then this will no longer be true.

READING

Taro Yamane *Statistics: An Introductory Analysis 3rd Ed.* Harper, pages 735–753.

EXERCISES

1(U) Using *tables*, evaluate the probabilities for $r = 0,1,2, \ldots 6$ in a Poisson distribution with $\lambda = 3.0$ and compare your results with those given in Section 10.6. If you find any discrepancies, explain why.

2(U) *Compute* the probabilities to four decimal places for a Poisson distribution with $\lambda = 1.0$ and hence obtain:

(a) $\Pr(R > 0)$; (b) $\Pr(R < 3)$; (c) $\Pr(R$ is 2 or more); (d) $\Pr(2 \leqslant R \leqslant 5)$.

3(U) Discuss which of the following variables are likely (or not likely) to have a Poisson distribution. *Hint*: consider whether the phenomena meet the requirements of a Poisson process (Section 10.8).

(i) The number of rainy days in July.
(ii) The number of students reporting sick per day during (a) examination weeks, (b) a mild month in winter.
(iii) The number of goals scored in an association football (soccer) match.
(iv) The number of breakdowns in a mile of motorway in a normal week.
(v) The number of defective screws produced per minute by a machine working with an output of 1,000 per minute.
(vi) The number of spelling mistakes per page of news in a newspaper.

4(U) Obtain the value of λ for which $p(1) = p(2)$ in a Poisson distribution. Check your answer with Exhibit 1.

5(P) (i) Take a sample of 50 full lines of print from a novel and obtain the frequency distribution of the number of times the letter "r" is used per line. Fit a Poisson distribution to the data, as in Section 10.7.

(ii) Check on Exercise 3 (iii) by obtaining the distribution of goals scored per match on a match day in either the English Football League or another league whose results are known to you.

6(E) Fit a Poisson distribution with the same total frequency to the following distribution of the number of breakdowns per day on B.B.C. television programmes.

No. of Breakdowns	*No. of Days*
0	100
1	70
2	45
3	20
4	10
5	5
	250

Comment in general on how well the Poisson distribution fits this data.

7(E) An Acceptance Sampling scheme for defective commutators bought in by a manufacturer of domestic electric appliances has the following rule. If the batch size exceeds 2,000 take a random sample of 120 commutators: accept the batch if 0 or 1 defectives are found. If 2 or more defectives are found return the batch to the supplier.

Draw a graph showing the probability of accepting batches that contain 0%, 0.5%, 1%, 2%, 3%, 4% defective. Assume that the Poisson distribution is an appropriate model to use.

8(E)* A company specialises in selling spares for pre-war cars. The company keeps records (for items costing more than £5) of the *number of weeks that elapse between consecutive demands*. For a cylinder head of a particular model their records were as follows:

1,	32,	0,	9,	13,	4,	4,	30,	9,	5,
27,	2,	2,	25,	9,	14,	1,	11,	5,	9,
12,	10,	12,	1,	10,	23,	5,	5,	0,	10.

(i) Using the above data, construct a frequency distribution of the following form:

No. of demands per *week*	No. of weeks in which stated no. of demands occurred
0	
1	
2	
3	

(ii) Fit a suitable theoretical distribution to the frequency distribution you have constructed, explaining your choice of model.

(iii) By means of tables estimate the probability that the current stock of 10 cylinder heads will be exhausted within a year. What assumptions are implicit within this estimate?

Other examination questions involving the Poisson distribution may be found in CU11.

SOLUTIONS TO EVEN NUMBERED EXERCISES

2.

r	0	1	2	3	4	5	6	7	8 and more
$p(r)$	0.3679	0.3679	0.1839	0.0613	0.0153	0.0031	0.0005	0.0001	$\simeq$ 0.0000

(a) $1 - 0.3679 = 0.6321$; (b) 0.9197; (c) 0.2642; (d) 0.2636.

4. $p(1) = \lambda e^{-\lambda}$ and $p(2) = \dfrac{\lambda^2}{2} e^{-\lambda}$

so $\lambda e^{-\lambda} = \dfrac{\lambda^2}{2} e^{-\lambda}$, from which $\lambda = 2$.

6.

x_i	f_i	$x_i f_i$	*Poisson Prob. (approx.)*	$\hat{f}_i$ (based on Poisson distribution) (to nearest whole number)
0	100	0	0.320	80
1	70	70	0.365	91
2	45	90	0.208	52
3	20	60	0.079	20
4	10	40	0.022	6
5 (or more)	5	25	0.005	1 (actually 1.25)
	250	285	0.999	250

$\bar{X} = \Sigma x_i f_i / \Sigma f_i = 285/250 = 1.14$.

The "fit" is not very good: in particular, the observed frequency distribution has a greater "spread" (higher variance).

8. With the interpretation given, the frequency distribution becomes:

(i) *No. of demands/week* x_i	*Frequency* f_i	$x_i f_i$	$x_i^2 f_i$
0	272	0	0
1	26	26	26
2	2	4	16
3	0	0	0
	300	30	42

Note: (1) The class $x_i = 3$ has zero frequency because there were no consecutive "0"s in the table given.
(2) The frequency of $x_i = 1$ could validly be stated as 27 – it depends on how the counting process is started off.

(ii) $\bar{X} = \sum x_i f_i / \sum f_i = 30/300 = 0.1$

$$s^2 = \sum x_i^2 f_i / \sum f_i - \bar{X}^2$$
$$= 42/300 - (0.1)^2$$
$$= 0.13$$

The Poisson distribution might be appropriate because:

(a) the process considered is one of counting the number of random events occurring per unit time (week);
(b) very approximately, the mean and variance of the distribution are equal.

Now $e^{-0.1} \simeq 0.9048$, leading to the following Poisson probabilities for $\lambda = 0.1$:

r	*p(r)*	*Expected Frequency* 300 $p(r)$	*Actual Frequency*
0	0.9048	271.44	272
1	0.9048 × 0.1 = 0.0905	27.15	26
2	0.0905 × 0.1/2 = 0.0045	1.35	2
3	0.0045 × 0.1/3 = 0.0002	0.06	0
	1.0000	300.00	300

An excellent fit!

(iii) In 52 weeks the mean demand will be (52 × 0.1) units = 5.2 units. Referring to Poisson tables with $\lambda = 5.2$ we find, Pr(10 or more) = 0.0397, or approx. 0.040.

COURSE UNIT 11 – THE NORMAL DISTRIBUTION AND CENTRAL LIMIT THEOREM

11.1 INTRODUCTION

The two probability distributions studied so far, the binomial and Poisson, are discrete distributions. We now turn our attention to the most important *continuous*[1] probability distribution, the so-called **normal** distribution, which far outstrips any other distribution in its wide range of applications.

The probability density function (p.d.f.) of the normal distribution is given by the following, somewhat formidable, formula.

$$f(x) = \frac{1}{\sigma\sqrt{2\pi}}\, e^{-\frac{1}{2}\left(\frac{x-\mu}{\sigma}\right)^2} \qquad (1)$$

μ is the mean of the distribution
σ is the standard deviation of the distribution
π is the constant 3.142 . . .
e is the constant 2.718 . . .

It may reassure the reader to know that it is rarely, if ever, necessary to evaluate probability densities of the normal distribution using this formula – tables, as we shall see, are available for most applications. But the analytical form of the p.d.f. does show that the normal is a two-parameter distribution – the parameters being μ (the mean) and σ (the standard deviation) – and consequently we can only draw the p.d.f. in the form of a graph for chosen values of these two parameters. Exhibit 1 shows some examples.

A careful examination of the p.d.f. formula (and/or the graphs in Exhibit 1) reveals the following properties of the normal distribution.

---

[1] Readers should be clear on the differences between a *discrete* and a *continuous* distribution before proceeding, CU8 went into the details, but in summary:

(a) a *discrete* distribution gives a set of probabilities corresponding to each possible value a (discrete) random variable can take;

(b) a *continuous* distribution is specified by a probability density function (p.d.f.) and may be depicted as a graph where the x-axis represents the measurement scale The *area* under the graph between two chosen measurement values represents the probability of the (continuous) random variable lying between those two values. The *total* area under the graph is unity.

EXHIBIT 1: Graphs of the p.d.f. of the Normal Distribution for Various μ and σ

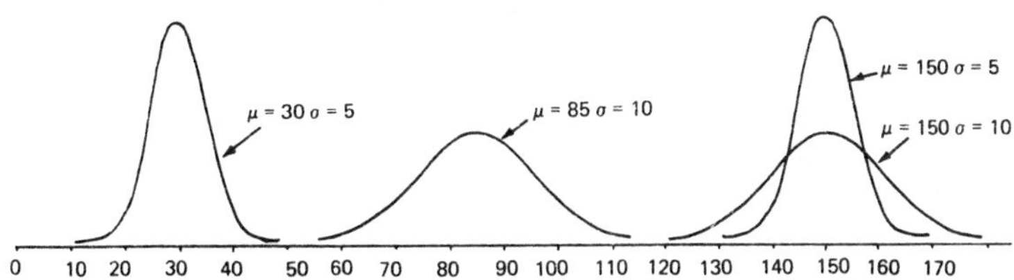

- The p.d.f. is symmetrical about the mean value μ. Consequently the mean, mode and median of the normal distribution have the same value.
- The "normal curve" (i.e. the p.d.f. plotted as a graph) has a form which may be described as bell-shaped, where the width of the bell mouth depends on the standard deviation σ.
- The p.d.f. is never *quite* zero for any measurement value, x – the bell mouth approaches the measurement axis asymptotically, never quite touching it.

In view of the symmetrical bell-shaped properties of the normal curve, the normal distribution finds application in describing the *population* of measured quantities in cases where a *sample* histogram displays similar bell-shaped characteristics. To see what we mean examine Exhibit 2 where superimposed on the *sample* histogram of the Series A data (taken from CU2) is the *population* histogram (assumed drawn up from the population of size 10,000 by using class widths of 1 therm). The population distribution has the normal curve "shape" and so the

EXHIBIT 2: Sample Histogram and "Guess" of Population Histogram for Series A Data

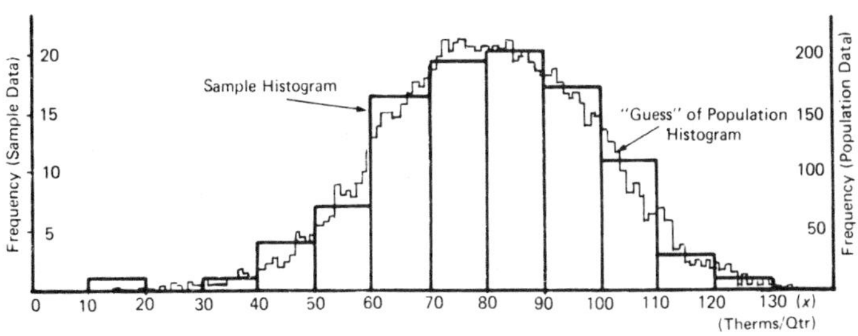

normal distribution can serve as a "model" for gas consumption of the central heating unit[1].

The normal distribution, with its p.d.f. expressed in the general form of Equation (1) suffers from a severe drawback. Being a two-parameter distribution (like the binomial, for instance) it is extremely cumbersome to tabulate – a separate table being required for each μ and σ combination. Instead of providing a whole bookful of such tables, it is the accepted convention to publish just *one* table for the particular normal distribution with zero mean ($\mu = 0$) and unit standard deviation ($\sigma = 1$) – the so-called **standard normal** distribution. Later (Section 11.4) we shall see how this table is used.

The p.d.f. of the *standard* normal distribution is obtained by substituting $\mu = 0$ and $\sigma = 1$ into Equation (1). In order to distinguish between this particular normal distribution and the distribution in general, it is accepted practice to use the symbol z, in place of x, as the measurement variable.

$$f(z) = \frac{1}{\sqrt{2\pi}} e^{-\frac{1}{2}z^2} \qquad (2)$$

π is the constant 3.142 . . .
e is the constant 2.718 . . .

The p.d.f. of the standard normal distribution is graphed in Exhibit 3, from which it can be seen that probability density becomes very small indeed at z values of more than +3 and less than −3.

Before we move on to the use of the standard normal distribution we first digress to consider the process of "standardisation" itself, in more general terms.

---

(1) We should not of course assume that simply because the shape looks right the normal distribution can be applied. Pending a formal "fitting" of the normal distribution to the Series A data in Section 11.6, the following quick check adds conviction to the argument. Consider the class 79.5–89.5, with frequency 20, of the Series A data. An estimate of the probability *density* (CU8) at the class mid-point 84.5 is given by (relative frequency/class width) = 0.20/10 = 0.02. The corresponding probability density of the normal distribution is given by

$$\frac{1}{19.02\sqrt{2\pi}} e^{-\frac{1}{2}\left(\frac{84.5 - 79.7}{19.02}\right)^2} = 0.0203,$$

a closely similar value.

EXHIBIT 3: The p.d.f. of the Standard Normal Distribution

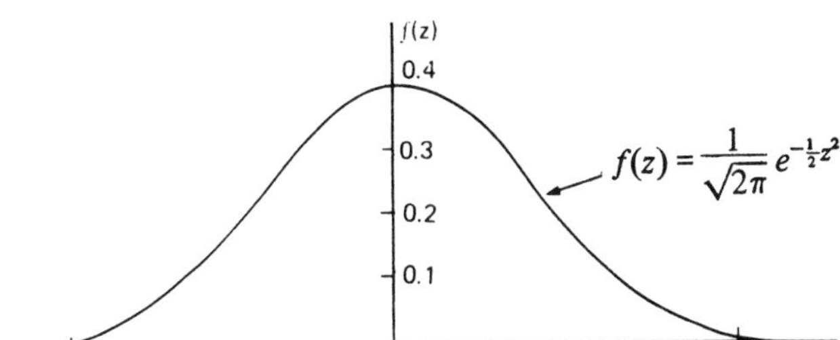

11.2 THE STANDARDISATION OF DATA

If we take *any* univariate set of data and calculate its mean and standard deviation, then we can express the data in "standardised" form by the following process: from each item we subtract the mean and divide the resulting deviations from the mean by the standard deviation. In analytical terms we are using the following linear transformation:

$Z = \dfrac{X - \mu}{\sigma}$ if we are dealing with population data, or

$Z = \dfrac{X - \bar{X}}{s}$ if we are considering sample data.

In either case, the X's are the original data and the Z's the data in standardised form. As an example consider our washing machine sales data (first introduced in CU4).

Month	*Jan*	*Feb*	*Mar*	*Apr*	*May*	*June*
Sales of washing machines (X)	2	0	4	6	2	3

This data has a mean of 2.83 ($\bar{X}$) and a standard deviation of 1.86 (s). Standardising this data in the manner indicated above gives:

Month	*Jan*	*Feb*	*Mar*	*Apr*	*May*	*June*
Sales of washing machines in standardised units, (Z)	−0.45	−1.52	0.63	1.70	−0.45	0.09

It can be seen that the process of standardisation amounts to expressing each item of data in terms of how many standard deviations it is, *more* than (or *less* than) the mean. The standardised January figure (–0.45), for example, indicates that the January sales were 0.45 standard deviations less than the mean sales of 2.83. The procedure for standardising data has the following important properties:

- It does not require any more information than is provided by the data itself. (Remember that the mean and standard deviation are themselves calculated directly from the data.)
- If we have any quantity expressed in standardised units, then this can be readily reconverted to the original measurement units by multiplication by the standard deviation and then adding the mean to the result, i.e.

  $X = \mu + Z\sigma$ (for population data) or

  $X = \bar{X} + Zs$ (for sample data)

- Standardised data has a mean of zero and a standard deviation, and hence variance, of unity. (You should be able to convince yourself of this by looking carefully at the general formula for standardisation, and also by checking that this is true for the standardised values of washing machine sales – a formal proof of this is given in CU12, Section 12.4)
- Classified data (i.e. a frequency distribution) can be standardised more conveniently by simply standardising the class limits (or mid-points). This effectively means changing the measurement scale of a histogram to one where the mean falls at zero and the standard deviation is unity.

Standardised "scores" find particular application in psychometric work. For many psychological tests the mean value and the standard deviation of test results have no particular significance: what is important is the score an individual obtains *in comparison* with those obtained by a large group of individuals. By expressing scores in standard units, ready comparison can be made of an individual's performance on a number of different tests. But for our present purposes it is the fact that standardised data has zero mean and unit standard deviation that is particularly important in the context of the normal distribution. If a measurement quantity is normally distributed then by working in standardised units the standard normal distribution can be employed for all probability calculations.

11.3 APPLICABILITY OF THE NORMAL DISTRIBUTION

The normal distribution has found wide application in describing the variability inherent in many natural (and man-made) phenomena. Heights of adult males, lengths of "identically" manufactured components, weekly demands of heavily consumed products – all can be adequately described by the normal distribution *in particular circumstances*. Even if the measurement variable itself is not normally distributed it is sometimes the case that by applying a suitable transformation to the measurement quantities (e.g. by taking logarithms of the observations), the transformed quantities are found to be approximately normally distributed. But it is fundamentally a matter of empirical testing as to whether the normal distribution forms an appropriate description in a particular case (see Section 11.6) – it is undoubtedly true that the normal distribution is all too often assumed to be appropriate without any testing, or even questioning, if it is reasonable.

There is however a theoretical reason why many observed quantities are found to have a normal distribution. The reason is related to a fundamental theorem of statistics called the **central limit theorem** (more of which in Section 11.5). For the present it is sufficient to state that if the variability in a measured quantity can be attributed to a large number of separate (independent) additive random effects then that quantity will be normally distributed. Consider, for instance, the height of an adult male. The height of his parents is one random influence, the quantity and quality of food consumed in his youth is another, the amount of exercise taken is yet another, and so on. If such influences are independent of each other and produce effects on height which are additive, then we will expect heights of adult males to be normally distributed. Similarly, the overall demand for a heavily consumed product in a particular week may be expressed as:

demand from customer 1
\+ demand from customer 2
\+ demand from customer 3
\+ etc.

Thus if there are a large number of customers, each with his own random pattern of ordering, and such customers order independently of each other, the total demand will be normally distributed (the effects are clearly additive in this case).

Such ideas, related to the Central Limit Theorem, enable us to see why the normal distribution occurs so frequently in practice; but they are of little use in helping us to decide if a particular measured quantity will be normally distributed because, in general, we will not know if the various assumptions – *independent additive* effects – hold true. The one major exception to this statement is where we specifically design a statistical investigation so that such assumptions *do* hold. Provided, for instance, sampling is carried out randomly, the mean values of large samples (of the same size) taken from the same population are sure to be normally distributed. In fact, the importance of the normal distribution lies in its applications to sampling processes rather than as a description of the variability observed in natural or man-made phenomena. Further details are given in Section 11.5.

11.4 NORMAL DISTRIBUTION TABLES – AN EXAMPLE OF THEIR USE

The standard normal p.d.f. is easily tabulated but its practical value is extremely limited simply because it gives probability *densities* rather than the more useful probabilities (which are areas under the p.d.f. curve). The most commonly (and usefully) tabulated quantity is the "area in the tail" (see Exhibit 4) of the standard normal p.d.f. curve which gives the probability of a standardised normal random variable Z *exceeding* a particular standardised value, z, i.e. $\Pr(Z > z)$.

EXHIBIT 4

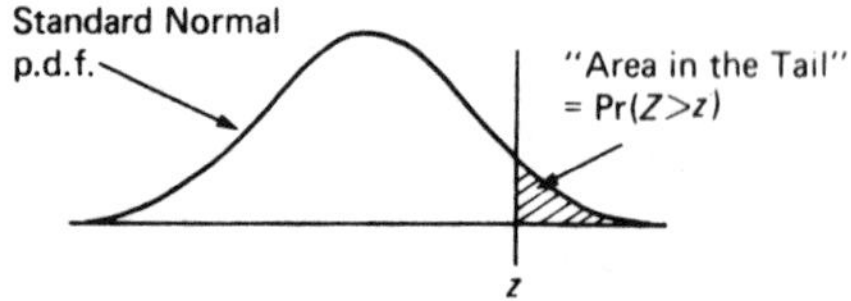

Areas in the tail of the standard normal distribution are given in tables[1] for z values of 0 to just over 3 (beyond which the probabilities become very small indeed). The tabulated value for $z = 0$ is 0.5000 which shows that the probability of a normal variate exceeding its mean value is one half – a fact which follows directly from the symmetrical nature of the normal distribution about its mean value. It should be noted that tabulated quantities are usually only given for positive values of z: the values for negative z's are obtained by appeal to the symmetry of the distribution. A number of probability evaluations, making use of tables, are shown in Exhibit 5.

---

[1] See Appendix 4. Not all normal distribution tables give "areas in the tail". Some tables give "areas under the normal curve between 0 and z". See Exercise 3, as an example.

EXHIBIT 5: Use of Normal Distribution Tables for Evaluating Probabilities

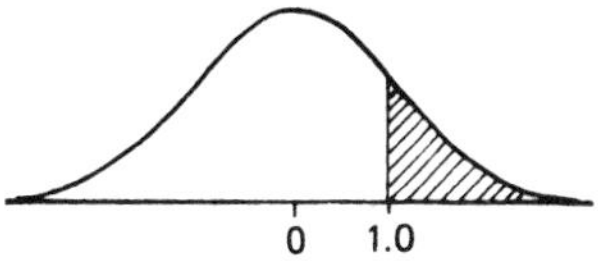

0 1.0

Pr(Z>1.0) = **0.1587**

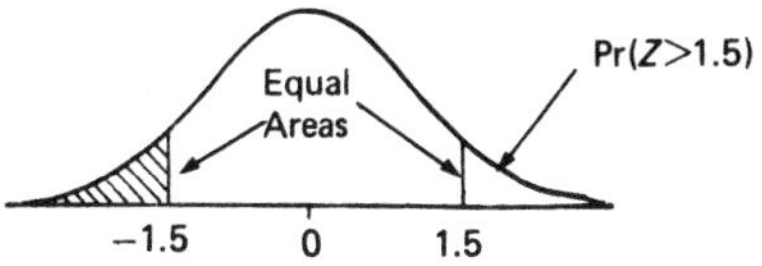

Equal Areas

Pr(Z>1.5)

−1.5 0 1.5

Pr(Z<−1.5) = Pr(Z>1.5)
= **0.0668**

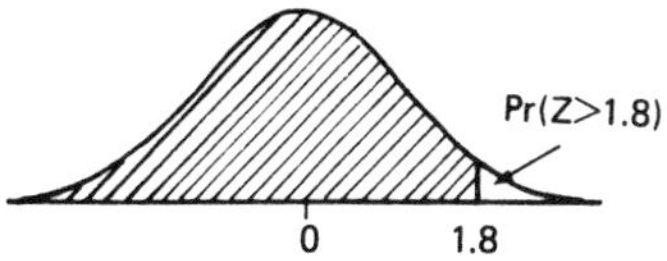

Pr(Z>1.8)

0 1.8

Pr(Z<1.8) = 1−Pr(Z>1.8)
= **1−0.0359**
= 0.9641

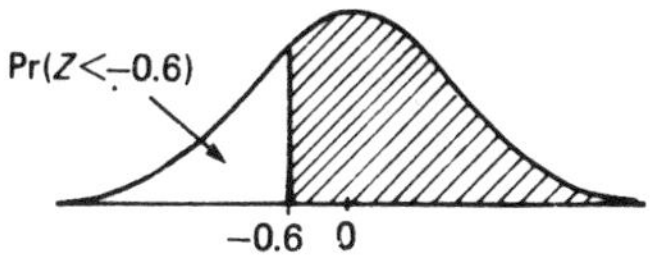

Pr(Z<−0.6)

−0.6 0

Pr(Z>−0.6) = 1−Pr(Z<−0.6)
= 1−Pr(Z>0.6)
= **1−0.2743**
= 0.7257

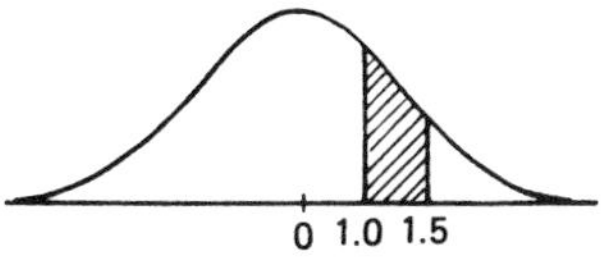

0 1.0 1.5

Pr(1.0<Z<1.5) = Pr(Z>1.0)−Pr(Z>1.5)
= **0.1587−0.0668**
= 0.0919

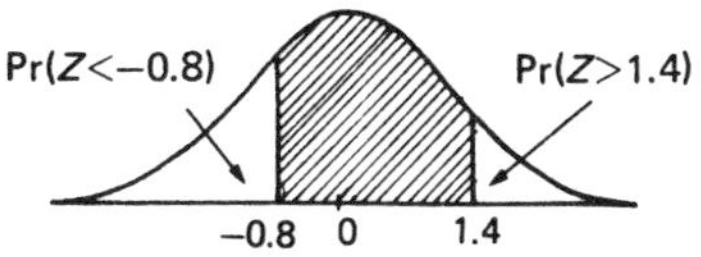

Pr(Z<−0.8)

Pr(Z>1.4)

−0.8 0 1.4

Pr(−0.8<Z<1.4)
= 1−Pr(Z>1.4)−Pr(Z<−0.8)
= 1−Pr(Z>1.4)−Pr(Z>0.8)
= **1−0.0808−0.2119**
= 0.7073

(The shaded area represents the required probability in each case. The figures shown in bold type are taken from tables)

We next illustrate the use of tables with a simple example. The highly technical nature of transistor manufacture results in products which differ widely in their measurable characteristics. One of the important measurable features of a transistor is its "leakage current" (X). Suppose in a particular case it is known that the leakage currents of transistors manufactured by a particular process is normally distributed with a mean value of 60 units (on a scale of an electronic measuring device) and a standard deviation of 12 units. Transistors with leakage currents of less than 50 units are classified as "Grade A", those with leakage currents of between 50 and 80 are "Grade B" and those with higher leakage currents have to be scrapped. What proportion of the output is expected to be (i) scrap, (ii) Grade A, and (iii) Grade B?

The first step in tackling such a problem is to draw a rough sketch of the normal distribution showing both x (measurement) and z (standardised) scales and indicating the mean value and the "end points" of the distribution – the latter, for all practical purposes, can be considered as z values of ± 3, that is, x values of $60 + 3 \times 12$ (= 96) and $60 - 3 \times 12$ (= 24). This sketch (Exhibit 6) enables us to gain a "feel" for the problem.

EXHIBIT 6: Distribution of Transistor Leakage Currents

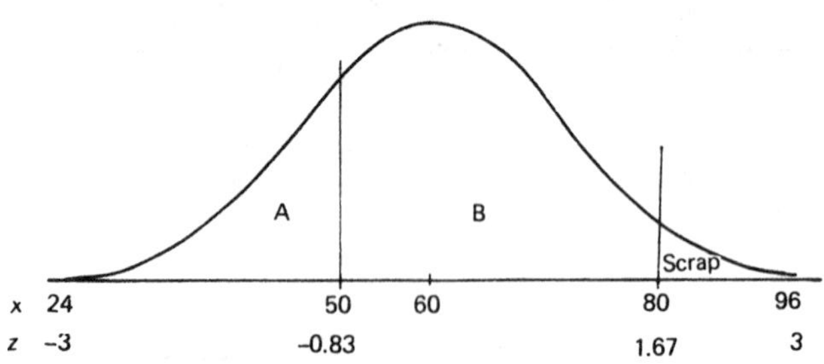

Next the measurement values of interest (50 and 80) are indicated appropriately and the corresponding z values calculated (see Section 11.2). So using tables we have:

(i) Proportion of scrap = $\Pr(X > 80) = \Pr(Z > 1.67) = 0.0475$

(ii) Proportion of Grade A = $\Pr(X < 50) = \Pr(Z < -0.83)$
$= \Pr(Z > 0.83) = 0.2033$

(iii) The proportion of Grade B is obtained by subtracting the proportion of scrap and the proportion of Grade A from unity, i.e. $1 - (0.0475 + 0.2033) = 0.7492$.

Having worked through this exercise the reader might be tempted to ask what service the normal distribution has performed. Presumably some sample data was available from which the mean and standard deviation of the leakage current were calculated; it would have been perfectly feasible to estimate the required proportions from the ogive of the sample data (as, for example, in Exercise 8, CU2). There are, in fact, three reasons why it is more desirable to base the estimates on the normal distribution (or, for that matter, any other theoretical distribution which can be shown to fit the data well) rather than on the sample data directly.

- The sample data will contain "irregularities" which are peculiar to that sample and are not strictly representative of the population. These irregularities are smoothed out by using the normal distribution (with the same mean and variance) – the assumption being that the normal distribution is a closer approximation to the *population* distribution than is the sample distribution itself.
- By using the normal distribution, we can take advantage of its many convenient theoretical properties (which are explored in later Course Units). We could, for example, evaluate the probability of the difference between the leakage currents of two randomly chosen transistors being within specified limits – a matter of some considerable importance in certain transistor applications.
- The normal distribution enables us to carry out various hypothetical experiments to determine the effect of changes of the production process on the characteristics of transistors produced, thus reducing the cost of development work. An example of this particular type of application is given below.

Suppose that it is believed that by employing stricter controls over the manufacturing process one could (at a cost) reduce the average leakage current of transistors produced. What reduction in the mean leakage current would be required to reduce the scrap level to 1% of output? Let us assume, for the moment, that any such controls applied to the manufacturing process result in changes in the mean leakage current but have no effect on the standard deviation of leakage currents (i.e. the standard deviation will remain unchanged at 12).

Referring to tables we see that a standardised value $z = 2.33$ results in a 1% (0.01) area in the tail of the normal distribution. So our scrap level of 80 would need to correspond to $z = 2.33$ under the new conditions of operation with mean leakage current μ', say. Exhibit 7 illustrates the problem.

EXHIBIT 7: A Problem with an Unknown Mean Value

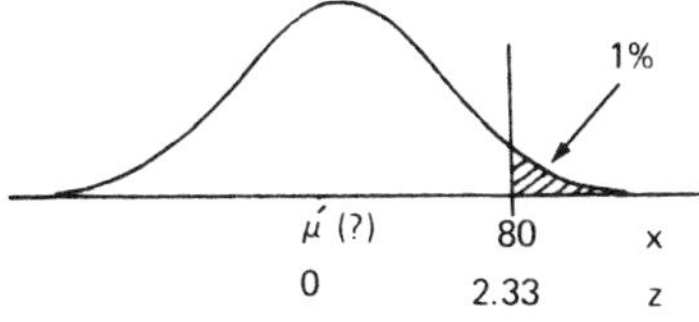

So we require $z = 2.33 = (80 - \mu')/12$ from which the value of μ' of 52 is readily obtained. Consequently under the stated assumptions we should need to reduce the mean leakage current from 60 to 52 in order to reduce the scrap proportion from 4.75% to 1%[1].

11.5 THE CENTRAL LIMIT THEOREM AND ITS APPLICATIONS

The Central Limit Theorem (C.L.T.) on which much statistical theory and application depends can be stated formally as follows:

The *sum* of a large number, n, of independent random variables is approximately normally distributed[2] (and the larger n is, the better the approximation) (3)

The distribution of *sums of variables* may appear an unfamiliar concept, but we have in fact already dealt with sums of variables in earlier Course Units without recognising them as such, as the following examples illustrate.

Suppose we take a simple random sample of size 30 from a large batch of components of which 20% have some minor defect. The

---

[1]Now suppose, to take the example one stage further, a number of trials were carried out on the manufacturing process and it was found that it was indeed possible to reduce the mean leakage current by the application of stricter controls, but in so doing, for any reduction in the mean leakage current, the standard deviation of leakage current was also reduced in the same proportion (i.e. our earlier stated assumption of an unchanged standard deviation is, in fact, untrue). If we reduced the mean value from 60 to 50, say, the SD would also change from 12 to $12 \times \frac{50}{60} = 10$. Under these new assumptions to what level must the mean be reduced in order to achieve a 1% scrap level?

Firstly, note that whatever the value of the new mean, μ', is the new standard deviation would be $\mu'/5$. So for a 1% scrap value we require:

$$z = 2.33 = \frac{80 - \mu'}{(\mu'/5)}$$

Solving this equation for μ' gives $\mu' = 54.6$. Thus under the new assumptions of the mean and standard deviation being reduced in proportion we only need to reduce the mean leakage current from 60 to 54.6 (rather than to 52, as under the old assumption of unchanged standard deviation). The reason for this is that any reduction in the standard deviation would in itself reduce the scrap level – consequently we need to reduce the mean value less than we had previously estimated.

[2]There are a few technical limitations on the validity of the C.L.T., but none need concern the reader. The main point is that the theorem holds irrespective of the form of the distribution of the individual random variables (they do not even need to have the same distribution). The key assumption is that the random variables have to be *independent*. The precise meaning of this is examined in CU12, but for the present it may be taken that random variables are independent if their values are obtained by a process of *random* sampling.

number of defects found in the sample can of course be described by a binomial distribution with $n = 30$, $\pi = 0.20$. (See Exhibit 8.) This has a normal curve shape. But we could equally consider this sampling experiment as being conducted by taking 30 separate (independent) samples of size 1 – each sample having either 0 or 1 defectives. So we can consider the number of defects found in our sample of size 30 to be *the sum of* the number of defects found in 30 samples each of size 1[1]. The fact that the binomial distribution with $n = 30$, $\pi = 0.2$ is approximately normally distributed is one particular demonstration of the truth of the C.L.T.

EXHIBIT 8

EXHIBIT 9

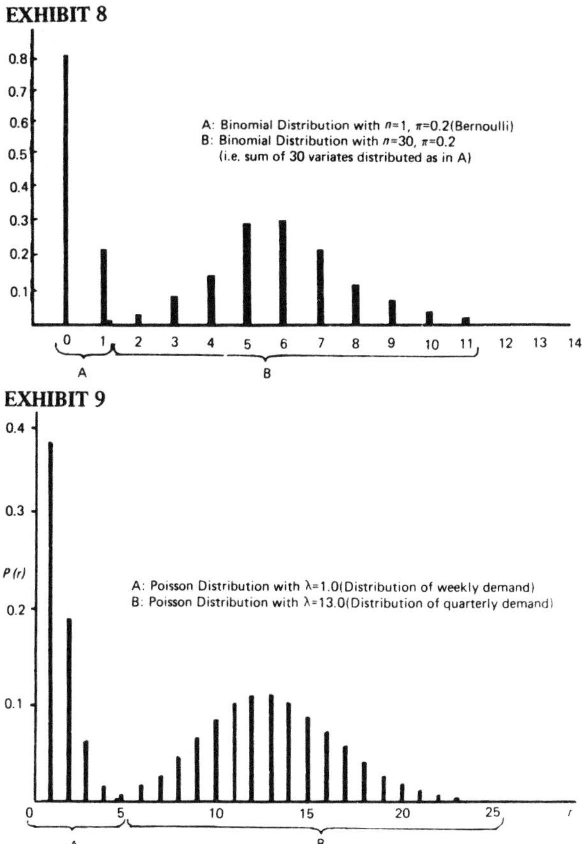

---

[1]This is a reiteration of what was stated formally in CU9, that a binomially distributed random variable (with n, π) is the sum of n independent Bernoulli random variables each with the same π.

As another example consider the case of demands for a product which occur randomly in time. Suppose the weekly demand (i.e. total number of units demanded per week) is Poisson distributed with mean of 1.0. The histogram of this distribution is shown in Exhibit 9(A) – it has a considerable positive skew. What can be said of the distribution of quarterly (13 week) demand? We might expect the quarterly demand to be Poisson distributed, because it still represents the number of (random) demands in another fixed time period, with mean value 13.0. Now we can certainly conceive of the demand in a quarter being *the sum of* 13 independent (because demands are random) weekly demands. As can be seen from Exhibit 9(B), the distribution of quarterly demand is *near* normally distributed – the distribution of *yearly* demand would be even closer to a normal distribution.

To show that the C.L.T. applies to the sum of independent random variables from *any* distribution is a complex theoretical task. One more example is now given in the hope that the reader will be prepared to take this on trust.

A company operates a "Flexitime" system whereby employees can, by working overtime, amass a number of days leave (within limits) owing to them, which can be taken at some convenient future time. Suppose it is known that as a result:

30% of the employees are owed 0 days' leave,
20% of the employees are owed 1 day's leave,
50% of the employees are owed 2 days' leave.

Now consider n employees chosen at random. What is the distribution of total leave (i.e. the *sum* of individual leaves) owing to them in the case of (a) $n = 1$ (b) $n = 2$, (c) $n = 3$ and so on. In the case of $n = 1$ the answer is simple (see Exhibit 10a). For higher values of n the distributions can be derived, quite simply, by using tree diagrams as shown in Exhibit 10b. Despite the fact that the distribution for the $n = 1$ case is markedly unsymmetrical, by the time $n = 5$ the distribution exhibits near symmetry and is (very) approximately normally distributed.

To return to our earlier examples involving the binomial and Poisson distributions, we can see that in appropriate conditions we could use the normal distribution as an approximation. It is often desirable to do this since the particular binomial (two-parameter distribution) or Poisson (one-parameter distribution) we need may not be tabulated, whereas the standard normal distribution, being a zero-parameter distribution, has a complete tabulation. The conditions under which it is often stated that a normal approximation to the binomial and Poisson distributions are valid are given on the following page.

EXHIBIT 10a:

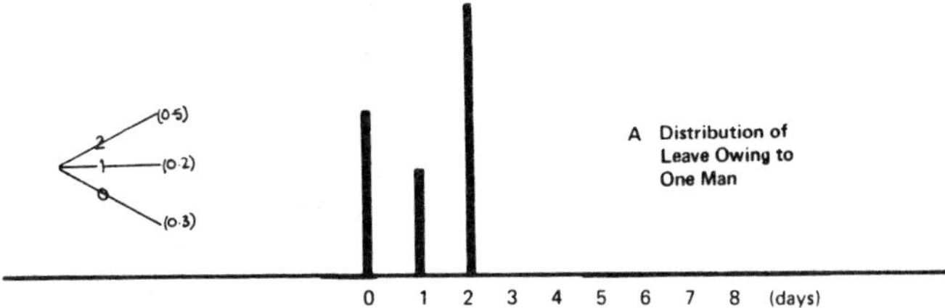

EXHIBIT 10b:

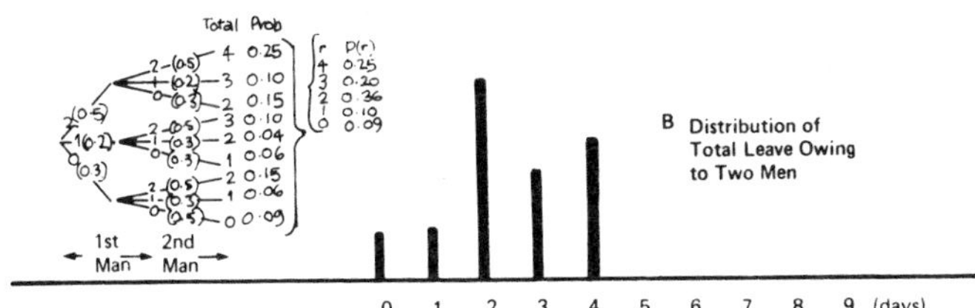

EXHIBIT 10c:

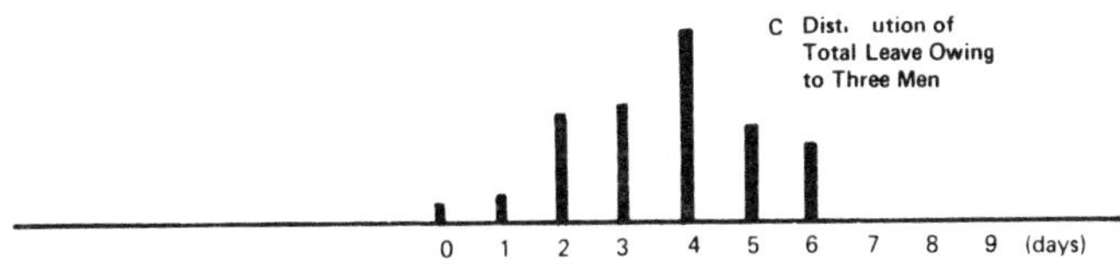

EXHIBIT 10d:

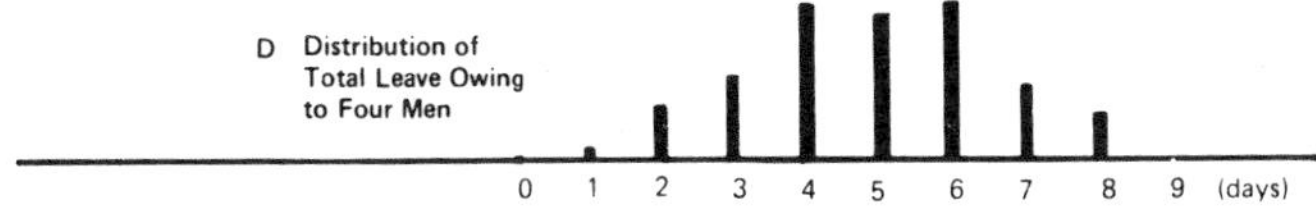

EXHIBIT 10e:

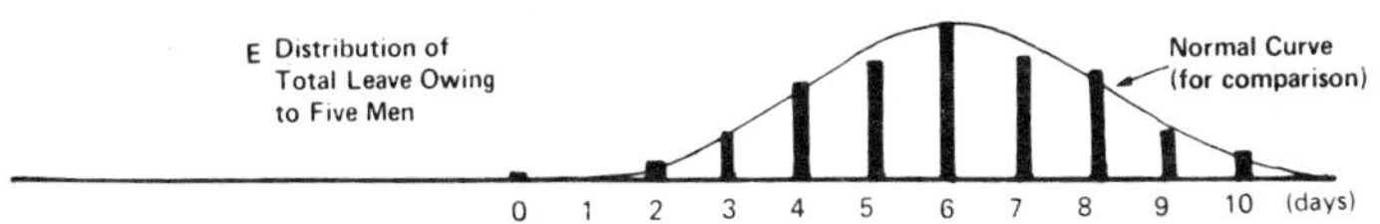

Binomial: for $0.1 \leqslant \pi \leqslant 0.5$[(1)] and $n\pi > 5$ use the normal approximation with $\mu = n\pi$ and $\sigma = \sqrt{n\pi(1-\pi)}$

Poisson: for $\lambda > 10$ use the normal approximation with $\mu = \lambda$ and $\sigma = \sqrt{\lambda}$

There is one fundamental problem which arises when using the normal approximation to the binomial and Poisson distributions. Because the binomial and Poisson are discrete distributions and the normal distribution is continuous, we have to correct for this fact by employing a "continuity correction" as illustrated in the example below.

Suppose we wish to approximate the binomial distribution with $n = 80$, $\pi = 0.16$ by the normal distribution. We note that our conditions for a valid approximation are satisfied, in that $0.1 < \pi < 0.5$ and $n\pi = 12.8 > 5$. Now we have

$$\mu = n\pi = 12.8$$

$$\sigma = \sqrt{n\pi(1-\pi)} = \sqrt{80 \times 0.16 \times 0.84} = 3.279$$

Exhibit 11 shows a sketch of the problem.

EXHIBIT 11: The Normal Approximation to the Binomial

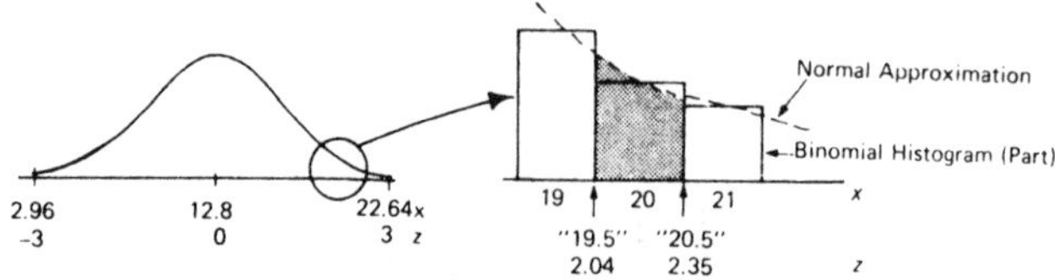

Let us evaluate the probability of exactly 20 "successes". We first draw a detailed sketch showing the binomial histogram (using bars rather than lines, in order to emphasise the "area under the graph" concept) and its normal approximation.

The shaded area (under the normal curve) is approximately equal to the area under the binomial histogram bar for $x = 20$. We call the class limits of this bar "19.5" – "20.5"[(2)] and evaluate their standardised z values. Hence using tables we have

---

[(1)]If $0.5 \leqslant \pi \leqslant 0.9$ we require $n(1-\pi) > 5$ for the approximation to hold. It should be noted that some other texts give other criteria for the approximation of greater or lesser stringency.

[(2)]The addition (and subtraction) of 0.5 to the integer value of $x = 20$ is the so-called continuity correction.

$\Pr(Z > 2.04) = 0.0207$, and

$\Pr(Z > 2.35) = 0.0094$.

Hence $\Pr(2.35 > Z > 2.04) = 0.0207 - 0.0094 = 0.0113$, which is our normal approximation to the binomial probability that $x = 20$. The correct binomial value is 0.0122 – the error being due partly to the approximation and partly to working to only two decimal places for the z values. (Had we simply required the probability of 20 *or more* successes, then this would have been approximated by $\Pr(X > 19.5) = \Pr(Z > 2.04) = 0.0207$.)

Before leaving the topic of the Central Limit Theorem, mention must be made of its most important application, previously alluded to in Section 11.3 – the distribution of sample means. Consider taking a simple random sample of n observations, $X_1, X_2, X_3, \ldots X_n$ and then forming their total,

$$T = X_1 + X_2 + X_3 + \ldots + X_n.$$

Each observation X_i is itself a random variable and independent of the other observations (because we *chose* to sample randomly). Consequently T will be approximately normally distributed for *large* samples, irrespective of the form of the distribution of the measurement quantity, X, itself. Furthermore, the sample mean, $\bar{X} = T/n$, will also be normally distributed since it is simply the sample total divided by the (constant) sample size. Now despite the fact that we usually only take a single sample and calculate from it just one sample mean, this result does demonstrate that the particular value of $\bar{X}$ obtained is just one of the many values of $\bar{X}$ that could be obtained from other samples of the same size from the same population: all such values come from normal distribution, and so we may use the properties of the normal distribution to evaluate the probability of different results for $\bar{X}$. This fundamental result will be used time and time again throughout later Course Units.

11.6 FITTING A NORMAL DISTRIBUTION TO SAMPLE DATA

The principles of fitting a normal distribution to data follow in outline the corresponding process for the binomial and Poisson distributions. Fitting a normal distribution to classified Series A data (Exhibit 12) we first estimate the parameters of the distribution (μ and σ) from the sample statistics $\bar{X} = 79.7$ and $s = 19.02$ obtained in CU4 and CU5 respectively. We next set about evaluating the normal probabilities in each class. To do this we standardise the *upper* class limits throughout (there is no need to standardise the lower class limits separately as the upper class limit of one class is automatically the lower class limit of the next), and then use normal distribution tables, taking particular care with the class which contains the mean, $\bar{X}$. Finally, the probabilities within each class are multiplied by the sample size ($n = 100$) to obtain expected frequencies. As can be seen from a comparison of columns (2) and (6), the normal distribution fits Series A data extremely well.

Because of the number of snares involved in this calculation, the reader is advised to study Exhibit 12 carefully and then attempt to reproduce it.

EXHIBIT 12: Fitting a Normal Distribution to Series A Data

(1) *Class Limits* x_L x_u	*(2)* *Observed Frequency* f	*(3)* *Standardised Value of Upper Class Limit* z_u	*(4)* *Area in the Tail* $Pr(Z > \lvert z_u \rvert)$	*(5)* *Normal Probability in Class*	*(6)* *Expected Frequency* $\hat{f}$ *(to 1 d.p.)*
9.5– 19.5 (below 19.5)	1	−3.166	0.0008	0.0008	0.1
19.5– 29.5	0	−2.639	0.0041	0.0033	0.3
29.5– 39.5	1	−2.114	0.0172	0.0131	1.3
39.5– 49.5	4	−1.588	0.0561	0.0389	3.9
49.5– 59.5	7	−1.062	0.1442	0.0881	8.8
59.5– 69.5	16	−0.536	0.2961	0.1519	15.2
69.5– 79.5	19	−0.011	0.4956	0.1995	19.9
79.5– 89.5	20	+0.515	0.3033	0.2011†	20.1
89.5– 99.5	17	1.068	0.1427	0.1606	16.1
99.5–109.5	11	1.563	0.0590	0.0837	8.4
109.5–119.5	3	2.093	0.0181	0.0409	4.1
119.5–129.5 (over 119.5)	1	∞	0	0.0181	1.8
Totals n =	100			1.0000	100.0

Notes: Col (1) As the normal distribution extends from $-\infty$ to $+\infty$, we interpret the lowest class as "below 19.5" and the highest class as "over 119.5" for comparison purposes.

Col (3) In general $z_u = (x_u - 79.7)/19.02$. For the highest class only we take $z_u = +\infty$ – see note on Col (1) above.

Col (4) Areas in the tail taken from standard normal tables with interpolation.

†Col (5) In general, these probabilities are obtained by subtraction of consecutive values in Col (4). In the case of class 79.5–89.5, *which includes the mean*, the z values of the upper and lower class limits differ in sign. In this case the probability is (1–0.4956 – 0.3033).

READING

Ronald E. Walpole *Introduction to Statistics* Collier-Macmillan, Chapter 7. (Note that Walpole tabulates $\Pr(0 < Z < z)$.)

EXERCISES

1(P) Series C[1] (of the Standard Data Sheets) gives 600 "random numbers". These values have been randomly selected from a **uniform** distribution (a distribution where each integer from 0 to 99 is equally likely).

(i) By forming a histogram from an appropriate number of these values, satisfy yourself that the population is in fact uniformly distributed.
(ii) Calculate the total (T) of each group (60 in all) of the 10 random numbers. Classify your results and draw up a histogram of the values of T.
(iii) Find the mean $\overline{T}$ and standard deviation s_T of T.
(iv) Fit a normal distribution with the same mean and standard deviation to your distribution of T, and compare the observed and expected frequencies. Do your results demonstrate the Central Limit Theorem? (You may need to complete Exercise 2 below before attempting this part.)

2(U) *Using Tables of the Normal Distribution*

(i) Show that the probability of Z, a standard normal variable, being
(a) greater than 1.25, $\Pr(Z > 1.25) = 0.1056$
(b) less than –0.78, $\Pr(Z < -0.78) = 0.2177$
(c) less than 0.62, $\Pr(Z < 0.62) = 0.7324$
(d) greater than –0.24, $\Pr(Z > -0.24) = 0.5948$
(e) between 0.20 and 1.30, $\Pr(0.20 < Z < 1.30) = 0.3239$
(f) between –0.90 and –0.30, $\Pr(-0.90 < Z < -0.30) = 0.1980$
(g) between –1.65 and 0.72, $\Pr(-1.65 < Z < 0.72) = 0.7147$

---

[1] See Appendix 1.

(ii) Show that:
(a) 5% of the normal distribution lies above $z = 1.645$
(b) 95% of the normal distribution lies within $z = \pm 1.960$
(c) 1% of the normal distribution lies below $z = -2.326$
(d) 99% of the normal distribution lies within $z = \pm 2.576$
(e) the quartiles of the normal distribution lie at $z = -0.6745$ and $z = 0.6745$.

(iii) In a large delivery of steel washers the average thickness is 0.400 cms and the standard deviation of thickness is 0.005 cm. Assuming that thickness is normally distributed, what percentage of washers are thicker than 0.403 cm?

(iv) The mean score in an apprentice aptitude test is 54 with a standard deviation of 6. Assuming a normal distribution, below which mark will we find the lowest 10% of apprentices?

(v) The number of arrivals at a supermarket between 9.00 a.m. and 9.30 a.m. has a mean of 230 and a standard deviation of 20. On what percentage of days will there be between 200 and 250 arrivals during this period? Assume a normal distribution.

(vi) The completed length of service of typists in an office typing pool has an average of 345 working days with a standard deviation of 80 days. Assuming a normal distribution of completed length of service, obtain
(a) the percentage of typists who have completed 400 working days,
(b) the value above which we will find the 25% longest serving typists,
(c) the proportion of typists who have completed between 150 and 450 working days.

(vii) An instant coffee manufacturer wishes to set his filling machine so that on average only 1 jar in 100 is filled with less than the nominal 500 gms. If the variability in the contents, based on previous experience, is a standard deviation of 12 gms, by what amount should the manufacturer overfill his jars on average?

(viii) As company Buyer you obtain quotations from two wire rope manufacturers. Company A quotes its ropes as having an average breaking strain of 32 kg, with a standard deviation of 2.5 kg. Company B, for the same diameter rope, quotes a mean of 30 kg, and a standard deviation of 1.5 kg. If the cost of the ropes is approximately equal, which would you recommend for buying when the minimum breaking strain required for the job on which the wires are to be used is 24 kgs?

3(E) The average annual earnings of a group of 10,000 unskilled engineering workers employed by firms in north-east England in 1971 was £1,000 and the standard deviation £200.
Assuming that the earnings were normally distributed, find how many workers earned:

(a) less than £1,000
(b) more than £600 but less than £800
(c) more than £1,000 but less than £1,200
(d) above £1,200.

Use the abstract from the table of areas under the normal curve given below where "Z" is the amount of the deviation from the mean and "O" the appropriate area:

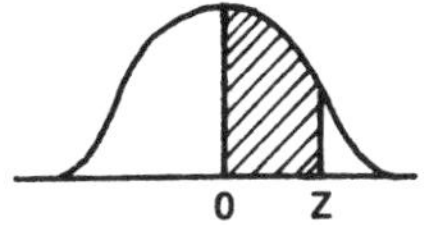

Z	O	Z	O
0.0	0.0000	1.3	0.4032
0.1	0.0398	1.4	0.4192
0.2	0.0793	1.5	0.4332
0.3	0.1179	1.6	0.4452
0.4	0.1554	1.7	0.4554
0.5	0.1915	1.8	0.4641
0.6	0.2258	1.9	0.4713
0.7	0.2580	2.0	0.4772
0.8	0.2881	2.1	0.4821
0.9	0.3159	2.2	0.4861
1.0	0.3413	2.3	0.4893
1.1	0.3643	2.4	0.4918
1.2	0.3849	2.5	0.4938

(ACCA December 1972)

4(E) The overtime earnings of 400 manual workers for the last week in June were

Earnings £p	No. of workers
0.25–0.75	30
0.75–1.25	44
1.25–1.50	28
1.50–1.75	62
1.75–2.00	86
2.00–2.25	84
2.25–2.50	46
2.50–2.75	20
	400

(a) Obtain the mean and the standard deviation of this distribution.

(b) Compare the proportion of workers earning between £1.75 and £2.25 in overtime, with the corresponding proportion given by the normal distribution with the same mean and standard deviation.

(DMS)

5(E) The normal or Gaussian distribution is widely used in statistics.

(a) Describe the main features of this distribution.

(b) Explain why many naturally occurring measurements are believed to come from this distribution.

(c) How would you investigate whether a given sample of observations (e.g. measurements of height on 1,000 adult males) do in fact come from a normal distribution?

(IPM November 1973)

6(E) As a control in a Market Research Survey designed to test the response to a new formula of detergent ("with added Albymes") a sample of 100 housewives, randomly selected, were supplied with two packets of detergent. Both packets contained the *same* detergent but on one packet no reference was made to the additive, whereas the other packet displayed the additive slogan boldly. If the packaging slogans have *no* effect on the choice made by the housewives, what is the probability that at least 60 housewives will prefer the packet with the slogan?

Part Question BA (Business Studies)

N.B. *Evaluate the probability using the normal approximation to the binomial distribution. Compare your result with that obtained directly from binomial tables.*

7(E) A complex television component has 1,000 joints soldered by a machine which is known to produce, on average, one defective joint in forty. The components are examined, and faulty soldering corrected by hand. If components requiring more than 35 corrections are discarded, what proportion of the components will be thrown away?

(ICWA December 1971)

8(E) A meat importer prepares fillet steak for supermarket sale in an appropriate wrapping. It is very difficult to control the variation in the weight of the steaks cut by an automatic slicer, and so in order to obviate the cost of individual weighing and pricing the importer labels all steaks with the same price and offers a £1 "compensation bonus" to any purchaser who finds that the weight of the steak falls short of 8 oz. (In order to obtain the compensation bonus the customer has to return the steak in its wrapping, but these returned steaks cannot be resold.)
During a previous month a production test was carried out where all steaks cut in that month were weighed and their packaging marked with an identifying symbol. The results were that the weights of steaks cut had an average of 8.94 oz. with standard deviation of 0.50 oz. (the weights being approximately normally distributed). The company found that 0.3% of that month's output was returned by customers claiming the compensation bonus.

(i) What proportion of steaks less than 8 oz. were returned?
(ii) If the production cost per steak of steaks with average weight x oz. is given by: $(2x + 5)$ pence, evaluate the total expected cost per steak (including the expected compensation bonus) for steaks with an average weight of 8 oz., 8.5 oz., 9.0 oz., etc. Hence determine the average weight per steak (to the nearest $\frac{1}{4}$ oz.) the company should aim for. (You may assume that the standard deviation of weight remains the same, i.e. 0.50 oz.)
(iii) What other assumptions have you made in answering (ii)?

BA (Business Studies)

The following two examination questions are only partly concerned with the normal distribution: they mainly provide revision material on the binomial and Poisson distributions.

9(E) (a) Describe the main features of the Binomial distribution.
(b) Give two examples of practical situations in which you would expect to encounter this distribution.
(c) In what way is the Binomial distribution related to the normal (or Gaussian)?

(IPM Summer 1973)

10(E) (a) Describe the main features of the Poisson distribution.
(b) Indicate the way in which the Poisson distribution is related
(i) to the Binomial distribution;
(ii) to the Normal (or Gaussian) distribution;
(iii) to a "random process".
(c) A random sample (of size 1) is drawn from a Poisson distribution with mean 49. Calculate the probability that the resulting value is in excess of 51.

(IPM June 1974)

SOLUTIONS TO EVEN NUMBERED EXERCISES

2. (iii) 27.43; (iv) $54 - 1.28 \times 6 = 46$; (v) 0.77% (vi) (a) 24.51, (b) 399, (c) 0.8976; (vii) 27.96 gms.
(viii) For company A $Z=3.2$, for company B $Z=4.0$. Hence company B's ropes are preferred since fewer will have breaking strains less than 24 kgs., even though the mean breaking strain is less.

4. (a) Using hand calculation methods we have:

x_i	f_i	d_i	$d_i f_i$		$d_i^2 f_i$	
0.500	30	−11	−330		3,630	
1.000	44	−7	−308		2,156	
1.375	28	−4	−112		448	$a = 1.875$
1.625	62	−2	−124		248	$c = 0.125$
1.875	86	0	−874		0	*Note unequal*
2.125	84	2		168	336	*class widths.*
2.375	46	4		184	736	
2.625	20	6		120	720	
$n = \Sigma f_i =$ 400				472	$8,274 = \Sigma d_i^2 f_i$	
				−874		
			$\Sigma d_i f_i =$	−402		

$$\bar{X} = a + \frac{c}{n}\sum d_i f_i$$

$$= 1.875 + \frac{0.125}{400} \cdot (-402)$$

$$\simeq 1.75\ (£)$$

$$s^2 = c^2\left[\frac{\Sigma d_i^2 f_i}{n} - \left(\frac{\Sigma d_i f_i}{n}\right)^2\right]$$

$$= (0.125)^2\left[\frac{8,274}{400} - \left(\frac{402}{400}\right)^2\right]$$

$$= (0.125)^2 \cdot 19.675$$

$$\therefore s = 0.125 \times \sqrt{19.675}$$

$$= 0.554\ (£)$$

(b) Proportion of workers earning between £1.75 and £2.25
$= (86 + 84)/400 = 0.425$
Standardised value of £1.75 = 0 (mean value)
Standardised value of £2.25 = $(2.25 - 1.75)/0.55 = 0.903$
Corresponding proportion from normal distribution

$$= \Pr(Z > 0) - \Pr(Z > 0.903)$$
$$= 0.5000 - 0.1833$$
$$= 0.3167,$$

considerably less than the observed proportion.

6. If the slogans have no effect on the choice, then the number of housewives preferring the packet with the slogan will be given by a binomial distribution $n = 100$, $\pi = 0.5$. We have

$$\mu = n\pi = 50; \ \sigma = \sqrt{n\pi(1-\pi)} = \sqrt{25} = 5.$$

$$z = \frac{(60-0.5) - 50}{5} \text{ (using the continuity correction)}$$

$$= 9.5/5 = 1.90$$

From normal tables $\Pr(Z > 1.90) = 0.0287$, which is close to the true binomial value of 0.0284.

8. Let weight of a steak be X oz.
Then during the production test the proportion of steaks less than 8 oz. is given by

$$\Pr(X < 8) = \Pr\left(Z < \frac{8.0 - 8.94}{0.5}\right)$$

$$= \Pr(Z < -1.88) = \Pr(Z > 1.88)$$

$$= 0.0301$$

(i) As 0.3% (i.e. 0.003) of total production was returned, the proportion of underweight steaks which were returned is $0.003/0.0301 \simeq 0.1$ (or 10%).

(ii)

(1) Mean Steak Weight $\bar{X}$	(2) Z value corresponding to 8 oz.	(3) Prop. of steaks less than 8 oz.	(4) Expected cost of bonus per steak†	(5) Average Production cost per steak	(6) Total expected cost per steak i.e. (4) + (5)
8.0	0	0.5000	5p	21p	26p
8.25	−0.5	0.3085	3.085p	21.5p	24.585p
8.50	−1.0	0.1587	1.587p	22p	23.587p
8.75	−1.5	0.0668	0.668p	22.5p	23.168p
9.00	−2.0	0.0227	0.227p	23p	23.227p
9.25	−2.5	0.0062	0.062p	23.5p	23.562p
9.50	−3.0	0.0014	0.014p	24p	24.014p

†for explanation see part (iii) below.

Hence company should aim for $\bar{X}$ = 8.75 oz. as that gives minimum expected cost.

(iii) We assume that 10% of underweight steaks will be returned irrespective of *average* weight. If, for example, $\bar{X}$ = 8.0, then 50% of all steaks will be underweight but only 10% of these (i.e. 5% of all production) will be returned, adding 5p on average to the cost of a steak.

COURSE UNIT 12* – TWO (OR MORE) INDEPENDENT RANDOM VARIABLES

12.1 INTRODUCTION

So far in the course we have been concerned with distributions (either *frequency* distributions compiled from empirical data, or theoretical *probability* distributions) of a single random variable. There are many instances in practice, however, where the physical nature of the problem means that one can observe the values of two (or more) distinct random variables on the same occasion (resulting in bivariate or multivariate observations). In this Course Unit we consider the case where such random variables are independent and where they *interact* with each other to produce an overall *composite* effect.[(1)] The following "problems" illustrate the issue under discussion.

Problem 1. Exhibit 1 illustrates how components A and B are fixed together lengthwise in a manufacturing process. If the length of A is X and the length of B is Y, then the combined length of A and B will be $X + Y$ (= Z, say). If we know the distribution of X and the distribution of Y, the question arises: what is the distribution of the combined length $Z = X + Y$?

EXHIBIT 1

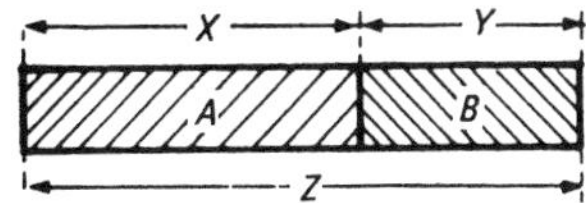

EXHIBIT 2

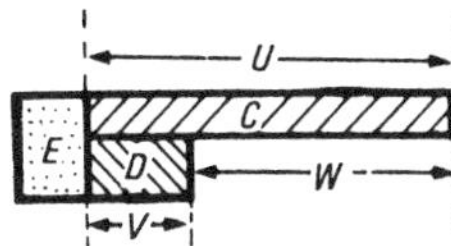

Problem 2. Consider another manufacturing process (Exhibit 2) in which two components C (of length U) and D (of length V) are abutted squarely to a third component E. An important measurable feature is the overlap between C and D, $W = U - V$. We need to know how the distribution of W is related to the distributions of U and V.

---

(1) Our interest in this this topic is partly theoretical – it forms the basis of sampling theory – and partly practical. Its practical importance arises from the fact that it is often desired to control a physically measurable quantity (e.g. length, weight or electrical resistance), the variability of which can be attributed to two (or more) separate *sources* of variability. Control is sometimes only possible over the separate sources. It is obviously of importance to know what overall effect such controls would have.

Problem 3. Suppose $X_1, X_2, X_3, \ldots X_n$ are a random sample of n items from a population with a known distribution.

Two questions of interest are: what is the distribution of

(a) $T = X_1 + X_2 + X_3 + \ldots + X_n$ (i.e. the sample total) and

(b) $\bar{X} = \frac{1}{n}(X_1 + X_2 + X_3 + \ldots X_n)$ (i.e. the sample mean)?

The answer to (b) has a particularly important application in conditions where the distribution of the population is *not* known, for it enables us to judge how close an observed value of $\bar{X}$ will be to the population mean, μ.

Note that we have asked in each problem how the *distribution* of the composite random variable (Z in Problem 1, W in Problem 2, T and $\bar{X}$ in Problem 3) relates to the *distributions* of the component random variables (X and Y in Problem 1, etc.).[(1)]

To answer such questions *in general* is a topic beyond the scope of this course – although we shall later consider some particular simple cases. In many practical situations, however, it is sufficient to know the mean (and variance) of the composite random variable in terms of the means (and variances) of the component random variables, and this, as we shall see, can be determined quite simply.

The results which we shall develop later in this Course Unit apply to both discrete and continuous random variables, although, for the sake of simplicity, they will be demonstrated using discrete random variables only. In order to emphasise the general applicability of the results the symbols X, Y and Z will be used for random variables, even when they are discrete.

12.2 THE MEANING OF INDEPENDENCE

It was stated earlier that our concern here is with *independent* random variables. (The analogous results for random variables which are not necessarily independent are to be given in CU13. The special case of independence is so important that we are devoting the whole of this

---

[(1)] In all three problems the composite random variable is related *linearly* to the component random variables.

$Z = X + Y$ (Problem 1)
$W = U - V$ (Problem 2)
$T = X_1 + X_2 + X_3 + \ldots + X_n$ (Problem 3)

Throughout, we shall restrict our attention to linear relationships.

Course Unit to it.) It is crucial that the meaning of "independence" be made clear before moving on to the analysis.

In CU7 we stated that two events A and B are independent if the occurrence of event A does not affect the chances of the occurrence of event B; that is, the probability of B occurring has the same value irrespective of whether A has occurred or not. In formal terms we have

$$\Pr(B|A) = \Pr(B|\bar{A}) = \Pr(B)$$

A consequence of this result is the simple "multiplication rule" for joint probabilities:

$$\Pr(A \text{ and } B) = \Pr(A) \times \Pr(B).$$

Expressing these results in terms of random variables we may state that for two independent random variables X and Y the probability of Y having a particular value (y, say) is not affected by the value taken by X. To put it another way, if we know the value of X then this does not help us to be any more precise about the value of Y than if we did *not* know the value of X.

The point is perhaps best made by quoting an example. If we are told a man's height (and nothing else about him) this does enable us to make a better guess at his weight – because tall men *tend to be*, although are not invariably, heavier. So height and weight are *not* independent. But knowledge of his height will probably not enable us to guess his income more precisely. So height and income are independent random variables.[(1)]

For random variables X and Y which are independent the multiplication rule for joint probabilities can be written as

$$\Pr(X = x \text{ and } Y = y) = \Pr(X = x) \times \Pr(Y = y)$$

12.3 THE SUM AND DIFFERENCE OF TWO INDEPENDENT RANDOM VARIABLES

The analysis of this topic is to be illustrated within the following setting (first introduced in CU11).

[(1)]We cannot even be *sure* that height and income are independent random variables. It may just be, for example, that very short men find difficulty in obtaining employment and thus tend to have incomes of less than average. If this were so height and income would not be independent. In fact independence between pairs of random variables related to human beings, companies or even countries is probably a pure abstraction – *some* dependency is always present. Nonetheless the notion of independence is a valuable one in that it provides a reference point from which "dependency" can be measured.

A parent company operates a "Flexitime" system, and on 1st March the distribution of days' leave owing to the employees is as follows:

30% of the parent company's employees are owed 0 days' leave,
20% of the parent company's employees are owed 1 day's leave,
50% of the parent company's employees are owed 2 days' leave.

A subsidiary of the parent company has also recently introduced the Flexitime system, but it has not been received as favourably as in the parent company as the following figures indicate:

70% of the subsidiary's employees are owed 0 days' leave,
30% of the subsidiary's employees are owed 1 day's leave.

Now consider choosing at random one employee from the parent company and also one employee from the subsidiary company. Denoting the number of days' leave owing to these two employees by the random variables X and Y respectively, we pose the question: what is the distribution of total leave owing to both employees, $(X + Y)$? This situation can be readily analysed by means of a tree diagram as shown in Exhibit 3. Because the selection of the employees is *at random* the random variables X and Y are independent and this results in a particularly simply constructed tree where, for example, the probability associated with the "$Y = 0$" branches is the same (0.7) in each case (i.e. irrespective of the value of X).

The distribution of the total leave owing to the two employees $(X + Y)$ can be taken directly from the joint probabilities listed on the right of tree diagram. The means and variances[(1)] of the component random variables X and Y and of the composite random variable $(X + Y)$ have also been calculated in Exhibit 3 using expected value techniques (CU8). The results are summarised below.

$$\begin{array}{llll} E(X) & = 1.2 & \text{Var}(X) & = 0.76 \\ E(Y) & = 0.3 & \text{Var}(Y) & = 0.21 \\ E(X+Y) & = 1.5 & \text{Var}(X+Y) & = 0.97 \end{array}$$

It can be seen immediately in our example that:

$$E(X + Y) = E(X) + E(Y), \quad \text{(i.e. } 1.5 = 1.2 + 0.3\text{), and}$$

$$\text{Var}(X + Y) = \text{Var}(X) + \text{Var}(Y), \quad \text{(i.e. } 0.97 = 0.76 + 0.21\text{).}$$

This is, in fact a completely general result for the sum of two *independent* random variables. So we have:

---

(1)The "operator" notation for the variance of X, Var (X), is used here in place of σ_X^2 in order to avoid the excessive use of suffices.

EXHIBIT 3: The Distribution, Mean and Variance of the SUM of Two Independent Random Variables

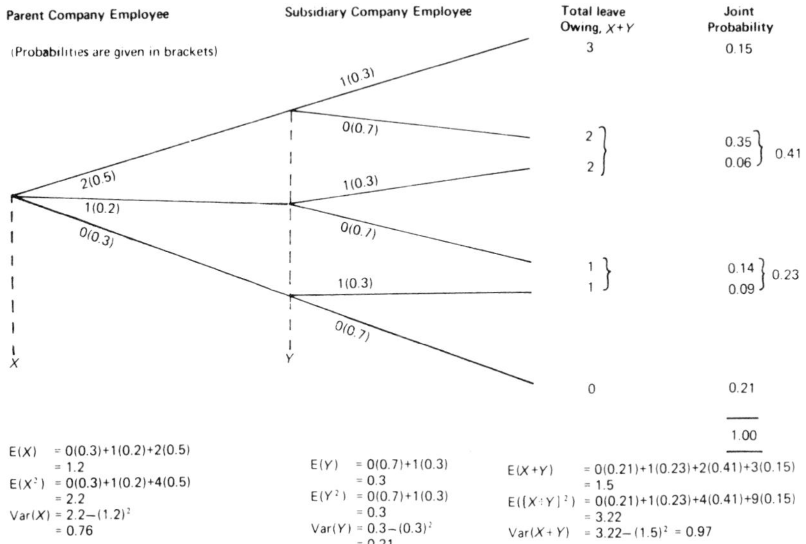

EXHIBIT 4: The Distribution, Mean and Variance of the DIFFERENCE Between Two Independent Random Variables

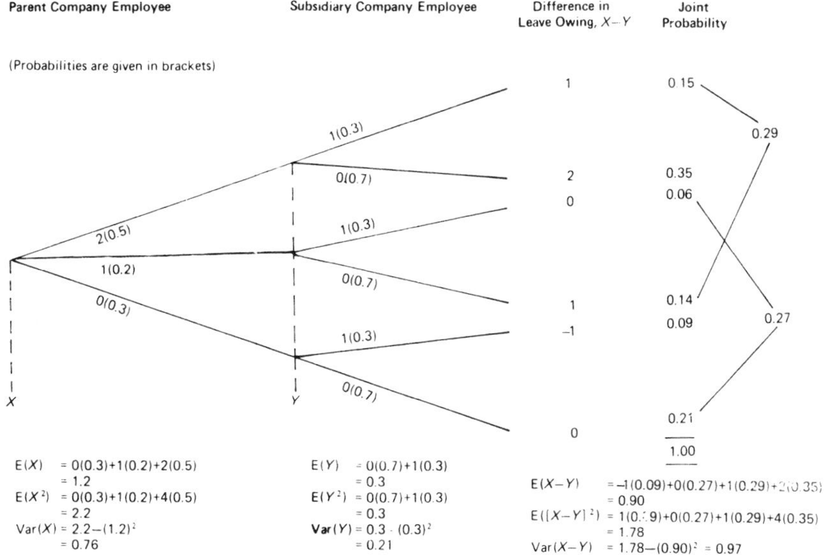

If X and Y are independent random variables

$$\begin{aligned} E(X+Y) &= E(X)+E(Y) \\ \text{Var}(X+Y) &= \text{Var}(X)+\text{Var}(Y) \end{aligned} \tag{1}$$

As a quite separate question we could ask: what is the distribution of the *difference* in days between the leave owing to the parent company's employee and that of the subsidiary's employee (i.e. the quantity $X - Y$)? Analysing this in a similar manner, as shown in Exhibit 4, we obtain the following results for the means and variances.

$$\begin{aligned} E(X) &= 1.2 & \text{Var}(X) &= 0.76 \\ E(Y) &= 0.3 & \text{Var}(Y) &= 0.21 \\ E(X-Y) &= 0.9 & \text{Var}(X-Y) &= 0.97 \end{aligned}$$

So it would appear that while the mean value of the difference between two independent random variables is equal to the difference between their respective mean values (i.e. $0.9 = 1.2 - 0.3$), the variance of the difference between the two variables is equal to the sum (*not* the difference) of their respective variances (i.e. $0.97 = 0.76 + 0.21$) – the same result as when the random variables are added. Again, this is a completely general result for independent random variables.

If X and Y are independent random variables

$$\begin{aligned} E(X-Y) &= E(X)-E(Y) \\ \text{Var}(X-Y) &= \text{Var}(X)+\text{Var}(Y) \quad \textbf{note the sign}^{(1)} \end{aligned} \tag{2}$$

[1]Why does this variance result still have a positive sign? Here are two plausible explanations. First, note that we could write

$$\text{Var}(X-Y) = \text{Var}(X+[-Y]) = \text{Var}(X)+\text{Var}(-Y),$$

using the result from Equations (1). But Var $(-Y)$ is simply the variance of the Y values with a minus sign in front of each. This process of "negativing" the Y values will not affect their variation (or spread) – although it will affect their average value – and so their variance will remain unchanged, i.e. Var $(-Y)$ = Var (Y). Consequently,

$$\text{Var}(X-Y) = \text{Var}(X)+\text{Var}(-Y) = \text{Var}(X)+\text{Var}(Y),$$

the stated result. Looking at the problem another way, if the correct relationship were Var $(X - Y)$ = Var (X) – Var (Y), then this would imply that Var $(X - Y)$ would be negative in cases where Var (Y) were greater than Var (X). But it is impossible (by definition) to have a negative variance and so Var $(X - Y)$ = Var (X) – Var (Y) must be incorrect.

12.4 A MORE GENERAL RESULT

Still keeping within the "Flexitime" setting, it has been left as an exercise for the reader (Exercise 3) to prove that the mean value and the variance of the composite random variable $(3X + 4Y + 5)$ are given by 9.8 and 10.2 respectively.

Now it can be shown (although the proof is outside the scope of this Course Unit) that a more general result for independent random variables X and Y is given by the following.

If X and Y are independent random variables and a, b, and c are constants then:

$$\begin{aligned} E(aX + bY + c) &= aE(X) + bE(Y) + c \\ \text{Var}\,(aX + bY + c) &= a^2\,\text{Var}\,(X) + b^2\,\text{Var}\,(Y) \end{aligned} \qquad (3)$$

Note carefully that on the right hand side of the variance relationship the constant multipliers are a^2 and b^2 (rather than a and b, as in the expected value relationship). The reason for this is that variance itself is a squared quantity. Note also that c does not appear on the right hand side. This is because the process of adding a constant quantity to a random variable does not affect its variability.

So using these results on the problem in Exercise 3, we obtain:

$$\begin{aligned} E(3X + 4Y + 5) &= 3E(X) + 4E(Y) + 5 \\ &= 3 \times 1.2 + 4 \times 0.3 + 5 \\ &= 9.8 \end{aligned}$$

and

$$\begin{aligned} \text{Var}\,(3X + 4Y + 5) &= 9\,\text{Var}\,(X) + 16\,\text{Var}\,(Y) \\ &= 9 \times 0.76 + 16 \times 0.21 \\ &= 10.2, \text{ as stated earlier.} \end{aligned}$$

Incidentally the general result in Equations (3) includes the special cases of the sum and difference of two independent random variables. If we take $a = 1$, $b = 1$, $c = 0$ then we obtain the result in Equation (1) for the sum of two variables:

$$E(X + Y) = E(X) + E(Y), \text{ and}$$

$$\text{Var}\,(X + Y) = \text{Var}\,(X) + \text{Var}\,(Y).$$

If we take $a = 1, b = -1$ and $c = 0$, we obtain

$$E(X - Y) = E(X) - E(Y)$$

$$\text{Var}(X - Y) = \text{Var}(X) + (-1)^2 \text{Var}(Y) = \text{Var}(X) + \text{Var}(Y),$$

which is the result given earlier in Equations (2).

One simple theoretical application of the general result in (3) is to the process of standardisation (CU11). For a random variable X with mean μ and variance σ^2, the standardised variable Z is related as follows

$$Z = \frac{X - \mu}{\sigma},$$

which can be written as $Z = X/\sigma - \mu/\sigma$. Recognising that σ and μ/σ are constants, we can write

$$E(Z) = E(X/\sigma) - \mu/\sigma = E(X)/\sigma - \mu/\sigma$$
$$= \mu/\sigma - \mu/\sigma = 0, \text{ and}$$
$$\text{Var}(Z) = \text{Var}(X/\sigma) = \text{Var}(X)/\sigma^2 = \sigma^2/\sigma^2 = 1,$$

proving that a standardised random variable (a **deviate**) has zero mean and unit variance.

12.5 APPLICATIONS TO SIMPLE PROBABILITY DISTRIBUTIONS

In this section we consider applications to the theoretical probability distributions treated in earlier Course Units. Although the practical value of the applications to the binomial and Poisson distributions is limited, the topic does serve to illustrate the link between material in this Course Unit and earlier material (CU's 9 and 10). The applications to the normal distribution, on the other hand, are of fundamental importance.

Binomial Distribution. Consider a large batch of items of which a proportion π have some specified defect. Suppose that *two* random samples are taken from the batch; the first sample of size n and the second (independent) sample of size m (see Exhibit 5). As with all binomial applications we assume that the samples are either taken with replacement or, more realistically, that the sample sizes n and m are a very small fraction of the total batch size. Let us represent the number of defective items found in the two samples by the random variables R and S respectively. Knowing that R and S will be binomially distributed we may state:

R is binomially distributed with mean, $E(R) = n\pi$, and variance, $Var(R) = n\pi(1 - \pi)$
S is binomially distributed with mean $E(S) = m\pi$, and variance $Var(S) = m\pi(1 - \pi)$.

EXHIBIT 5: Two Samples from a Batch

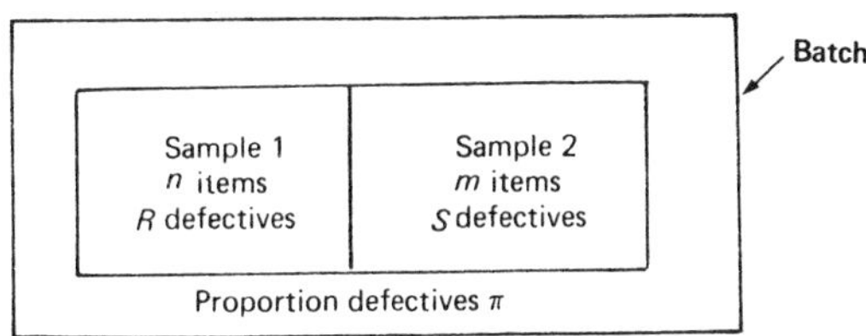

What can be said of the distribution of the total number of defectives in both samples (i.e. $R + S$)? Using Equations (1), we can state:

$$E(R + S) = E(R) + E(S)$$
$$= n\pi + m\pi = (n + m)\pi,$$

and also

$$Var(R + S) = Var(R) + Var(S)$$
$$= n\pi(1 - \pi) + m\pi(1 - \pi) = (n + m)\pi(1 - \pi).$$

These results should come as no surprise as they are the same as the mean and variance of the number of defectives in one single sample of size $(n + m)$. This provides an illustration of the fact that it is valid to "pool" the results from two (or, in fact, more than two) attribute samples provided that the samples are themselves independent and taken from populations with the same value of the parameter π – a result which we shall use later (CU18) when testing hypotheses concerning the parameter π.

Poisson Distribution. We take an illustration from the baking industry. A large quantity of dough, into which is mixed weighed quantities of currants and sultanas, is used to make small buns. The weights of currants and sultanas used are sufficient to produce buns which on average have 4 currants and 3 sultanas each. What proportion of buns will have no currants nor sultanas if the dough has been properly mixed?

Under the stated conditions (proper mixing) we may assume that the number of currants, C, and the number of sultanas, S, in a bun will both be Poisson distributed with mean values of 4 and 3 respectively. Consequently (see CU10) we have:

$$\Pr(C = 0) = e^{-4} = 0.0183$$

$$\Pr(S = 0) = e^{-3} = 0.0498$$

Now as C and S are independent under conditions of thorough mixing we may state:

$$\Pr(C = 0 \text{ and } S = 0) = \Pr(C = 0) \times \Pr(S = 0) = 0.0183 \times 0.0489$$
$$\simeq 0.0009,$$

a result which shows that on average less than 1 bun in 1000 will be without any dried fruit.

This example raises the question of whether it is necessary to consider the currants and sultanas separately, as we are basically only interested in the distribution of $(C + S)$. Are we justified in saying that $(C + S)$ will be Poisson distributed? Intuitively this must be the case as the statistician has no need to distinguish between currants and sultanas – they are both dried fruit, the pieces of which are randomly positioned within the dough. This intuitive result may be confirmed, remembering that for the Poisson distribution the mean and variance are equal, by calculating $E(C + S)$ and $\text{Var}(C + S)$ we have:

$$E(C) = \text{Var}(C) = 4$$

$$E(S) = \text{Var}(S) = 3$$

Now

$$E(C + S) = E(C) + E(S) = 7$$

and

$$\text{Var}(C + S) = \text{Var}(C) + \text{Var}(S) = 7$$

Thus the mean and variance of $(C + S)$ are equal, as for a Poisson distribution. Re-working the example on this basis we have:

$$\Pr(C + S = 0) = e^{-7} \simeq 0.0009, \text{ as before.}$$

Normal Distribution. A result of fundamental importance for later work can be stated as follows:

If X and Y are independent *normally distributed* random variables, then the composite random variables

(i) $X + Y$
(ii) $X - Y$, and more generally (4)
(iii) $aX + bY + c$ (where a, b and c are constants)
are also normally distributed

We shall illustrate this result, and at the same time make use of the results in (1) and (2) earlier, using Problem 1 and Problem 2 set out in the Introduction to this Course Unit. (The statements of these problems should now be re-read before proceeding.) We shall assume that the random variables X and Y in Problem 1 (and U and V in Problem 2) are independent. Physically this means that the two components A and B (and C and D) may be conceived as being selected randomly from large batches of such components before being assembled.

Referring now to Problem 1, suppose the means and standard deviations of the lengths X and Y of the two components are as follows:

Component	A	B
Mean length (cm)	3.210	1.730
SD of length (cm)	0.003	0.004

The combined length of the two components, $Z = X + Y$, will have mean value $E(Z) = E(X) + E(Y) = 3.210 + 1.730 = 4.940$ (cm). The variance of the combined length, $\text{Var}(Z) = \text{Var}(X) + \text{Var}(Y) = (0.003)^2 + (0.004)^2 = 0.000025$. Hence the *standard deviation* of the combined length $Z = \sqrt{0.000025} = 0.005$ (cm). It is worth noting that this SD of combined length is greater than the SD's of length of both A and B but less than the sum of their individual SD's (i.e. $0.003 + 0.004$). Mathematically, the reason for this is that it is the variances, *not* the standard deviations, which possess the additive property. As a practical explanation we would not expect the addition of two independently variable quantities (X and Y) to exhibit less variability than each exhibits separately, but neither would we expect their individual variations to reinforce each other totally.[1]

So far no mention has been made of the probability distribution of X or Y. Our analysis to this point is valid for *any* probability distribution. But if we now confine our attention to the case of X and Y both being normally distributed, the result in (4) immediately tells us that the combined length must be normally distributed too. On this basis we can now calculate, for example, the proportion of all assembled components with combined length greater than an upper limit to the accept-

---

[1]To be a little more precise, as X and Y are independent random variables then we would observe, with roughly equal frequencies, the following four conditions:

(a) a relatively large X with a relatively large Y
(b) a relatively small X with a relatively small Y
(c) a relatively large X with a relatively small Y
(d) a relatively small X with a relatively large Y

In conditions (a) and (b) the variations in X and Y *reinforce* each other; in conditions (c) and (d) the variations tend to cancel each other out. The net effect is one of *partial* reinforcement.

EXHIBIT 6: Analysis of Problem 1

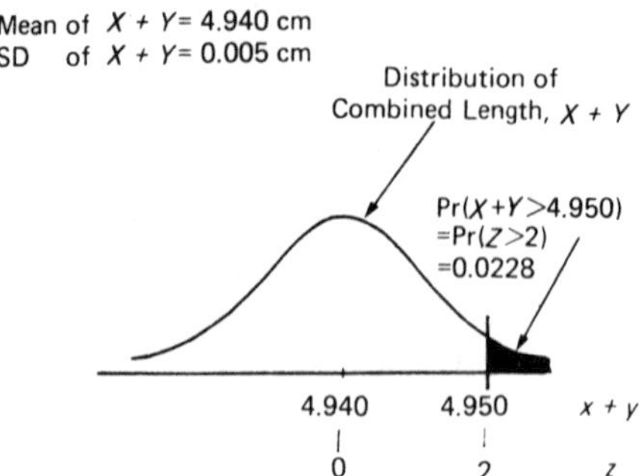

able length of say 4.950 cm. As shown in Exhibit 6, this works out to about 2.3%.

We can analyse *Problem 2* in a similar manner. Suppose U and V are both normally distributed with means and SD's as follows.

Component	C (Length U)	D (Length V)
Mean length (cm)	10.480	4.210
SD of length (cm)	0.012	0.005

The overlap between the two components, $W = U - V$, will have mean value, $E(W) = E(U) - E(V) = 10.480 - 4.210 = 6.270$ (cm). The variance of the overlap, $\text{Var}(W) = \text{Var}(U) + \text{Var}(V) = (0.012)^2 + (0.005)^2 = 0.000169$. Hence the standard deviation of overlap is given by $\sqrt{0.000169} = 0.013$ (cm). Using the result from (4) that $W(= U - V)$ will be normally distributed we can show, in a similar matter to our analysis of Problem 1, that some 6.2% of assembled components will have an overlap of *less than* 6.250 cm, for example. (Verify this figure for yourself.)

12.6 APPLICATIONS MAINLY TO SAMPLING

In a certain industry a recent census revealed that the number of working days lost through illness last year per employee was distributed as shown in Exhibit 7. The mean value of this distribution is 10 days and the standard deviation 5 days.

Now consider the process of selecting *at random* an employee in the industry and determining the number of days lost through illness last year (X) by that person. Clearly, resulting from the *random* selection, we may say that X is distributed as shown in Exhibit 7 and that:

EXHIBIT 7: Distribution of Working Days Lost

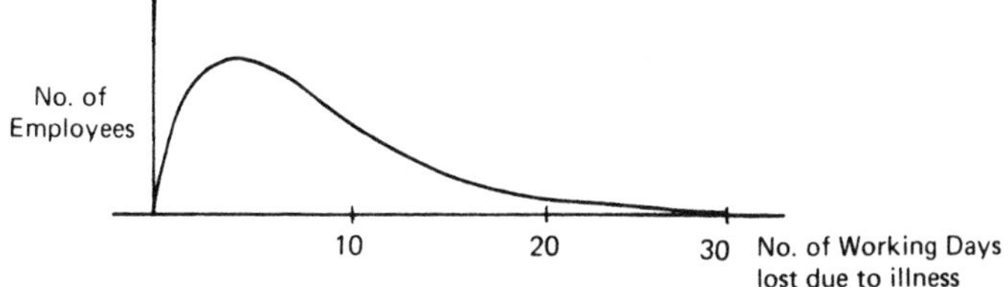

$$E(X) = 10 \text{ and } Var(X) = 5^2 = 25.$$

(Note that were we to consider a deliberately non-random selection, e.g. *deliberately* choosing an employee over the age of 55 or earning over £3,000 a year, such statements would not be true.)

Now consider choosing *another* employee (again at random) and similarly determining the number of days lost (Y) by him. Again we may state:

$$E(Y) = 10 \text{ and } Var(Y) = 5^2 = 25.$$

So far we have considered (in painstaking detail!) the process of selecting a sample of *two* items from the population whose distribution is shown in Exhibit 7. The point of this is to emphasise two properties of X and Y. Firstly X and Y are independent random variables as both employees were considered to be selected at random. Secondly (provided the population is large or the sampling is considered with replacement) there is no difference in *principle* between X and Y – no significance can be attached to the fact that X represents the first item and Y the second item in the sample. Now if we actually carried out this sampling procedure we would probably obtain different values for X and Y, so we cannot say that X and Y are equal. We can say, however, that X and Y are **identically distributed** and one consequence of this is that their mean values and variances are equal.

What is the mean value and variance of the total (T) number of days lost to both employees? We have $T = X + Y$. So

$$\begin{aligned} E(T) &= E(X) + E(Y) & Var(T) &= Var(X) + Var(Y) & \text{(i)} \\ &= E(X) + E(X) & &= Var(X) + Var(X) & \text{(ii)} \\ &= 2\,E(X) & &= 2\,Var(X) & \text{(iii)} \\ &= 2 \times 10 & &= 2 \times 25 & \\ &= 20 & &= 50 & \end{aligned}$$

Line (i) is true because X and Y are independent. Line (ii) because X

and Y are identically distributed. The constant "2" in line (iii) is clearly a consequence of there being just two items in the sample.

It is a short step from here to the general result for a random sample of n items.

If a random sample[1] of n items, $X_1, X_2, X_3, \ldots X_n$ is taken from a population with mean μ ($=\mathrm{E}(X)$) and variance σ^2 ($=\mathrm{Var}(X)$), then the sample *total*,

$$T = X_1 + X_2 + X_3 + \ldots + X_n$$

has a expected value (5)

$$\mathrm{E}(T) = \mu_T = n\mu$$

and variance, $\mathrm{Var}(T) = \sigma_T^2 = n\sigma^2$

Thus considering a random sample of 100 employees in the industry, the total number of days lost last year (T) would have a mean value of 1,000 (= 100 × 10) and a variance of 2,500 (= 100 × 25), (and hence an SD of 50). Furthermore as the sample size itself is large, T will be approximately normally distributed despite the fact that each item in the sample is far from normally distributed (as is clear from Exhibit 7). This follows directly from the Central Limit Theorem (CU11).

Armed with these results we can tackle such questions as: what is the probability that 100 randomly chosen employees will have lost a total of more than 1,120 working days? The answer, as the reader should verify from normal distribution tables, is 0.008.

The results given in (5) have applications beyond what are normally considered as sampling processes. One example is from the field of wire rope manufacture. Wire rope is made up of a number of strands of equal diameter and the breaking strain properties (in fact, the distribution of breaking strains) of individual strands are well known from laboratory tests. On the assumption that the breaking strain of the rope will be equal to the sum of the breaking strains of the individual strands, the manufacturer can readily predict the breaking strain properties of ropes made up from various numbers of strands. Statistically one can consider a rope as consisting of a "sample" of strands. Returning to sampling proper, the majority of sample surveys are con-

---

[1] Throughout this section we assume that the sampling is either with replacement or that the sample size is a small fraction of the population size in order to meet the requirement that all items be identically distributed.

ducted to determine not the sample total[(1)] but the sample mean, or the proportion (or percentage) of items in the sample with a specified attribute. We treat these situations briefly here and we will return to consider them in greater depth in CU15.

Sampling for Means. Considering a random sample of n items, the sample mean $\bar{X}$ can be expressed as:

$$\bar{X} = \frac{1}{n} \sum_{i=1}^{n} X_i = T/n,$$

where T denotes sample total. So

$$\begin{aligned} E(\bar{X}) &= E(T)/n, && \text{using the result from (3)} \\ &= n\mu/n, && \text{using the result from (5)} \\ &= \mu \end{aligned}$$

Thus the average of all possible sample means is equal to the population mean – a useful result which can be expressed in formal terms by saying that $\bar{X}$ is an unbiased estimator of μ. Also

$$\begin{aligned} \mathrm{Var}(\bar{X}) &= \mathrm{Var}(T)/n^2, && \text{using the result from (3)} \\ &= n\sigma^2/n^2, && \text{using the result from (5)} \\ &= \sigma^2/n. \end{aligned}$$

This result has an intuitive appeal. It shows that if n is large then the variation in the value of the sample mean from one sample to another is small; in other words large samples tend to give more accurate results (accuracy being defined, for our purposes, as closeness of $\bar{X}$ to the population value μ). It is also worth noting a special case of this result: if we consider samples of just one item each, then, because the sample means are in each case equal to the observations themselves, the variance of the sample means must be equal to the population variance, a result which is confirmed by putting $n = 1$ in the variance relationship. Summarising we have:

If a random sample of n items is taken from a population with mean μ and variance σ^2, then the sample mean $\bar{X}$, will have

an expected value $E(\bar{X}) = \mu_{\bar{X}} = \mu$ (6)

and variance $\mathrm{Var}(\bar{X}) = \sigma^2_{\bar{X}} = \sigma^2/n$

[(1)] Before leaving the notion of a sample total altogether, consider the following neat theoretical application. A binomial random variable R can be expressed as the total of n independent Bernoulli random variables. A Bernoulli random variable S has mean value π and variance $\pi(1 - \pi)$ – a result given in CU9. Consequently $E(R) = nE(S) = n\pi$. Also $\mathrm{Var}(R) = n\ \mathrm{Var}(S) = n\pi\ (1 - \pi)$. This is probably the simplest proof of the binomial mean and variance.

As for the sample total, if n is large then $\bar{X}$ will be normally distributed (C.L.T.). Using these results on our "days lost" example, the distribution of sample means for samples of size 100 is shown in Exhibit 8. It can be seen that it is extremely unlikely to obtain a sample mean of less than $8\frac{1}{2}$ days or more then $11\frac{1}{2}$ days.

EXHIBIT 8: Distribution of $\bar{X}$

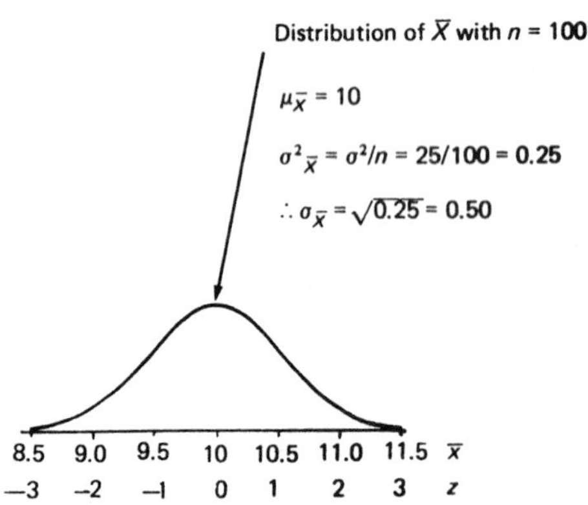

Sampling for Proportions. Let the random variable R represent the number of "successes" in a random sample of n items taken from a population with a proportion of π "successes". (R, then, is a binomially distributed random variable.) The proportion of successes in the *sample*, P, can be written as

$$P = R/n$$

so

$$\begin{aligned} \mathrm{E}(P) &= \mathrm{E}(R)/n, && \text{using the result in (3)} \\ &= n\pi/n, && \text{using the binomial mean, } n\pi \\ &= \pi. \end{aligned}$$

So on average the sample proportion is equal to the population proportion, a comforting result. Similarly

$$\begin{aligned} \mathrm{Var}(P) &= \mathrm{Var}(R)/n^2, && \text{using the result in (3)} \\ &= n\pi(1-\pi)/n^2, && \text{using the binomial variance, } n\pi(1-\pi) \\ &= \pi(1-\pi)/n \end{aligned}$$

Summarising, we have:

If the proportion of "successes" in a random sample of n items, taken from a population with an overall proportion π, is denoted by P, then

$$E(P) = \pi \tag{7}$$

and

$$\text{Var}(P) = \pi(1 - \pi)/n$$

If n is large (and π not too small) then P will be approximately normally distributed with this mean and variance (C.L.T.)

As an example, if an industry has 80% men and 20% women employees, then in a random sample of 100 employees, the proportion of those who are men (P) has:

$$E(P) \quad = \pi = 0.80 \ (= 80\%)$$

$$\text{Var}(P) = \pi(1 - \pi)/n = 0.80 \times 0.20/100 = 0.0016$$

Hence the SD of P is 0.04 (i.e. 4%). Using the normal approximation we see that the vast majority of such samples will contain a proportion of between 0.68 (i.e. $0.80 - 3 \times 0.04$) and 0.92 (i.e. $0.80 + 3 \times 0.04$) of men: that is between 68% and 92% of men.

Incidentally it is somewhat easier to work such examples throughout in terms of percentages. Denoting percentage values by a % subscript, we can write $\pi_{\%} = 100\pi$ and $P_{\%} = 100P$. The results in (7) become:

$$E(P_{\%}) \quad = \pi_{\%}$$

$$\text{Var}(P_{\%}) = \pi_{\%}(100 - \pi_{\%})/n.$$

Reworking the example on this basis, we have

$$E(P_{\%}) = 80\%$$

$$\text{Var}(P_{\%}) = (80 \times 20/100 = 16$$

So

$$\text{SD of } P_{\%} = \sqrt{16} = 4\%.$$

READING

T.H. & R.J. Wonnacott *Introductory Statistics 2nd Ed*, Wiley, Chapter 5.

EXERCISES

1(P) Consider a trial of rolling two dice simultaneously, one with the left hand, the other with the right, Let X and Y denote the faces shown by left-hand and right-hand dice respectively.

(i) Explain why X and Y are independent random variables.

(ii) Perform the trial 20 times and note the two faces, x and y, shown in each case. Also calculate the sum of the faces shown, $u = x + y$, and the difference between the two faces shown, $v = x - y$, for each trial. Set out your results in the following tabular form

Trial	x	y	$u = x + y$	$v = x - y$	
1	2	3	5	−1	←(Example)
2					
3					
.					
.					
.					
20					

(iii) Calculate the means and variances of x, y, u and v. Do you find that:

$$\bar{u} = \bar{x} + \bar{y} \text{ and } s_u^2 = s_x^2 + s_y^2,$$

$$\bar{v} = \bar{x} - \bar{y} \text{ and } s_v^2 = s_x^2 + s_y^2?$$

If not, explain why. Under what conditions would you expect these relationships to hold exactly?

(iv) Calculate, using expected values, the means and the variances of the random variables X and Y. Hence find the means and variances of the quantities $(X + Y)$ and $(X - Y)$. Compare these (population) results with the sample results in (iii) by drawing up a suitable table and inserting the results.

2(U) State whether you consider the following pairs of variables to be independent or not. Give reasons in each case:

(i) Heights of two randomly chosen males.

(ii) Heights of randomly chosen male twins.

(iii) Age and income of a randomly chosen male employee in the computer industry.

(iv) Distance from London and annual profit of a randomly chosen British company.

(v) Sales of roofing tiles by a large building contractor on a randomly chosen day in May, and the sales the *following* day.

(vi) Sales of roofing tiles by a large building contractor in a randomly chosen month in 1973 and the sales the *following* month.

3(U) Using the information given in Section 11.2 of the text (related to the company operating a "Flexitime" system) draw up a tree diagram (as in Exhibit 3) and use this to determine the mean and variance of the composite random variable:

$$3X + 4Y + 5$$

Check your answers with those given in Section 11.4 of the text.

4(E) (i) If the weights of all men travelling by air from Manchester airport have a mean of 162 lbs. with an SD of 20 lbs., what is the probability that the combined gross weight of 49 men on a plane departing from Manchester is more than 8,330 lbs.?

(ii) A sample of 256 employees is chosen at random from among the employees of a very large company whose mean I.Q. is 100 with an SD of 18.0. What is the probability that the mean of the I.Q.s in the sample will exceed 101.0?

(iii) 25% of employees in a large company have joined the company pension scheme. What is the probability that more than 90 employees out of a random sample of 300 will be members of the pension scheme?

HND (Business Studies)

5(E) Rope is made up from strands which each have a mean breaking strain of 100-lb. and SD of breaking strain of 10-lb. How many strands are needed to make a rope which has less than a 1 in 10,000 chance of breaking under a strain of 1,000-lb.?

BA (Business Studies)

6(E) Suppose that as a member of a company manufacturing a branded product X you have gained access to some results of sample surveys carried out by a rival company. This rival company produces a product Y which is in direct competition with your product X. The survey is known to relate to the percentage of consumers interviewed who preferred product Y to product X. The results are as follows:

Survey Date	*% of consumers interviewed preferring product Y to X*
1st Jan. 1974	32
3rd Feb. 1974	36
4th Mar. 1974	38
1st Apr. 1974	30
29th Apr. 1974	36
10th May 1974	40
2nd June 1974	34
1st July 1974	32
7th Aug. 1974	36
1st Sep. 1974	36

Make the following assumptions:

(a) the true proportion of consumers preferring product Y to product X has remained constant over the period 1st Jan. 1974 to 1st Sept. 1974.

(b) the rival company employed a simple random sampling scheme with the same sample size for all surveys.

Calculate the mean and variance of the percentage of customers preferring product Y to X and hence estimate the "true" (i.e. population) percentage. Use the values you have calculated to make an estimate of the sample size used by the rival company. Discuss in general terms how your estimate would differ if it were known that assumption (a) and/or assumption (b) were untrue.

7(E) A building is erected in two stages. A and B are the prefabricated side walls of the first stage, C and D are the prefabricated side walls of the second stage.

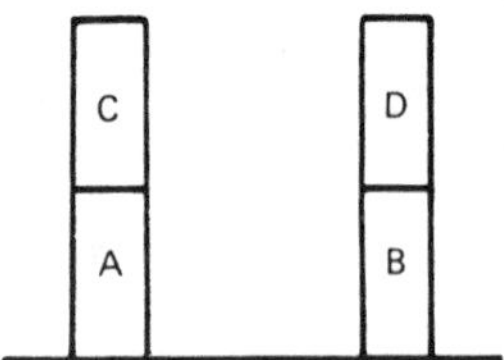

Units A and B both have mean heights of 3 m and SD's of 0.002 m. Units C and D have mean heights 2.5 m and SD's 0.003 m. If the foundations of the buildings are level, calculate the proportion of such buildings where the total heights of their two side walls differ by more than 0.01 m. (State any assumptions made.) If this proportion were unacceptable, suggest ways of reducing it.

BA (Business Studies)

8(E) One production line of a company produces resistors with a nominal resistance of 100 ohms. In fact the resistance of the resistors is normally distributed with mean 100 ohms and standard deviation 5 ohms. An operator sits by a conveyor belt down which the resistors flow and tests randomly selected resistors *in pairs*. If a pair of resistors differ in their resistance by at most 2 ohms then they are said to be "matched" and they are put aside for separate use. Now suppose that the operator takes two resistors (A and B) from the conveyor belt.

(i) If A has a resistance of 106 ohms, find the probability that B will "match" A.

(ii) If A has a resistance of 101 ohms, find the probability that B will "match" A.

(iii) What is the probability that a randomly chosen A and B will be "matched"?

(iv) If A and B do not match then the operator can do one of two things. He can either (a) replace both resistors on the conveyor belt and pick another two, or alternatively (b) return *one* resistor to the conveyor belt and pick another *one* in the hope that it will match the one not replaced. Using your results from (i), (ii) and (iii), demonstrate that it is sometimes better to do (a) and sometimes better to do (b). Suggest the form (but not the detail) of a decision rule to determine when to do (a). Also indicate *how* the decision rule could be worked out.

BA (Business Studies)

SOLUTIONS TO EVEN NUMBERED EXERCISES

2. (i) Independent because the males are chosen randomly.

(ii) Twins tend to be of similar height, so not independent despite the fact that they are chosen randomly.

(iii) Income tends to increase with age, so not independent.

(iv) Probably independent. Note that this would not be so if we consider companies anywhere in the world because some countries tend to have more profitable companies than others.

(v) Independent because sales on one day are unlikely to be affected by sales the previous day if there are a large number of customers.

(vi) Not independent because building is a seasonal trade i.e. low sales would be expected in winter months and high sales in the summer months.

4 (i) Let $T = \Sigma_{i=1}^{49} X_i$, where X_i is weight of an individual

$$\mu_T = n\mu = 49 \times 162 = 7{,}938$$

$$\sigma_T = \sigma\sqrt{n} = 20\sqrt{49} = 140$$

For a value of T equal to 8,330, we have,

$$z = (8{,}330 - 7{,}938)/140 = 2.8$$

$$\Pr(T > 8{,}330) = \Pr(Z > 2.8) = 0.00256$$

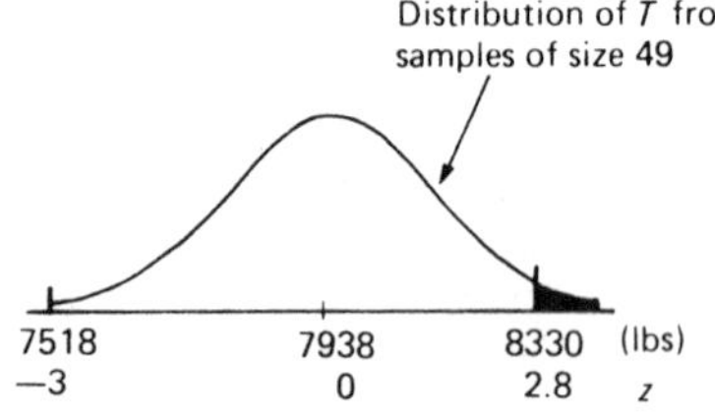

(ii) $\mu_{\bar{X}} = \mu = 100$

$$\sigma_{\bar{X}} = \sigma/\sqrt{n} = 18.0/\sqrt{256} = 18.0/16 = 1.125$$

So for $\bar{X}$ of 101,

$$z = (101 - 100)/1.125 = 0.89$$

So

$$\Pr(\bar{X} > 101) = \Pr(Z > 0.89) = 0.1867$$

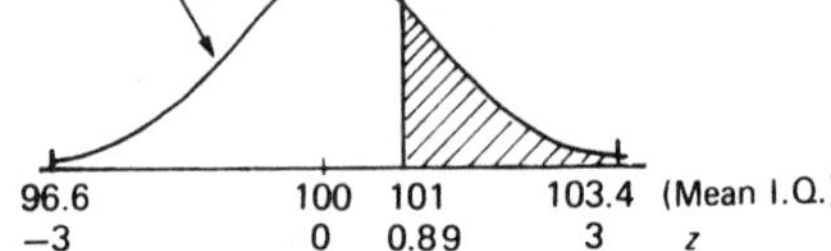

(iii) Working in percentages we have $\pi_{\%} = 25\%$. Let sample % be $P_{\%}$
So

$$E(P_{\%}) = \pi_{\%} = 25\%$$

$$\text{Var}(P_{\%}) = \pi_{\%}(100 - \pi)/n = 25 \times 75/300 = 6.25$$

So

$$\text{SD of } P_{\%} = \sqrt{6.25} = 2.5\%$$

Our value of

$$P_{\%} = \left(\frac{90}{300} \times 100\right)\% = 30\%$$

So

$$z = (30 - 25)/2.5 = 2$$

$$\Pr(P_{\%} > 30\%) = \Pr(Z > 2) = 0.0225$$

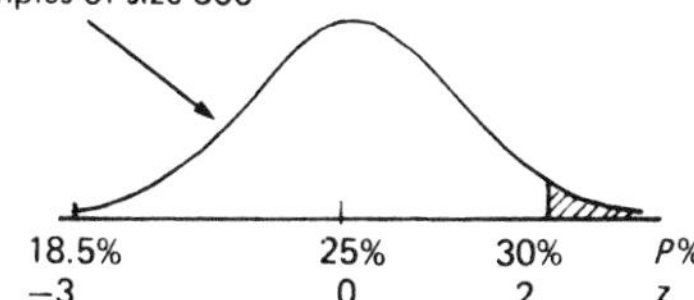

6. Denoting the percentage figures by $P_{\%}$, we have

$$\bar{P}_{\%} = 350/10 = 35\%,$$

$$s^2_{P_{\%}} = 82/10 = 8.2$$

But we know

$$E(P_{\%}) = \pi_{\%} \text{ (the population percentage),} \qquad (1)$$

$$\text{Var}(P_{\%}) = \pi_{\%}(100 - \pi_{\%})/n. \qquad (2)$$

So from (1) we can say that $\pi_{\%} \simeq 35\%$. But as $s_P^2 \simeq \text{Var}(P_{\%})$, we have from (2)

$$\pi_{\%}(100 - \pi_{\%})/n = 8.2$$

(Strictly we should correct s_P^2 by Bessel's correction.) Substituting for $\pi_{\%}$ ($\simeq 35\%$), we obtain

$$35 \times 65/n = 8.2$$

or

$$n = 35 \times 65/8.2 = 277$$

This is our estimate of the sample size.
If (a) is untrue n could be much larger than that estimated as differences in $P_{\%}$ could simply have arisen by the changes in the "true" proportion preferring Y to X.
If (b) is untrue the differences in $P_{\%}$ could have arisen by faulty sampling alone and so n could again be larger than estimated.

8. (i) $\Pr(\text{A matches B}) = \Pr(104 < B < 108) = \Pr(0.8 < Z < 1.6)$
$= 0.2119 - 0.0548$
$= 0.1571$

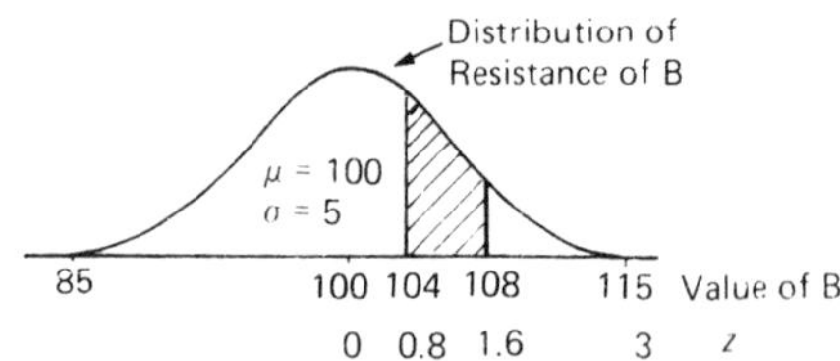

(ii) $\Pr(\text{A matches B}) = \Pr(99 < B < 103) = \Pr(-0.2 < Z < 0.6)$
$= 1 - (0.4207 + 0.2743)$
$= 0.3050$

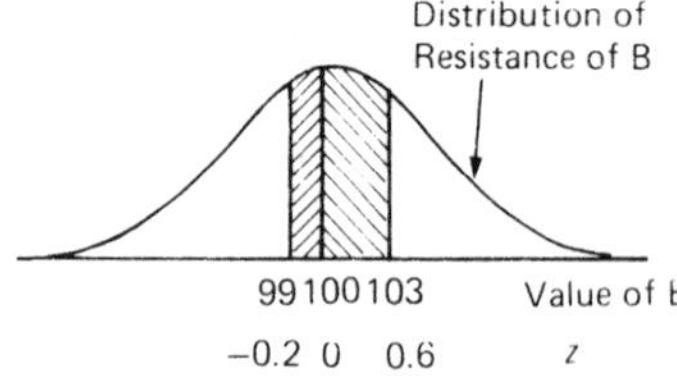

(iii)

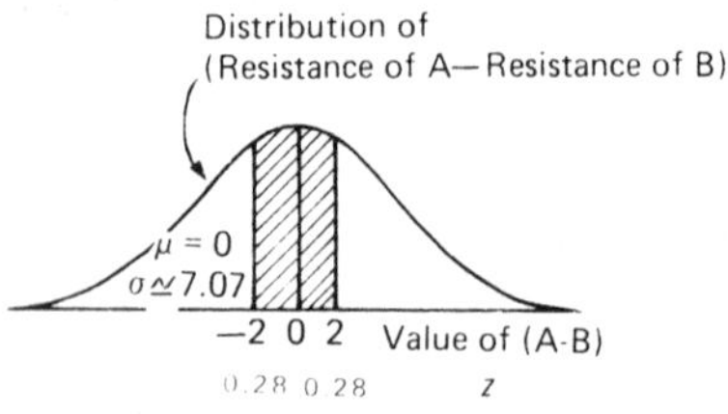

We have

$$E(A - B) = E(A) - E(B) = 0$$

$$\text{Var}(A - B) = \text{Var}(A) + \text{Var}(B) = 2 \times 5^2;$$

hence

$$SD = 5\sqrt{2} \simeq 7.07$$

$$\begin{aligned}\Pr(\text{A matches B}) &= \Pr(-2 < (A - B) < 2) \\ &= \Pr(-0.28 < Z < 0.28) = 1 - 2(0.3890) \\ &= 0.2220\end{aligned}$$

(iv) We note (i) < (iii) < (ii), so if $A = 106$ (i.e. (i)), then the probability of matching is increased by replacing both resistors, (a). If $A = 101$ (i.e. (ii), the probability of matching is increased by retaining A, (b).

Rule Replace both resistors (a) if the resistance of *both* resistors is outside the range $100 \pm r$. (r is the critical value to be determined.)

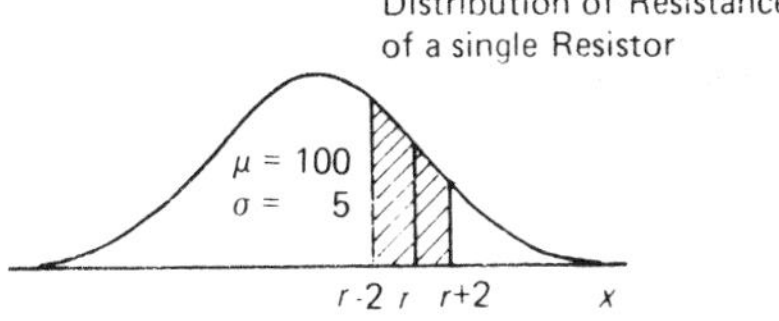

r is obtained from $\Pr(r - 2 < X < r + 2) = 0.2220$

COURSE UNIT 13 – COVARIANCE AND CORRELATION

13.1 INTRODUCTION

In CU12 we studied some properties of pairs of random variables (X and Y) which are independent. In this Course Unit we consider the case where X and Y are *not* independent. In terms of practical usefulness this is by far the more important case, as the main reason for studying two variables jointly is to investigate their dependency on each other. If it can be established, for example, how the effective life (Y) of a frying pan depends on the thickness (X) of a "non-stick" coating then this information can be put to good use in assisting production and marketing decisions. Similarly if we know how computer programming ability (Y, measured in some specified manner) is related to the score on a standard psychological test (X) such information would be useful to a personnel department in helping select entrants for the data processing division of a company.

As for other statistical topics, the idea of dependency (or **association**, as it is normally called in this context) between pairs of random variables can be approached in two ways – theoretically, leading to a definition of a population parameter measuring the association, or empirically in the development of a sample statistic. In this Course Unit we take the unusual step of mixing up the two approaches. Firstly (Section 13.2) we consider some empirical results which serve to illustrate graphically the meaning of association.

In Section 13.3 we work towards the definition of a population parameter[(1)], returning in Sections 13.4, 13.5 and 13.6 to a further consideration of sample results and the calculation and interpretation of some sample statistics.

13.2 SCATTER DIAGRAMS

In CU12 we argued loosely that the height and income of randomly chosen males are independent random variables whereas height and weight are not independent. In our current terminology we could say that there is an association between height and *weight*, but not between height and *income*.

[(1)]Section 13.3 may be omitted by the reader whose main concern is with sample results provided he is prepared to forgo the insights into the nature of the statistics provided by a study of the population parameters.

An *indication* of whether there is an association between a pair of random variables can be provided empirically be means of a **scatter diagram**. Suppose for example, that the height and weight of a man are 175 cm and 70.0 kg. These values (which constitute a single bivariate observation) can be plotted as a single point on a graph with axes height (x) and weight (y), as shown as point A in Exhibit 1.

Other bivariate observations (taken from a random sample of men)

EXHIBIT 1: Direct (or Positive) Association

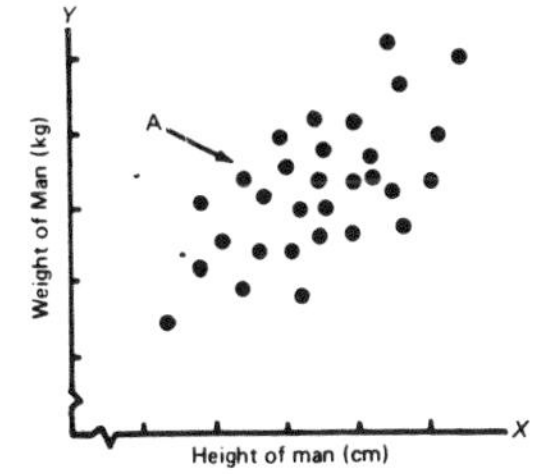

EXHIBIT 2: No Association

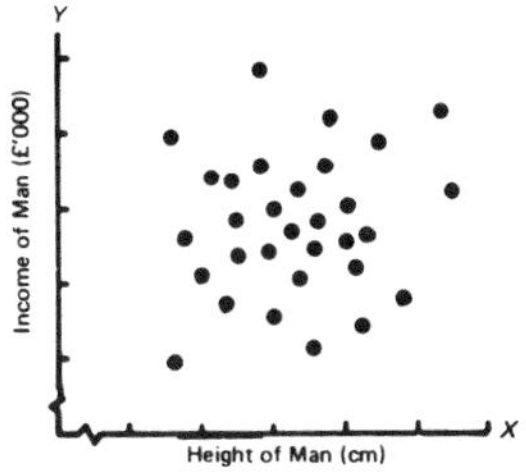

EXHIBIT 3: Inverse (or Negative) Association

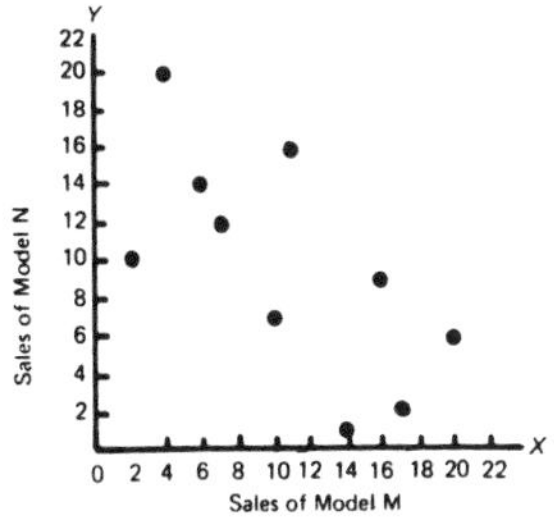

EXHIBIT 4: Some Association but not Direct or Inverse

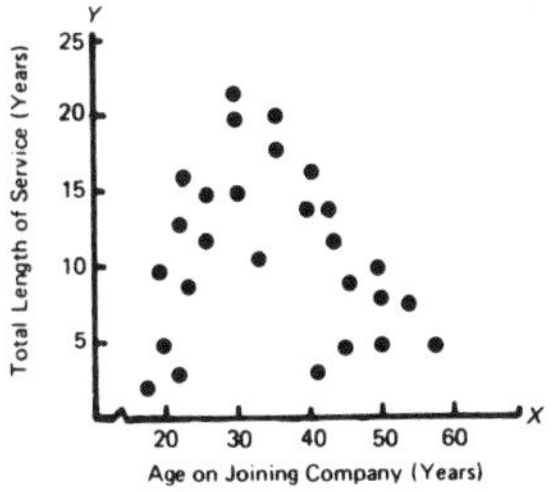

EXHIBIT 5: Perfect Positive Linear Association

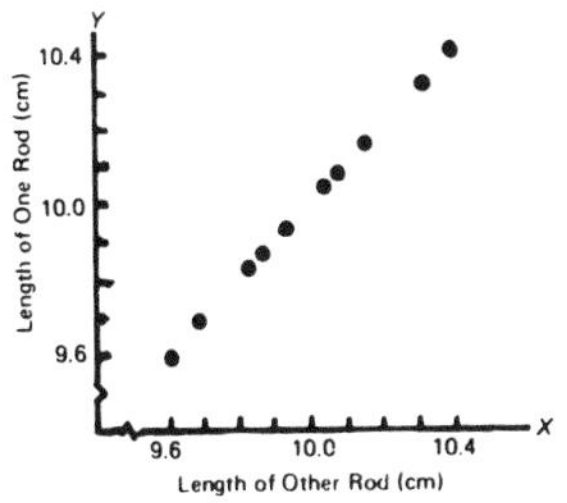

EXHIBIT 6: Perfect Negative Linear Association

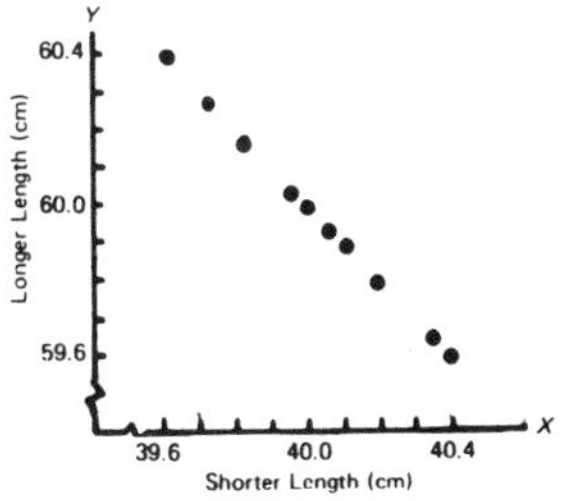

can be plotted in a similar manner and the result is a scatter diagram with a swarm of points forming the "cigar shaped" pattern as shown. Compare this with the scatter diagram of height and income (Exhibit 2) where there is no discernible pattern ("random scatter"). It is the pattern in Exhibit 1 which provides an indication that height and weight are associated – tall men tend to be heavier – and the absence of pattern in Exhibit 2 which indicates that height and income are independent.

Now for two business examples. A car distributor sells two models of cars (which we shall call model N and model M) which are believed in the trade to be in direct competition with each other. The sales of these two models in the first ten weeks of the year are used to construct the scatter diagram in Exhibit 3 – each point representing the sales of the two models in a particular week. Is there any evidence of association here? It is difficult to tell as there is a fair amount of "scatter", but the indications are that high sales of model M are associated with low sales of N and vice versa. Thus there is some rather weak evidence of *inverse* or *negative* association. (Compare with Exhibit 1 where there is a stronger indication of *direct* or *positive* association.)

Association between variables is not always "direct" (a tendency for large X to be associated with large Y and small X with small Y) or "inverse" (where large X tends to be associated with small Y, and small X with large Y). Consider the scatter diagram in Exhibit 4 constructed from the records of the last 40 employees to leave a company well known for its employee loyalty. A pattern can be seen, but it certainly is not a simple one. Thus there is some indication of association between "age on joining the company" and "total length of service" – it is certainly not a case of random scatter, but nor is the association of the form where we could call it direct or inverse.

Scatter diagrams, perhaps more than any other pictorial aid to statistical analysis, are fraught with difficulties of interpretation. There are three main areas of difficulty.

Randomness of the Sample. A scatter diagram can, of course, be drawn up from any (bivariate) sample of observations, but unless the sample is randomly selected any indications of association provided by the scatter diagram will have limited validity. It is easy to imagine, for example, that, in a study of the relationship between height and income if we use a faulty sampling technique which tends to choose men who are tall and rich and/or those who are short and poor then the scatter diagram will indicate a positive association. Such a result would give an erroneous impression of the relationship between height and income in general.

Small Sample Effects. It has already been seen in earlier Course Units that small samples (even small *random* samples) can convey misleading impressions of the population: this is particularly true in the area of bivariate analysis. A scatter diagram drawn up from a small sample of some five observations may show no evident pattern – the pattern only "emerging" when a much larger sample is used. Conversely, and possibly more misleadingly, a small sample may show a distinct pattern which "disappears" when the sample size is increased.

Causal Relationships. The fact that two variables are associated does not *prove* that there is a cause-effect relationship at work. To take an example, for most companies there is an association between sales revenue in a given year and the advertising expenditure in that year. A scatter diagram constructed for sales revenue and advertising expenditure (each point representing a particular year) would possess an upward sloping pattern, providing an indication of positive association, and as such act as a source of comfort to the Sales Manager. One possible interpretation is that higher sales *result from* higher advertising expenditure, but that is not the only interpretation. It could be that in years when the sales were high, and the company prosperous, a higher advertising budget was allocated. Yet again it could be that high sales and high advertising expenditure both resulted from a general feeling of business confidence. Nonetheless, scatter diagrams giving indications of association can lead one to *suspect* a (perhaps hitherto unknown) causal relationship and, as such, lead to a series of investigations designed specifically to find out whether the effect is a causal one or not.[(1)]

13.3* TWO EXTREME EXAMPLES OF LINEAR ASSOCIATION

We now touch on a theoretical development of the subject. In CU12 it was established that for two independent random variables X and Y,

$$E(X + Y) = E(X) + E(Y), \text{ and} \tag{1}$$

$$\text{Var}(X + Y) = \text{Var}(X) + \text{Var}(Y). \tag{2}$$

We might enquire whether these relationships hold for pairs of variables which are associated (i.e. not independent): if not, the extent to which these relationships are violated might be used to *measure* the association. The fact of the matter is that although result (1) still holds for

[(1)] The reader may recall the debate concerning the association between smoking and lung cancer in this context.

pairs of variables which are associated, result (2) does not. Nonetheless result (2) can be generalised for *any* X and Y as follows:

$$\mathrm{Var}(X+Y) = \mathrm{Var}(X) + \mathrm{Var}(Y) + \text{“correction term”}. \qquad (3)$$

The interesting feature of this "correction term"[(1)] is that it has a value which is:

- *zero* in the case of independent variables – in which case, of course, result (3) is identical with result (2),
- *positive* in cases where X and Y are directly or positively associated,
- *negative* in cases where X and Y are inversely or negatively associated.

So this "correction term" can play a key role in the measurement of association. We illustrate this point by considering two examples where the association takes an extreme form.

Example 1. A do-it-yourself enthusiast wants to cut a large number of rods to a length of 10 cm. He does this by clamping a pair of rods together, squaring up their ends, marking out and sawing through both rods at the same time. Now although neither his marking nor sawing are particularly accurate this process does at least ensure that pairs of rods are the same length (but with each pair differing slightly in length from the other pairs). Let X represent the length of one rod *of a pair* and Y the length of the other (Exhibit 8).

EXHIBIT 8: Cutting Two Rods to the Same Length

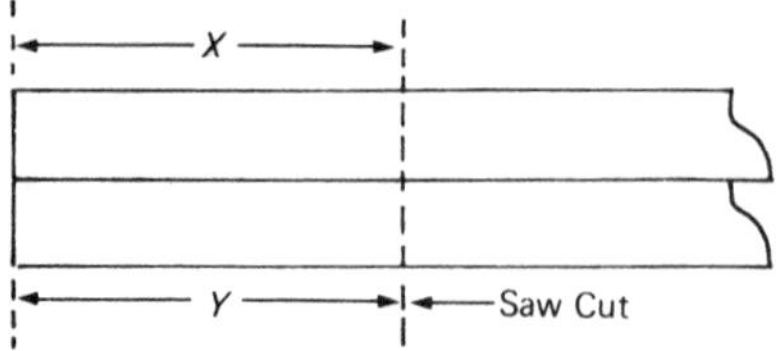

Now X and Y are clearly not independent: in fact X and Y will always be equal, although not necessarily quite 10 cm. If we were to collect

(1) We have met other "correction terms" in other areas of analysis. Recall that if A and B are mutually exclusive events then $\Pr(A \text{ or } B) = \Pr(A) + \Pr(B)$. In general, however, $\Pr(A \text{ or } B) = \Pr(A) + \Pr(B) - \Pr(A \text{ and } B)$. Clearly $\Pr(A \text{ and } B)$ can be regarded as a "correction term" allowing for the possibility that events A and B are not mutually exclusive.

some data from the D-I-Y man and form a scatter diagram we would achieve a result similar to Exhibit 5 where all the points lie on an upward sloping straight line. This is an example of a *perfect positive linear* association. It is perfect because from a knowledge of the value of X we will know with certainty the corresponding value of Y; it is positive and linear because the points fall on an upward sloping straight line.

Returning to the "correction term", we note that as in this example $X = Y$, $\text{Var}(X)$ must equal $\text{Var}(Y)$. So

$$\text{Var}(X) + \text{Var}(Y) = \text{Var}(X) + \text{Var}(X) = 2\,\text{Var}(X).$$

But

$$\text{Var}(X + Y) = \text{Var}(X + X) = \text{Var}(2X) = 4\,\text{Var}(X),$$

by using the result in CU13 that $\text{Var}(aX) = a^2\,\text{Var}(X)$.

By subtraction we obtain the value of the correction term, $2\,\text{Var}(X)$. This is a positive quantity, confirming the fact that there exists a positive association between X and Y.

Example 2. The same enthusiast as in Example 1 has a large number of 100 cm rods which he wishes to cut into two unequal lengths, 40 cm and 60 cm. He does this by (imprecisely) marking out and cutting the rods one at a time. For a given rod let X denote the shorter length and Y the longer length obtained (Exhibit 7).

EXHIBIT 7: Cutting a Rod into Two Unequal Lengths

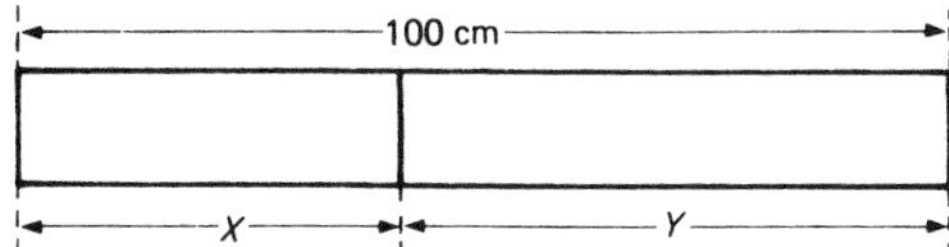

Plotting the results from a number of rods would give a scatter diagram similar to that of Exhibit 6, showing that the points fall on a backward sloping straight line. This is an example of a perfect negative linear association.

Now for each millimetre oversize (i.e. greater than 40 cm) X is, the corresponding value of Y will be the *same amount* undersize (less than 60 cm) and consequently $\text{Var}(X) = \text{Var}(Y)$.

So $\text{Var}(X) + \text{Var}(Y) = \text{Var}(X) + \text{Var}(X) = 2\,\text{Var}(X)$. Now although X and Y are variables, their sum $(X + Y)$ will be constant at 100 cm (or rather 100 cm less the width of the saw cut): consequently $\text{Var}(X + Y)$

= 0. From these results we can easily see that the correction term = -2 Var(X), the negative value confirming that the association is itself negative in form.

In both these examples it can be seen that the magnitude of the correction term will depend upon the values of Var(X) and Var(Y); that is, on the precision of the saw cuts, and also on the units (cm or inches) in which length is measured. But whatever the cutting precision and the units used the association will still be *perfect* in both cases. Consequently the correction term itself is not a very useful measure of association; a better measure is obtained by "cancelling out" the effects of Var(X) and Var(Y) in the following fashion:

$$\rho = \frac{\frac{1}{2}\,(\text{correction term})}{\sqrt{\text{Var}(X) \cdot \text{Var}(Y)}} \tag{4}$$

For Example 1 we obtain $\rho = \text{Var}(X)/\sqrt{\text{Var}(X) \cdot \text{Var}(X)} = 1$ and for Example 2, $\rho = -\text{Var}(X)/\sqrt{\text{Var}(X) \cdot \text{Var}(X)} = -1$. In fact these two values of ρ (± 1) are the extreme values that ρ can take. In general $-1 \leqslant \rho \leqslant +1$; the *sign* of ρ tells us whether the association is positive or negative, the numerical value of ρ tells the *extent to which* the association can be considered as being linear (the value 1 indicating a *perfect* linear association). If ρ = o, then there is no *linear* association: this is true of variables X and Y which are independent, but it is also true of variables which are associated in such a way that with no stretch of the imagination could we consider the association to be linear in form.[(1)]

Variables are said to be **correlated** if they have a degree of linear association (*highly* correlated if the numerical value of ρ is close to 1). Consequently the parameter ρ is called the **correlation coefficient** (or more precisely, the product-moment correlation coefficient).

What we have so far referred to as the "correction term" is in fact *twice*[(2)] a quantity known technically as the **covariance** of X and Y, written Cov(X, Y) or σ_{XY}. Covariance then, unlike variance, can be either positive or negative. The definition of covariance is given in expected value notation as:

$$\text{Cov}(X, Y) = \sigma_{XY} = \text{E}(XY) - \text{E}(X)\,\text{E}(Y); \tag{5}$$

that is, "the expected value of the product of X and Y less the product of the expected values of X and Y".

---

(1)If X and Y are related in such a form that $X^2 + Y^2 = 1$ then $\rho = 0$. Here X and Y are perfectly associated, but the association is by no means linear: in fact the scatter diagram would show a circular pattern.

(2)See footnote on p. 233 opposite.

Note that the covariance of X and X *itself* is the variance of X because $\text{Cov}(X, X) = \text{E}(X^2) - [\text{E}(X)]^2 = \text{Var}(X)$. (This agrees with our finding in Example 1 that the correction term is $2\text{Var}(X)$.)

Translating result (5) into form suitable for computation (for discrete variables) we get:

$$\text{Cov}(X, Y) = \sigma_{XY} = \sum_{\text{all } x, y} xyp(x, y) - \mu_X\mu_Y, \tag{6}$$

where $p(x, y)$ is the *joint* probability $\Pr(X = x \text{ and } Y = y)$.

Finally we can rewrite results (3) and (4) in their conventional form as:

$$\text{Var}(X + Y) = \text{Var}(X) + \text{Var}(Y) + 2\text{Cov}(X, Y) \tag{7}$$

---

and

$$\rho = \frac{\text{Cov}(X, Y)}{\sqrt{\text{Var}(X) \cdot \text{Var}(Y)}} = \frac{\sigma_{XY}}{\sigma_X \sigma_Y} \tag{8}$$

---

(2)*Continued from previous page.*

The reason for "twice" is as follows:

$$(X + Y)^2 = X^2 + Y^2 + 2XY$$

So

$$\text{E}(X + Y)^2 = \text{E}(X^2) + \text{E}(Y^2) + 2\text{E}(XY) \qquad \text{(i)}$$

Also

$$\begin{aligned}[\text{E}(X + Y)]^2 &= [\text{E}(X) + \text{E}(Y)]^2 \\ &= [\text{E}(X)]^2 + [\text{E}(Y)]^2 + 2\text{E}(X)\,\text{E}(Y) \qquad \text{(ii)}\end{aligned}$$

Subtracting (ii) from (i), we obtain

$$\begin{aligned}\text{Var}(X + Y) &= \text{E}(X + Y)^2 - [\text{E}(X + Y)]^2 \\ &= \text{E}(X^2) - [\text{E}(X)]^2 \qquad (= \text{Var}(X)) \\ &\quad + \text{E}(Y^2) - [\text{E}(Y)]^2 \qquad (= \text{Var}(Y)) \\ &\quad + 2\text{E}(XY) - 2\text{E}(X)\,\text{E}(Y)\end{aligned}$$

It can be seen that the correction term is

$$2[\text{E}(XY) - \text{E}(X)\,\text{E}(Y)] \text{ or } 2\text{Cov}(X, Y).$$

13.4 THE CORRELATION COEFFICIENT AS A SAMPLE STATISTIC, *r*

In Section 13.3 we developed a population parameter ρ for measuring the extent of linear association between two variables X and Y: in this section we develop a sample statistic r, the sample correlation coefficient, to do the same job. Here we are mainly concerned with the ideas: Section 13.5 gives a streamlined method of computation for r.

Consider the following sample of bivariate observations.

Observation No.	i	1	2	3	4	5	6	7
	X_i	7	2	7	9	11	2	4
	Y_i	7	5	9	13	11	2	9

As a preliminary to the analysis we calculate the sample mean and standard deviation for the X values and Y values separately.

$$\bar{X} = \sum X/n = 42/7 = 6 \qquad \bar{Y} = \sum Y/n = 56/7 = 8$$

$$s_X^2 = \sum X^2/n - \bar{X}^2 \qquad s_Y^2 = \sum Y^2/n - \bar{Y}^2$$

$$= 324/7 - 6^2 = 10.29 \qquad = 530/7 - 8^2 = 11.71$$

$$\therefore\ s_X = \sqrt{10.29} = 3.21 \qquad \therefore\ s_Y = \sqrt{11.71} = 3.42$$

Exhibit 8 shows the scatter diagram for this data on which two auxiliary axes have been drawn; one, parallel to the y axis, at $X = \bar{X} = 6$, and the other parallel to the x axis at $Y = \bar{Y} = 8$. These auxiliary axes divide the scatter diagram into four regions: if the observations lie predominantly in regions 1 and 3 (as in our case) this provides evidence of positive association; if they lie predominantly in regions 2 and 4 then this indicates a negative association.

We can develop a *measure* of association as follows. Consider observation 2 which is $X_2 = 2$, $Y_2 = 5$: this observation is at the corner of the shaded rectangle (formed with the auxiliary axes) shown on the scatter diagram, the area of which is 12 (= 4 × 3). In general, for any observation, we can calculate such an area from:

$$\text{“Signed area”} = (X_i - \bar{X})(Y_i - \bar{Y})$$

By using this expression to evaluate the areas, a positive value will be obtained for those observations in regions 1 and 3 and a negative quantity for those in regions 2 and 4.

EXHIBIT 8: Measuring Association in a Sample

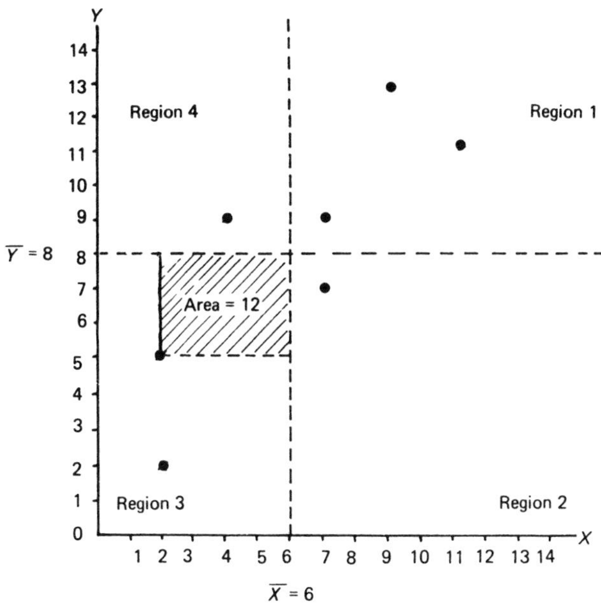

	X_i	Y_i	$(X_i - \overline{X})$	$(Y_i - \overline{Y})$	"Signed Area" $(X_i - \overline{X})(Y_i - \overline{Y})$
	7	7	1	−1	− 1
	2	5	−4	−3	12
	7	9	1	1	1
	9	13	3	5	15
	11	11	5	3	15
	2	2	−4	−6	24
	4	9	−2	1	− 2
Totals	42	56	0	0	67 − 3 = 64

$\overline{X} = 42/7$ $\quad$ $\overline{Y} = 56/7$
$= 6$ $\quad$ $= 8$

$$s_{XY} = \frac{1}{n}\sum (X_i - \overline{X})(Y_i - \overline{Y}) = 64/7 = 9.14$$

$$r = \frac{s_{XY}}{s_X s_Y} = \frac{9.14}{(3.21)(3.42)} = 0.83$$

So if the *average* "signed area" (of all observations) is positive this corresponds to a positive association; if negative, a negative association.

This average "signed area" is called the **sample covariance**,

$$s_{XY} = \frac{1}{n} \sum (X_i - \bar{X})(Y_i - \bar{Y}), \tag{9}$$

which has the positive value 9.14 in our example (Exhibit 8), indicating a positive association.

The sample covariance, s_{XY}, is itself a measure of association but not a very useful one, as the following argument demonstrates. The numerical values of the "signed areas" depend on the units used to measure X and Y. Consequently, to take a specific example, the sample covariance of height and weight of a random sample of men will have a certain value if height is measured in cm. and weight in Kg; it will have a quite different value if height is measured in inches and weight in pounds. Any sensible measure of association must be independent of the measurement units. Such a measure is a "descaled" version of the sample covariance obtained by dividing it by the standard deviation of X and standard deviation of Y[1]:

$$r = \frac{s_{XY}}{s_X s_Y} \tag{10}$$

The statistic r is known as the **sample correlation coefficient**, or more ponderously, the sample product-moment (PM) correlation coefficient[2], and it measures the *degree of linear association* in a sample; that is, the extent to which the points on a scatter diagram cluster around a sloping straight line.

In general r (like ρ – Section 13.3[3]) lies between its extreme values of -1 (*perfect negative* linear association) and $+1$ (*perfect positive* linear association).

[1] This quantity is in fact the sample covariance of the standardised value (CU12) of the observations. Readers should note that this expression for r is exactly analogous to that for the population parameter ρ given in result (8).

[2] Also known as Pearson's correlation coefficient (after its originator).

[3] r in fact can be used to estimate the value of ρ, provided it is remembered that it may be an unreliable estimate as the result of a small sample size or a non-random sampling procedure.

In our example we have:

$$r = \frac{s_{XY}}{s_X s_Y} = \frac{9.14}{(3.21)(3.42)} = 0.83,$$

a result which indicates that the association is positive and "strong" (the value is quite close to +1), but not perfect.

As a guide to the interpretation of r the reader should study the scatter diagrams in Exhibits 9–16, where approximate values of r are

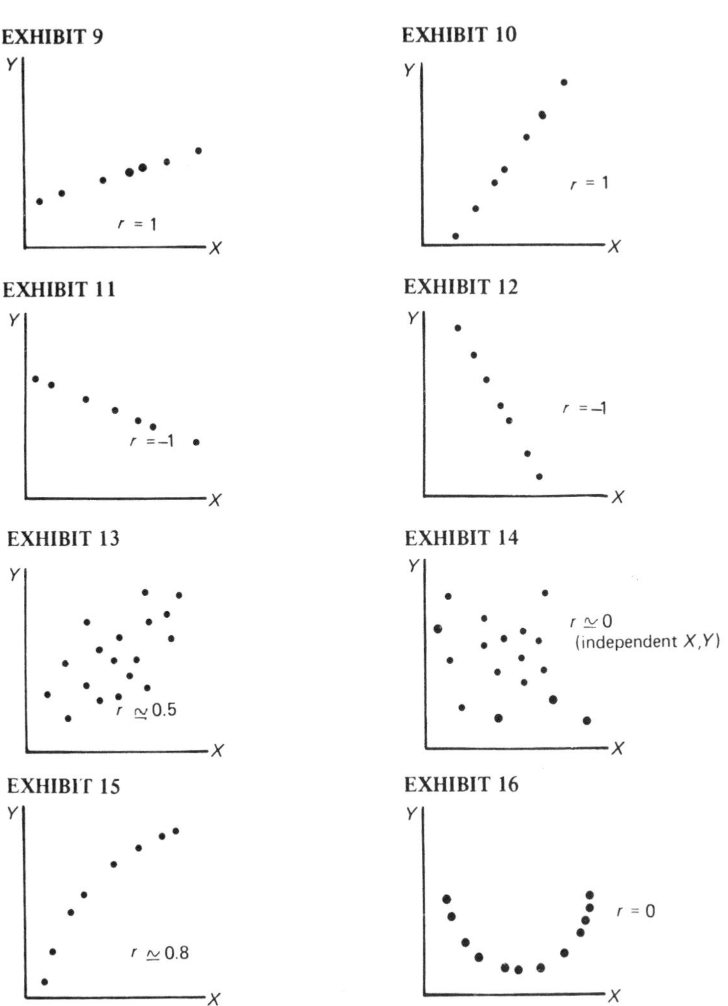

shown in each case. Note that where a perfect linear association is shown (Exhibits 9–12) the value of r is $+1$ or -1 ($+1$ for a direct association, -1 for an inverse association) irrespective of the slope of the line on which the points lie: this is simply a consequence of the independence of scale of r. Note too that in Exhibits 15 and 16 a perfect association exists but not a *linear* association; consequently r is neither $+1$ nor -1. In Exhibit 15 the association is nonetheless direct and hence r is positive, but in Exhibit 16 the association is neither direct nor inverse and so $r = 0$.

13.5 PRACTICAL CALCULATION OF CORRELATION COEFFICIENTS

We first develop an improved method for calculating r. The sample covariance, given in result (9) can be expressed equivalently and more conveniently as:

$$s_{XY} = \frac{1}{n}\sum XY - \bar{X}\bar{Y}$$

while the sample variances are given by:

$$s_X^2 = \frac{1}{n}\sum X^2 - \bar{X}^2$$

and

$$s_Y^2 = \frac{1}{n}\sum Y^2 - \bar{Y}^2$$

(Note the similarity in form between the expressions for the variance and the covariance). So we have

$$r = \frac{s_{XY}}{s_X s_Y} = \frac{(\Sigma XY)/n - \bar{X}\bar{Y}}{\sqrt{[(\Sigma X^2)/n - \bar{X}^2]\,[(\Sigma Y^2)/n - \bar{Y}^2]}}\,.$$

Multiplying the numerator and denominator by n, we arrive at the usual form for computation:

$$r = \frac{\Sigma XY - n\bar{X}\bar{Y}}{\sqrt{[\Sigma X^2 - n\bar{X}^2]\,[\Sigma Y^2 - n\bar{Y}^2]}}\,. \qquad (11)$$

where n is the number of paired observations.

Although appearing quite formidable the evaluation of r can be broken down into simple steps. Exhibit 17 shows the calculation of r for the data from which the scatter diagram in Exhibit 3 was drawn (concerning the sales of two competing models of cars, M and N). The resulting value of −0.62 shows that there is an inverse association, for the sample, between the sales of model M and model N (which we had concluded earlier from a visual inspection of Exhibit 3). But in this example it is not easy to extend the interpretation beyond this simple fact, as the result of the non-random selection of the data (sales in ten *consecutive* weeks). In other cases, however, where the sampling may be considered at random, it sometimes is possible to draw conclusions about association in the *population* by examining the value of r (CU20).

Equation (11) is a formula for calculating r with *unclassified* bivariate data. As with other statistics alternative forms can be developed for use with classified data, but these are more formidable still! The reader is advised to use Equation (11) even when the data is classified, by considering what the various summations mean in terms of the classified data (see Exercise 8).

Because the value of r (and ρ) is independent of the units of measurement of X and Y the values of X and the values of Y may be coded (see CU4) in any convenient linear fashion to simplify the calculations, and this process will still result in the same value for r.

EXHIBIT 17: Calculation of P.M. Correlation Coefficient, r

	Weekly Sales				
Week	*(Model M)* X	*(Model N)* Y	X^2	Y^2	XY
1	7	12	49	144	84
2	14	1	196	1	14
3	11	16	121	256	176
4	4	20	16	400	80
5	2	10	4	100	20
6	16	9	256	81	144
7	10	7	100	49	70
8	20	6	400	36	120
9	17	2	289	4	34
10	6	14	36	196	84
Totals	107	97	1,467	1,267	826

$n = 10$, $\bar{X} = 107/10 = 10.7$, $\bar{Y} = 97/10 = 9.7$

$$r = \frac{\Sigma XY - n\bar{X}\bar{Y}}{\sqrt{[\Sigma X^2 - n\bar{X}^2][\Sigma Y^2 - n\bar{Y}^2]}}$$

$$= \frac{826 - 10 \times 10.7 \times 9.7}{\sqrt{[1{,}467 - 10(10.7)^2][1{,}267 - 10(9.7)^2]}}$$

$$= \frac{-201.2}{\sqrt{322.1 \times 326.1}} = \frac{-201.2}{324.0} = -0.62$$

EXHIBIT 18: Calculation of Rank Correlation Coefficient, r'

X	Y	x'	y'	x'^2	y'^2	$x'y'$	$(x' - y')^2$
7	12	4	7	16	49	28	9
14	1	7	1	49	1	7	36
11	16	6	9	36	81	54	9
4	20	2	10	4	100	20	64
2	10	1	6	1	36	6	25
16	9	8	5	64	25	40	9
10	7	5	4	25	16	20	1
20	6	10	3	100	9	30	49
17	2	9	2	81	4	18	49
6	14	3	8	9	64	24	25
		55	55	385	385	247	276

$n = 10, \bar{x}' = 55/10 = 5.5, \bar{y}' = 55/10 = 5.5$

$$r' = \frac{\Sigma x'y' - n\bar{x}'\bar{y}'}{\sqrt{[\Sigma x'^2 - n\bar{x}'^2][\Sigma y'^2 - n\bar{y}'^2]}}$$

$$= \frac{247 - 10(5.5)^2}{\sqrt{[385 - 10(5.5)^2][385 - 10(5.5)^2]}}$$

$$= \frac{-55.5}{82.5} = -0.67$$

Alternatively,

$$r' = 1 - \frac{6\Sigma(x' - y')^2}{n(n^2 - 1)}$$

$$= 1 - \frac{6 \times 276}{10 \times 99}$$

$$= -0.67$$

13.6 RANK CORRELATION COEFFICIENT

In CU4 we made use of the idea (for the purpose of calculating the median) of arranging a set of data in increasing value (i.e. forming an ordered set). When the data is ordered in this way the first item (the smallest) is said to have *rank* 1, the second item rank 2, and so on up to the last item (the largest) which has rank n (where n denotes the sample size).

For small samples we can work out the ranks without physically rearranging the data. In Exhibit 18 we show the ranks of X, denoted by x', and the ranks of Y, denoted by y', for the same data (sales of competing cars) as was used for the calculation of the correlation coefficient r in Exhibit 17. In Exhibit 18, however, we calculate a correlation coefficient between the *ranks* of the data, x' and y'. The resulting statistic is known as the **rank correlation coefficient** and denoted by r' to avoid confusion with r. Many texts use the symbol r_s in recognition of Spearman its originator.

The rank correlation coefficient has a number of useful properties:

- Provided that the scatter diagram is *roughly* "cigar-shaped" its value approximates to the P.M. correlation coefficient (in our example $r' = -0.67$ while $r = -0.62$). This is useful because there is a short-cut method for calculating r', given by[1]

$$r' = 1 - \frac{6\Sigma(x' - y')^2}{n(n^2 - 1)} \tag{12}$$

  As shown in Exhibit 18 this enables us to calculate r' (and hence an approximate value for r) in a fraction of the time needed to calculate an exact value for r.
- In cases where the scatter diagram is decidedly *not* cigar-shaped the values of r and r' may differ substantially. Consider the scatter diagram of Exhibit 15 where we said that $r \simeq 0.8$: the rank correlation coefficient is exactly 1.0 since, for each observation $x' = y'$, a perfect correlation between the ranks. Thus r' is a more useful measure of association when the form of association is slightly non-linear, as may be the case in many industrial applications.
- We may use r' to examine how good a judge is in comparison with some known objective standard. Thus a wine taster could be asked to rank a number of wines in order of increasing alcohol content. These ranks may then be compared (by calculating r') with those obtained from a chemical analysis. (Note that it would only be possible to calculate r if the judges were given the much more difficult task of assessing the percentage of alcohol in each wine.)
- We may also use r' to compare the extent of agreement between

---

(1)The short-cut method is based on the property that $\Sigma x' = \Sigma y'$ and $\Sigma y'^2 = \Sigma x'^2$, as can be seen in Exhibit 18. It is an *exact* method for calculating r' provided there are no *tied* ranks (see later), but it is often used even if there are some ties.

two judges on some issue for which no measurement scale exists. In this way we might compare the judgement of car designers on the visual appeal of a number of model cars, by asking them to rank body styles in order of preference.

Finally, one difficulty frequently occurs in the calculation of r': consider the problem of ranking the following set of data:

(X) 10 14 6 8 6 4 8 8.

We see that the value 6 occurs twice and the value 8 three times. Where such *ties* occur we assign an *average* rank to the tied values in the following manner

(X)	10	14	6	8	6	4	8	8
(x')	7	8	2.5	5	2.5	1	5	5

Thus the two sixes are each given the rank $(2 + 3)/2 = 2.5$ and the three eights, $(4 + 5 + 6)/3 = 5$.

Ranking methods are becoming increasingly popular in modern statistical work. Not only do they have a wider range of application than conventional methods – it is often possible, for instance, to rank order individuals or objects where direct measurement is impossible or prohibitively expensive – but they are often simpler computationally. Another particularly important reason for their adoption is discussed in CU22.

REFERENCE

G.W. Snedecor & W.G. Cochran *Statistical Methods 6th Ed.* Iowa State, Chapter 7.

EXERCISES

1(P) Consider the following game:
Toss two coins and observe the number of heads. (Let X denote the variable "the number of heads shown": it can take values 0, 1 or 2.) Suppose that x heads are shown. Next, toss x coins and again count the number of heads shown (Y).

(i) Are the variables X and Y associated? What form will the association take and why?
(ii) Repeat the game 50 times, recording your results. Depict

your results graphically (you will have to overcome a fundamental difficulty in some way).

(iii) Calculate r and interpret the value you obtain.

(iv) Draw up a (bivariate) frequency distribution of your results by filling in the following "boxes" with their observed frequencies.

y \ x	0	1	2
0			
1			
2			

(v) By suggesting your own notation, produce a formula for r for bivariate data *classified* as in (iv). Check your answer by recalculating r from the table in (iv).

(vi) Draw up a tree diagram for the game and work out all possible joint probabilities (see CU7). Calculate the *expected* frequencies of all possible outcomes of the game and compare with the observed frequencies of part (iv). Comment on your results.

(vii)* From your tree diagram derive the probability distribution of X, the (marginal) probability distribution of Y and the probability distribution of $X + Y$. Evaluate the mean and variance of X, Y, and $X + Y$. Hence find the value of the "correction term" (Section 13.3) and show that the population correlation coefficient ρ is approximately 0.577. Compare this value with that of r you obtained from your sample.

(viii)* Evaluate ρ directly from equation (6) in the text.

2(U) Plot the following bivariate observations on separate scatter diagrams.

(i)		(ii)		(iii)		(iv)		(v)	
X	Y	X	Y	X	Y	X	Y	X	Y
4	3	1	3	3	2	1	0	0	6
2	2	3	2	1	2	4	6	5	1
1	2	3	1	4	2	3	4	6	0
1	1	1	4	2	2	2	2	1	5
3	4	2	3	0	2	5	8	3	2

Describe in words any indications of association you see. Guess the value of r in each case, and check your results by calculation.

3(E) A comparison of an Index of Wage Rates and Index of Productivity for 10 forms over a period of years gave the following results for 1966:

Index of Wage Rates	*Index of Productivity*
110	112
105	108
112	114
113	114
108	108
116	111
110	106
120	115
105	108
108	106

Calculate the "Product Moment" Correlation Coefficient between the two series.

Part question HND (Business Studies)

4(E) (a) What do you understand by the terms positive (direct) and negative (inverse) correlation? Give two illustrations of the use of a correlation coefficient in business.

(b) Calculate the Pearson, product-moment, correlation coefficient between the two series below, which relate to the aptitude test scores of 10 apprentices and their mark in a practical examination.

Apprentice	*Aptitude Test (Mark %)*	*Practical Examination Mark (%)*
A	66	58
B	73	67
C	67	72
D	74	82
E	70	70
F	79	84
G	68	66
H	75	90
I	70	69
J	80	80

(c) Comment on your findings.

HND (Business Studies)

5(E) The table below shows a traffic-flow index and the related site costs in respect of eight service stations of Universal Garages Ltd. You are required to

(a) calculate the coefficient of *rank* correlation for this data, and
(b) discuss briefly what your calculations indicate.

UNIVERSAL GARAGES LTD

Site Number	Traffic Flow Index	Site Cost £(000)
1	100	100
2	110	115
3	119	120
4	123	140
5	123	135
6	127	175
7	130	210
8	132	200

ACCA June, 1973

6(E) Fifteen pairs of non-twin brothers and sisters were given a "psychological inventory" questionnaire to complete. Their scores on the introversion-extroversion scale are given below.

Introversion-Extroversion Score

Brother:	110,	98,	116,	85,	113,	131,	89,	98
Sister:	108,	90,	110,	90,	107,	135,	93,	98

Brother:	107,	94,	116,	102,	120,	115,	88
Sister:	111,	103,	110,	106,	122,	117,	80

(a) Calculate the rank correlation coefficient between the brother's and sister's scores.

HND (Business Studies)

7(E) It is suggested that the performance of a particular task is likely to be improved by better education.

(a) Given for each of a large sample of adults a measurement of performance and the person's age of finishing full-time education, explain how you would analyse these data to throw light on this statement.
(b) Suppose that you find clear evidence of a positive association between education and performance, how would you answer the sceptic who suggests that both factors are really only indirect evidence of innate intelligence and that any attempt to improve educational levels in a particular community is therefore likely to be fruitless?

IPM November, 1973.

8(E) Evaluate the sample correlation coefficient r from the data in the following table.

	No. of Industrial Accidents in 1973/74		
	0	1	2 or more
Age of employee			
15–40	270	30	0
40–65	90	10	0

Number of Employees

Part question BA (Business Studies)

9(E) (a) If two variables are recorded for a set of objects, what is meant by the following statements:

(i) the two variables are positively associated, and
(ii) there exists a causal relation between the two

(b) As a result of standardised interviews, an assessment was made of the IQ and the attitude to the employing company of a group of six workers. The IQ's were expressed as whole numbers within the range 50–150 and the attitudes were assigned to five grades labelled 1, 2, 3, 4 and 5 in order of decreasing approval. The results obtained are summarised below.

Employee	A	B	C	D	E	F
IQ	127	85	94	138	104	70
Attitude score	2	4	3	1	2	5

Is there evidence of an association between the two attributes?

IPM November, 1972.

SOLUTIONS TO EVEN NUMBERED EXERCISES

2. (i) $r \simeq 0.77$ (ii) $r \simeq -0.88$ (iii) $r = 0$ (iv) $r = +1$ (v) $r = -1$

4.

Apprentice	*Aptitude Test* X	*Practical Mark* Y	XY	X^2	Y^2
A	66	58	3,828	4,356	3,364
B	73	67	4,891	5,329	4,489
C	67	72	4,824	4,489	5,184

4. (continued)

Apprentice	*Aptitude Test* X	*Practical Mark* Y	XY	X^2	Y^2
D	74	82	6,068	5,476	6,724
E	70	70	4,900	4,900	4,900
F	79	84	6,636	6,241	7,056
G	68	66	4,488	4,624	4,356
H	75	90	6,750	5,625	8,100
I	70	69	4,830	4,900	4,761
J	80	80	6,400	6,400	6,400
	722	738	53,615	52,340	55,334

$n = 10$

$\bar{X} = \Sigma X/n = 722/10 = 72.2$

$\bar{Y} = \Sigma Y/n = 738/10 = 73.8$

$$r = \frac{\Sigma XY - n\bar{X}\bar{Y}}{\sqrt{(\Sigma X^2 - n\bar{X}^2)(\Sigma Y^2 - n\bar{Y}^2)}}$$

$$= \frac{53{,}615 - 10 \times 72.2 \times 73.8}{\sqrt{(52{,}340 - 10 \times 72.2^2)(55{,}334 - 10 \times 73.8^2)}} = 0.77$$

N.B. It is possible to reduce the volume of calculation when no machine is available by choosing an "assumed average" in either or both variables. We suggest you subtract 72 from each X value and 74 from each Y value and rework the example. You will find that you have less arithmetic to do and you will obtain exactly the same answer.

6.

Brother X	x'	*Sister* Y	y'	$x' - y'$	$(x' - y')^2$
110	9	108	9	0	0
98	$5\frac{1}{2}$	90	$2\frac{1}{2}$	3	9
116	$12\frac{1}{2}$	110	$10\frac{1}{2}$	2	4
85	1	90	$2\frac{1}{2}$	−1.5	2.25
113	10	107	8	2	4
131	15	135	15	0	0
89	3	93	4	−1	1
98	$5\frac{1}{2}$	98	5	0.5	0.25
107	8	111	12	−4	16
94	4	103	6	−2	4
116	$12\frac{1}{2}$	110	$10\frac{1}{2}$	2	4
102	7	106	7	0	0
120	14	122	14	0	0
115	11	117	13	−2	4
88	2	80	1	1	1
				$\Sigma(x' - y')^2 =$	49.50

$$r' = 1 - \frac{6\Sigma(x' - y')^2}{n(n^2 - 1)} = 1 - \frac{6 \times 49.50}{15(225 - 1)} = +0.9116$$

8. The proportions of employees in the age ranges 15–40 and 40–65 who have had one accident are 30/(270 + 30) = 0.1 and 10/(90 + 10) = 0.1, respectively. As these proportions are equal there is no association between age (as defined) and the number of accidents i.e. $r = 0$. We can show this formally by calculating r. Denoting the number of accidents by y and the age group 15–40 by the coded value of $x = 0$, and for age 40–65 by $x = 1$, then:

$$\text{“}\Sigma XY\text{”} = \Sigma xyf_{xy} = 0 \times 0 \times 270 + 0 \times 1 \times 30 + 1 \times 0 \times 90 + 1 \times 1 \times 10 = 10$$

$$\text{“}\Sigma X^2\text{”} = \Sigma x^2 f_x = 0 \times 360 + 1 \times 40 = 40$$

$$\text{“}\Sigma Y^2\text{”} = \Sigma y^2 f_y = 0 \times 300 + 1 \times 100 = 100$$

$$\text{“}\Sigma X\text{”} = \Sigma xf_x = 0 \times 360 + 1 \times 40 = 40$$

$$\text{“}\Sigma Y\text{”} = \Sigma yf_y = 0 \times 300 + 1 \times 100 = 100$$

So

$$\bar{X} = 40/100 = 0.1$$

and

$$\bar{Y} = 100/400 = 0.25,$$

and

$$r = \frac{\Sigma XY - n\bar{X}\bar{Y}}{\sqrt{(\Sigma X^2 - n\bar{X}^2)(\Sigma Y^2 - n\bar{Y}^2)}}$$

$$= \frac{10 - 400 \times 0.1 \times 0.25}{\sqrt{(40 - 400 \times 0.1^2)(100 - 400 \times 0.25^2)}} = 0$$

COURSE UNIT 14 – LINEAR REGRESSION

14.1 INTRODUCTION

In CU13 we saw that if the correlation coefficient r of a set of bivariate data was close to +1 (or −1) then the points on a scatter diagram would cluster around a straight line But so far we have not attempted to draw the straight line itself on a scatter diagram. This prompts two questions:

- Is there a particular straight line one can plot on a scatter diagram around which the points cluster more closely than any other?
- What is the physical interpretation and use of such a straight line?

It suits our purposes to answer the second question immediately (on the assumption that the answer to the first is "yes") and return to the first question later.

Consider the straight line in Exhibit 1, superimposed on a scatter diagram drawn up for a sample of annual maintenance costs during various "years of life" of cars in a company's fleet (all the same model). We note in particular that only one point actually lies on the straight line in this case.

EXHIBIT 1: Annual Maintenance Costs vs. Year of Life

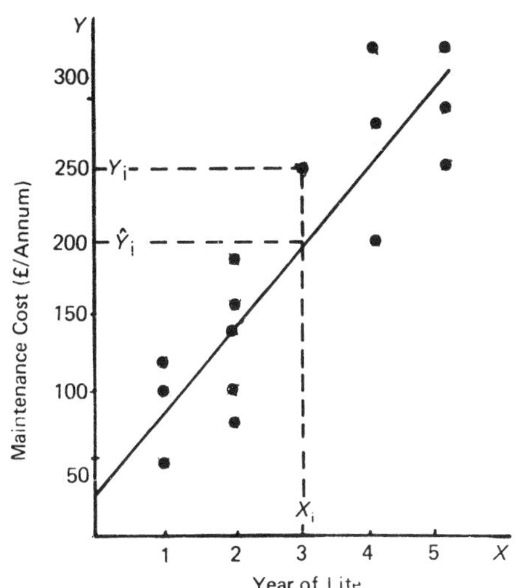

A tentative interpretation of the straight line is that it is an estimate of the line on which all the points *would lie* if it were possible to eliminate all random effects on maintenance costs arising as the result of variations in mileage and driver skills, manufacturing variations in quality, and so on. It should be clear that, given this interpretation, the straight line could be used for the purposes of *predicting* the maintenance costs of another car in the fleet (or a similar car yet to be bought) at each year of its life, in conditions where we have no knowledge of the influence of such random effects.

Consider, for example, predicting the maintenance costs of a car in the third year of its life ($X = 3$). The "predicted" value from the straight line, $\hat{Y}$, is £200. This is in marked contrast with the single sample result in the same year of life ($X_i = 3$) which gave an actual maintenance cost $Y_i = 250$. It is as if our *belief* in the existence of a straight line relationship between maintenance cost and year of life has enabled us to ignore the only directly comparable observation in our sample[(1)]. That being so, we can interpret the *difference* between the actual and predicted values of maintenance costs for each observation ($= Y_i - \hat{Y}_i$ the so-called "Y-error") as being due to the previously mentioned random effects.

Thus for our observation $X_i = 3$, $Y_i = 250$, we have a Y-error of £50 (= 250 − 200) which might be attributable to a higher than average mileage, a worse than average driver, a "poor" car, etc. Using the convention that we always calculate the Y-errors in the form $(Y_i - \hat{Y}_i)$, points *above* the straight line, having *higher* than predicted maintenance costs, will have *positive* Y-errors; those points below the line, with *lower* maintenance costs than predicted, will have *negative* Y-errors.

A line drawn on a scatter diagram in this way is called a **regression line**, and the associated analysis, **regression analysis**. We shall confine our attention to straight lines (*linear* regression), although in certain circumstances curves might be more appropriate[(2)] (*curvilinear* regression). Again, we shall deal only with *bivariate* regression (predicting Y from X) rather than the more general *multivariate* regression (predicting Y from a number of other variables as, for example, predicting maintenance costs from age *and* mileage).

Prediction is the main use of regression analysis. By using a regres-

---

(1) We have not ignored this observation completely, of course; it has played its part in forming the overall pattern on the scatter diagram to which the straight line has been "fitted".

(2) In a study of the relationship between quantity sold and price a *linear* regression would be inappropriate because it would predict *negative* quantities sold at a high price level!

sion line we can make predictions (of Y values) when the amount of *strictly comparable* data (i.e. having the same X values) is limited, and even when there is no strictly comparable data at all – we could, for instance, have predicted the maintenance costs of a three year old car even if there were no three year old car in our sample. Naturally, such uses of regression are not without their attendant risks, but we leave a discussion of these until later.

14.2 "GUESSES" AND "LEAST-SQUARES" LINES

Of the many straight lines which *could* be drawn on a scatter diagram for the purposes we have discussed, which one do we choose? It is obviously desirable to have some objective standard by which such lines are drawn in order that different workers, using the same set of (bivariate) data, all arrive at the same result, and that just comparisons may be made between different sets of data.

We approach this problem by taking a deliberately exaggerated example where the answer to "which line" is not at all obvious. A mail order firm receives order forms from customers, each form listing the number of items ordered (X). For each order, a packer gathers together the various items, carries out various administrative procedures (checking the arithmetic and the details on the enclosed payment, etc.), and packs the items in a parcel. The length of time (Y, in minutes) taken to complete an order is given for a sample of ten orders as follows:

X (No. of items in order)	8	11	8	11	4	6	12	9	6	5
Y (Time taken)	6	8	8	9	5	4	9	6	8	7

The mean of the X values, $\bar{X} = 80/10 = 8.0$; the mean of the Y values, $\bar{Y} = 70/10 = 7.0$. The data is displayed as a scatter diagram in Exhibit 2 on which the bivariate mean ($\bar{X} = 8.0$, $\bar{Y} = 7.0$) has been indicated by a circled point.

Our first clue, as to the "best" line, is given by considering the following question: "For an X value equal to the sample mean of X (8.0), what would one expect the predicted value of Y to be?" The answer must be the sample mean of Y, $\bar{Y} = 7.0$, if only because it is very difficult to find a convincing argument for the predicted value of Y being any other value (given that we are relying on *sample* evidence). The implication of this is that our regression line must pass through the bivariate mean – Exhibit 3 shows a few examples. Such lines have an interesting property: for any line passing through the bivariate mean the algebraic sum of the Y-errors will be zero; that is, the extent of over-prediction for some points (below the line) will exactly balance

EXHIBIT 2: Scatter Diagram Showing Bivariate Mean

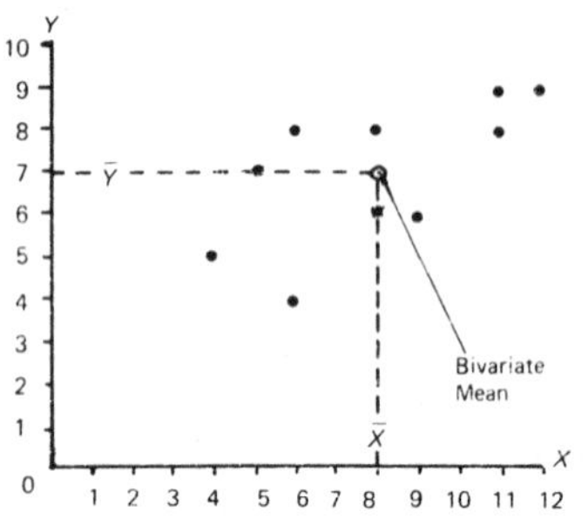

EXHIBIT 3: Which Line is "Best"?

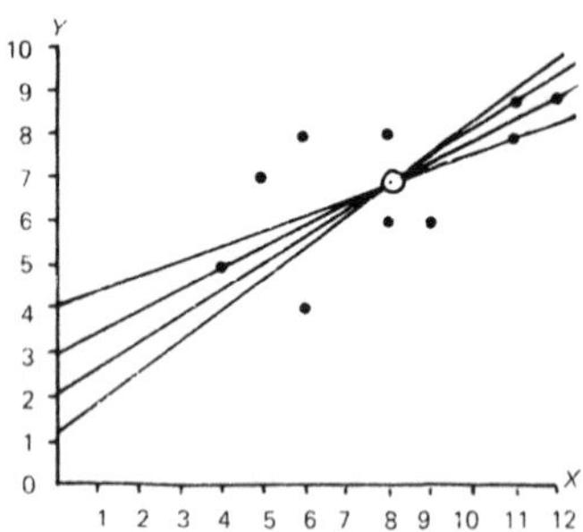

EXHIBIT 4: The Authors' Guess (and *Y*-errors)

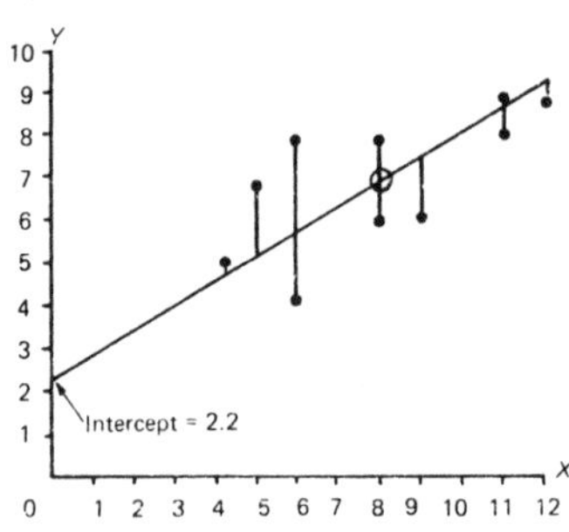

EXHIBIT 5: The Authors' Guess (and Squared *Y*-errors)

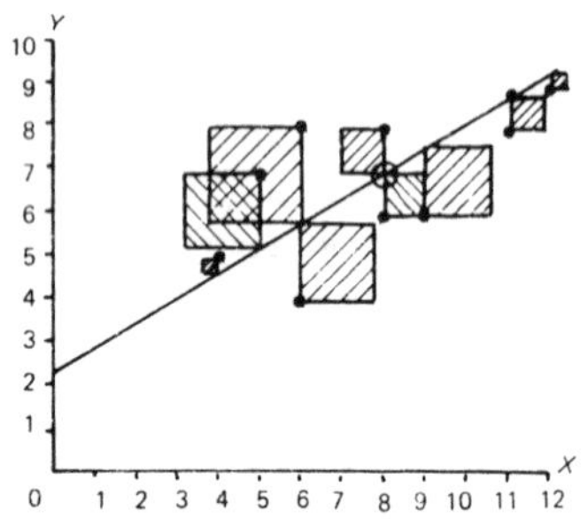

EXHIBIT 6: "Least Squares" Line (and *Y*-errors)

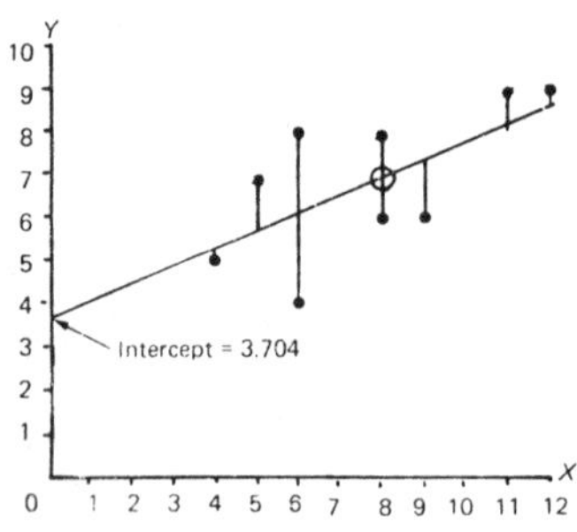

EXHIBIT 7: "Least Squares" Line (and Squared *Y*-errors)

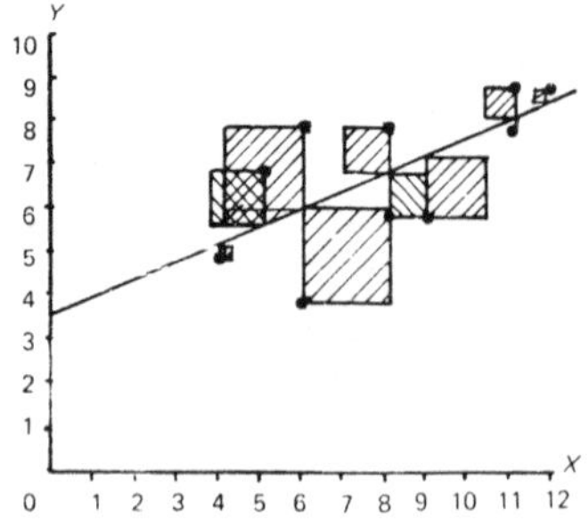

out the under-prediction of the others (above the line) throughout the sample[(1)]. But this idea does not help us a great deal since there are still an infinite number of lines which pass through the bivariate mean: which one of *these* is "best"? In the rest of this section we consider two straight lines – one, the authors' guess, fitted "by eye", and the other, the so-called "least squares" regression line, in order to make some important comparisons.

The authors' guess of the best-fitting line passing through the bivariate mean is shown in Exhibit 4 (the vertical lines represent Y-errors). We note that the intercept on the y-axis is 2.2 and the "slope" of the line, given by the increase in Y values corresponding to a *unit* increase in X values, is 0.6[(2)]. The equation of the line is therefore $\hat{Y} = 2.2 + 0.6X$. (This may be checked readily by substituting $X = \bar{X} = 8$, giving $\hat{Y} = 2.2 + 0.6(8) = 7$, the value of $\bar{Y}$.)

In Exhibit 8(a), we conduct an "error analysis" on the authors' guessed line. Firstly, the predicted value $\hat{Y}_i$ is calculated for each observation by using the regression equation in the form $\hat{Y}_i = 2.2 + 0.6X_i$ (an alternative would be to read the values off an accurately drawn graph). The Y-errors are then calculated and their squares – squared Y-errors – formed. As expected, we find that the sum of the Y-errors is zero. The

---

(1) Representing the regression line by $\hat{Y} = a + bX$, where a is the intercept on the y-axis and b the "slope", the requirement that the line must pass through the bivariate mean implies that:

$$\bar{Y} = a + b\bar{X}$$

or

$$\frac{1}{n}\Sigma Y_i = a + \frac{b}{n}\Sigma X_i$$

or

$$\Sigma Y_i = na + b\Sigma X_i$$

or

$$\Sigma(Y_i - a - bX_i) = 0 \qquad \text{(i)}$$

But the predicted value of Y, $\hat{Y}_i$, corresponding to an X_i is given by $\hat{Y}_i = a + bX_i$, and so (i) can be rewritten as:

$$\Sigma(Y_i - \hat{Y}_i) = 0,$$

giving the result that the sum of the Y-errors is zero.

(2) The line passes through the bivariate mean (8,7) and the intercept on the y-axis has co-ordinates (0, 2.2). Hence an increase in Y values of 7 – 2.2 = 4.8 corresponds to an increase in X values of 8 – 0 = 8. The slope is therefore

$$\frac{7 - 2.2}{8 - 0} = \frac{4.8}{8} = 0.6.$$

squared Y-errors, whose values are represented graphically as the shaded areas in Exhibit 5, sum to 16.88. This figure provides a measure, for a particular set of data, of how well the line "fits" the data. (The reader should convince himself of this by drawing up a diagram similar to Exhibit 5 with a line which is obviously a very bad fit to the data, and comparing the shaded areas formed.)

EXHIBIT 8: Error Analysis of "Guess" and "Least-Squares" Lines

		(a) "Guess": $\hat{Y} = 2.2 + 0.6X$			*(b)* "Least-Squares": $\hat{Y} = 3.704 + 0.412X$		
X_i	Y_i	$\hat{Y}_i$	*Y-error* $(= Y_i - \hat{Y}_i)$	*Squared Y-error* $(Y_i - \hat{Y}_i)^2$	$\hat{Y}_i$	*Y-error* $(= Y_i - \hat{Y}_i)$	*Squared Y-error* $(Y_i - \hat{Y}_i)^2$
8	6	7.00	−1.00	1.00	7.00	−1.00	1.00
11	8	8.80	−0.80	0.64	8.24	−0.24	0.06
8	8	7.00	1.00	1.00	7.00	1.00	1.00
11	9	8.80	0.20	0.04	8.23	0.77	0.58
4	5	4.60	0.40	0.16	5.35	−0.35	0.12
6	4	5.80	−1.80	3.24	6.18	−2.18	4.75
12	9	9.40	−0.40	0.16	8.65	0.35	0.12
9	6	7.60	−1.60	2.56	7.41	−1.41	1.99
6	8	5.80	2.20	4.84	6.18	1.82	3.31
5	7	5.20	1.80	3.24	5.76	1.24	1.54
		Totals	0.00	16.88		0.00	14.47

We now pose the final question of which line, drawn through the bivariate mean, makes the total squared Y-errors as small as possible; and hence the "fit" as good as possible. The answer is provided by the so-called "least-squares regression line of Y on X" (the reason for "Y on X" is that we are using the line to predict Y *from* X – a matter to be taken up again in Section 14.4). Leaving aside details of the calculations involved until Section 14.3, we quote the result for the equation of this regression line:

$$\hat{Y} = 3.704 + 0.412X.$$

An error analysis of this regression line is given in Exhibit 8(b), alongside that for the "guessed" line, and is illustrated graphically in Exhibits 6 and 7. The sum of squared Y-errors has the value 14.47; this is smaller than that given by the authors' guess; it is also smaller than any other straight line, whether drawn through the bivariate mean or not. By common assent, and a considerable amount of theoretical justification, the least-squares regression line of Y on X is the "best" regression line for predicting Y from X in statistical work.

14.3 CALCULATING AND INTERPRETING THE REGRESSION LINE

In more mathematically oriented texts it is proved that the slope b, of the least-squares regression line (of Y on X), $\hat{Y} = a + bX$, is given by:

$$b = \frac{\text{Sample Covariance of } X \text{ and } Y}{\text{Sample Variance of } X} = \frac{s_{XY}}{s_X^2}. \qquad (1)$$

Following the procedure adopted in Section 13.5 we can re-write this in the following convenient form for computation:

$$b = \frac{\Sigma XY - n\bar{X}\bar{Y}}{\Sigma X^2 - n\bar{X}^2}, \qquad (2)$$

the slope of the regression line of Y on X.

Note that the slope b has "units" Y/X – a fact that assists in the memorising of the formula[1].

EXHIBIT 9: Layout for Regression Calculation

X	Y	X^2	Y^2	XY
8	6	64	36	48
11	8	121	64	88
8	8	64	64	64
11	9	121	81	99
4	5	16	25	20
6	4	36	16	24
12	9	144	81	108
9	6	81	36	54
6	8	36	64	48
5	7	25	49	35
$\Sigma X = 80$	$\Sigma Y = 70$	$\Sigma X^2 = 708$	$\Sigma Y^2 = 516$	$\Sigma XY = 588$

N.B. The Y^2 column is not used for evaluating b, but is included in this table for later use.

---

[1] Another memory aid, which has a mathematical justification, is as follows:

Write down the regression equation:	$Y = a + bX$	(i)
Multiply (i) by X:	$XY = aX + bX^2$	(ii)
"Operate" on (i) and (ii), using Σ giving	$\Sigma Y = na + b\Sigma X$	(iii)
and	$\Sigma XY = a\Sigma X + b\Sigma X^2$	(iv)

Finally, solve equations (iii) and (iv) simultaneously for b, giving the result shown in the text.

Exhibit 9 shows the calculations of the various summations required for our mail-order data, which by substitution into (2) give:

$$b = \frac{588 - 10 \times 8.0 \times 7.0}{708 - 10 \times 8.0^2} = \frac{28}{68} = 0.412$$

To obtain a, the intercept on the y-axis, we remember that the regression line must pass through the bivariate mean ($\bar{X} = 8$, $\bar{Y} = 7$), and so $\bar{Y} = a + b\bar{X}$. Solving for a we obtain:

$$a = \bar{Y} - b\bar{X} = 7 - 0.412 \times 8 = 3.704.$$

Thus our regression equation is $\hat{Y} = 3.704 + 0.412X$, which we earlier quoted in Section 14.2. It is most conveniently drawn on the scatter diagram (Exhibits 6 and 7) by joining the bivariate mean to the intercept on the y-axis.

We can interpret this regression line in the following way. The intercept $a = 3.704$ (mins.) represents, for our mail-order problem, that component of the total time involved in processing an order which is independent of the number of items ordered, and corresponds to such activities as gathering the packaging materials, writing the address label, weighing the completed package and affixing the appropriate postage. The slope $b = 0.412$ (mins.) represents the additional time involved *per item* in the order and attributable to the time spent physically selecting an item from stock, checking that the correct price has been entered on the order form, etc. Using the regression line to predict the time taken to complete an order involving, for example, 10 items, we obtain

$$\hat{Y} = 3.704 + 0.412(10) \simeq 7.8 \text{ (mins.)}$$

The procedure we have given for calculating the regression equation (by hand) involves some unwieldy arithmetic if the X_i and Y_i values are large or given to more than two significant figures. Some simplification can be brought about by subtracting a constant from each X_i and another (maybe different) constant from each Y_i. The value of b is unchanged by this, as the *slope* of the regression line is unaffected by a simple change of origin. The value of a, however, must be calculated using the mean values ($\bar{X}$ and $\bar{Y}$) of the original observations.

14.4* ANOTHER REGRESSION LINE AND ADDITIONAL INTERPRETATIONS OF THE CORRELATION COEFFICIENT

Suppose, in our mail order example, that an order had just been made up, taking 8 minutes, and that we wished to estimate the number of items it contained (X). (This may seem rather a strange request; but it does not sound too unreasonable if we assume that the customer's order form is in the completed package and thus there is no remaining record of the number of items it contained.) Although we *could* use the regression line of Y on X for this purpose, it would be quite inappropriate. The reason is that we are now attempting to predict X from Y (rather than Y from X) and we are concerned with the X-errors involved (rather than Y-errors), as illustrated in Exhibit 10.

EXHIBIT 10: Regression Line of *X* on *Y* Showing *X*-Errors

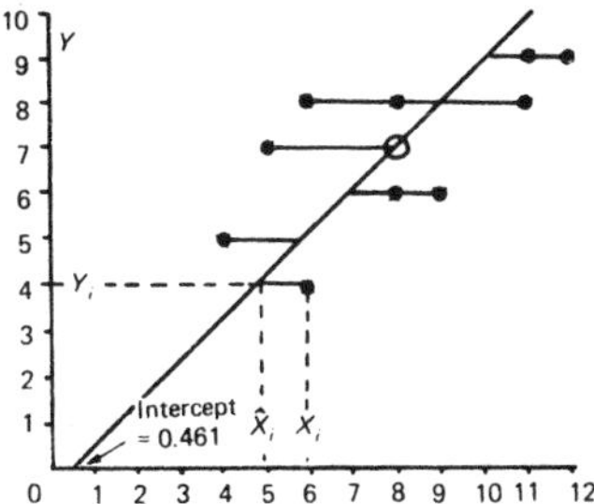

EXHIBIT 11: The Three Regression Lines Compared

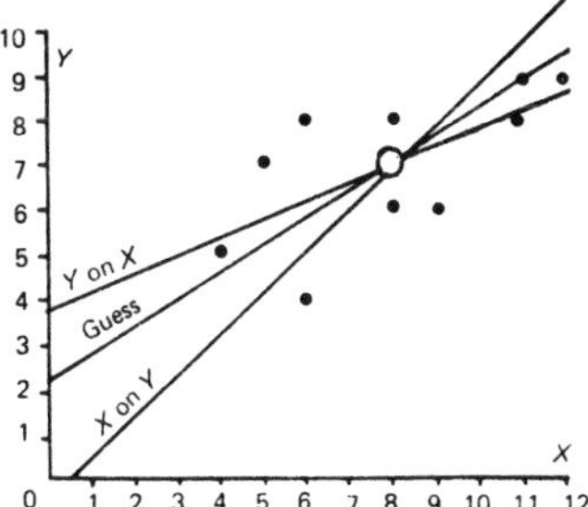

What is needed for this purpose is the regression line of X *on* Y, which (to avoid confusion with the regression line of Y on X) we write as

$$\hat{X} = a' + b'Y,$$

which minimises the sum of the squared X-errors. (Note that a' is the intercept on the x-axis and b' the slope with respect to the y-axis.) The value of b' is obtained quite simply by interchanging X and Y in Formulae (1) and (2), giving:

$$b' = \frac{\text{Sample Covariance of } X \text{ and } Y}{\text{Sample Variance of } Y} = \frac{s_{XY}}{s_Y^2}, \tag{3}$$

and

$$b' = \frac{\Sigma XY - n\overline{X}\overline{Y}}{\Sigma Y^2 - n\overline{Y}^2}, \tag{4}$$

the slope (with respect to the y-axis) of the regression line of X on Y.

Numerically, we have

$$b' = \frac{588 - 10 \times 8.0 \times 7.0}{516 - 10 \times 7.0^2} = 1.077.$$

As the regression line passes through the bivariate mean we have $\bar{X} = a' + b'\bar{Y}$, and so $a' = (\bar{X} - b'\bar{Y}) = (8 - 1.077 \times 7) = 0.461$. Finally then, our regression line is $\hat{X} = 0.461 + 1.077Y$ and it is this line that has been drawn on Exhibit 10, by joining the bivariate mean to the intercept on the x-axis. We can see that the predicted number of items in a package that took 8 minutes to make is given by:

$$\hat{X} = 0.461 + 1.077 \times 8 = 9.077,$$

which we would interpret realistically as 9 items. (Compare this with the prediction that would be made from the regression line of Y *on* X, 10.43.)

In Exhibit 11 we compare the regression of Y on X, X on Y and the authors' guess. (It is true in general that the regression line of X on Y makes a steeper angle with the x-axis than the regression line of Y on X.) We can now perhaps see why the authors' guess falls between the two "least-squares" lines. In making their guess, the authors have been concerned to achieve a good "fit", but have not considered the use to which the line would be put (predicting Y from X or alternatively, X from Y). Hence their line is in some sense a compromise between the two calculated regression lines, each one having a distinct use.

It is instructive to enquire under what conditions the regression lines of Y on X and X on Y coincide. As both lines pass through the bivariate mean they would simply need to have the same slope (with respect to the x-axis) to do this. Now we can re-write the regression line of X on Y, $\hat{X} = a' + b'Y$, as $Y = (-a'/b') + (1/b')X$, showing that the slope with respect to the x-axis is $(1/b')$. So the two regression lines coincide if $b = (1/b')$ or $bb' = 1$. Quite generally, the product of b and b' is given by:

$$bb' = \frac{s_{XY}^2}{s_X^2 s_Y^2} = r^2, \tag{5}$$

where r is the *correlation* coefficient[1] (CU13). So we see that the regression lines coincide only in cases where $r^2 = 1$. This is hardly a surprising result, since we found in CU13 that $r = \pm 1$ (and hence $r^2 = 1$)

[1] In our example we have $b = 0.412$ and $b' = 1.077$. Hence $r^2 = (0.412)(1.077) = 0.444$. When taking the square root of this value, in order to obtain the correlation coefficient, we must take care with the sign. We see from the scatter diagram that the association is *positive*, so $r = +\sqrt{0.444} = 0.666$.

only if the points on a scatter diagram lie *exactly* on a straight line. This idea leads us to an interpretation of r^2 (called the **coefficient of determination** – as distinct from r, the correlation coefficient) as a measure of the "goodness of fit" of the least-squares regression lines – the extreme value of $r^2 = 1$ indicating a perfect fit and $r^2 = 0$[1], a worthless fit.

Another interpretation of r^2 is given by a consideration of the amount of variation "explained" by the regression: here we shall confine our argument to the regression of Y on X. Taking the Y_i values alone, the "sum of squared deviations from the mean, $\bar{Y}$" is given by:

$$\sum (Y_i - \bar{Y})^2 = \sum Y_i^2 - n\bar{Y}^2 = 516 - 10 \times 7.0^2 = 26,$$

in our example. This quantity is a measure of "total variation" in the Y values, and which by division by n gives the sample variance. Now consider the following identity (true for *any* value of r^2):

$$\underset{\text{Total Variation}}{\sum(Y_i - \bar{Y})^2} = \underset{\text{"Explained" Variation}}{r^2\sum(Y_i - \bar{Y})^2} + \underset{\text{Residual Variation}}{(1 - r^2)\sum(Y_i - \bar{Y})^2} \qquad (6)$$

Evaluating the "residual variation", we obtain

$$(1 - r^2)\sum(Y_i - \bar{Y})^2 = (1 - 0.444)(26) = 14.46$$

This value is equal (apart from rounding errors) to the sum of the squared Y-errors of the regression line of Y on X (see Exhibit 8) and corresponds to the variation (about the regression line) that remains after regression. It is for this reason that the other term, $r^2\,\Sigma(Y_i - \bar{Y})^2 = 11.54$, is called the "explained" variation – it is the amount by which the total variation is *reduced* by the regression. Clearly then, r^2 has the interpretation of the proportion of variation explained by the regression: if $r^2 = 1$, it is *all* explained; if $r^2 = 0$, none is explained; in our example, 44.4% (= 0.444) of the total variation is explained.

14.5* POPULATION RESULTS (Y ON X ONLY)

Given a *population* of bivariate observations we could obtain the "true" regression line, $\hat{Y} = \alpha + \beta X$, in a manner similar to that used for a sample:

---

[1] If $r^2 = 0$ the two regression lines are parallel to the respective axes and perpendicular to each other. This shows that the regression is meaningless in such a case.

$$\beta = \frac{\sigma_{XY}}{\sigma_X^2} \text{ and } \alpha = \mu_Y - \beta\mu_X.$$

How close are the *sample* regression coefficients, the a and b of $Y = a + bX$, to the true values of α and β? To begin to answer such questions we need to assume a particular model for the relationship between X_i and Y_i. A commonly used model is as follows:

$$Y_i = \alpha + \beta X_i + e_i,$$

where the e_i's are *independent* "errors", all drawn from the *same* distribution, with mean value zero and variance[1] σ_e^2. The simplest interpretation of this model is that the errors arise in connexion with the Y_i values, not the X_i values (which, it is assumed, can be measured without error). Although the model makes no assumptions about the *cause* of the errors – they may arise from imprecise measurement of the Y_i's and/or a number of other random influences – the assumption that they are independent (they arise, as it were, from a process of random sampling) is particularly important.

Examining these assumptions in the light of our mail order example, it is certainly reasonable that the X_i values (number of items in the order) can be measured without error. The "errors" in the Y_i values will arise out of the greater or lesser difficulty (and hence, time) involved in collecting the individual items from stock and packing them. But is it reasonable to assume, as the model requires, that such errors are independent and "drawn from the same distribution"? Both assumptions seem rather suspect: it could be that customers placing *large* orders (high X_i's) will have a greater tendency to order some "unusual" items which may be stocked in less accessible places; and this alone would mean that the errors tend to increase in size as X_i increases. (The only satisfactory way of examining these assumptions is to carry out a number of tests on the errors $(Y_i - \hat{Y}_i)$ from a large sample regression, but such tests are beyond the scope of this course.)

---

(1) We can *estimate* σ_e^2 from a sample regression as:

$$\hat{\sigma}_e^2 = \frac{1}{n-2}\Sigma(Y_i - \hat{Y}_i)^2$$

(The "$n - 2$" division gives an unbiased estimate.)
Instead of evaluating $\Sigma(Y_i - \hat{Y}_i)^2$ directly, it is easier to calculate the equivalent residual variation, $(1 - r^2)\,\Sigma(Y_i - \bar{Y})^2$, giving

$$\hat{\sigma}_e^2 = (1 - r^2)\,\Sigma(Y_i - \bar{Y})^2/(n-2)$$

In our mail order example this becomes $\hat{\sigma}_e^2 = (1 - 0.444)(26)/8 = 14.46/8 = 1.807$.

If the assumptions we have discussed *are* valid we can identify two separate sources of inaccuracy in the sample regression line – the extent to which a and b differ from α and β respectively. We may assess the magnitude of such errors by the use of the following theoretical results:

$$E(a) = \alpha \quad \text{or } \hat{\alpha} = a \tag{5}$$

$$E(b) = \beta \quad \text{or } \hat{\beta} = b \tag{6}$$

$$\text{Var}(a) = \sigma_e^2 \left\{ \frac{1}{n} + \frac{\bar{X}^2}{\Sigma(X_i - \bar{X})^2} \right\} \tag{7}$$

$$\text{Var}(b) = \frac{\sigma_e^2}{\Sigma(X_i - \bar{X})^2} \tag{8}$$

Results (5) and (6) show that the sample results a and b are unbiased estimates of the true population values – so "on average" the sample regression line is accurate. Results (7) and (8) show, in particular, that the larger the spread of the X_i values, and hence the larger the value of $\Sigma(X_i - \bar{X})^2$, the more accurate a and b will be. (This makes sense since it is clearly impossible to obtain a regression line at all if the X_i values are all the *same*.) We shall make use of these results in later Course Units (CU 16 and 20) when making judgements about the accuracy of regression lines and the accuracy of predictions made with their use.

14.6 SOME PRACTICAL POINTS ON REGRESSION

Before we carry out a regression analysis on a bivariate set of data, we must first decide which is the predict*or* variable (the variable from which the predictions will be made) – let us call this variable X. We shall then be interested in the regression of Y on X.

The next step is to plot the data in the form of a scatter diagram: the purpose of this is to satisfy ourselves that it is reasonable to employ *linear* regression. If the scatter diagram suggests a *non-linear* relationship between the variables, it is nonetheless sometimes still possible, by suitably "transforming" one (or both) of the variables before carrying out the analysis, to retain the technique of *linear* regression. Exhibit 12 shows a typical example where the scatter diagram indicates a non-linear relationship between the variables X and Y. When the X_i values are transformed by taking their logarithms (Exhibit 13), a linear relationship becomes apparent, and so we could proceed with a linear regression analysis of Y on $Log(X)$ with a fair measure of confidence.

EXHIBIT 12: A Non-Linear Relationship

EXHIBIT 13: The Same Data with Transformed X Values

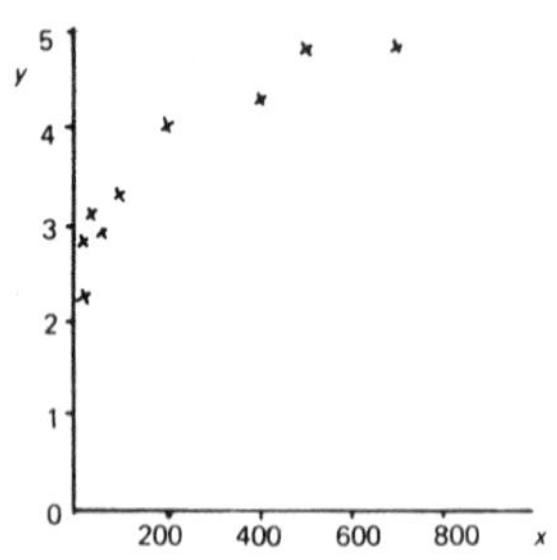

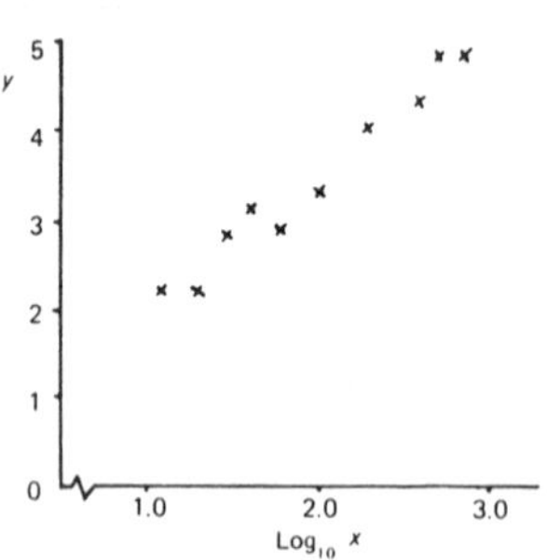

When using regression for the purposes of prediction we run a number of risks, some of which may be summarised as follows:

(a) The regression line $\hat{Y} = a + bX$ is based on a bivariate *sample* and serves as an estimate of the "true" population regression line $\hat{Y} = \alpha + \beta X$ (Section 14.5). If our sample happens to be unrepresentative of the population (as the result of non-random sampling, or simply by chance) this will lead to inaccurate predictions. In CU16 we shall see how we can make allowances for such sample effects under certain conditions.

(b) The predictions that are made are effectively of the *expected value* of the Y variable corresponding to a given X value. Thus our prediction, in Section 14.3, that an order consisting of 10 items would take 7.8 mins. to complete is a statement about the *average* time taken for all orders involving 10 items. Although this is the best prediction that can be made of the time taken for the *next* order involving that number of items, the individual characteristics of that order (random effects) will mean that the actual time taken may be more or less than 7.8 mins. (Again CU16 gives details of how we may allow for this.)

(c) We have concentrated on *linear* regression – straight lines. The "true" relationship between X and Y may not be linear at all; but yet we may not be able to detect this on the basis of a small sample. Hence our predictions may be in error as the result of using the *wrong model*.

The risk mentioned in (c) brings us to an important distinction in the use of regression. The point is that the relationship between two variables X and Y may often be *considered* linear over the restricted range of X values of the sample data (see Exhibit 13). Provided that our predictions are made for X values *within the same range* (interpolation) the risk of using the wrong model may well be acceptably low. If we move outside this range of X values (extrapolation) the risk increases immeasurably. Such predictions should be treated with the utmost caution.

EXHIBIT 13: Interpolation and Extrapolation

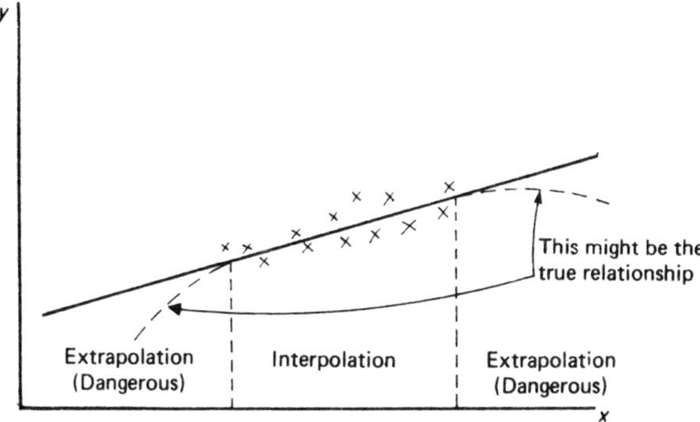

Finally, we give two examples of the use of regression analysis for decision making. The regression analysis of annual maintenance costs (Section 14.1) enabled us to predict the (average) cost during each year of life of cars in a fleet. By a similar analysis we could predict the annual depreciation (change in resale value) during each year of life. These two costs, maintenance and depreciation, represent the two costs that can be *controlled* by a policy of replacing fleet cars at a chosen age. Thus the two analyses would enable the company to decide what the "best" replacement age is – perhaps by choosing a policy designed to reduce the average *total* annual cost to a minimum value.

Referring to the mail-order problem (Sections 14.2 and 14.3) the regression analysis enables us to predict the (average) time required to complete an order according to the number of items it contains. Now suppose the company decided to encourage some customers to group together in placing their orders (by offering, for instance, free postage and packing, or a discount, on large orders) and that it is estimated, from a market survey, that this would result in the average order size increasing from the current level of 8 items to 10 items per order. It is assumed that this would take place without affecting the total number of *items* ordered per day. Exhibit 14 shows that, for a current level of 10,000 items per day this policy would result in a reduction of (8,750 – 7,800) = 950 minutes of packaging time per day.

EXHIBIT 14: Comparison of Total Packaging Time

	No. of Items dispatched per day	*Average No. of Items/Order*	*Average No. of Orders/Day*	*Packaging Time per Order (mins.)*	*Total Packaging Time/Day (mins.)*
Current situation	10,000	8.0	1,250	7.0	8,750
Possible future situation	10,000	10.0	1,000	7.8	7,800

Using this information the company could readily assess whether the policy suggested would bring about an overall reduction in cost.

READING

K.A. Yeomans, *Introducing Statistics*, Statistics for the Social Scientist: Volume One, Penguin, Chapter 5.

EXERCISES

1(P) Obtain the height (H) and weight (W) of a number of friends of the same sex (preferably at least 10, using consistent measurement units throughout).

(i) Plot the data on a scatter diagram and indicate the bivariate mean on the graph.
(ii) Derive the least squares regression lines of H on W and W on H and draw these on the graph.
(iii) Use the regression lines to predict (a) the average weight of individuals of the same height as yourself and (b) the average height of individuals of the same weight as yourself. Comment on the validity of these predictions.

2(U) Consider the following four bivariate observations:

X_i	1	1	3	3
Y_i	2	3	3	4.

(i) Plot these observations on a scatter diagram and indicate the bivariate mean.
(ii) *Without* performing any calculations, determine the least-squares regression lines of Y on X and X on Y and draw these

on the graph. State the equations of these lines.

(iii) Check your answers in (ii) by calculation.

(iv) Determine the value of r^2. (See Equation (5) in the text.)

(v) For the regression line of Y on X carry out an error analysis and hence find $\Sigma(Y_i - \hat{Y}_i)^2$.

(vi) Check your result in (v) by evaluating $(1 - r^2)\Sigma(Y_i - \bar{Y})^2$.

3(E) (a) From the following information draw a scatter diagram and by the method of least squares draw the regression line of best fit.

Volume of sales (in thousands of units)	5	6	7	8	9	10
Total expenses (£'000's)	74	77	82	86	92	95

(b) What will be the expected total expenses when the volume of sales is 7,500 units?

(c) If the selling price per unit is £11, at what volume of sales will the total income from sales equal the total expenses?

(ICMA 1973)

4(U) The data below, which relates to the number of vehicles licensed and road accident casualties in the United Kingdon, is taken from the "Annual Abstract of Statistics".

(i) Obtain the formula for the straight line ($\hat{Y} = a + bX$) of best fit regression equation of casualties (Y) upon number of vehicles (X). Enter the points and your line on a graph.

Year	*Vehicles (m)*	*Casualties (000)*	*Year*	*Vehicles (m)*	*Casualties (000)*
1958	7.9	300	1962	10.5	342
1959	8.6	333	1963	11.4	356
1960	9.4	348	1964	12.3	385
1961	9.9	350	1965	12.9	400

(ii) From your formula, estimate the number of casualties when the number of vehicles reaches 15 millions.

(iii) From the current Abstract, enter the latest information on your graph. Comment on the extrapolated line and the latest data.

5(E) The following data has been collected over eight periods:

Period	Units of Output	Total Cost £
1	10,000	32,000
2	20,000	39,000
3	40,000	58,000
4	25,000	44,000
5	30,000	52,000
6	40,000	61,000
7	50,000	70,000
8	45,000	64,000

Draw a scatter diagram and by the method of least squares draw a straight line which best fits the data.
Give the equation of the line and estimate the cost likely to be incurred at the output levels of 26,000 units and 48,750 units.

(Hint: Work in more convenient units.) (ICMA 1973)

6(E) A company marketing men's toiletries calculated its pre-Christmas advertising expenditure in the Main I.T.V. areas as a percentage of total expenditure by all companies in the product field. These percentages were compared with the corresponding sales figures for the product field (in £ value).

% Expenditure in I.T.V. Advertising and % Value of Sales

I.T.V. Area	*% Expenditure*	*% Sales*	*I.T.V. Area*	*% Expenditure*	*% Sales*
London	15.1	16.0	Northern	20.5	18.1
Southern	18.2	17.2	T.W.W.	10.6	14.2
Anglia	14.2	15.8	Central Scottish	12.4	15.3
Midlands	22.1	18.6	Westward	18.7	17.4

(a) Estimate the equation of the straight line $\hat{Y} = a + bX$ which best estimates the relationship between % Expenditure (X) and % Sales (Y).

(b) Plot the points on a scatter diagram and enter the line you have calculated.

(HNC Business Studies)

7(E) The production of small cigars in the United States 1961–70 is shown in the following table.

Year	*Number of Small Cigars (Millions)*
1961	98.2
1962	92.3
1963	80.0
1964	89.1
1965	83.5
1966	68.9
1967	69.2
1968	67.1
1969	58.3
1970	61.2

You are required to:

(a) graph the data shown above,
(b) calculate the equation of a least squares trend line fitting the data, say what the trend value is at 1961 and 1970, and explain your result, and,
(c) estimate the production of small cigars for the year 1971.

(ACCA 1971)

8(E) The production Foreman discovered accidentally that the addition of a small percentage of a new silicone compound to a vat of paint mix reduced the time required for stirring to obtain the required consistency.
A trial under production conditions was carried out with varying percentages of the additive. The results are as follows:

% Additive	*Reduction in Stirring Times (mins)*
.02	7
.04	12
.06	16
.08	22
.10	26
.12	30
.14	34
.16	40

Obtain the straight line regression equation of the form $\hat{Y} = a + bX$ which best fits the data, and shows the dependence of reduction in stirring time on the % additive.
Discuss briefly the interpretation of the two constants a and b

and comment on the possibility of increasing the % additive still further.

Part question (BA Business Studies)

9(E) A mail order company has two separate groups of packers (A and B) who have been trained to use different *procedures* when packing the goods. For a random sample of orders the packing time was noted together with the number of items each order contained. The results were classified according to whether the packer came from group A or B, and are as follows:

No. of Items	*Time (mins)*	*Group*	*No. of Items*	*Time (mins)*	*Group*
18	4.8	B	20	6.0	B
10	3.6	B	18	6.8	A
4	2.0	A	16	6.0	A
5	5.0	B	13	4.0	B
1	2.4	B	2	0.8	A
19	8.0	A	24	6.8	B
12	5.8	A	3	2.4	B
2	1.8	A	10	4.4	A
8	3.2	B	14	4.8	B

By using regression analysis answer the following question as *quantitatively* as possible: Is there any benefit to be derived from assigning individual orders to one group of packers rather than the other? (Consider the practical aspects of any scheme you suggest.)

(BA Business Studies)

SOLUTIONS TO EVEN NUMBERED EXERCISES

2. (i) and (ii)

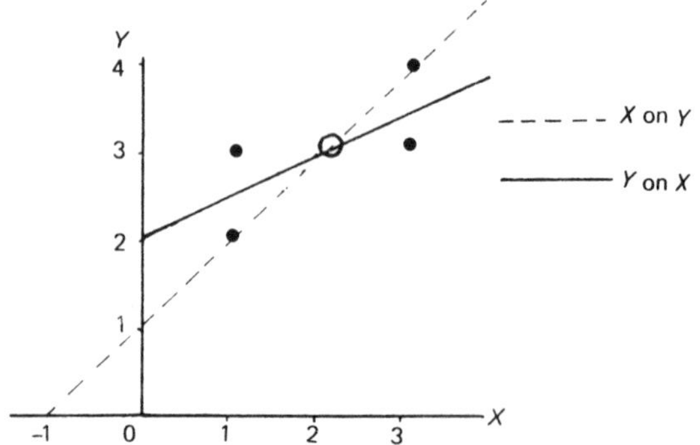

Y on X is $\hat{Y} = 2 + 0.5X$
X on Y is $\hat{X} = -1 + Y$.

(iii)

X	Y	XY	X^2	Y^2	
1	2	2	1	4	
1	3	3	1	9	
3	3	9	9	9	
3	4	12	9	16	
8	12	26	20	38	Totals

$n = 4; \bar{X} = 8/4 = 2; \bar{Y} = 12/4 = 3$

Y on X

$$b = \frac{\Sigma XY - n\bar{X}\bar{Y}}{\Sigma X^2 - n\bar{X}^2} = \frac{26 - 4 \times 2 \times 3}{20 - 4 \times 8^2} = \frac{2}{4} = 0.5$$

$$a = \bar{Y} - b\bar{X} = 3 - 0.5 \times 2 = 2$$

Hence

$$\hat{Y} = 2 + 0.5X$$

X on Y

$$b' = \frac{\Sigma XY - n\bar{X}\bar{Y}}{\Sigma Y^2 - n\bar{Y}^2} = \frac{26 - 4 \times 2 \times 3}{38 - 4 \times 3^2} = \frac{2}{2} = 1.0$$

$$a' = \bar{X} - b'\bar{Y} = 2 - 1 \times 3 = -1$$

Hence

$$\hat{X} = -1 + Y.$$

(iv) $r^2 = bb' = 0.5 \times 1.0 = 0.5$ ($\therefore r = 0.71$).

(v) Using $\hat{Y}_i = 2 + 0.5X_i$, we have:

X_i	Y_i	$\hat{Y}_i$	$(Y_i - \hat{Y}_i)$	$(Y_i - \hat{Y}_i)^2$
1	2	2.5	−0.5	0.25
1	3	2.5	−0.5	0.25
3	3	3.5	0.5	0.25
3	4	3.5	0.5	0.25
		Totals	0.0	1.00

So

$$\Sigma(Y_i - \hat{Y}_i)^2 = 1.00.$$

(vi) $(1 - r^2)\Sigma(Y_i - \bar{Y})^2 = (1 - 0.5)(38 - 4 \times 3^2) = 1.0.$

4. (i) From the data given we obtain the following values:

$$\Sigma X = 82.9,\ \Sigma Y = 2{,}814,\ \Sigma XY = 29{,}514.9,\ \Sigma X^2 = 880.65$$

As $n = 8$, we have $\bar{X} = 82.9/8 = 10.3625$; $\bar{Y} = 2{,}814/8 = 351.75$. Now

$$b = \frac{\Sigma XY - n\bar{X}\bar{Y}}{\Sigma X^2 - n\bar{X}^2}$$

$$= \frac{29{,}514.9 - 8 \times 10.3625 \times 351.75}{880.65 - 8 \times 10.3625^2} = 16.428$$

And $a = \bar{Y} - b\bar{X} = 351.75 - 16.428 \times 10.3625 = 181.5148$.
So $\hat{Y} = 181.51 + 16.43X$ is the required regression line.

(ii) When $X = 15$, $\hat{Y} = 181.51 + 16.43 \times 15 = 427.96$.

6(a) $\Sigma X = 131.8$, $\Sigma Y = 132.6$, $\Sigma XY = 2{,}226.73$, $\Sigma X^2 = 2{,}285.36$. As $n = 8$, we have $\bar{X} = 131.8/8 = 16.475$; $\bar{Y} = 132.6/8 = 16.575$. Now

$$b = \frac{2{,}226.73 - 8 \times 16.475 \times 16.575}{2{,}285.36 - 8 \times 16.475^2} = 0.3698$$

$$a = \bar{Y} - b\bar{X} = 16.575 - 0.3698 \times 16.475 = 10.4825$$

Hence approximately $\hat{Y} = 10.48 + 0.37X$.

8. Denoting % Additive by X and Reduction in Stirring Time by Y, we have $\Sigma X = 0.72$, $\Sigma Y = 187$, $\Sigma XY = 20.70$, $\Sigma X^2 = 0.0816$. As $n = 8$, $\bar{X} = 0.72/8 = 0.09$; $\bar{Y} = 187/8 = 23.375$. So

$$b = \frac{\Sigma XY - n\bar{X}\bar{Y}}{\Sigma X^2 - n\bar{X}^2} = \frac{20.70 - 8 \times 0.09 \times 23.375}{0.0816 - 8 \times 0.09^2} = 230.357,$$

$$a = \bar{Y} - b\bar{X} = 23.375 - 230.357 \times 0.09 = 2.643$$

Hence approximately $\hat{Y} = 2.64 + 230.4X$.
The constant a is the intercept on the y-axis, and so is the predicted value for $X = 0$. In this example the "true" intercept should be zero as there will be no reduction in stirring time when no additive is used. So a differs from zero simply as the result of sampling errors. (The *particular* sample used.)
The constant b is the slope of the regression line and corresponds to the predicted reduction in stirring time per unit increase in % additive.
If more additive is used then the stirring time may be reduced still further, but clearly the relationship cannot continue to be linear

indefinitely or eventually a negative stirring time would be reached! The relationship seems likely to be of the form shown.

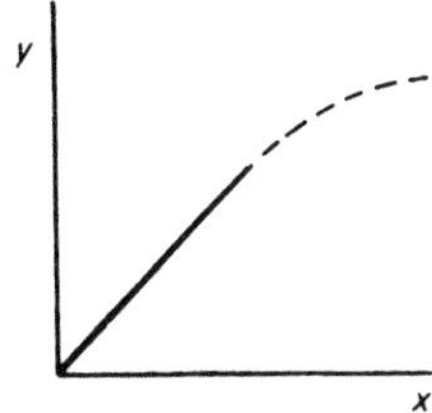

COURSE UNIT 15 – INTRODUCTION TO SAMPLING AND CONFIDENCE INTERVALS

15.1 INTRODUCTION

The statistical techniques we have discussed so far fall under the general heading of *descriptive statistics* – we have been concerned with *describing* (and summarising) the properties of samples. But throughout, the reader should have been aware that the underlying purpose has been to find out about the population from which a sample is taken. If sample results are to be used for decision making in industry we must consider seriously how reliable (or, more to the point, how unreliable) such results are. After all, it would be singularly unfortunate for a company to embark on a major reorganisation on the basis of the ten employees whose views were sought reacting favourably, if, in fact, these employees represented a distinct minority view.

The development of techniques used to assess the reliability of sample results is an area of study called the statistics of *inference*. Our starting point is an examination of the reasons for sampling and the consideration of a method of sampling which enables us to derive the required quantitative relationships.

At first sight it may seem desirable to base a decision on a complete count or measurement of all members of the population, e.g. a marketing manager would like a census of all housewives on their attitudes to his product. But it is not difficult to see that a complete census of millions of people is not feasible for every marketing decision that has to be taken. What we shall now show is that a census is not only impractical but is also unnecessary. We shall first consider why a sample may be preferable, and then discuss the principles of selecting a simple random sample. This will enable us to discover the inferences that may be drawn validly about a population parameter from a statistic calculated from such a sample.

15.2 WHY WE SAMPLE

Cost – The inspection cost of fewer items (people, occasions) is obviously less than if a 100% inspection (census) is carried out. But this reduction in cost is only justifiable if the results are sufficiently reliable.

Time – Even if cost were not a consideration, information if often required within a specified time so that a decision can be made and action taken. It is invariably better to have reliable sample data in time rather than more complete data several weeks after the decision has been taken.

Accuracy – Absolute accuracy may not be required. The size of a sample can be worked out so that the resulting accuracy is sufficient for the decision to be made: if a larger sample is taken then resources are being wasted. But samples can sometimes be more accurate than a complete count. With a smaller volume of work to be performed more skilled, better trained and better supervised labour may be employed; inspection fatigue may be reduced or eliminated, missing items of data may be traced. A thorough check on a small sample may be more useful than a more superficial inspection on a larger number.

Resource Allocation – By the use of samples it is possible to carry out several studies concurrently. Thus a work study practitioner may cover a number of projects by sampling different processes at varying times each day during a week, an inspector can check on several production lines, a market research team run various surveys.

The Test may be Destructive – In quality control many tests destroy the product – from T.V. tubes to H-bombs. No products would be left unless samples were used.

This last point illustrates that we are not interested in the sample members except in so far as they may be used to draw inferences about the population from which they were selected. The market researcher wishes to draw inferences about 20,000,000 or so housewives in Great Britain, not the 2,000 he has had interviewed. The inspector is interested in the 60,000 components produced each hour, not the 20 he has torn to pieces. The work study practitioner wishes to make inferences about the year's production, not the 100 occasions on which he has timed the process.

15.3 SIMPLE RANDOM SAMPLING

Simple random sampling refers to a *method* of sampling a population in which sample members are chosen one at a time, and where, at each selection, each *eligible* member stands the same chance of being chosen. When sampling *with* replacement all population members are eligible throughout; when sampling without replacement only those not previously chosen are eligible. As far as the theoretical methods of statistics are concerned we specify which of the two methods we are using: in applied statistics it is commonly assumed that we are sampling without replacement, since this is by far the more usual case. The adjective "simple" is used to distinguish this basic method from more complex methods which are also random, the more important of which are discussed in SU5.

A common form of simple random sampling (SRS) is the lottery, where numbered discs or tickets are placed in a box, thoroughly mixed and the winning numbers drawn one at a time for each available prize. A winning ticket is not usually returned to the box to allow it to win a second prize and so the sampling is without replacement. Note that it is the *thorough mixing* which makes the sample (the prizewinners) "random".

What this lottery example makes clear – all too clear, if one fails to win a prize – is that if we engage in sampling operations there is some uncertainty in the answers obtained. The way in which we shall handle the uncertainty is to quote our results in terms of probabilities, making use where appropriate of the probability distributions – binomial, Poisson and normal – that we have already studied.

Whether a sample can be termed random or not is determined solely by the *way* the sample was selected, and it is not related in any way to the *properties* of the resulting sample. That a sample of men contains too many who use electric razors (i.e. it is not "representative" of the shaving population) does not prove that the sample was not randomly selected – it may or may not have been. We would expect SRS's with varying degrees of "unrepresentativeness" to occur with certain probabilities. A very unrepresentative result will occur by chance with a small probability and indeed it should occur with a relative frequency equal to this probability when a large number of similar samples are taken.

In commercial applications (sampling housewives or shops in marketing studies, components in quality control inspection, invoices for auditing etc.) it is not convenient – to say the least – to have a large box full of numbered tickets to help us determine which members of the population should be selected for the sample. It is more practical to use tables of **random numbers** in conjunction with a list of the members of the population, known as a **sampling frame**, which enables us to identify the sample member by using a random number as a *reference*.

Tables of random numbers have been constructed by a number of different ingenious devices over the years, latterly by electronic scrambling machines such as ERNIE (Electronic Random Number Indicating Equipment). For our purposes the essential feature of these tables is that each digit from 0 to 9 is found an equal number of times in a large section of the tables, with no more repetitiveness than should properly occur by chance, and with no tendency for the numbers to form repeating patterns.[1] There are a number of statistical tests which may be carried out to determine if a series of numbers is truly random.

Random numbers may be used in any sequence – for convenience they are often used down the columns, or across the rows. The numbers may be used singly if the population has up to 10 members, in pairs if the population has up to 100 members, and so on. In random sampling for consumer market research in the United Kingdom the Electoral Register is most often used as a sampling frame. The electors for a Polling District are listed according to name and address and each elector has a number, starting from 1 for each Polling District. Hence the Electoral Register is rather like a large school register usually listing over 1,000 individuals entitled to vote. So if a particular register has 2,404 electors, each elector will have a number from 1 to 2,404; we would then use random number tables to give us reference numbers between 0001 to 2404. Typically we would take a table of random numbers and read off the numbers in groups of four, e.g.

1204 0917 9687 4825 1398 0246.

Our first sample member would be elector "1,204" on the Electoral Register, the second, elector 917. The next two random numbers, 9687 and 4825, would then be ignored since our population size is only 2,404. So the third sample member is elector 1,398. We carry on in this manner until we have uniquely identified the required number of people for our sample by name and address.

In other applications, such as when sampling shops, customer invoices, employee records, boxed manufactured components, a certain amount of ingenuity often has to be used in order to obtain a method which is both practicable and random. *Simple* random sampling is not always the best method to use. Other forms of random sampling are frequently more convenient, cheaper or more accurate. However, the SRS is the standard against which these other methods are evaluated. The SRS plays a central role in the theory of statistics. In particular, the technique of randomisation ensures the validity of the techniques to be discussed; so unless samples are randomly selected, experiments performed in random sequence etc. the techniques of setting confidence intervals and significance testing do not lead to valid conclusions. If other sampling methods are used then the effect on the validity of the findings has to be evaluated. Where another form of *random* sampling is used (e.g. Stratified Random Sampling, SU5) then the confidence

---

(1) More formally the individual digits are Uniformly distributed, $p(r) = 0.1$, $0 \leqslant r \leqslant 9$, and the correlation coefficient of a digit with the preceding digit (and with the one before that, and so on) is zero.

intervals and signficance levels will differ (sometimes being greater, sometimes smaller) from those predicted from SRS theory. If *non*-random methods are used then the validity of all inferences is open to question.

In surveys of the human population, "non-response" (persons selected not completing the questionnaire) will similarly throw doubt on the validity of the findings even if the remainder of the sample has been randomly selected – the non-respondents may differ in any number of known or unknown ways from the respondents. What particularly bedevils such surveys is that the non-response is often concentrated in the very rich and very poor sections of the community, and such people may well have different opinions from the rest of the population.

15.4 SUMMARY OF SAMPLING THEORY

Some aspects of sampling theory have already been covered in earlier Course Units (CU11 and CU12); here we attempt to provide a coherent summary of the argument, and at the same time introduce a few technical terms used in the context of sampling.

The process we are about to describe is not usually carried out in practice[(1)]. Consider taking a large number of SRS's, all the same size, from a given population[(2)] and calculating from each the value of the *same* sample statistic[(3)]. To make it easier to visualise, the reader may care to think in terms of calculating the mean value of each sample, but the argument is not confined to this statistic.

We start with some definitions.

- The variations in value of the statistic from one sample to another are known as **sampling fluctuations** or **sampling errors** (see Exercise 1, CU1). It is important to realise that such sampling errors are an unavoidable phenomenon and are not to be confused with "mistakes" which might occur in the sampling procedure itself.
- These values of the statistic could be formed into a frequency distribution and depicted as a histogram, just as if they were observations in a single sample. The distribution of the values of a statistic is called a **sampling distribution**.

---

(1)The exceptions being "replicated sampling" methods (beyond the scope of this course) but the number of replications is usually not large.

(2)As in earlier Course Units we (temporarily) impose the restriction that the sampling takes place *with* replacement, or alternatively that the population is effectively infinite. We turn briefly to sampling without replacement from finite populations in Section 15.5.

(3)Or consider taking *all possible* samples: See Exercise 1(ii).

- The values of the statistic could be summarised by calculating *their* mean value and *their* standard deviation. If the mean value of the sample statistic, from *all possible* samples, is equal to the corresponding parameter of the population from which the samples were drawn, we say that the statistic is "unbiased". The standard deviation of the values of the statistic is given the special name **standard error** (SE) in order to avoid confusion with the standard deviation of the population.

In terms of practical usefulness, by far the most important statistics calculated from samples are mean values and proportions (or percentages). It is singularly fortunate that these two statistics have very convenient sampling properties. Firstly, both statistics are unbiased. So if, for example, from SRS's we estimated that (a) the average expenditure by housewives on frozen foods per week is £1.80, and (b) that 80% of women are married, these estimates are the "best" single-figure estimates of the true population values – we have no need to "correct for bias" for these two statistics (see Section 12.6). Secondly, the sampling distributions of means and proportions can be taken to be normally distributed over a wide range of practical applications (CU11). Finally, the standard errors of these statistics are particularly simple to calculate.

We next consider some further aspects of sampling for means and sampling for proportions separately.

Sampling for Means. If the population from which the sample is drawn is itself normally distributed then the sampling distribution of means will also be normally distributed for samples of *any* size. The simplest illustration of this is to consider samples of size 1; in this case the sampling distribution of means is identical with the population distribution. If on the other hand the population is *not* normally distributed, but it is nonetheless nearly symmetrical, sample sizes of 4 or 5 are sufficient to make the sampling distribution of means normal, and this fact is utilised in the construction of Statistical Quality Control charts for averages (SU4). With skewed populations, sample sizes of the order of 30 or more are usually considered adequate to obtain a normal sampling distribution for means, as a result of the Central Limit Theorem (CU11).

The standard error of a sample mean (taken from a SRS) has a particularly simple relationship to the standard deviation of the population (σ):

$$\text{Standard Error (SE) of mean} = \sigma_{\bar{X}} = \frac{\sigma}{\sqrt{n}},$$

where n is the sample size from which the *mean* is calculated. (This result was proved in CU12).

To take a numerical example, if the standard deviation of the number of hours per week that children watch television is 4.0, then with a SRS of 25 children, the SE of the mean is $4.0/\sqrt{25} = 4.0/5 = 0.80$ hrs. Similarly, for an SRS of 100 children the SE of the mean is $4.0/\sqrt{100} = 4.0/10 = 0.40$. Thus by increasing the sample size by a factor of 4 we have reduced the SE by a factor of 2.

The implication of these results is that the sampling distribution of means is less spread out than that of the population from which the samples are selected. This phenomenon is due to the fact that observations (from the two tails of the distribution) tend to balance each other out when the mean is calculated. Thus the plus and minus deviations about the population mean, μ, will tend to cancel out, leaving the sample mean closer to the population mean than any of the extreme values found in the sample. The larger the sample size, the more opportunity there is for the cancelling process to operate, and hence the closer the sample means will tend to cluster around the population mean. Eventually, if we let the sample size equal the population size (i.e. we take a census), then the standard deviation of the sampling distribution will become zero – there will be no sampling error. These effects are illustrated graphically in Exhibit 1, for the special case of a normal population.

EXHIBIT 1: Sampling Distributions of Means

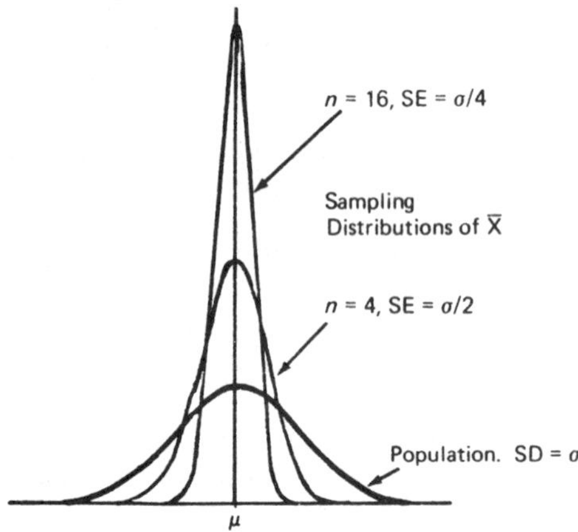

We can summarise the most general (and useful) result from the foregoing discussion as follows:

If a simple random sample of size n ($n \geqslant 30$) is taken from a population with mean μ and standard deviation σ, the sampling distribution of the sample mean $\bar{X}$ is normally distributed with mean μ and standard error $\dfrac{\sigma}{\sqrt{n}}$. (1)

Sampling for Proportions (or Percentages). We obtain a proportion from a sample of n items by counting the number of items r with the required property (such as the number of women in a sample who are married) and forming the quotient r/n (= P). Similarly a sample percentage is obtained from $100r/n$ (= $P_{\%}$). Now if the true (population) proportion is π, the number r in a SRS will be binomially distributed (CU9) with mean $n\pi$ and standard deviation $\sqrt{n\pi(1-\pi)}$. Consequently the proportion $P = r/n$ will have a mean value $n\pi/n = \pi$ (showing that P is an unbiased statistic) and standard deviation (i.e. standard error) of $\sqrt{n\pi(1-\pi)}/n = \sqrt{\pi(1-\pi)/n}$. So,

$$\text{SE of a proportion } P = \sqrt{\frac{\pi(1-\pi)}{n}}$$

and the equivalent result for a percentage is given by

$$\text{SE of a percentage } P_{\%} = \sqrt{\frac{\pi_{\%}(100-\pi_{\%})}{n}},$$

where $\pi_{\%}$ is the population *percentage*, and where the standard error itself is expressed as a percentage.

Example. If we toss a true coin ($\pi_{\%} = 50\%$) 100 times, the SE of the percentage of heads shown is $\sqrt{50 \times 50/100} = 5\%$. If we toss it 400 times the SE becomes $\sqrt{50 \times 50/400} = 2.5\%$. This shows, as is the case when sampling for means, that a fourfold increase in sample size results in the SE being reduced by a factor of two. Generally, and again a similar result as for the mean[(1)], the larger the sample the more closely will the the sample statistic P or $P_{\%}$ cluster around the population value π or $\pi_{\%}$.

[(1)] The parallel between sampling for means and sampling for proportions is very close. Suppose 7 women in a sample of 10 were married; the sample could be represented as:

1 1 1 0 1 1 0 0 1 1

where "1" indicates married, and "0", not married (Bernoulli variables). The *mean* value of these 0's and 1's is 7/10 which is itself the sample proportion. Now the SE of the *mean* is $\sigma/\sqrt{n}$ which by substitution of the Bernoulli standard deviation, $\sigma = \sqrt{\pi(1-\pi)}$, gives $\sqrt{\pi(1-\pi)/n}$, the SE of *proportion*.

Sample proportions and percentages have a normal sampling distribution for *large* samples provided that π is not too small and not too large, i.e. the same conditions that apply when approximating the binomial distribution by the normal distribution. In general the sample sizes required for normality are larger than those required when sampling for means; but precisely how large our samples need to be is a matter we shall come to presently.

For the moment we summarise the argument as follows:

If a "large" simple random sample of size n is taken from a population with a percentage $\pi_{\%}$ of items having a certain property then the sampling distribution of the percentage of items in the sample, $P_{\%}$, having this property is approximately normally distributed with mean $\pi_{\%}$ and standard error (2)

$$\sqrt{\frac{\pi_{\%}(100 - \pi_{\%})}{n}}.$$

Large Sample Procedures. Provided that our samples meet the conditions we have outlined – whether for means or proportions – we are able to use the so-called "large sample" methods. These enable us to make statements about the probability of a sample mean or proportion differing in value by more than a certain amount from the population mean or proportion, by reference to normal distribution tables. For smaller samples the normal distribution is not applicable and we have to resort to the "small sample" methods, some of which are discussed later in the course.

Before we turn to the matter of drawing inferences from sample data, a fundamental problem has to be overcome. That is, in order to determine the standard error of a sample mean or proportion, we need to know the value of the relevant population parameter involved (σ if we are sampling for means, π if we are sampling for proportion). Yet the reason *why* we are sampling is that we do not know the values of these parameters! This problem is usually overcome in practice by approximating the value of the population parameter needed in the standard error formula by the value of the corresponding statistic obtained from the sample. So

$$\text{SE of } \bar{X} = \sigma_{\bar{X}} = \sigma/\sqrt{n} \simeq s/\sqrt{n}$$

$$\text{SE of } P = \sigma_p = \sqrt{\pi(1 - \pi)/n} \simeq \sqrt{P(1 - P)/n}$$

$$\text{SE of } P_{\%} = \sigma_{p\%} = \sqrt{\pi_{\%}(100 - \pi_{\%})/n} \simeq \sqrt{P_{\%}(100 - P_{\%})/n}$$

These approximations will obviously introduce errors into our later calculations of confidence intervals but, provided that the samples are "large", they may be taken to be sufficiently accurate for most practical purposes. Indeed this is the reason, when sampling for proportions (or percentages) in particular, that larger samples are needed than would be thought necessary on the basis of the normal approximation to the binomial. Cochran (Reference) recommends the minimum sample sizes shown in Exhibit 2 for various values of P.

EXHIBIT 2: Minimum Sample Sizes Needed for the Normal Approximation

P	*Minimum Sample Size*
0.5	30
0.4 (or 0.6)	50
0.3 (or 0.7)	80
0.2 (or 0.8)	200
0.1 (or 0.9)	600
0.05 (or 0.95)	1,400

15.5 CONFIDENCE INTERVALS AND SAMPLE SIZE

Two problems arising from the joint consideration of the size and precision of a simple random sample are:

- What is the degree of accuracy in the result based on a sample? The answer to this problem is given, in probability terms, by a **confidence interval**, a range of values within which we can be reasonably sure that the population parameter lies.
- Bearing in mind the sampling error, how large a sample should be taken in order to obtain a certain degree of accuracy, or, stating the problem statistically, what size sample is needed to achieve a confidence interval of some predetermined size?

The principles outlined below are common to most "large sample" confidence intervals (and significance tests) where it is assumed that the statistic in question has a normal distribution centred on the true (but unknown) population value.

EXHIBIT 3: The Development of Confidence Intervals

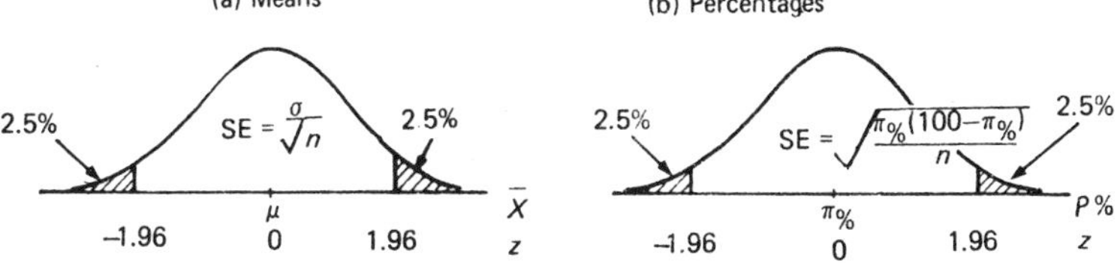

Referring to Exhibit 3(a) we may predict (approximately, because of the approximate form of the standard error) that:

95% of sample means will lie between the two values $\mu \pm 1.96\ s/\sqrt{n}$.

As we are interested in the unknown value of μ, we may rewrite this statement as:

For 95% of $\bar{X}$'s the population mean μ will be within $1.96\ s/\sqrt{n}$ units

or yet again:

For 95% of $\bar{X}$'s the population mean μ will lie within the range $\bar{X} \pm 1.96\ s/\sqrt{n}$.

Now suppose we take a *single* sample, and calculate $\bar{X} + 1.96\ s/\sqrt{n}$ and $\bar{X} - 1.96\ s/\sqrt{n}$. These two values are known respectively as the upper and lower *Confidence Limits* on μ, and the range of values between them as the 95% *Confidence Interval* for μ. Clearly the true value of μ will either lie between these limits or it will not: all we are saying is that, over a large number of such samples, we shall be right 95% of the time.

The factor 1.96 used in the above statements is taken from normal distribution tables and corresponds to $2\frac{1}{2}$% of the distribution in *each* tail (95% in the "centre"). It is commonplace to use a 95% Confidence Interval, but by no means universal. As the reader should verify, a 90% Confidence Interval is given by a factor of 1.645 and 99% by 2.576. Further light on which level of percentage confidence we should choose will be given when we discuss Significance Tests in CU17, but for the present we simply point out that by increasing the z-factor we *widen* the confidence interval. In other words we may increase our "confidence" by making a less precise statement (and *vice versa*).

Example: An SRS of housewives spent on average £1.80 per week on frozen foods (with a standard deviation of £0.60). How accurate would this average be if (a) the sample size were 100, (b) the sample size were 225.

We are 95% confident that true value lies in the range:

for (a), $1.80 \pm \dfrac{1.96 \times 0.60}{\sqrt{100}} = £1.80 \pm 0.118$

(i.e. between £1.68 and £1.92)

and for (b), $1.80 + \dfrac{1.96 \times 0.60}{\sqrt{225}} = £1.80 \pm 0.078$

(i.e. between £1.72 and £1.88).

The differences between these results reflect the increase in accuracy brought about by a larger sample size, for a given level of confidence.

The argument we have used for means applies equally to proportions (or percentages), and may be summarised as:

For 95% of $P_{\%}$'s the population value $\pi_{\%}$ will lie within the range $P_{\%} \pm 1.96 \sqrt{\dfrac{P_{\%}(100 - P_{\%})}{n}}$.

Example: An SRS of 900 women gave 80% as being married. How accurate is this figure?

We are 95% confident that the true percentage lies in the range:

$$80 \pm 1.96 \sqrt{\frac{80 \times 20}{900}} = 80 \pm 2.6\% \text{ (i.e. between 77.4\% and 82.6\%).}$$

It should be noted that in some applications we do not need to specify a Confidence Interval as such, but rather a *single* Confidence Limit. If we wished to estimate simply the *maximum* likely percentage of women who are married (for the purposes of some planning exercise) we would quote 82.6%. This figure is a "one-sided" Confidence Limit, not at 95% confidence, but at 97.5% because we are dealing with just one tail of the distribution: clearly there is only a 2½% chance of our being wrong.

We next turn to the second question posed at the beginning of this section – how large a sample? We must distinguish between two situations; firstly where something is known about the population (e.g. as a result of an earlier study), and secondly where nothing is known.

Where something is known about the population. Suppose, in our frozen food example that a *previous* study had revealed that the standard deviation of amount spent on such goods was £0.60. How large a sample should be taken to estimate the true mean expenditure to within £0.10? Working to 95% confidence we can write $0.10 = 1.96(0.60)/\sqrt{n}$, which gives the necessary sample size of $n = 138$. In general then, to estimate a mean value to within $\pm x$ units we set

$$x = \frac{zs}{\sqrt{n}} \quad \text{giving } n = \left(\frac{zs}{x}\right)^2,$$

where z is the normal deviate for the desired level of confidence.

Similarly, if we were to hazard a guess that about 80% of women were married, but wished to estimate the true percentage to within 2%, we would require $2 = 1.96\sqrt{80 \times 20/n}$, giving $n = 1537$. In general, to estimate a percentage to within $\pm x\%$ we set

$$x = z\sqrt{\frac{P_{\%}(100 - P_{\%})}{n}} \quad \text{giving } n = \left(\frac{z}{x}\right)^2 \cdot P_{\%}(100 - P_{\%})$$

In both cases a smaller sample (than that calculated) would lead to a less accurate result than required and a larger sample, although giving a more accurate result, would be wasteful of resources.

Where nothing is known about the population. In many situations when sampling for means no results from a previous survey are available; and when sampling for proportions, we may be hesitant to guess the value for the purposes of determining the necessary sample size. In such cases the rigorous approach is to conduct a small scale "pilot" study in order to obtain preliminary estimates and hence determine the full-scale sample size required. If this is not feasible we may nonetheless proceed to obtain the required sample sizes in the following ways.

From normal distribution tables we see that 99.7% of the distribution will lie between z values of ±3. So there will be about *six* standard deviations' difference between the lowest and highest values in a large population of normally distributed data. Now if we were to guess that the great majority of housewives spend between nothing and £4 per week on frozen foods, we may crudely estimate the standard deviation of expenditure (assuming it is *roughly* normally distributed) as £4/6 = £0.67. This value could then be used to determine the required sample size.

When sampling for proportions there is an easier way. It can be shown that $\sqrt{P_{\%}(100 - P_{\%})/n}$ has a *maximum* value when $P_{\%} = 50\%$. So provided we determine the sample size on the basis of $P_{\%} = 50\%$, we are bound to get results which meet our accuracy requirements. Again, if we wanted to ensure that our estimate of the percentage of women who are married was accurate to within 2% (with 95% confidence) we could set $n = (1.96/2)^2 \times 50 \times 50 = 2{,}400$. This is an increase in sample size of 863 (= 2,400 – 1,537) over the case where we substituted our guess of $P_{\%} = 80\%$, and corresponds to the penalty to be paid for having no background information. Now if, on the basis of a sample of size 2,400, we subsequently obtain $P_{\%} = 80\%$ it will be true that the accuracy of our result (with 95% confidence) is greater than our needs of ±2%, and so we will have spent money on collecting data that is more accurate than is necessary for the problem in hand. However without some background information there is no way of avoiding this waste of resources.

15.6* FINITE POPULATION CORRECTIONS

The results we have quoted for standard errors apply to the process of sampling *with* replacement, or, alternatively, without replacement from an infinite population. When the sample size n is a sizeable fraction of population size N (i.e. when $n > N/20$, say) finite population correction (fpc) factors should be applied. The results when sampling for means are shown below – corresponding corrections may be made for the standard error of other statistics.

$$\text{SE of } \bar{X} = \sigma_{\bar{X}} = \frac{\sigma}{\sqrt{n}} \sqrt{\left(\frac{N-n}{N-1}\right)}, \text{ where the term } \sqrt{\left(\frac{N-n}{N-1}\right)}$$

is the fpc factor.

It can be seen that when $n = 1$ the fpc = 1, underlining the fact that for samples of size 1 there is no difference between sampling with and without replacement. If we take a census and $n = N$ the fpc = 0, showing that the standard error is zero.

Another form of the fpc, more convenient for advanced work, is incorporated into the following standard error expression:

$$\text{SE of } \bar{X} = \sigma_{\bar{X}} = \frac{\sigma}{\sqrt{n}} \sqrt{\left(1 - \frac{n}{N}\right)}$$

This form relies on a slight redefinition of the population variance, but this is of no practical significance. We see that at the limit of our criterion for applying the fpc ($n = N/20$), i.e. where the sample is 5% of the population size, we have

$$\sqrt{\left(1 - \frac{n}{N}\right)} = \sqrt{1 - 0.05} = \sqrt{0.95} \simeq 0.975$$

So if we fail to apply the fpc at this level of sampling our standard errors will be about $2\frac{1}{2}$% too large and Confidence Intervals about $2\frac{1}{2}$% too wide.

READING

A useful discussion on sampling may be found in:
Richard Goodman *Teach Yourself Statistics*, E.U.P., Sections 7.1–7.6.
A pictorial treatment of confidence intervals is given in:
T.H. & R.J. Wonnacott *Introductory Statistics 2nd Ed.*, Wiley, Section 7.1.

REFERENCE

William G. Cochran *Sampling Techniques 2nd Ed.*, Wiley, Table 3.3.

EXERCISES

1(P) (i) Suppose the Post Office wished to determine the percentage of telephone subscribers who would prefer "push-button" dialing. Explain in detail:

(a) How you would decide on the sample size,
(b) How you would select your sample.

Using the London telephone directories[1] as your sampling frame select the first ten members of your sample – explaining precisely how this is achieved. Comment on the use of the London directories as a sampling frame.

(ii) By considering the 36 possible results for the total score when rolling two dice simultaneously, obtain a histogram for the sampling distribution of $\overline{X}$, the average score of the two dice. Compare with the histogram of the score from throwing a single die.

2(E) (i) Define a simple random sample.

(ii) The Register of Electors of a certain ward (a particular geographical area) is to be used for the purpose of selecting a simple random sample of 100 electors from that ward. The ward is divided into four districts as follows:

District	*No. of Electors in District*
KA	3,658
KB	1,306
KC	2,417
KD	948
Total	8,329

The electors within each district are arranged by house number and street – the electors being numbered consecutively from 1 to the maximum number of electors in the district. (e.g. in district KC the electors are numbered

[1] For those readers who do not have access to the London directories, proceed with a local directory but also consider the special problems involved with using the London ones; the following details apply: pages of subscribers (*not* advertising) are numbered consecutively from 1 in each volume: there are four volumes having the following number of subscriber pages each: A–D, 827; E–K, 785; L–R, 833; S–Z, 672.

1,2,3, . . ., 2,416, 2,417). A small section of the register for district KA is shown below:

		HILLTOP ROAD		
	2,625	SMITH, VIOLET M.	1	
	2,626	CLUMP, ADA	3	
	2,627	JONES, ARTHUR J.	5	
	2,628	JONES, SARAH E.	5	
Code	2,629	LINTEL, EVA F.	7	
numbers	2,630	LINTEL, SIDNEY A.	7	House
for	2,631	HELP, AUDREY J.	9	Numbers
electors	2,632	HELP, THOMAS W.	9	
	2,633	WILLIAMS, LYDIA M.	11	
	2,634	BRANCH, IVY E.	13	
	2,635	BRANCH, RACHEL	13	
	2,636	BRANCH, WILLIAM D.	13	

Describe in detail (using examples as appropriate) how you would select the required sample.

(iii)* Indicate *how* you could use the sample to estimate the distribution of the *number of electors per household*.

BA (Business Studies)

3(E) A large industrial company wishes to construct a "profile" of its hourly-paid work force, with particular reference to items of information such as family structure, housing conditions and financial commitments. Assuming that all the activities of the company are located in a single geographical area:

(i) design a sample survey based upon a simple random sample to obtain the necessary information.

(ii) how would you determine the appropriate sample size for this purpose?

(iii) what would be the advantages and disadvantages of a sample survey of this kind as opposed to a complete census?

(IPM June, 1974)

4(U) (i) A random sample of 625 families in a certain town reported an average income of £2,800 with a standard deviation of £400. Quote 95% confidence limits for the *total* income of the town's 100,000 families.

(ii) A random sample of 480 T.V. viewers on a certain time and day revealed that 160 were tuned to a B.B.C. channel. What is the 99% confidence interval of the true proportion of viewers watching B.B.C. at that time?

5(E) (a) What is the standard error of a proportion?

(b) In a sample of 500 beer drinkers, 50 were women. You are required to estimate the population proportion at the 95% confidence level, and explain your answer.

(ACCA December, 1972)

(c) What is a sampling distribution of the means?

(d) For a random sample of the annual salaries of 50 salesmen drawn from a population of 1,000 salesmen the mean salary was £1,500 and the standard deviation £320. Calculate the standard error of the mean and explain the use of this measure.

(ACCA December 1973)

6(E) (i) In a sample survey of 200 union members in the electricity industry 140 were in favour of accepting a proposed productivity deal. Find the 95% confidence interval for the proportion π of all union members in the industry who would be in favour.

(ii) How large would the sample need to be in order to estimate the proportion of union members in favour to within ±0.05 (i.e. 5%) with 99% confidence? Distinguish between the two cases of (a) the survey result in (i) being available, and (b) no previous estimates being available.
Comment briefly on your results.

(DMS)

7(E) As personnel manager you are asked to assess the industrial accident position in an organisation and you have access to personal records for each employee giving details of each accident reported to management.

(a) How would you summarise this information to provide a succinct description of the situation and to what limitations would your analysis be subject?

(b) Assuming that conditions remain stable, what form of distribution would you expect the total numbers of accidents in successive years to follow?

(c) By using an appropriate approximation to this distribution, derive 95% confidence limits for the number of accidents in the current year if the average annual accident rate is known to be 36.

(IPM Summer 1973)

8(E) A company manufacturing paint employed 36 amateur decorators to test a new formulation of non-drip paint. Each decorator was given a 500 ml can of paint and asked to paint a large flat surface using the paint as if he were redecorating his own home. The total area covered by each decorator (using the whole can) was measured. The results were an average area of 6.50 sq. metres and a standard deviation of 0.60 sq. metres. Suppose the company is prepared to risk no more than a one-in-a-hundred chance of being wrong in its advertising claims, what is the maximum value the company can use for x and y in the two following advertising claims.

(a) this can covers on average x sq. metres.
(b) this can will cover *at least* y sq. metres.

(DMS)

9(E) The table shown below was published by the Decimal Currency Board and gives the approved conversion procedure from £sd to £p. Note that only the old penny (d) to new penny (p) conversion is shown, as shillings convert to an exact number of new pence. The table also shows the effect of the "rounding up" and "rounding down" of the conversion procedure, e.g. an item priced 3d (or 1s.3d, 2s.3d. etc.) before conversion will cost 0.6d less after the approved conversion to 1p (or 6p, 11p etc.).

Old £sd	New £p		Decimal value is:	
1	$\frac{1}{2}$		0.2d more	
2	1		0.4d more	
3	1			0.6d less
4	$1\frac{1}{2}$			0.4d less
5	2			0.2d less
6	$2\frac{1}{2}$ same value			
7	3		0.2d more	
8	$3\frac{1}{2}$		0.4d more	
9	4		0.6d more	
10	4			0.4d less
11	$4\frac{1}{2}$			0.2d less
		Totals	1.8d more	1.8d less
			RESULT:	ALL SQUARE

Two random samples of goods sold in a particular supermarket were taken just prior to decimalisation. Sample A was a simple random sample of prices of goods bought by customers taken

from the cash register tally roll. Sample B was a sample of the prices of products on the shelves (in this case each product line had the same chance of inclusion in the sample).
The results have been tabulated according to the number of old pence in the price, as follows:

Old pence (d) part of price	*Number of items having price as stated* *Sample A*	*Sample B*
0	23	21
1	30	30
2	40	40
3	10	20
4	25	30
5	20	30
6	47	49
7	25	20
8	30	20
9	70	50
10	30	50
11	50	40
No. in sample	400	400

Analyse this data with a view to commenting on the Decimal Currency Board's statement "**result: all square**". What further information would be required in order to estimate the percentage price rise (or fall) resulting from the conversion procedure? Also briefly discuss the problem of how one would place confidence limits on the % price rise (or fall) figure.

BA (Business Studies)

SOLUTIONS TO EVEN NUMBERED EXERCISES

2. (ii)

District	*No. of Electors*	*Cumulative No. of Electors*	*Random Number Groups*
KA	3,658	3,568	0,001–3,658
KB	1,306	4,964	3,659–4,964
KC	2,417	7,381	4,965–7,381
KD	948	8,329	7,382–8,329
	8,329		

Procedure – Select a 4 digit random number (ignore 0000 and any number greater than 8,329) e.g. 5,432.
Locate elector by cumulative number i.e. elector 5,432 lives in district KC and has number (5,432 – 4,964) = 468 in the register for that district.

– Repeat the procedure 100 times, using random numbers in any *consistent* manner (across rows or down columns).

(iii) For each elector in the sample note the number of electors in his/her household. Now although electors were sampled randomly the households will not be. For example a particular household with 3 electors has three times the probability of being included in the sample than a particular household with one elector (and thus on average there will be three times as many such households in the sample in comparison with a SRS of households). We can correct for this as shown in the following hypothetical example.

No. of Electors in Household	*No. of Electors in Sample coming from Household of stated size*	*Estimated Frequency of Household size*
1	8	8/1 = 8
2	20	20/2 = 10
3	36	36/3 = 12
4	20	20/4 = 5
5	10	10/5 = 2
6	6	6/6 = 1
	100	

4. (i) $\bar{X} = 2{,}800$ (£)
SE of $\bar{X} \simeq s/\sqrt{n} = 400/\sqrt{625} = 400/25 = 16$ (£)
So 95% CI on μ is given by $2{,}800 \pm 1.96 \times 16 \simeq 2{,}800 \pm 31$
Consequently 95% CI on *total* income is given by $100{,}000(2{,}800 \pm 31)$.
So the total income lies between £276.9m and £283.1m with 95% confidence.

(ii) $P = 160/480 = 1/3$
SE of $P = \sqrt{P(1-P)/n} = \sqrt{\frac{1}{3} \cdot \frac{2}{3}/480} = \sqrt{0.000463}$
$= 0.0214$
So 99% CI on π is given by $0.3333 \pm 2.58 \times 0.0214$
$= 0.333 \pm 0.055$
So π lies between 0.278 and 0.388.
(Incidentally, this exercise shows that it would have been preferable to work with percentages rather than proportions and so avoid the risk of errors with very small decimals).

6. (i) $P_{\%} = (140/200) \times 100 = 70\%$
SE of $P_{\%} \simeq \sqrt{P_{\%}(100 - P_{\%})/n} = \sqrt{70 \times 30/200} = 3.24\%$
CI for $\pi_{\%}$ is given by $(70 \pm 1.96 \times 3.24)\% = (70 \pm 6.4)\%$

(ii) (a) The largest SE occurs with $P_{\%} = 50\%$

So $5 = 2.58\sqrt{50 \times 50/n}$, where n is the required sample size

$$\therefore \quad n = \left(\frac{2.58}{5}\right)^2 \cdot 2{,}500 = 665$$

(b) Using $P_{\%} = 70\%$, we obtain similarly

$$n = \left(\frac{2.58}{5}\right)^2 (70 \times 30) = 559$$

The difference in the value of n between (a) and (b) reflects the price one pays for the lack of background information (i).

8. $\bar{X} = 6.50$; SE of $\bar{X} = s/\sqrt{n} = 0.60/\sqrt{36} = 0.10$
We are dealing with a 99% *one sided* confidence limit and so the appropriate z value is 2.33 (normal distribution tables).

(a) We must quote a value less than $\bar{X}$.
So $x = 6.50 - 2.33 \times 0.10 = 6.50 - 0.23 = 6.27$.

(b) To derive the value of y we must consider the population distribution, not the sampling distribution of means, so we use the standard deviation in place of the standard error.

$$y = 6.50 - 2.33 \times 0.60 = 6.50 - 1.40 = 5.10.$$

COURSE UNIT 16* – FURTHER LARGE SAMPLE CONFIDENCE INTERVALS

16.1 INTRODUCTION

In CU15 we found that for means and proportions of large samples a 95% Confidence Interval (CI) for the population parameter is given by:

Value of the Sample Statistic ±1.96 SE's of the Sample Statistic (1)

The essential justification for this is that the statistics considered ($\bar{X}$ and P) have a *normal* sampling distribution centred around the true (population) value provided that (a) simple random sampling is employed, and (b) the samples are "large". But this statement is also true of a number of other statistics (but not all) calculated from samples.

Unfortunately there is no common measure of agreement as to what is meant by a "large" sample – it is a matter of degree. In Exhibit 1 we quote "commonly stated" values of the sample size, n, which lead to normal sampling distributions, together with the formulae for their standard errors, for some statistics in frequent use.

What happens if the samples employed are "small" by this criterion? Strictly, this is a topic for a later Course Unit (CU20), but a few comments here are not out of place in order to give some perspective to the course as a whole. With small samples we shall need to assume that the *population* from which the samples are taken is normally distributed: with our large samples there is generally no need to make this assumption as the Central Limit Theorem[(1)] (CU11) comes to our aid. Provided then that this normality assumption is valid the process of setting confidence intervals with small samples is similar to that given by (1), but with one exception – the 95% factor of 1.96 has to be increased. To take just one example, if we were setting a confidence interval on a population mean from a sample of size 10 the appropriate 95% factor would be 2.23 in place of 1.96.

This Course Unit is arranged as a series of examples designed to illustrate some of the pitfalls involved in setting confidence limits.

[(1)] That is the sampling distribution of the statistic if normal even though the population distribution is not.

EXHIBIT 1: Standard Errors – Results for Large Samples

Statistic	*Expected Value of Statistic*	*Standard Error*	*Estimate of SE*	*Conditions for Normal Sampling Distribution*
Sample Mean, $\overline{X}$	Population Mean μ	$\sigma/\sqrt{n}$	$s/\sqrt{n}$	$n \geqslant 30$ for any distribution of X, but *any n* provided X is normally distributed and σ is known and not estimated.
Sample Median	Population Median	$\simeq 1.25\sigma/\sqrt{n}$	$= 1.25s/\sqrt{n}$	$n \geqslant 30$ but standard error is only accurate for normal populations.
Sample Variance s^2	$\sigma^2\left(\frac{n-1}{n}\right)$	$\sigma^2\sqrt{\frac{2}{n}}$	Solve for σ^2 (see Example 2)	$n \geqslant 50$ but standard error is only accurate for normal populations.
Proportion† ($P = r/n$)	Population proportion π	$\sqrt{\frac{\pi(1-\pi)}{n}}$	$\sqrt{\frac{P(1-P)}{n}}$	$n \geqslant 30$ if $0.4 < \pi < 0.6$. For larger or smaller π larger sample sizes required (see CU15).
Difference between means of independent samples $\overline{X}_1 - \overline{X}_2$	$\mu_1 - \mu_2$	$\sqrt{\frac{\sigma_1^2}{n_1} + \frac{\sigma_2^2}{n_2}}$	$\sqrt{\frac{s_1^2}{n_1} + \frac{s_2^2}{n_2}}$	Provided conditions are met for $\overline{X}_1$ and $\overline{X}_2$ separately.

Difference between proportions† of independent samples $P_1 - P_2$	$\pi_1 - \pi_2$	$\sqrt{\frac{\pi_1(1-\pi_1)}{n_1} + \frac{\pi_2(1-\pi_2)}{n_2}}$	$\sqrt{\frac{P_1(1-P_1)}{n_1} + \frac{P_2(1-P_2)}{n_2}}$	Provided conditions are met for P_1 and P_2 separately.
Correlation Coefficient r	ρ	$\simeq \frac{1-\rho^2}{\sqrt{n}}$	$\frac{1-r^2}{\sqrt{n}}$	$n \geqslant 50$ provided $\lvert\rho\rvert$ is "small", say $\leqslant 0.5$. $n \geqslant 100$ provided ρ is not too close to ± 1.
Regression Coefficient b in $\hat{Y} = a + bX$	β in $Y = \alpha + \beta X$	$\sigma_e \sqrt{\frac{1}{\Sigma(X_i - \bar{X})^2}}$	Put $\hat{\sigma}_e^2 = \frac{1}{n-2} \Sigma(Y_i - \hat{Y}_i)^2$	$n \geqslant 30$, provided standard regression model applies (CU14).
$\hat{Y}_0 = a + bX_0$	$Y_0 = \alpha + \beta X_0$	$\sigma_e \sqrt{\frac{1}{n} + \frac{(X_0 - \bar{X})^2}{\Sigma(X_i - \bar{X})^2}}$	$= \frac{(1-r^2)}{n-2} \Sigma(Y_i - \bar{Y})^2$	$n \geqslant 30$, provided standard regression model applies (CU14).
An individual value of Y_0 for a given X_0	$Y_0 = \alpha + \beta X_0$	$\sigma_e \sqrt{1 + \frac{1}{n} + \frac{(X_0 - \bar{X})^2}{\Sigma(X_i - \bar{X})^2}}$		$n \geqslant 30$, and the distribution of errors about the regression line is normal.

†When working with percentages rather than proportions simply replace π by $\pi_{\%}$ and $(1 - \pi)$ by $(100 - \pi_{\%})$ in the SE formula.

16.2 EXAMPLES

For the first two examples we use the statistics calculated from our Series A data viz:

Mean ($\bar{X}$)	=	80.0
Median	=	80.5
SD (s)	=	19.0
(Sample size n = 100)		

Example 1. Calculate a 95% CI for the population mean and median.

$$\text{The CI for the mean } \mu \text{ is given by } 80.0 \pm 1.96 \frac{19.0}{\sqrt{100}} = 80.0 \pm 3.7$$

Using the standard error from Exhibit 1 we have:

$$\text{CI for the population median is given by } 80.5 \pm \frac{1.96 \times 1.25 \times 19.0}{\sqrt{100}} = 80.5 \pm 4.7$$

We see that the median has a wider CI than the mean – the median is a less *efficient* statistic. Indeed, to ensure the same degree of reliability in the median we would need to use a sample size which is $(1.25)^2 = 1.56$ times as large as that used for the mean. However, the process of setting a CI on a median (not on a mean) in this manner depends on the population being normally distributed (Series A *is* normal). The reason is that the factor 1.25 in the calculation of the SE is appropriate only to a normal population – for populations which have longer "tails" than the normal distribution (for the same SD) this factor becomes less than 1.25, and may in extreme cases be less than 1.0 (resulting in the median being a *more* efficient statistic than the mean). This argument ties up with the comment made in CU4 that the median is less affected by outlying observations in the tails of the distribution than is the mean.

Example 2. Calculate a 95% CI for the population variance.

Incorrect method. The SE of $s^2 = \sigma^2 \sqrt{\frac{2}{n}}$ for large n ($\geqslant$50).

We approximate[1] σ^2 by $s^2 = (19.0)^2$, so the SE of $s^2 \simeq (19.0)^2 \sqrt{\frac{2}{100}} \simeq 51.1$.

Thus the CI for σ^2 is given by $(19.0)^2 \pm 1.96 \times 51.1$
$\simeq 361 \pm 100$ (i.e. $261 < \sigma^2 < 461$).

Correct method. For the two confidence limits we have

$$\sigma^2 = s^2 \pm 1.96\sigma^2 \sqrt{\frac{2}{n}}$$

Solving for σ^2 gives $\sigma^2 = \dfrac{s^2}{1 \pm 1.96\sqrt{\frac{2}{n}}} = \dfrac{361}{1 \pm 1.96\sqrt{\frac{2}{100}}}$.

So $283 < \sigma^2 < 499$, a confidence interval which, unlike those previously calculated, is *not* symmetrically disposed about the sample statistic $s^2 = 361$.

It is important to note that we cannot similarly derive a CI for the standard deviation since it is only the variance, and not the standard deviation, which is normally distributed for large samples.

Example 3. A company specialising in personality testing used a SRS of 100 students who were found to take an average of 12 mins. (SD of 2 mins.) to complete a "standard" test. Another independent SRS of 400 students took an average of 10 mins. (SD of 3 mins.) to complete the same test, but these students, unlike the first 100, had been given some practice on similar material beforehand. Calculate a 99% CI on the reduction in average time taken brought about by the practice.

Incorrect method. For the first 100 students the CI for μ_1 is given by:

$$12 \pm 2.57 \cdot \frac{2}{\sqrt{100}} = 12 \pm 0.51$$

i.e. $11.49 < \mu_1 < 12.51$

---

[1]Because s^2 is a biased estimator of σ^2 (CU5), we could have "corrected for bias" by using

$$\hat{\sigma}^2 = s^2\left(\frac{n}{n-1}\right) = 361 \times \frac{100}{99} = 364.6$$

in place of s^2 (= 361) in our calculations. But this correction is inconsistent with a normal sampling distribution and is rarely used with large samples. It is necessary, however, with small samples (CU20).

For the 400 students the CI for μ_2 is given by:

$$10 \pm 2.57 \cdot \frac{3}{\sqrt{400}} = 10 \pm 0.39$$

i.e. $9.61 < \mu_2 < 10.39$

So the *greatest* reduction in average time is $12.51 - 9.61 = 2.90$ mins., and the *smallest* reduction in average time is $11.49 - 10.39 = 1.10$ mins. Consequently the CI for reduction in average time is:

$$1.10 < \mu_2 - \mu_1 < 2.90.$$

This method is incorrect because we cannot simply "join up" two (valid) 99% CI's in this way without affecting the percentage confidence of the end result. In fact the quoted answer gives a 99.97% CI.

Correct method. We have observed the value of the (composite) sample statistic, $\bar{X}_1 - \bar{X}_2$ (= 12 − 10 = 2), and we wish to quote a 99% CI for the value of the population parameter, $\mu_1 - \mu_2$. To do this we seek a value for the SE of the sample statistic in the following way:

$\text{Var}(\bar{X}_1 - \bar{X}_2) = \text{Var}(\bar{X}_1) + \text{Var}(\bar{X}_2)$, because $\bar{X}_1$ and $\bar{X}_2$ are independent (CU12)

$$= (\text{SE of } \bar{X}_1)^2 + (\text{SE of } \bar{X}_2)^2$$

$$= \frac{\sigma_1^2}{n_1} + \frac{\sigma_2^2}{n_2}$$

$$\text{So SE of } (\bar{X}_1 - \bar{X}_2) = \sqrt{\frac{\sigma_1^2}{n_1} + \frac{\sigma_2^2}{n_2}} \simeq \sqrt{\frac{s_1^2}{n_1} + \frac{s_2^2}{n_2}}$$

$$= \sqrt{\frac{4}{100} + \frac{9}{400}} = 0.25$$

Hence the CI for $\mu_1 - \mu_2$ is given by $2 \pm 2.57\,(0.25)$

$$= 2 \pm 0.64$$

or $1.36 < \mu_1 - \mu_2 < 2.64$, a narrower band than that given by the incorrect method.

Example 4. A polling organisation took a SRS[1] of 600 electors and found that 40% would vote for party X. A week later they took

[1] Polling organisations do not in fact use simple random samples because of the high cost involved. They tend to use "representative quota samples" (SU5). For such samples a valid standard error cannot be evaluated but it may be taken that the standard error is at least as great as that from an SRS.

another (independent) SRS of 500 electors and found that 45% of these would similarly vote for party X. Quote a 95% CI for the improvement of party X's percentage vote.

Let $P_{\%1} = 40\%$ $\quad n_1 = 600$

$P_{\%2} = 45\%$ $\quad n_2 = 500$

We have observed the value of $P_{\%2} - P_{\%1} = 5\%$ and wish to quote a 95% CI for the population parameter $\pi_{\%2} - \pi_{\%1}$.
Now Var $(P_{\%2} - P_{\%1})$ = Var $(P_{\%2})$ + Var $(P_{\%1})$, as $P_{\%2}$ and $P_{\%1}$ are independent (CU12)

$$= (\text{SE of } P_{\%2})^2 + (\text{SE of } P_{\%1})^2$$

$$= \frac{\pi_{\%2}(100 - \pi_{\%2})}{n_2} + \frac{\pi_{\%1}(100 - \pi_{\%1})}{n_1}$$

$$\text{So SE of } (P_{\%2} - P_{\%1}) \simeq \sqrt{\frac{P_{\%2}(100 - P_{\%2})}{n_2} + \frac{P_{\%1}(100 - P_{\%1})}{n_1}}$$

$$= \sqrt{\frac{40 \times 60}{600} + \frac{45 \times 55}{500}}$$

$$= \sqrt{8.95} = 2.99$$

So the CI for $\pi_{\%2} - \pi_{\%1}$ is given by $5 \pm 1.96\ (2.99) = 5 \pm 5.86$ or $-0.86 < \pi_{\%2} - \pi_{\%1} < 10.86$.

Note that this range includes zero; that is, we *cannot* be confident that the true support for party X has actually increased – beware of opinion polls! Actually the problem of determining whether there has been a real change in a population parameter is the province of *significance testing* (CU17 and 18), the techniques for which are closely similar to setting confidence intervals.

Example 5. The correlation coefficient r of 100 pairs of observations (X, Y) is calculated. Quote a 95% CI for the population parameter ρ for the two cases where (a) $r = 0.2$, (b) $r = 0.8$.

As the conditions set out in Exhibit 1 are met, we may state;

$$\text{SE of } r = \frac{1 - \rho^2}{\sqrt{n}} \simeq \frac{1 - r^2}{\sqrt{n}}$$

So for (a) SE of $r = \dfrac{1 - (0.2)^2}{\sqrt{100}} = 0.096$: hence for ρ the CI is givenby

$0.2 \pm 1.96\ (0.096) = 0.2 \pm 0.19$, i.e. $0.01 < \rho < 0.39$, and so on this basis we can be 95% confident (just!) that the population is correlated, and that our value of $r = 0.2$ is not simply the result of sampling fluctuations.

For (b) we have SE of $r = \dfrac{1-(0.8)^2}{\sqrt{100}} = 0.036.$

$$\text{giving } 0.8 \pm 1.96\,(0.036) = 0.8 \pm 0.07 \text{ (i.e. } 0.73 < \rho < 0.87)$$

We see that the CI is narrower for high values of r than for low values of r. Why do you think this is? (Consider the limiting cases of $r = 0$ and $r = 1$.)

The method employed here is strictly only valid in cases where both X and Y are normally distributed, but it may be taken to give reasonably accurate results provided that the scatter diagram is roughly "cigar shaped". In addition the method is one that *cannot* be adopted for use with small samples – in fact the sampling distribution of r is not normal for small samples, even if the population itself is normally distributed. To cope with this other methods of setting CI's on correlation coefficients have been proposed. David's *Tables of the Correlation Coefficient* (Biometrika Office, University College, London) provide a convenient alternative method, as does Fisher's Z-transformation method, the details (and tables) for which can be found in most advanced texts. Using the Fisher Z-transformation on our example the following values are obtained:

$$\text{(a) } 0.00 < \rho < 0.38; \qquad \text{(b) } 0.72 < \rho < 0.86,$$

results which differ little from ours.

Example 6. A bivariate sample of 40 observation resulted in the following statistics:

$$\sum X = 292,\ \sum Y = 768,\ \sum X^2 = 2{,}190,\ \sum Y^2 = 15{,}648,\ \sum XY = 5{,}790.$$

(i) Calculate the regression line of Y on X, $\widehat{Y} = a + bX$.
(ii) Quote a 95% CI for the value of β, the *population* regression coefficient in $\widehat{Y} = \alpha + \beta X$.
(iii) Obtain $\widehat{Y}_0$ for $X_0 = 9.0$ and quote a 95% CI for the value of $\widehat{Y}_0$ that would be obtained from the population regression.
(iv) Establish a 95% *Prediction Interval* for Y_0 the observed value associated with an individual $X_0 = 9.0$.

In such examples it is well worthwhile to calculate initially the value of r, the *correlation* coefficient, even though it is not specifically asked for. The reasons are threefold: it enables us to check the strength of the association between X and Y; all the intermediate calculated quantities can be used later in the analysis; the value of r itself is used in the determination of the various standard errors.

We have $\bar{X} = 292/40 = 7.3$ and $\bar{Y} = 768/40 = 19.2$. So

$$r = \frac{\sum XY - n\bar{X}\bar{Y}}{\sqrt{[\sum X^2 - n\bar{X}^2][\sum Y^2 - n\bar{Y}^2]}}$$

$$= \frac{5{,}790 - 40(7.3)(19.2)}{\sqrt{[2{,}190 - 40(7.3)^2][15{,}648 - 40(19.2)^2]}}$$

$$= \frac{183.6}{\sqrt{58.4 \times 902.4}} = 0.800,$$

showing a high measure of positive linear association (CU13).

We now proceed with the solution:

(i) $b = \dfrac{\sum XY - n\bar{X}\bar{Y}}{\sum X^2 - n\bar{X}^2} = \dfrac{183.6}{58.4} = 3.144$ (by simply taking the values from the calculation of r, above) and $a = \bar{Y} - b\bar{X} = 19.2 - 3.144(7.3) = -3.751$. So $\hat{Y} = -3.751 + 3.144X$.

(ii) For the purpose of calculating all CI's we need to assume that the "standard" regression model is valid (CU14). To evaluate all the standard errors required we first need to estimate σ_e^2, the residual variance (of the Y–errors about the regression line). We *could* evaluate this directly from CU14:

$$\sigma_e^2 = \frac{1}{n-2}\sum(Y_i - \hat{Y}_i)^2,$$

if we had access to the individual Y_i values. But in most cases it is more convenient to use the equivalent formula (also from CU14):

$$\hat{\sigma}_e^2 = \frac{1}{n-2} \cdot (1 - r^2)\sum(Y_i - \bar{Y})^2 = \frac{(1 - 0.800^2)(902.4)}{38}$$

$$= 8.549$$

So

$$\hat{\sigma}_e = \sqrt{8.549} = 2.924.$$

Note that we have again taken $\Sigma(Y_i - \bar{Y})^2 = \Sigma Y^2 - n\bar{Y}^2 = 902.4$ from our earlier calculations of r (this is also true of $\Sigma(X_i - \bar{X})^2$ in the following calculation).

Now the SE of $b = \sigma_e \sqrt{\dfrac{1}{\Sigma(X_i - \bar{X})^2}} \simeq 2.924 \sqrt{\dfrac{1}{58.4}} = 0.383$

So 95% CI for β is given by $3.144 \pm 1.96\ (0.383) = 3.144 \pm 0.751$, or β lies between 2.393 and 3.895 with 95% confidence.

(iii) For $X_0 = 9.0$ we obtain $\hat{Y}_0 = a + bX_0 = -3.751 + 3.144\ (9.0) = 24.545$.

This, of course, is the predicted value of $\hat{Y}_0$ based on the sample regression. We quote a 95% CI for the value of $\hat{Y}_0$ from the *population* regression, $\hat{Y}_0 = \alpha + \beta X_0$, as follows:

$$\text{SE}^{(1)} \text{ of } \hat{Y}_0 = \sigma_e \sqrt{\frac{1}{n} + \frac{(X_0 - \bar{X})^2}{\Sigma(X_i - \bar{X})^2}}$$

$$= 2.924 \sqrt{\frac{1}{40} + \frac{(9.0 - 7.3)^2}{58.4}}$$

$$= 2.924\sqrt{0.025 + 0.0495}$$

$$= 2.924\sqrt{0.0745} = 0.798$$

So a 95% CI for the population prediction of Y_0 is given by:

$$24.545 \pm 1.96(0.798) = 24.545 \pm 1.564.$$

(iv) If we wish to estimate a range of values for an *individual* Y_0 (corresponding to $X_0 = 9.0$) we have to take into account not only the errors in the regression equation itself, but also the residual error about the regression line. Such a range of values is called a prediction interval, not a confidence interval. For a valid prediction interval we need to know the *distribution* of residual Y–errors about the true regression line: we assume a normal distribution.

$$\text{SE of individual } Y_0 = \sigma_e \sqrt{1 + \frac{1}{n} + \frac{(X_0 - \bar{X})^2}{\Sigma(X_i - \bar{X})^2}}$$

$$= 2.924\sqrt{1 + 0.0745} = 3{:}031$$

---

(1)Taking the prediction $\hat{Y}_0 = a + bX_0$ and adding $\bar{Y} - a - b\bar{X}$ (= 0) to the right-hand side we arrive at $\hat{Y}_0 = \bar{Y} + b(X_0 - \bar{X})$.
Taking variances by using the rules of CU12 ($\bar{Y}$ and b are independent statistics),

$$\text{Var}\ (Y_0) = \text{Var}\ (\bar{Y}) + (X_0 - \bar{X})^2 \cdot \text{Var}\ (b)$$

$$= \frac{\sigma_e^2}{n} + (X_0 - \bar{X})^2 \cdot \frac{\sigma_e^2}{\Sigma(X_i - \bar{X})^2} = \sigma_e^2 \left\{\frac{1}{n} + \frac{(X_0 - \bar{X})^2}{\Sigma(X_i - \bar{X})^2}\right\}$$

Hence the formula for the SE.

So 95% Prediction Interval for Y_0 (X_0 = 9.0)
= 24.545 ± 1.96 (3.031) = 24.545 ± 5.941.
The solutions to this example are illustrated in Exhibit 2.

EXHIBIT 2: Applications of Confidence Intervals to Regression

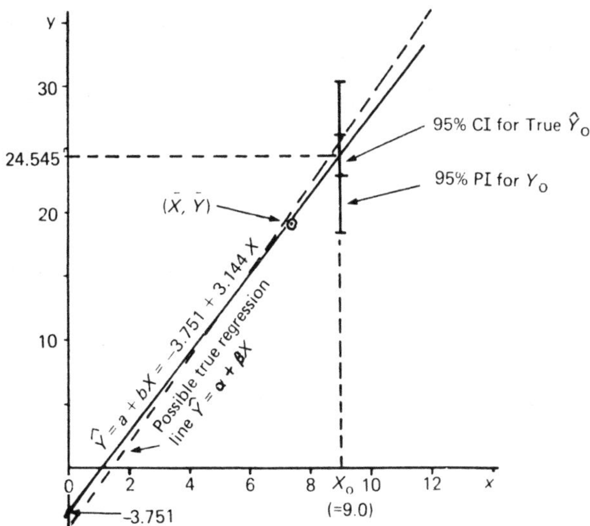

EXERCISES

1(U) The following statistics relate to a SRS of 100 *bivariate* observations (X_i, Y_i):

$$\sum X = 980; \sum Y = 3{,}012; \sum X^2 = 10,948; \sum Y^2 = 101,209;$$
$$\sum XY = 31,849.$$

Show that with 95% confidence:

(i) the population mean of the Y values lies between 28.11 and 32.13.

(ii) the population variance of the Y values lies between 82.11 and 145.09.

(iii) the population correlation coefficient ρ lies between 0.50 and 0.74.

(iv) the population regression coefficient β lies between 1.30 and 2.17.

(v) the predicted value $\hat{Y}_0$ for $X_0 = 9.80$ lies between 28.53 and 31.71; and for $X_0 = 15.0$ it lies between 36.38 and 41.90. Explain why the first interval (for $X_0 = 9.80 = \bar{X}$) is narrower than that given in part (i); and why the second interval (for $X_0 = 15.0$) is wider than the first.

In each case state any assumptions you are making about the population from which the sample is drawn.

2(E) A survey among a SRS of men and women gave the following results.

	Male	Female
Smokers	30	40
Non-Smokers	70	160

Place a 95% CI on the percentage difference in smokers between males and females. Comment on your results.

Part question BA (Business Studies)

3(U) Taking the results from Exercise 2 for men only, calculate a 95% CI for the percentage of men who smoke, in the usual way i.e.

$$\pi_{\%} = P_{\%} \pm 1.96\sqrt{\frac{P_{\%}(100 - P_{\%})}{n}}.$$

Also calculate this CI *without* approximating $\pi_{\%}$ by $P_{\%}$, by solving

$$\pi_{\%} = P_{\%} \pm 1.96\sqrt{\frac{\pi_{\%}(100 - \pi_{\%})}{n}}$$

for $\pi_{\%}$. Comment on your results.

4(E) The following sketch illustrates how two components A and B, manufactured on separate production lines, are assembled.

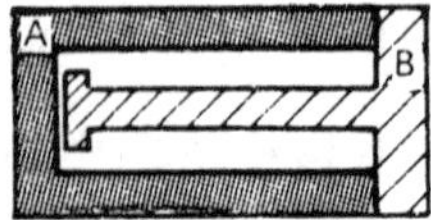

X is the depth of the cavity in A

Y is the length of the plunger B

The critical dimension of the assembled components is the clearance between the plunger base and the base of the cavity (i.e. $X - Y$).

(i) A sample of 100 A components has mean value of the X dimension 1.0050" and SD 0.0020". A sample of 25 B components has mean value of the Y dimension of 0.9950" with SD 0.0010". Calculate 99% confidence limits for the mean clearance of all assembled components.

(ii) The process of manufacturing A and B is to be controlled by taking hourly samples of A and B (a total of 5 components in all, A's and B's, is to be selected from the production lines each hour). In order to obtain the most accurate information on the current clearance value, how many A's and how many B's should be taken?

(Note: do not attempt to use the calculus in answer to part (ii). You may assume that both X and Y are normally distributed.)

BA (Business Studies)

SOLUTIONS TO EVEN NUMBERED EXERCISES

2. Working in percentages we have:

$$\text{Sample \% of males who smoke} = \frac{30}{100} \times 100 = 30\% = P_{\%1}$$

$$\text{Sample \% of females who smoke} = \frac{40}{200} \times 100 = 20\% = P_{\%2}$$

$$\text{SE of difference in \%'s} \simeq \sqrt{\frac{P_{\%1}(100 - P_{\%1})}{n_1} + \frac{P_{\%2}(100 - P_{\%2})}{n_2}}$$

$$= \sqrt{\frac{30 \times 70}{100} + \frac{20 \times 80}{200}}$$

$$= \sqrt{29} = 5.385\%$$

Observed difference in sample %'s = 30 – 20 = 10%.
Hence 95% CI for true difference in %'s is given by

$$10 \pm 1.96 \times 5.385 = (10 \pm 10.55)\%.$$

This range includes zero and so there may be no real difference between the percentages of men and women who smoke. Strictly this conclusion (and the method used) should be drawn by conducting a significance test (CU18) – the arithmetic is then *slightly* different.

4. We are interested in the accuracy of $\bar{X} - \bar{Y}$ (= 0.0100″).

(i) $$\text{SE of } (\bar{X} - \bar{Y}) \simeq \sqrt{\frac{s_X^2}{n_X} + \frac{s_Y^2}{n_Y}}$$

$$= \sqrt{\frac{0.000004}{100} + \frac{0.000001}{25}} = 0.00028$$

99% CI is given by (0.0100 ± 2.58 × 0.00028)″ ≃ (0.0100 ± 0.0007)″

(ii) "The most accurate information" is given when the SE of $\bar{X} - \bar{Y}$ has a minimum value.

Sample Constitution		*SE of $(\bar{X} - \bar{Y})^2$ in units of 0.000001*
5 of A		Not known as no B components taken
4 of A	1 of B	$\frac{4}{4} + \frac{1}{1} = 2$
3 of A	2 of B	$\frac{4}{3} + \frac{1}{2} = 1.83$
2 of A	3 of B	$\frac{4}{2} + \frac{1}{3} = 2.33$
1 of A	4 of B	$\frac{4}{1} + \frac{1}{4} = 4.25$
5 of B		Not known as no A components taken

Hence the sample should consist of 3A and 2B components. (Note that the sampling distribution of $\bar{X} - \bar{Y}$ is normal even for small samples because X and Y are both normally distributed).

COURSE UNIT 17 – INTRODUCTION TO SIGNIFICANCE TESTING: STATISTICAL HYPOTHESES, TYPE I AND II ERRORS

17.1 A LEGAL ANALOGY

CU15 and CU16 have been concerned with the process of estimating the values of population parameters. We now turn to an allied problem – that of making statistical judgements based on sample data. The mechanics of the approach used are very similar to those used in the construction of confidence intervals but the underlying philosophy is quite different.

In business we are constantly called upon to make judgements based partly on numerical information. We may be asked for example to consider whether a recently introduced change in the way we conduct our business has led to an increase in output, greater demand for our products or improved labour relations. In answer to such questions we cannot normally reply in purely probabilistic terms: we are instead required to answer either "yes" or "no" (albeit suitably qualified) or at least plead that further data must be collected before a well-founded judgement can be made. It is of course of paramount importance to understand, and to communicate, precisely what is implied by making a judgement in this way.

There are a number of different approaches used to make statistical judgements, all of which are termed "significance testing". The approach used here is the so-called "classical" approach, although hints of another approach will be given in Section 17.6. The manner in which classical significance testing is carried out has a number of important features in common with the English legal system, a brief sketch of which is now given.

If it were possible to present a court with a *complete* and accurate statement of *all* the facts relating to a particular crime the establishment of guilt or innocence would become a very much easier task. Unfortunately in law, as in Statistics, we have to pass a judgement on the basis of partial information, some of it possibly misleading. A cardinal principle of English law is exercised when the jury is instructed to assume initially that the accused man is innocent. Only if the evidence presented indicates "beyond reasonable doubt" the guilt of the accused can the jury bring in a verdict of "guilty". As a consequence of this principle it is indeed possible, in cases where the evidence presented is sketchy, to find an accused man "not guilty" when in fact he had committed the crime. Statistical significance testing proceeds in a similar manner: *before* assessing the importance of the evidence provided by a sample statistic we make an assumption (called the "null hypothesis") that the sample has been drawn from some known, or conveniently postulated, population (like the assumption of innocence). Only if the evidence presented by the sample indicates "beyond reasonable doubt" that the sample could not come from the postulated population do we reject our original assumption (in statistical terms the result is then "statistically significant"). Just as in the legal case, if the evidence is insufficient to do this, we cannot necessarily assume that our assumption is correct: it is simply that the evidence is not strong enough to prove otherwise. (The reader may be aware that, in contrast to the English system, the Scottish legal system permits a "not proven" verdict which allows for a retrial if further evidence should come to light. Later, we shall find an analogous procedure in statistical significance testing.)

Let us consider one more point of similarity. We have already noted that there is a chance, as the result of insufficient evidence, of arriving at a verdict of "not guilty" for a man who has committed the crime for which he has been tried. It is also possible in cases where the circumstantial evidence is great to bring in a verdict of "guilty" for a truly innocent man. The statistical counterpart of this occurrence is where, as a rare chance event, we just happen to examine a sample which is unrepresentative of the population from which it is drawn and consequently conclude (wrongly) that it must have come from some other population.

The reader is advised to treat this Course Unit rather differently from the rest – it should be read with a view to understanding the *principles* of significance testing, rather than the details of the calculations. The detail can be taken in during CU18, or at a later re-reading.

17.2 THE NULL HYPOTHESIS AND TYPE I ERRORS

In Sections 17.2–4 of this Course Unit the classical approach to significance testing will be developed making use of an example drawn from marketing where a decision had to be taken on the evidence of a sample of consumers' reactions to a trial batch of a new product. From time to time we shall refer back to our legal analogy.

A food manufacturing company is considering introducing a new formulation of meat paste to replace the existing product. The new formulation is intended to be identical with the old one, but it is cheaper to manufacture and hence has a greater profit margin. However, it will not be introduced if the public can detect any difference, since the "goodwill" build up under the old brand name will be lost.

A trial batch is prepared and 20 housewives are selected at random from regular buyers of the product over the past six months. The housewives are told that one of the products is a new one, they are then given two sandwiches to taste and afterwards are asked which of the two sandwiches contained the new product. From the results of this test a decision has to be made whether to proceed with the new formulation or not.

The company would ideally wish to know which of the following two statements (which we shall refer to as "states of nature") represents the true state of affairs:

(1) The new product (N) is really not distinguishable from the old product (O), or
(2) N is in fact distinguishable from O.

If (1) were known to be true the company could proceed with the new formulation with confidence; if (2) were known to be true they would stick to the old formulation. In either event, by the company's own lights, the correct decision would be taken.

Unfortunately the company does not have access to such precise information – the best they can do is to examine the evidence presented by the sample and draw their own conclusions. They will either:

(i) conclude that (1) is true (and hence (2) is false), or
(ii) conclude that (2) is true (and hence (1) is false)

and act accordingly. That being so, we have to admit the possiblity that, resulting from the random sampling errors ("guesses" by the housewives), the company might draw the wrong conclusion and hence take the wrong decision.

Conceptually speaking then, we have four distinct possibilities to consider – the two states of nature, (1) and (2) combined with each of the two possible outcomes of the evidence, (i) and (ii). We can conveniently summarise these possibilities in a table as shown on p. 310 in Exhibit 1.

It can be seen that in only two of these four cases is the "correct decision" taken[(1)] – in the other two the decision would be incorrect (like finding a truly innocent man guilty or acquitting a man who is in

[(1)]The phrase "correct decision" should be interpreted in its narrowest sense, i.e. in agreement with the previously expressed view of the company that it would only wish to proceed with the new formulation if N and O were in fact indistinguishable. Whether that view is in the best interests of the company is a separate issue, and in any case we shall need to modify that "view" to take account of various risks later.

EXHIBIT 1: Decision Table for Marketing Problem

Sample Evidence (and decision) \ *State of Nature*	*(1) O and N Indistinguishable*	*(2) O and N Distinguishable*
(i) Conclude that O and N are indistinguishable (hence manufacture N)	Correct decision taken	Incorrect decision taken
(ii) Conclude that O and N are distinguishable (hence manufacture O)	Incorrect decision taken	Correct decision taken

fact guilty). From this logical framework we may built up an approach to tackling the problem in decision making.

First, we put forward a hypothesis. The hypothesis is usually constructed as a **null hypothesis**, "null" meaning "no change" or "no difference". The formulation of this null hypothesis is important since it must be capable of being tested, and in order to test it we must be able to derive the sampling distribution of the statistic obtained from our experiment.

In our case, the null hypothesis (denoted by the symbol H_0) can be stated as: "There is no distinguishable difference between the products". So if H_0 is true the sampling distribution of the number of persons correctly identifying the new product may be assumed to be binomial with $n = 20$ and $\pi = 0.5$.[(1)] That is to say, the probabilities in Exhibit 2 are those associated with obtaining 0, 1, 2, 3, . . ., 20 correct identifications of N *by chance* if the products are truly indistinguishable. (Note that it would be incorrect to adopt the hypothesis: "There is a distinguishable difference between the products" because this hypothesis is not sufficiently precise for the purpose of deriving a sampling distribution unless we are prepared to hypothesise further just *how* distinguishable the products are.)

The mean number of correct identifications (if H_0 is true) over a large number of such tasting experiments is of course 10 (= $n\pi$) although we might find the actual number in one experiment to be somewhat greater or less than 10 on account of sampling fluctuations (in accordance with the binomial probabilities). But if we find the actual number to be much greater than 10, then this must cast doubt on the truth of

[(1)] π is interpreted here as the proportion of all consumers who would correctly identify N under similar conditions.

H_0. In that case we might entertain the possibility that the products can truly be distinguished by at least some consumers.

We now make use of the probabilities in Exhibit 2 (and the histogram drawn from them in Exhibit 3) to assist in the construction of a suitable decision rule to determine what action should be taken on the outcome of the sampling experiment.

EXHIBIT 2: Individual Terms of the Binomial Distribution $n = 20$, $\pi = 0.5$

r		$p(r)$	
0		.0000	
1		.0000	
2		.0002	
3		.0011	
4		.0046	
5		.0148	
6		.0370	0.9423 (r = 0 to 13)
7		.0739	
8		.1201	
9		.1602	
10		.1762	
11		.1602	
12		.1201	
13		.0739	
14		.0370	
15		.0148	
16		.0046	
17	0.0207 (r = 15 to 20)	.0011	0.0577 (r = 14 to 20)
18		.0002	
19		.0000	
20		.0000	
	Total	1.0000	

EXHIBIT 3: Histogram of Binomial Distribution $n = 20$, $\pi = 0.5$

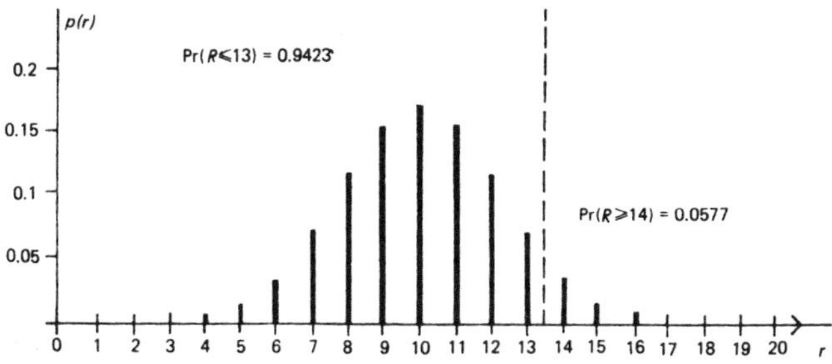

Number of correct identifications of N if H_0 is true

Consider for the sake of argument, the following decision rule: "If between 14 and 20 persons correctly identify N we will admit that the products are distinguishable, that is we will reject the null hypothesis. If 0–13 persons correctly identify N we will accept the null hypothesis".[1]

It can be seen from Exhibit 2 that the consequence of this decision rule is that there is a probability of 0.0577 of rejecting H_0 *when it is in fact true*. On those occasions when, as the result of sampling fluctuations, we do wrongly reject H_0 we say that we have committed a Type I error (and use the symbol α to denote the probability of this occurring). The corresponding error in our legal analogy occurs when an innocent man is found guilty.

Consequently in a long series of such tests (with H_0 true in each) we would falsely reject H_0 on 5.77% of occasions. But note we are able to determine this probability in advance of the test. If we consider a 5.77% risk of a wrong decision of this kind to be too high we may alter our decision rule in one of two ways.

(a) We may amend the rule to read, say:
"Reject H_0 if the number of persons correctly identifying N is between 15 and 20 inclusive; otherwise accept H_0".
The probability of obtaining 15–20 correct choices when H_0 is true is shown to be 0.0207 or 2.07% – a considerable reduction in α over the previous rule.

(b) Alternatively we may alter the sampling distribution. One possibility is by "doubling up", taking a sample of 40 housewives and rejecting H_0 if 28–40 of them are correct. As the reader may care to show, for $n = 40, \pi = 0.5$, this leads to the lower value of $\alpha = 0.0088$.

However, returning to point (a), we can see that we could eliminate our Type I error entirely by changing the decision rule to: "Accept H_0 if between 0 and 20 correct decisions are made". In this case we are making the same decision whatever the outcome of the experiment since no alternative outcomes exist, and the experiment need not be performed at all. (In law this is equivalent to finding all defendants "not guilty" irrespective of the evidence given at the trial).

---

[1] In consequence of the way in which the decision rule is worded there is one rejection region 14–20, and one acceptance region, 0–13. Thus we reject H_0 if the statistic falls in one particular tail (in this case the right hand tail) of the sampling distribution. Such a procedure is known as a one-sided or one-tailed test to distinguish it from the two-sided tests we shall encounter later.

17.3 TYPE II ERRORS

We have seen that a Type I error occurs when we wrongly reject the null hypothesis. There is another type of error, called a Type II error, we can commit: this occurs in conditions where the null hypothesis is in fact false and yet we accept it, like finding a man "not guilty" when he has actually committed the crime.

Consider again our example of the extreme decision rule (accept H_0 if between 0 and 20 correct identifications are made) which always accepts the null hypothesis. Although, by this means, we have eliminated the risk of a Type I error we stand a high chance of making a Type II error. As we would always accept that there is no difference between the two products, we should always be wrong when a truly distinguishable product is being tested. This illustrates the general principle that as we decrease the chance of a Type I error in a given situation we automatically increase the chance of a Type II error. In practice we have to decide on acceptable levels of risk in connexion with these two types of error (and the wrong decisions based on them), and design our experiment and decision rule in such a way that we satisfy both probabilities simultaneously. For experiments involving such simple statistics as the sample mean or proportion, the required sampling distribution may be obtained by choosing the sample size appropriately.

So a Type II error occurs when we accept H_0 to be true when in fact it is false. The symbol β is often used to denote the probability of this occurring. Unfortunately we cannot obtain one single value for β since the probability will depend on how false the null hypothesis is. If the hypothesis is only just false, say if only 5% of housewives can really distinguish the new product, then our Type II error will have a certain probability; if on the other hand 60% of housewives can really distinguish N from O then the null hypothesis is clearly wrong and the probability of accepting it will be much lower.[1]

Taking the broadest view, the true "state of nature" may be anything in the range of between 0–100% of housewives being able to distinguish between the two products.[2] The associated probabilities of making a Type II error consequently form a continuous curve over this

---

(1) Note that if 5% of housewives can truly distinguish the new product (as distinct from correctly identifying N by "guessing") the proportion (π) of all housewives who would correctly identify N for whatever reason is given by 5% + $\frac{1}{2}$ (95%) = $52\frac{1}{2}$%. Similarly if 60% of housewives can truly discriminate then π = 0.80 (i.e. 60% + $\frac{1}{2}$ (40%) = 80%).

(2) In view of footnote (1) above, this statement is equivalent to saying that all states of nature are accommodated by values of π in the range 0.5 to 1.0.

range. By changing the decision rule from accepting H_0 if 0–13 correctly identified N to accepting H_0 if 0–14 correctly identified N we reduce the probability of committing our Type I error from 5.77% to 2.07% but in so doing, we will have increased the probability of our Type II errors over the whole range of possible states of nature.

17.4* THE OPERATING CHARACTERISTIC CURVE

Having formulated our null hypothesis and fixed on our decision rule (of accepting H_0 if 0–13 persons correctly identify N) we may now proceed to calculate the Type II errors. The values in Exhibit 4 were derived from tables of the binomial distribution (all with n = 20, our sample size).(1)

EXHIBIT 4: Calculation of Type II Errors

π	$Pr(R \geqslant 7)$	π	$Pr(R \leqslant 13) = Pr(H_0 \text{ accepted})$
0.50	0.9423	0.50	0.9423
0.45	0.8701	0.55	0.8701
0.40	0.7500	0.60	0.7500
0.35	0.5834	0.65	0.5834
0.30	0.3920	0.70	0.3920
0.25	0.2142	0.75	0.2142
0.20	0.0867	0.80	0.0867
0.15	0.0219	0.85	0.0219
0.10	0.0024	0.90	0.0024
0.05	0.0000	0.95	0.0000

There is, for example, a probability of 0.7500 of accepting H_0 when the true value of π is 0.60.

The results from Exhibit 4 can be plotted as an "Operating Characteristic Curve" (Exhibit 5) for the decision rule we have used. The function of the O.C. curve is to enable us to study the behaviour of the decision rule in conditions where the null hypothesis is false. So if we have a certain proportion of housewives who *can* discriminate between the two products (and hence π is no longer 0.5) the O.C. curve enables us to determine the probability of (wrongly) accepting the null hypothesis that the housewives cannot discriminate.

(1)The column headed Pr($R \geqslant$ 7) is taken directly from cumulative binomial tables. As we are only interested in values with $\pi > 0.50$, (and only values with $\pi \leqslant 0.50$ are given in the tables) we need to derive these using the complementarity property of the binomial distribution. As an example, Pr($R \leqslant$ 13) for $n = 20$, $\pi = 0.60$ is equal to Pr($R \leqslant$ 7) for $n = 20$, $\pi = 0.40$ – the result of interchanging "success" and "failure" in the interpretation of the binomial distribution.

We can interpret this probability as implying that if the proportion of *all* consumers who would correctly identify N under similar conditions (π) is 0.60, say, then in a large series of such tests we would expect to accept H_0 ($\pi = 0.5$) in 75 out of every 100.[1]

Exhibit 6 shows graphically the sampling distributions of the number of correct identifications for various values of π. This shows vividly how the probability of accepting H_0 decreases as π departs further away from 0.5. The O.C. curve, of course, simply and economically summarises the properties of all such sampling distributions.

EXHIBIT 5: Operating Characteristic (O.C.) Curve

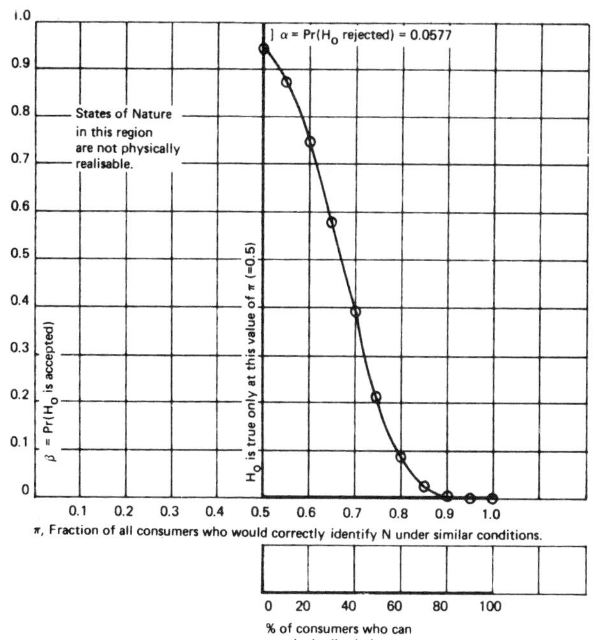

Decision Rule: $n = 20$; Accept H_0 if $R \leqslant 13$ where R is the number of correct identifications made.

---

(1)If it appears to the reader that our rule performs very poorly in its ability to distinguish between alternative states of nature, he is quite correct! The rule we have developed was chosen purely for illustrative purposes: in practice a much larger sample size would be employed and this would give a far improved performance. Also in practice tasters would be asked to pick the "odd man out" from three (or more) sandwiches arranged so that in some cases only one sandwich contains the new formulation, and in others only one contains the old formulation. This device improves the performance of the test further.

EXHIBIT 6: Sampling Distributions for Various Values of π

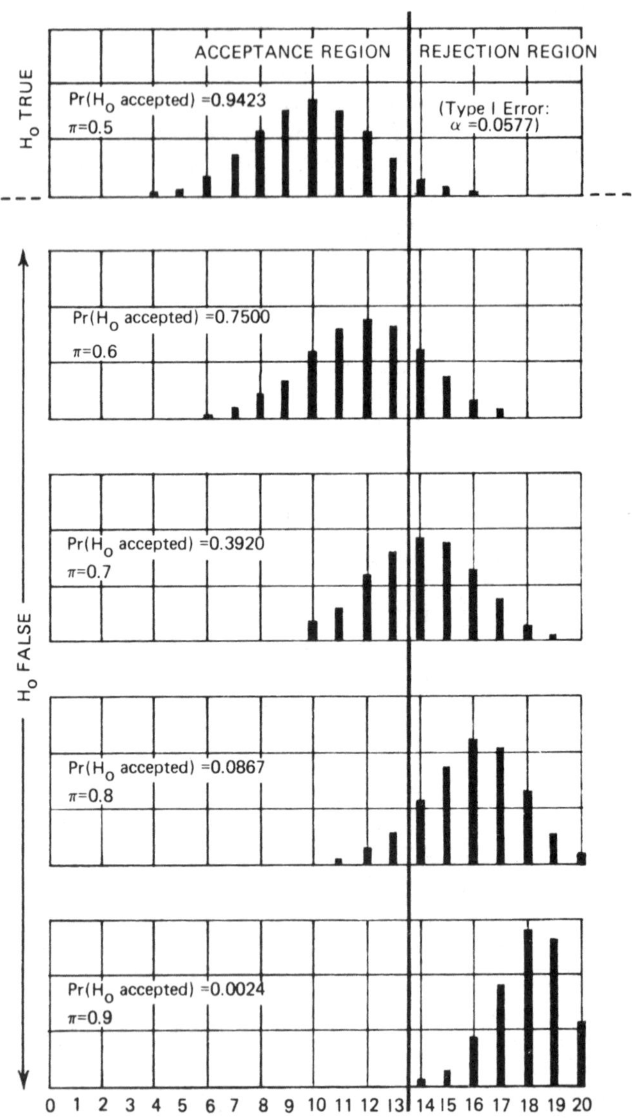

17.5 THE ALTERNATIVE HYPOTHESIS

In the example we have been considering we would reject H_0 only if the number of correct identifications were 14 or over. If on the other hand we observed a very small number of correct identifications (say 4), then although this would be interpreted as a very unlikely event, it would not give us grounds for rejecting H_0. Thus the test we have described is **one-sided** because rejection of H_0 occurs only on one side of the sampling distribution. If we reject H_0 what we are in fact doing is to accept an **alternative hypothesis** (H_1) that $\pi > 0.5$. In formal terms we have, as a statement of our test

$$H_0 : \pi = 0.5$$

and

$$H_1 : \pi > 0.5$$

Often we argue the other way around: having specified the form of the alternative hypothesis, this decrees whether a one-sided or two-sided test is appropriate. Alternative hypotheses formulated with $>$, $<$ or = signs imply one-sided tests; alternative hypotheses using $\neq$ signs imply two-sided tests.

For our next example we consider a two-sided test. To demonstrate the wide applicability of significance testing this example is set in a production, instead of a marketing context and the test concerns mean values rather than proportions.

The example relates to a Statistical Quality Control chart, the construction of which is covered in some detail in SU5. Here it suffices to say that a control chart is set up to detect if the mean value of a process variable under study is different from the value laid down for the process, namely the mean value set by the design engineer. The technique operates by taking a small sample of production, measuring the same dimension of each item, and calculating the mean value of these measurements. If this sample mean is found to lie outside either of two "control limits"[(1)] (one set above the designed mean, one below it) then the process is said to be "out of control" and appropriate action is taken.

We assume, in what follows, that the measured quantities are near normally distributed, and their variance is known.

The mean value (μ) of the thickness of a spindle 2" from a flange is designed to be 0.120". Over a long period the standard deviation (σ) of the process is found to be 0.003". Random samples of n = 4 spindles are taken every 20 minutes: if the sample mean of the spindle thick-

---

[(1)] To simplify the problem we shall use only what are referred to as "inner" or "warning" limits (SU4).

nesses is found to be outside the limits given by $\mu \pm 1.96\ \sigma/\sqrt{n}$, then the machine is switched off and reset.(1)

From our knowledge of the normal distribution we know that 95% of sample means (from a stable process) will fall between the limits calculated according to the above formula, as shown in Exhibit 7.

EXHIBIT 7

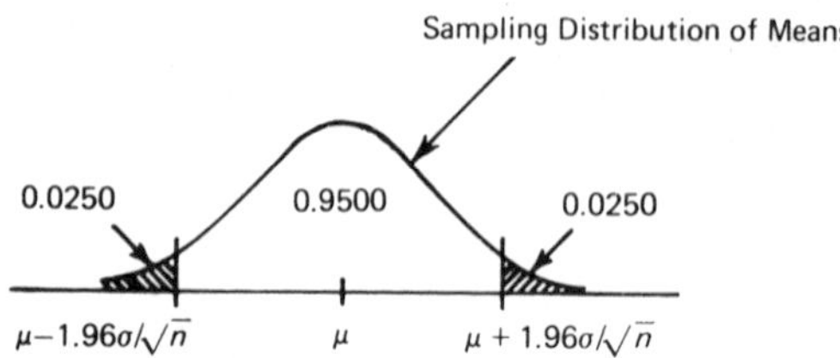

In significance testing terms, the null hypothesis is that the process mean at time t, μ_t, is no different from μ_0 (the design mean), i.e.

$$H_0 : \mu_t = \mu_0 \quad \text{or} \quad \mu_t - \mu_0 = 0$$

The probability of a type I error associated with the above decision rule is clearly 0.025 + 0.025 = 0.05 (5%) because we reject the null hypothesis in both tails of the sampling distribution.

In our particular example the control limits specifying the decision rule, work out to be

$$0.120'' \pm \left(1.96 \cdot \frac{0.003}{\sqrt{4}}\right)'' \simeq 0.120'' \pm 0.003'' \text{(i.e. } 0.117'' \text{ and } 0.123'')$$

Our decision rule points to the fact that it is undesirable to have a process mean which is either greater than or less than the design mean: a spindle will be considered unsatisfactory if it is either too large or too small. Hence in rejecting the null hypothesis we will be accepting the alternative hypothesis:

$$H_1 : \mu_t \neq \mu_0 \quad \text{or} \quad \mu_t - \mu_0 \neq 0^{(2)}$$

---

(1) As a matter of fact the accepted procedure is more subtle than this. The machine would only be switched off and reset if the sample mean fell outside *these* limits on two consecutive samples, but the principle still stands. (See SU4.)

(2) It would be quite in order, statistically speaking, to formulate the alternative hypothesis in a one-sided fashion, but the interpretation of the decision rule would then be quite different. If we had $H_1 : \mu_t > \mu_0$, then we would only have one (upper) control limit and this would imply that we were only interested in guarding against spindles which are too thick. Alternatively if we used $H_1 : \mu_t < \mu_0$ then we would only have a lower control limit to protect against spindles which are too thin. Note that in either case H_0 and H_1 taken together, do not cover all possible states of nature – only those states of nature in which we are interested.

For the purposes of comparison with our earlier marketing problem Exhibits 8 and 9 show some sampling distributions and the O.C. curve, respectively, for this example, on the assumption that the standard error of the sample mean remains constant at $0.003/\sqrt{4} = 0.0015''$. It should be noted that in consequence of conducting a two-sided test, the O.C. curve is itself "two-sided".

EXHIBIT 8: Sampling Distributions for Various Values of μ_t'

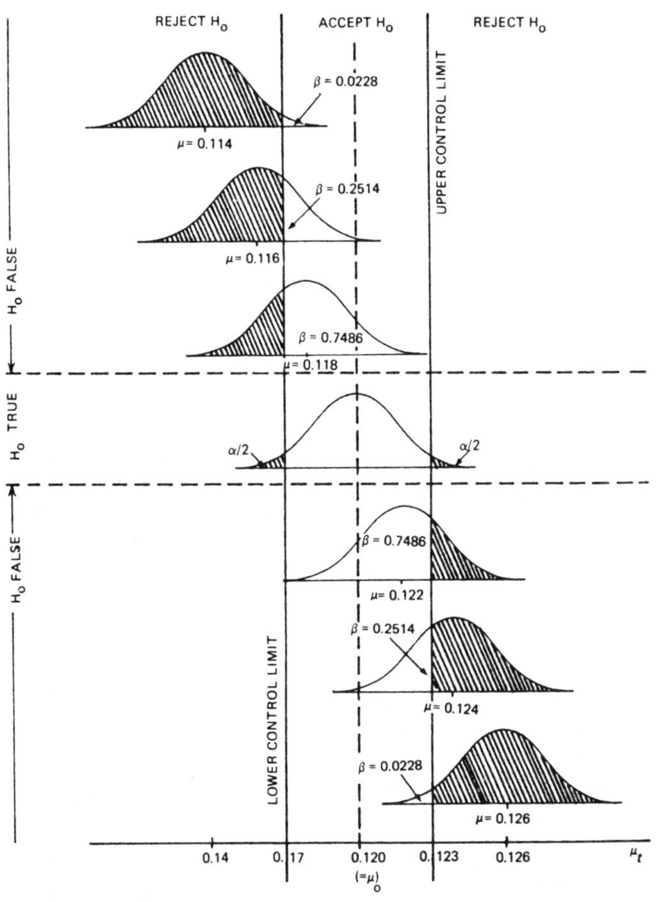

EXHIBIT 9: Operating Characteristic Curve

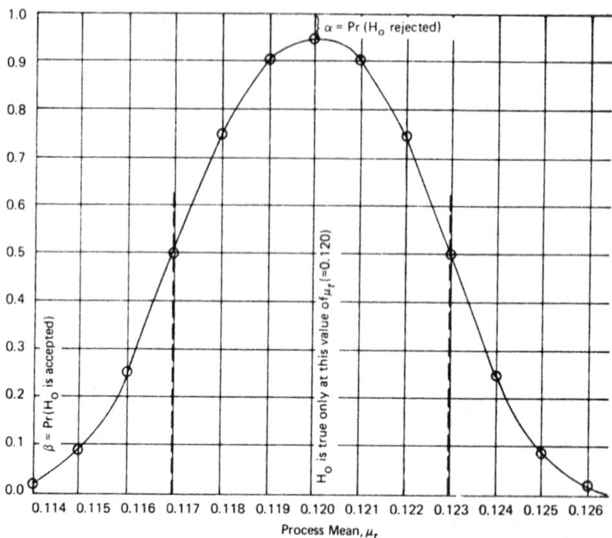

Decision Rule: Accept H_0 if $0.117 < \bar{X} < 0.123$.

17.6 FURTHER DISCUSSION ON TYPE I AND II ERRORS

Significance Testing and Confidence Intervals. The quality control example of Section 17.5 should have made it clear that there is a close connexion between carrying out a significance test and constructing a confidence interval. Suppose that on one occasion the sample mean ($\bar{X}$) was (a) 0.122" and on another (b) 0.124". We would have accepted H_0 for (a) but rejected it for (b). (See Exhibit 5). Alternatively we could have calculated a 95% confidence interval for the population mean in each case from $\mu = \bar{X} \pm 1.96\ \sigma/\sqrt{n}$, as follows:

(a) $\mu = 0.122 \pm 1.96 \times 0.003/\sqrt{4} \simeq 0.122 \pm 0.003$, which gives an interval (0.119 – 0.125) which *includes* the hypothesised value $\mu_0 = 0.120$.

(b) $\mu = 0.124 \pm 1.96 \times 0.003/\sqrt{4} \simeq 0.124 \pm 0.003$, which gives an interval (0.121 – 0.127) which *excludes* the value $\mu_0 = 0.120$.

So if a confidence interval includes/excludes the value of the population parameter specified by H_0, the *corresponding* significance test will accept/reject H_0. By "corresponding" we mean that if the Type I error of the test is α, (e.g. 0.05) the confidence interval has a level of confidence of 100 $(1 - \alpha)\%$, (i.e. 95%).[1]

Choosing α. In *classical* significance testing no formal guidance is given on the value of α one should choose, but the following ideas often prove useful:

- If, before carrying out the test, we have strong reasons to believe H_0 to be true for other, non-statistical reasons, we shall take a good deal of persuading that it is in fact false. Consequently we should choose a *small* value for α to reflect the fact that we are prepared to take only a small risk of H_0 being wrongly rejected.
- If the cost of wrongly rejecting H_0 is high – the cost of needlessly persisting with the more expensive formulation of meat paste or that of resetting the production machine without due cause – we shall again choose a small value for α to keep such costs down.[2]

In another approach to significance testing – the Bayesian approach – such considerations are taken into account explicity. *Prior* probabilities are evaluated for each state of nature (including H_0) and the test designed to minimise total expected costs – sampling costs and those associated with type I and II errors. Classical statisticians object to the difficulties of obtaining prior probabilities; Bayesians counter this objection by pointing to the arbitrary way α is chosen in classical tests.

Power of a Test. While the concept of an Operating Characteristic Curve has found favour in the field of business statistics, especially in Quality Control applications, a related measure of the **power** of a test is more frequently used when comparing alternative significance tests that could be applied to the same set of data.

Denoting, as previously, the probability of a Type II error as β, we define:

$$\begin{aligned}\text{Power} &= 1 - \beta \\ &= 1 - (\text{probability of accepting } H_0 \text{ when it is false}) \\ &= \text{Probability of } \textit{rejecting } H_0 \text{ when it is false.}\end{aligned}$$

So the higher the power of a test the more likely it is to reject a false null hypothesis in favour of the true alternative. In many practical

(1) Put in the most general terms a confidence interval is the set of *all* acceptable hypotheses.

(2) This statement ignores the cost of committing a Type II error – the risk of which will be increased by a small value for α; but read on!

situations there are a number of different tests one could employ[1]: for the same value of α, the test with the highest power (i.e. the most "powerful") should be chosen. Just as the power is defined by $1 - \beta$, the "power curve" is an inverted form of the O.C. curve, as shown for our two examples in Exhibits 10 and 11.

EXHIBIT 10:
Power Curve for the Marketing Problem

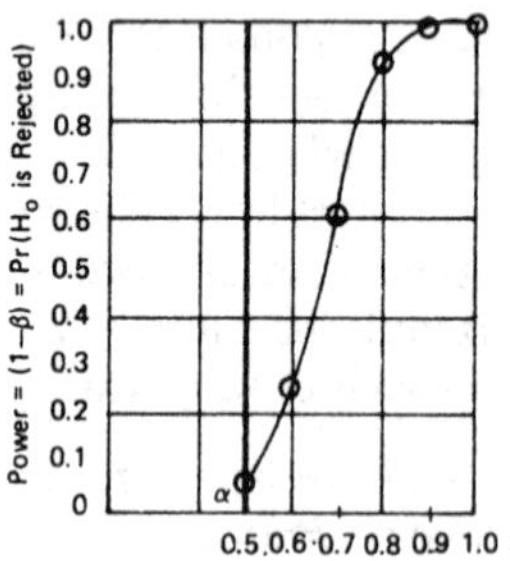

EXHIBIT 11:
Power Curve for the Production Problem

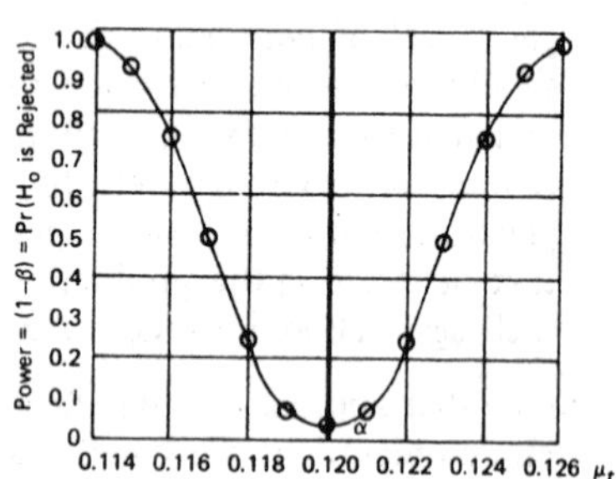

READING

Further discussion on the interpretation of significance tests is given in: Russell Langley *Practical Statistics*, Pan, Chapter 6.

A Bayesian treatment of a control chart for sample means is to be found in:

F.E. Croxton, D.J. Cowden & B.W. Bolch *Practical Business Statistics 4th Ed.*, Prentice-Hall, Chapter 13.

REFERENCE

D.E. Morrison & R.E. Henkel *The Significance Test Controversy* Butterworths.

A collection of 31 papers, of special interest to social science students, mainly non-mathematical discussions on the philosophy and appropriateness of significance testing.

---

[1] For example, in CU22 we conduct a *non-parametric* test on the same data as is used for a *parametric* test in CU20.

EXERCISES

1(P) In this exercise we develop a simple procedure for testing whether a pack of playing cards has been well shuffled. Consider taking the top 10 cards from a well shuffled pack and counting the number of black cards (r). It is a relatively simple matter[(1)] to evaluate the probability distribution for r. This is shown below:

r	$p(r)$	r	$p(r)$
0	0.0003	5	0.2735
1	0.0051	6	0.2176
2	0.0321	7	0.1081
3	0.1081	8	0.0321
4	0.2176	9	0.0051
		10	0.0003

(a) Formulate a null and alternative hypothesis for this problem and decide on a level of significance. Determine acceptance and rejection regions for the test.

(b) Take a pack of playing cards (without the joker) and arrange them so that all the black cards are at the top of the pack, all the red cards at the bottom. (This ensures that the pack is very badly shuffled.)

(c) *Gently* shuffle the pack in an overhand fashion. Three or four shuffling actions are sufficient – the aim is *not* to shuffle them well immediately. Cut the pack and count the number of black cards in the top ten. Record your result and apply your statistical test to it.

(d) Repeat step (c) until your test accepts the null hypothesis on five consecutive occasions. Tabulate your results as follows on the next page:

---

(1) At first sight this might be thought to be a binomial distribution. In fact, this is not so because the "sampling", is taking place *without* replacement from a finite (52 card) population. For example, the probability of finding no black cards in the top 10 of a well shuffled pack is given by:

$$\frac{26}{52} \cdot \frac{25}{51} \cdot \frac{24}{50} \cdot \frac{23}{49} \cdot \frac{22}{48} \cdot \frac{21}{47} \cdot \frac{20}{46} \cdot \frac{19}{45} \cdot \frac{18}{44} \cdot \frac{17}{43} = 0.000336 = p(0).$$

The distribution is given the special name of "hypergeometric".

After Shuffle Number	*Number of Black Cards in Top 10*	*Conclusion (Accept or Reject H_0)*
1		
2		
3		
4		
.		
.		
.		

(e) Comment on your results critically. (If you have accepted the null hypothesis on every occasion, you have probably been shuffling too vigorously: repeat from step (b)).

(f) How "good" a test do you think this is? Comment on any weakness you can see in the test. Can you think of a "better" statistic to test than the number of black cards in the top 10. State your reasons.

2(U) (i) Discuss formulating the following everyday decisions by making use of the concepts of a null hypothesis and Type I and Type II errors.

(a) On taking an umbrella in case it rains.
(b) On taking "comprehensive" as opposed to "third party" car insurance.
(c) On betting that Leeds United will be next year's league champions.
(d) On deciding to offer trading stamps on the sales of your product.

(ii) Sketch a "perfect" O.C. curve for the marketing problem discussed in the text. Say why it is (a) impractical and (b) undesirable to devise a test with this O.C.

3(E) (a) Define the terms "size"[(1)] and "power" in relation to a statistical test.

(b) Describe in detail *two* situations in the practice of personnel management in which the concept of testing an alternative hypothesis against a null hypothesis is appropriate. (Neither of your examples should involve topics covered in other questions on this examination paper.)

(IPM 1972)

N.B. *The reader will be in a better position to answer part (b) of this question after completing CU18.*

[(1)] "Size" is a term sometimes used for the probability of a Type I error, α.

4(E)* A company has production rights on a new lightweight ladder system for the building trade. Assuming there are 80,000 registered building companies in the U.K. the company estimates that it will show a moderate profit on the venture if 28% of the building firms buy the ladders. If only 20% buy them, however, a moderate loss would be incurred.

The company management decide to initiate a market research survey among building firms. They decide they will accept a 5% risk of the sample under-estimating a buying level of 28% when in fact it is as high as this, and a 1% risk of the sample overestimating a buying level of 20%.

Based on these risks obtain the sample size that should be used for the sample, assuming simple random sampling. Obtain the critical value (%) at which the decision is made – to go into production or not.

Plot the O.C. curve for the decision rule over the range 18% to 30%.

BA (Business Studies)

5(E) In the theory of hypothesis testing what is meant by the terms:

(i) null hypothesis;
(ii) alternative hypothesis;
(iii) errors of the first and second kinds?[(1)]

Part question (IPM 1973)

SOLUTIONS TO EVEN NUMBERED EXERCISES

2. (i)

Null Hypothesis, H_0	*Type I Error (Reject true H_0)*	*Type II Error (Accept false H_0)*
(a) It will not rain.	Needlessly carry umbrella.	Get wet!
(b) No accident will occur for which the owner is held responsible.	Needlessly insure comprehensively.	Pay for damage to own car.
(c) L.U. will not be league champions.	Wasted money placed as bet.	Miss the opportunity of winning by not placing a bet.
(d) Trading stamps have no effect on sales	Money spent on trading stamps wasted.	Forego the increase in sales as the result of not offering stamps.

[(1)]That is, Type I and II errors.

(ii)

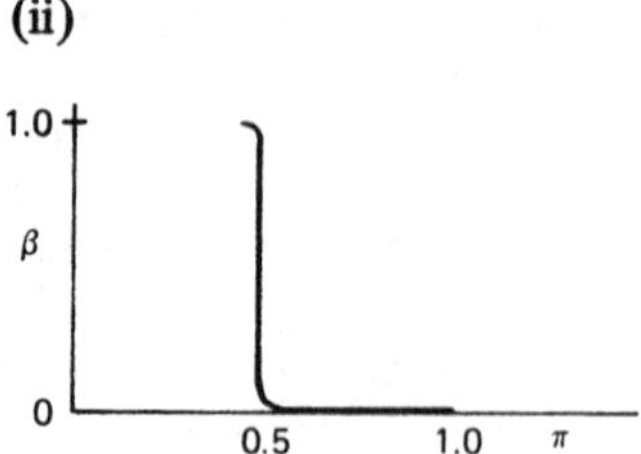

(a) Too costly; all consumers would have to carry out the tasting experiment.

(b) We would not wish to keep to the more expensive formulation (0) simply because a *very small* proportion of consumers could distinguish between the two products.

4. The percentages of interest are not too small and so, provided that the sample size is large, we may use the normal approximation to the binomial distribution as a suitable model.

Consider $H_0 : \pi_{\%} = 28\%$
against $H_1 : \pi_{\%} = 20\%$

(N.B. The hypotheses could be the other way around without affecting the solution.)
The problem may be illustrated diagramatically as follows, showing the two sampling distributions based on the unknown sample size *n*. *R* is the critical (%) value at which the production decision is made.

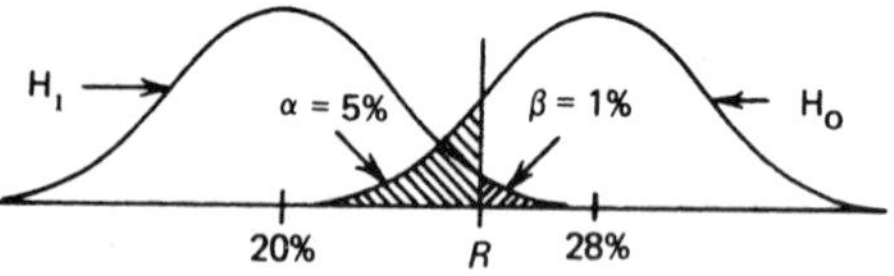

With

$$\alpha = 5\%, \quad z_\alpha = 1.645$$

$$\beta = 1\%, \quad z_\beta = 2.33$$

Also

$$\pi_{\%0} = 28\%, \; \pi_{\%1} = 20\%$$

Hence

$$R = 20 + 2.33\sqrt{\frac{20(100-20)}{n}} = 28 - 1.645\sqrt{\frac{28(100-28)}{n}}$$

So

$$n = \left(\frac{1.645\sqrt{20 \times 80} + 2.33\sqrt{28 \times 72}}{8}\right)^2 = 436.$$

Hence

$$R = 20 + 2.33\sqrt{\frac{20 \times 80}{436}} \text{ or } 28 - 1.645\sqrt{\frac{28 \times 72}{436}} = 24.46\ (\%)$$

The shape of the O.C. curve is as shown and will pass through the points (20%, 1%), (28%, 95%) and (24.46%, 50%).

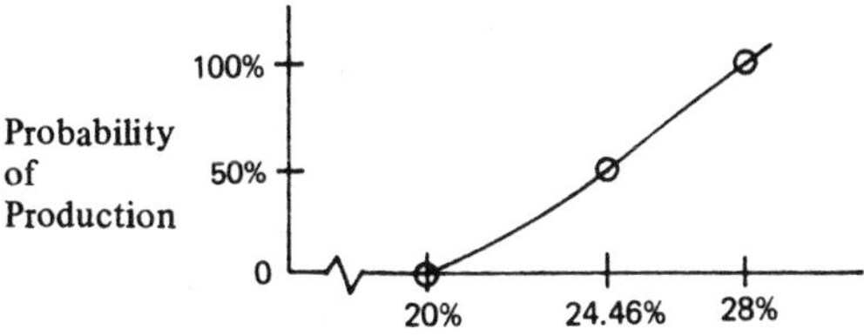

% of Buyers in the Population ($\pi_{\%}$)

The calculations for the O.C. curve can be obtained by application of the formula:

$$Z = (\pi_{\%} - 24.46) \Big/ \sqrt{\frac{\pi_{\%}(100 - \pi_{\%})}{436}}$$

where $\pi_{\%}$ is the percentage of buyers in the population. Values of $\pi_{\%}$ over the range 18% to 30% may be used until sufficient points are obtained to indicate the path of the O.C. curve.

COURSE UNIT 18 – SIGNIFICANCE TESTING II. FURTHER DISCUSSION AND LARGE SAMPLE TESTS

18.1 SIGNIFICANCE TESTS IN GENERAL

In CU17 we developed the principles of significance testing by examining two particular problems. These examples were chosen because the appropriate sampling distributions are ones with which the reader should be thoroughly familiar. But the principles have a quite general validity, and it would now seem appropriate to attempt a general description of significance testing without reference to any particular examples. At the same time the opportunity will be taken to introduce some new terminology.

We have seen that a statistical test of significance is concerned with the acceptance or rejection of the null hypothesis. If we accept the null hypothesis we are not necessarily saying that the null hypothesis is true, but that the evidence is insufficient to prove otherwise. If, on the other hand we reject the null hypothesis we are taking the view (rightly or wrongly) that the null hypothesis is *not* true.[(1)] If we *wrongly* reject the null hypothesis we have committed a Type I error and the probability that this can occur is called the **level of significance** of the test and is denoted by α.

We have also seen that the form of the significance test is to set up the acceptance region within which there is a high probability $(1 - \alpha)$ that the sample statistic will fall if the null hypothesis is true. If the sample statistic falls outside this region then the null hypothesis is rejected with a small probability α that this result could have arisen from sampling errors. If the null hypothesis is rejected we often say that the sample result is **significantly different** from the null hypothesis. Note that this "significant" difference may or may not be of practical importance: we are merely saying that we do not believe that such a large difference could have arisen from chance sampling errors.

There are some occasions where although we would formally accept the null hypothesis (because there is insufficient evidence to reject it) we may have other (non-statistical) reasons to believe it to be untrue. In such cases we leave the verdict open ("not proven" – as in the Scottish legal system) for further evidence to be accumulated. A classic case is in the development of a new drug where although laboratory tests have

---

(1) In view of the difference in character between acceptance and rejection of the null hypothesis it is common statistical practice to formulate the null hypothesis in such a way that it is a statement of what we wish to disprove.

proved promising, the clinical trials have failed to indicate a statistically significant improvement over existing drugs – maybe as the result of the small numbers used in the trials.

In many research reports it is found that significance testing is carried out without any attempt being made to calculate the Type II errors of the test. This is especially true of sample survey work where a large number of significance tests are carried out in the search for a significant difference. The calculation of the O.C. Curves for all of these tests would be prohibitively expensive. In these circumstances it is assumed that the probability of a Type II error for a large (practical) difference is small enough to be ignored.[(1)]

The most commonly used significance tests use the following form for the null hypothesis:

$$H_0 : \theta = \theta_0,$$

where θ is some parameter (e.g. the mean, μ) of the population from which the sample has been drawn. θ_0 is a constant, our hypothesised value of θ. The alternative hypothesis is not usually specified exactly but takes one or other of the following forms:

$$H_1 : \theta \neq \theta_0 \text{ (a two-sided test)}$$

or

$$H_1 : \theta > \theta_0 \text{ (a one-sided test)}$$

or

$$H_1 : \theta < \theta_0 \text{ (a one-sided test)}$$

Note that the form of the alternative hypothesis specifies whether a one or a two-sided test is appropriate.

The level of significance, α, of the test may in principle be chosen as any value, excepting 0 and 1. When dealing with a discrete sampling distribution (e.g. binomial) we have a limited number of possibilities to choose from, but for a continuous sampling distribution it is common practice to choose 0.01 or 0.05 (1% or 5%). (If we reject the null hypothesis in a test with $\alpha = 0.01$ we often say that the *result is significant at the 1% level.*) The validity of the procedure holds as long as the level of significance is selected *in advance* of seeing the survey results.

[(1)]The justification for this is that in practice the most commonly used tests are those which are most "powerful" (i.e. from the tests available in a given situation, the one with the smallest probability of a Type II error would be chosen).

A test of significance may be carried out on any sample statistic assuming its sampling distribution can be found or may be approximated by some known distribution. In elementary Statistics it is usually assumed that we are either sampling from a normal population, or that the sampling distribution of the statistic, with the size of sample we are employing, is approximately normally distributed. In what are known as "large sample" tests we can make the latter assumption in most (but not all) cases, even though the population from which the sample is drawn is not a normal distribution (the Central Limit Theorem). Sections 18.3 and 18.4 of this Course Unit give examples of large sample tests. If neither[(1)] of the above assumptions can be made, so-called "distribution-free" tests may be used (CU22).

Significance tests may be employed on sample statistics (of like kind) from any number of samples simultaneously. So far we have been concerned with single sample tests, but tests on two or more (k, say) samples only involve a simple extension of the basic ideas. As an example, in the two-sample test of means we may test that the two sample means $\bar{X}_1$ and $\bar{X}_2$ come from populations with the same population mean, in which case we would use the null hypothesis:

$$H_0 : \mu_1 = \mu_2$$

or as it is usually expressed,

$$H_0 : \mu_1 - \mu_2 = 0.$$

This second way of expressing the null hypothesis is to be preferred since it highlights the fact that the statistic $(\bar{X}_1 - \bar{X}_2)$ will have an expected value of zero if H_0 is true. Suitable alternative hypotheses are:

$$H_1 : \mu_1 \neq \mu_2 \text{ (or } \mu_1 - \mu_2 \neq 0) \quad \text{(a two-sided test)}$$

or

$$H_1 : \mu_1 > \mu_2 \text{ (or } \mu_1 - \mu_2 > 0) \quad \text{(a one-sided test)}$$

or

$$H_1 : \mu_1 < \mu_2 \text{ (or } \mu_1 - \mu_2 < 0) \quad \text{(a one-sided test)}$$

Whether we are using a one, two or k sample test the procedure takes the same form:

[(1)]Course Unit 20 deals with cases where we *can* make the assumption of a normal population but not the assumption of a normal sampling distribution – "small sample" tests.

(1) A statement of the null hypothesis
(2) A statement of the alternative hypothesis
(3) A statement of the significance level of the test, α
(4) The calculation of the acceptance/rejection regions
(5) A comparison of the empirical result with the acceptance/rejection region
(6) A statement accepting, rejecting or witholding judgement about the null hypothesis.

18.2 A PICTORIAL PRESENTATION

The argument presented in the previous section was, in the interests of generality, somewhat involved. For those readers who find it easier to understand or remember a pictorial presentation the following treatment is offered. Throughout, we assume that the sampling distribution of the statistic concerned is normally distributed.

The fundamental problem of significance testing is illustrated in Exhibit 1.

EXHIBIT 1

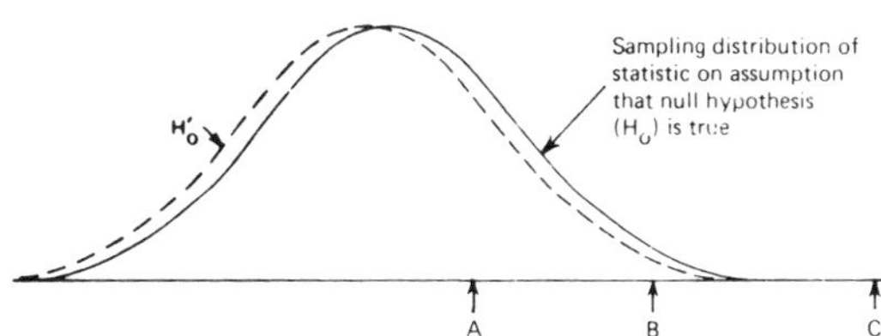

If the statistic observed is at A: then it is *consistent* with H_0 (but H_0 is not necessarily true – H_0' may be the truth). We say "accept H_0", *not* "H_0 true".
If the statistic observed is at C: then it is inconsistent with H_0. In the case drawn we could confidently conclude that H_0 were false. We say "reject H_0", or the result is "statistically significant".

But what do we conclude if the statistic is observed at B? To answer this we need to specify an acceptance and rejection region which in turn defines α, the probability of a Type I error (Exhibit 2).

EXHIBIT 2

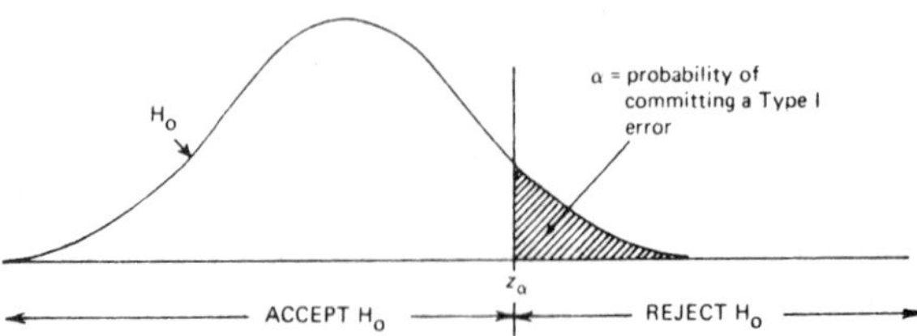

If the statistic falls in the rejection region (i.e. its standardised value $Z > z_\alpha$) we say the result is "significant at the α level".

Suppose we can also put forward a specific alternative hypothesis (H_1) whose sampling distribution can be evaluated. This also defines β, the probability of a Type II error (Exhibit 3).

EXHIBIT 3

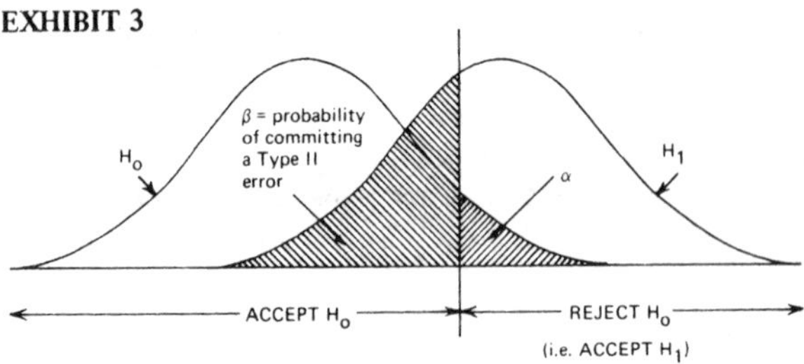

We are unable in most cases to put forward a specific alternative hypothesis: instead, we use a general statement. Suppose that θ is the population parameter corresponding to the statistic we are testing. Then one possible formulation is

$$H_0 : \theta = \theta_0$$

$$H_1 : \theta > \theta_0,$$

which is illustrated in Exhibit 4.

EXHIBIT 4

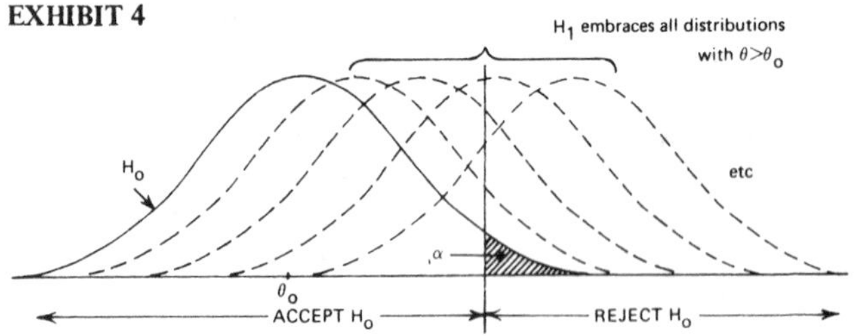

We cannot measure β directly, but only express it as a function of the population parameter, θ (Exhibit 5).

EXHIBIT 5: The O.C. Curve for a One-Sided Test

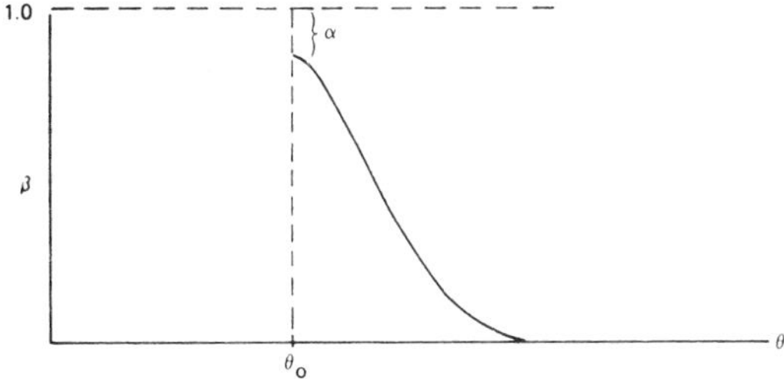

Another formulation is:

$$H_0 : \theta = \theta_0$$

$$H_1 : \theta < \theta_0$$

The pictorial treatment of this case is a lateral inversion of the above case (H_1: $\theta > \theta_0$). Both these tests are one-sided. The two-sided test has the formulation:

$$H_0 : \theta = \theta_0$$

$$H_1 : \theta \neq \theta_0$$

In this case we would accept H_0 if the standardised value of the test statistic lies *between* the critical values $-z_{\alpha/2}$ and $+z_{\alpha/2}$ (Exhibit 6).

EXHIBIT 6

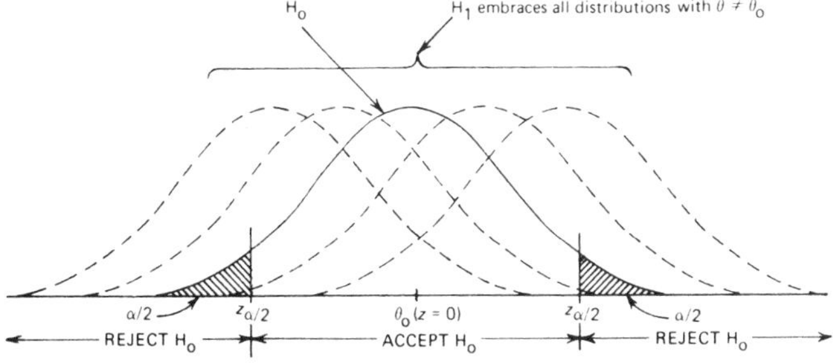

The corresponding O.C. Curve is shown in Exhibit 7.

EXHIBIT 7: O.C. Curve for Two-Sided Test

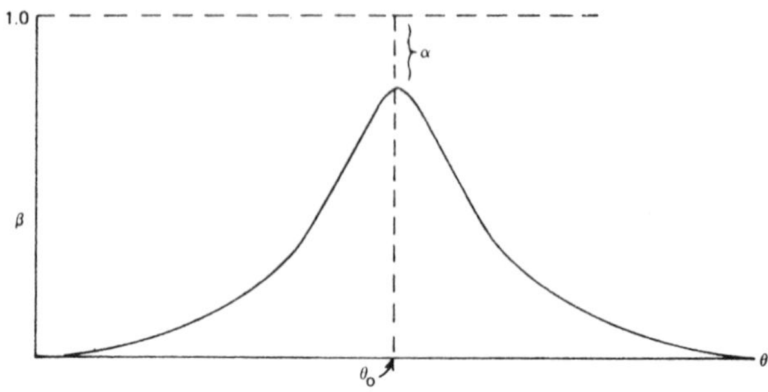

The reader should now be in a position to apply large-sample tests to sample means and proportions. Exercise 2 should now be attempted and checked with the solution given.

18.3 INDEPENDENT LARGE TWO-SAMPLE TEST OF MEANS

In this section (and the next) we demonstrate the use of significance testing in comparing statistics from *two independent large samples*. Such tests are the most complex large sample tests the reader is likely to encounter in an introductory Statistics course – they also have considerable practical value. The fact that the samples are large, as has been stated previously, means that we are able to approximate the sampling distribution of some statistics (in particular, the sample mean and proportion) by the normal distribution.

In some texts the tests we are about to demonstrate would fall under the general heading of "*Z*-tests" – *Z* being the symbol used for the standardised normal deviate used in the test.

Consider the following example:

In the West Country it was found that the last 200 female assembly workers to leave a company had an average completed length of service of 101.4 weeks (SD of 13.0 weeks). In the Midlands branch of the same company, for the corresponding group of 250 workers, the figures were an average of 98.4 weeks and SD of 12.0 weeks. Is the average completed length of service significantly different at the 1% level?

We have observed the sample means and variances from two independent large samples for which we will use the notation

	Sample Size	*Sample Mean*	*Sample Variances*
West Country (1)	$n_1 = 200$	$\bar{X}_1 = 104.1$	$s_1^2 = (13.0)^2$
Midlands (2)	$n_2 = 250$	$\bar{X}_2 = 98.4$	$s_2^2 = (12.0)^2$

We assume that $\bar{X}_1$ comes from a population of mean μ_1 and that $\bar{X}_2$ comes from a population of mean μ_2.

Our null hypothesis is that these two population means are equal so,

$$H_0 : \mu_1 = \mu_2 \text{ (or } \mu_1 - \mu_2 = 0)$$

As we are only interested in whether the population means are different (rather than one particular one being greater than the other) we formulate the alternative hypothesis as,

$$H_1 : \mu_1 \neq \mu_2 \text{ (or } \mu_1 - \mu_2 \neq 0)$$

and this decrees that a two-sided test is appropriate.

The statistic we are about to test is the difference between the sample means, $\bar{X}_1 - \bar{X}_2$ (= 104.1 – 98.4 = 5.7 in our example). As the sample sizes are large enough for us to assume that both $\bar{X}_1$ and $\bar{X}_2$ are normally distributed, we may also assume that the sampling distribution of their difference, $\bar{X}_1 - \bar{X}_2$, is similarly normally distributed.[1]

The mean value of $\bar{X}_1 - \bar{X}_2$ is $\mu_1 - \mu_2$ which, according to the null hypothesis, is zero (that is precisely the reason why we formulated the null hypothesis as we did – to enable us to evaluate the sampling distribution of $\bar{X}_1 - \bar{X}_2$).

The information deduced so far enables us to draw up a sketch of the sampling distribution as shown in Exhibit 8. Th^ acceptance and

EXHIBIT 8

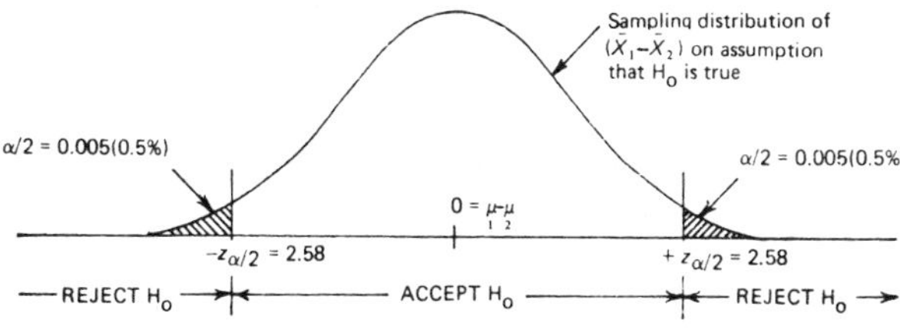

(1) The difference between two independent normally distributed variables is also normally distributed (CU12).

rejection regions have been added, based on the given value of α of 1%, the requirement of a two-sided test and the use of the normal distribution tables.

The standard deviation of the sampling distribution is the standard error of $\bar{X}_1 - \bar{X}_2$, which for independent samples is given by[1]

$$\text{SE of } (\bar{X}_1 - \bar{X}_2) = \sqrt{\frac{\sigma_1^2}{n_1} + \frac{\sigma_2^2}{n_2}}$$

where σ_1^2 and σ_2^2 are the population variances of the lengths of service. As the sample sizes are large we may approximate σ_1^2 by s_1^2 and σ_2^2 by s_2^2. Hence the

$$\text{SE of } (\bar{X}_1 - \bar{X}_2) \simeq \sqrt{\frac{s_1^2}{n_1} + \frac{s_2^2}{n_2}}$$

$$= \sqrt{\frac{169}{200} + \frac{144}{250}}$$

$$= \sqrt{1.421} = 1.19$$

Consequently the Z value[2] of our observed $\bar{X}_1 - \bar{X}_2$ is

$$Z = \frac{5.7 - 0}{1.19} = 4.78.$$

This value of Z falls in the rejection region (i.e. outside the "critical" value for acceptance of H_0, $z_{\alpha/2} = 2.58$) and hence we reject H_0 and accept H_1. The difference between the sample means is therefore significant at the 1% level.

Note that in the operation of this test we have had no need to assume that the populations of lengths of service are themselves normally distributed, nor even that the two populations have the same

---

(1)These matters were gone into in CU17. To summarise we have:

$$\text{Var}(\bar{X}_1 - \bar{X}_2) = \text{Var}(\bar{X}_1) + \text{Var}(\bar{X}_2), \text{ for independent } \bar{X}_1, \bar{X}_2,$$

$$= \frac{\sigma_1^2}{n_1} + \frac{\sigma_2^2}{n_2}$$

Hence the square root of this expression gives the standard error of $\bar{X}_1 - \bar{X}_2$.

(2)Summarising the argument we have:

$$Z \simeq \frac{(\bar{X}_1 - \bar{X}_2) - (\mu_1 - \mu_2)}{\sqrt{\dfrac{s_1^2}{n_1} + \dfrac{s_2^2}{n_2}}},$$

where $(\mu_1 - \mu_2) = 0$ under the null hypothesis.

variance. For this reason we might describe the test as being "robust"; that is, not requiring many untested assumptions for its success. This is in marked contrast to a similar test (see CU20) on *small* samples, and provides a demonstration of the fact that the benefits of employing large samples are not simply a question of the probabilities of Type I and II errors.

18.4 INDEPENDENT LARGE TWO-SAMPLE TESTS OF PROPORTIONS

Consider the following example. In one of the factories referred to in the last section it is found that 28% of the 400 male manual workers and 22% of the 200 female workers join the voluntary pension scheme. Is the proportion for men significantly greater than for women at the 5% level?

Let us use the following notation:

	Sample Size	*Proportions (%)*
Men (1)	$n_1 = 400$	$P_{\%1} = 28$
Women (2)	$n_2 = 200$	$P_{\%2} = 22$

We assume that the *percentage* of the population[1] of male and female manual workers who would join the scheme are $\pi_{\%1}$ and $\pi_{\%2}$ respectively. Immediately we formulate the null hypothesis as:

$$H_0 : \pi_{\%1} = \pi_{\%2} \text{ (or } \pi_{\%1} - \pi_{\%2} = 0)$$

As the question asks whether the proportion is *greater* for men than for women (not just different) our alternative hypothesis is:

$$H_1 : \pi_{\%1} > \pi_{\%2} \text{ (or } \pi_{\%1} - \pi_{\%2} > 0).$$

Clearly a one-tailed test is appropriate.

The statistic we are about to test is the difference between the sample percentages, $P_{\%1}-P_{\%2}$ (= 6%, in our example).

The values of the n's and $P_{\%}$'s enable us to assume a normal sampling distribution for $P_{\%1}$ and $P_{\%2}$, and hence for $P_{\%1}-P_{\%2}$. In addition, the mean value of $P_{\%1}-P_{\%2}$ is $\pi_{\%1}-\pi_{\%2}$ which is itself zero if H_0 is true. This enables us to sketch the sampling distribution and indicate

[1] See the note at the end of this section on the interpretation of "population" in this context.

the acceptance and rejection regions related to the stated value of α of 5%, as shown in Exhibit 9.

EXHIBIT 9

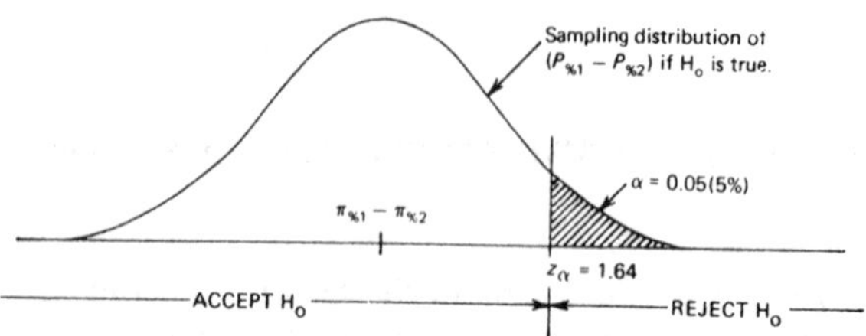

In general terms the standard error of $(P_{\%1}-P_{\%2})$ for *independent* samples is given by:

$$\sqrt{\frac{\pi_{\%1}(100-\pi_{\%1})}{n_1}+\frac{\pi_{\%2}(100-\pi_{\%2})}{n_2}}$$

But according to H_0, $\pi_{\%1} = \pi_{\%2}$ ($= \pi_{\%}$, say) and so we can rewrite this as

$$\sqrt{\pi_{\%}(100-\pi_{\%})\left(\frac{1}{n_1}+\frac{1}{n_2}\right)}\ (\%)$$

We need to estimate $\pi_{\%}$ *jointly* from the two samples in the following manner:

Total number of men and women in pension scheme

$$=\frac{n_1P_{\%1}+n_2P_{\%2}}{100}$$

$$=\frac{(400\times 28)+(200\times 22)}{100}$$

$$= 156$$

% of men and women in pension scheme

$$=\frac{n_1P_{\%1}+n_2P_{\%2}}{n_1+n_2}$$

$$=\frac{15{,}600}{600}$$

$$= 26\%\ (= \hat{\pi}_{\%})$$

Hence the standard error of $P_{\%1} - P_{\%2}$

$$= \sqrt{26(100-26)\cdot\left(\frac{1}{200}+\frac{1}{400}\right)}\ (\%)$$

$$= \sqrt{14.43}\%$$

$$= 3.80^{(1)}\%$$

So finally our test statistic[2]

$$Z = \frac{6-0}{3.80} = 1.59.$$

This value is less than z_α and so the null hypothesis is accepted. Consequently we conclude that the proportion for men is not significantly greater than for women at the 5% level.

It is important to note that in the above example (and in the example given in Section 18.3) we have used as "samples" the total number of persons involved (i.e. *all* the manual workers in a particular category). In this type of situation we are thinking of the "population" in some hypothetical sense, that is, all the workers we *could* employ in the West Country or in the Midlands.

One difference between the examples given in Section 18.3 and in this section is that in the former case we used a two-sided test and in the latter a one-sided test. We stated that it was the alternative hypothesis in each case that determined which test was appropriate. But consider the example in this section further. We may speculate that by common observation – or previous research – the person studying the pension position has discounted the possibility of there being a higher percentage of women than men with pensions. The point we wish to emphasise is that it is the research problem (and hence the researcher) that formulates the hypothesis to be tested; the statistical method (or the statistician) which provides the appropriate procedure.

---

(1) Some texts erroneously calculate the standard error of $P_{\%1}-P_{\%2}$ in the form

$$\sqrt{\frac{P_{\%1}(100-P_{\%1})}{n_1}+\frac{P_{\%2}(100-P_{\%2})}{n_2}}\quad (= 3.69).$$

This value is very little different from our own but the method of calculation is inconsistent with the null hypothesis, to say the least.

(2) Again, summarising, we have

$$Z \simeq \frac{(P_{\%1}-P_{\%2})-(\pi_{\%1}-\pi_{\%2})}{\sqrt{\hat{\pi}_{\%}(100-\hat{\pi}_{\%})\left(\frac{1}{n_1}+\frac{1}{n_2}\right)}},$$

where $\pi_{\%1} = \pi_{\%2}$ under the null hypothesis.

EXERCISES

1(U) Consider carrying out a significance test on a sample mean $\overline{X}$ of 100 observations randomly selected from a population the standard deviation of which is known to be 5 units. Show, by drawing sketches and using normal distribution tables, that the following alternative formulations of the test imply the acceptance regions (expressed in terms of standardised values) as given:

Test	*Accept H_0 If*
(i) $H_0 : \mu = 110$, $H_1 : \mu = 112$, $\alpha = 0.05$	$Z < 1.64$ $(= z_\alpha)$
(ii) $H_0 : \mu = 110$, $H_1 : \mu > 110$, $\alpha = 0.01$	$Z < 2.33$ $(= z_\alpha)$
(iii) $H_0 : \mu = 110$, $H_1 : \mu = 108$, $\alpha = 0.005$	$Z > -2.58$ $(= -z_\alpha)$
(iv) $H_0 : \mu = 110$, $H_1 : \mu < 110$, $\alpha = 0.02$	$Z > -2.05$ $(= -z_\alpha)$
(v) $H_0 : \mu = 110$, $H_1 : \mu \neq 110$, $\alpha = 0.10$	$-1.64 < Z < 1.64$
(i.e. between	$-z_{\alpha/2}$ and $+z_{\alpha/2}$)

2(E) A town has been chosen to test-market a new table-top freezer. As it is hoped that the results in this town will provide an indication of the sales potential throughout the country some initial checks have to be carried out to establish that the town is in fact "typical" of the country as a whole.

A simple random sample of 400 households in the town gave the following statistics:

(a) Average weekly net income of households = £53
(b) Percentage of households in the AB social class = 14.5%

Compare these results with the national figures (obtained by a recent large-scale survey) of an average household net income of £50 per week (and SD £20 per week) and 12% of households being AB's. Carry out the appropriate significance tests at the 5% level. Is the town typical? If not, comment on the likely effects on the validity of the test-marketing operation.

(HNC Business Studies)

3(E) (i) A manufacturer produces cables with a mean breaking strength of 2,000 lbs. and a standard deviation of 100 lbs. By using a new technique the manufacturer claims that the breaking strength can be increased. To test this claim a sample of 50 cables produced by the new technique is tested and the mean breaking strength found to be 2,050 lbs. Can the manufacturer's claim be supported at a .01 level of significance?

Table	
Level of significance	0.01
Critical value of z for a one-tail test	−2.33
or	2.33
Critical value of z for a two-tail test	−2.58
and	2.58

(ICMA 1972)

(ii) A market research company wishing to determine whether it is more usual in the North for working men to come home to a mid-day meal than in the South interviews two random samples each of 500 men. In the sample from the North 330 men come home mid-day compared with 280 in the sample from the South.
Is the difference significant?

(ICMA 1971)

4(E) In the development of the Avenger, market research tests on the interior showed that 46% of a sample of 250 motorists preferred the design to that of the Escort. The interior was restyled and another test showed that 53% of a sample of 250 preferred the Avenger. Test at the 2% level if the increase is statistically significant.
Indicate how you would modify your test if the same 250 motorists had been questioned on both occasions.[1]
State the assumptions underlying the test you have used.

(BA Business Studies)

5(E) The internal auditor of your company has reported to you as management accountant as
"From 1,000 sample postings made before metrication 210 errors were discovered whereas 250 errors were found in 1,000 sample postings made after metrication. It would appear that although the staff has not changed, the postings before metrication were more accurate than they are after. Special steps should be taken to regain the accuracy previously enjoyed."

(a) What statistical hypothesis does this statement imply?
(b) Evaluate this hypothesis and state if it should influence you as management accountant in your decision on whether to take special action.

(ICMA 1974)

---

[1] This part should not be attempted until CU19 is read.

6(E) In 1970 a survey gave the average number of hours per week overtime in a large industry as having an arithmetic average of 12.20 hours with a standard deviation of 4.10 hours, based on a sample of 400 manual workers. In 1973, the survey was repeated, using an independently selected sample, with the following results for 500 manual workers:

No. of hours overtime	*No. of workers*
0– 3	24
3– 6	58
6– 9	78
9–12	100
12–15	90
15–18	87
18–21	63
	500

(a) Calculate the arithmetic mean and standard deviation of the hours overtime worked for the 1970 sample.
(b) Test the significance of the difference of the two arithmetic means at the .05 (5%) level.

(HND Business Studies)

7(E) In a commercial organisation it is asserted that computer staff are in general better qualified than clerical staff and that a salary differential should be introduced. The qualifications of all the 25 members of the computer staff and all the 64 members of the clerical staff are assessed on a comparable basis using a standard scoring system. Within each population you can assume that the scores follow a Normal or Gaussian distribution. Given that the computer staff scores are denoted $x_1\, x_2, \ldots, x_{25}$, the clerical staff scores are denoted $y_1, y_2, \ldots, y_{64}$ and that

$$\bar{x} = 1.414, \bar{y} = 1.0, \sum_{i=1}^{25} x_i^2 = 80.0, \sum_{i=1}^{64} y_i^2 = 172.0,$$

what conclusions would you draw about the assertion? (Your answer should include a description of the basis of any statistical techniques which you use, as well as your calculations.)

(IPM 1974)

N.B. *The sample size of computer staff (25) does not strictly meet our criterion for a "large" sample. But as a useful exercise the reader may apply large sample methods to the question and later, having read CU20, compare the solution with that of an exact "small sample" test.*

8(E) In January 1972 a survey carried out by a Management Consultant Group showed that 55% of members of the sample of 400 firms had a larger order book than in January of the previous year. The survey was repeated in January 1973 with a larger sample of 1,000 firms of which 48% claimed to have a larger order book than in the previous January.
Test at both the 5% and 1% significance levels the hypothesis that there is no difference between the two sample results. Give an interpretation of your results.
As part of the same series of enquiries the average salaries of the Managing Directors of the Companies was £6,750 and £7,200 for 1972 and 1973 with standard deviations of £1,400 and £1,200 respectively. Is the average salary for 1973 greater than in 1972 at the 5% level of significance?

(BA Business Studies)

SOLUTIONS TO EVEN NUMBERED EXERCISES

2. The sample size (400) is large enough to employ "Large-sample" (normal distribution) methods.
The test of incomes can be formulated as:

$H_0 : \mu = 50$ (£)

$H_1 : \mu \neq 50$ (£), which indicates a two-sided test.

The sample mean $\bar{X}$ will, if H_0 is true, be normally distributed with mean $\mu = 50$ and standard error of $\sigma/\sqrt{n} = 20/\sqrt{400} = 1$ (£). Consequently the observed value of $\bar{X}$ (= £53) has the standardised value $Z = (53 - 50)/1 = 3$. With $\alpha = 0.05$, $z_{\alpha/2} = 1.96$ and so we reject the null hypothesis and conclude that the mean income in the town is significantly different at the 5% level from the national average.
The test of percentage of **AB**'s can be formulated as follows:

$H_0 : \pi_{\%} = 12$

$H_1 : \pi_{\%} \neq 12$ (a two-sided test)

If H_0 is true the sample percentage $P_{\%}$ will be normally distributed with mean $\pi_{\%} = 12\%$ and standard error of

$$\sqrt{\pi_{\%}(100 - \pi_{\%})/n} = \sqrt{12 \times 88/400} = 1.625\%.$$

Consequently the observed value of $P_{\%}$(= 14.5%) has Z-value of $(14.5 - 12)/1.625 = 1.538$ which is less than $z_{\alpha/2} = 1.96$.

We accept the null hypothesis and conclude that the difference in percentage AB's may well be attributable to sampling errors.

4. $n_1 = 250 \qquad n_2 = 250$

$P_{\%1} = 46\% \qquad P_{\%} = 53\%$

We wish to test if the *increase* in percentage is significant, indicating that a one-sided test is appropriate;

$H_0 : \pi_{\%1} = \pi_{\%2}$ against $H_1 : \pi_{\%2} > \pi_{\%1}$.

As $\alpha = 0.02$, $z_\alpha = 2.06$.
Pooling the percentages we obtain

$$\hat{\pi}_{\%} = \frac{n_1 P_{\%1} + n_2 P_{\%2}}{n_1 + n_2} = \frac{250 \times 46 + 250 \times 53}{500} = 49.5\%$$

So

$$Z = \frac{53 - 46}{\sqrt{49.5 \times 50.5\left(\frac{1}{250} + \frac{1}{250}\right)}} = \frac{7}{4.47} = 1.57,$$

and as $Z < z_\alpha$ we would accept the null hypothesis that there is no significant increase in preference on this evidence.

6. The earlier survey results are given:

$\bar{X}_1 = 12.20 \quad s_1^2 = (4.10)^2 = 16.81 \quad n_1 = 400$

(a) Using hand calculation methods on the frequency distribution we obtain (CU4 and CU5).

x_i	d_i	f_i	d_if_i		$d_i^2f_i$
1.5	−3	24		−72	216
4.5	−2	58		−116	232
7.5	−1	78		−78	78
a = 10.5	0	100		−266	0
13.5	1	90	90		90
16.5	2	87	174		348
19.5	3	63	189		567
		$\Sigma f_i = n =$ 500	453		$\Sigma d_i^2 f_i =$ 1,531
			−266		
			$\Sigma d_if_i =$ 187		

$$\bar{X} = a + \frac{c}{n}\sum d_i f_i = 10.5 + \frac{3}{500} 187 = 11.62$$

$$s_2^2 = c^2 \left[\frac{\sum d_i^2 f_i}{\sum f_i} - \left(\frac{\sum d_i f_i}{\sum f_i}\right)^2\right] = 3^2 \left[\frac{1,531}{500} - \left(\frac{187}{500}\right)^2\right] = 26.30$$

So

$$\bar{X}_2 = 11.62 \quad s_2^2 = 26.30 \quad n_2 = 500$$

(b) The following two-sided test is appropriate:

$$H_0 : \mu_1 = \mu_2 \ \ (\text{or } \mu_1 - \mu_2 = 0)$$
$$H_1 : \mu_1 \neq \mu_2 \ \ (\text{or } \mu_1 - \mu_2 \neq 0)$$

with

$$\alpha = 0.05, \qquad z_{\alpha/2} = 1.96$$

Out test statistic is

$$Z = \frac{(\bar{X}_1 - \bar{X}_2) - (\mu_1 - \mu_2)}{\sqrt{\dfrac{s_1^2}{n_1} + \dfrac{s_2^2}{n_2}}}$$

$$= \frac{(12.20 - 11.62) - 0}{\sqrt{\dfrac{16.81}{400} + \dfrac{26.30}{500}}} = \frac{0.58}{0.3076} = 1.89.$$

As $Z < z_\alpha$ we accept the null hypothesis and conclude that the observed difference is not significant and may be due to sampling errors.

8. The samples are large and the percentage values being tested are around 50%. Hence the large sample test of proportions may be used, i.e.

$$H_0 : \pi_{\%1} = \pi_{\%2} \text{ against } H_1 : \pi_{\%1} \neq \pi_{\%2} \text{ (a two-sided test)}$$

At the 5% and 1% levels $z_{\alpha/2} = 1.96$ and 2.58 respectively.
We first pool the sample percentages,

$$\hat{\pi}_{\%} = \frac{55 \times 400 + 48 \times 1{,}000}{400 + 1{,}000} = 50\%$$

to evaluate the test statistic:

$$Z = \frac{(P_{\%1} - P_{\%2}) - (\pi_{\%1} - \pi_{\%2})}{\sqrt{\hat{\pi}_{\%}(100 - \hat{\pi}_{\%})\left(\dfrac{1}{n_1} + \dfrac{1}{n_2}\right)}} = \frac{55 - 48}{\sqrt{50 \times 50\left(\dfrac{1}{400} + \dfrac{1}{1{,}000}\right)}} = 2.37$$

Hence the difference is found significant at the 5% level but not at the 1% level. Thus if we are prepared to take a 5% (1 in 20) chance of wrongly rejecting the null hypothesis, we would in fact reject it. If we are only prepared to take a 1% chance (1 in 100) then we avoid this risk by accepting the null hypothesis. Thus the conclusion differs depending on the attitude to the risk.

The test of mean salaries is one-sided since the object is to detect if the average salary for 1973 is significantly greater than in 1972. Hence we would use, for $\alpha = 5\%$, $z_\alpha = 1.645$.

Our test statistic is

$$Z = \frac{(\bar{X}_1 - \bar{X}_2) - (\mu_1 - \mu_2)}{\sqrt{\dfrac{s_1^2}{n_1} + \dfrac{s_2^2}{n_2}}}$$

which, under the null hypothesis of $\mu_1 = \mu_2$, becomes

$$Z = \frac{7{,}200 - 6{,}750}{\sqrt{\dfrac{1{,}400^2}{400} + \dfrac{1{,}200^2}{1{,}000}}} = \frac{450}{79.62} = 5.65$$

Hence the difference is statistically significant at the 5% level.

COURSE UNIT 19 – CHI-SQUARED TESTS

19.1 A PROBLEM IN SIGNIFICANCE TESTING

The results of market and social research surveys in particular are often presented in the form of tables. One example is shown in Exhibit 1, compiled for the purpose of assisting decisions on marketing policy. There are a number of difficulties (as we shall see) involved with applying our previously discussed tests of significance – large sample tests of means and proportions – to such tables. In this Course Unit we introduce a new test which is better suited to these applications and which, unlike our earlier tests, does not use a normal sampling distribution.

EXHIBIT 1: Car Ownership (Number of Households)

Social Class	*Number of Cars* 0	1	2 or more	*Totals*
Upper and Middle (ABC_1)	14	26	20	60
Skilled Working (C_2)	16	54	10	80
Unskilled Working (D)	30	20	10	60
Totals	60	100	40	200

Exhibit 1 is a *two-way* table, the classifying variables being social class (an attribute variable) and the number of cars owned (a discrete variable). The body of the table consists of 9 "cells", each cell showing the number of households of a particular social class owning a particular number of cars. The cells are mutually exclusive – each household will be identified with one (and only one) cell.

It is clear that we could employ our previously discussed tests on *parts* of this table. We could, for instance, calculate the mean number (and standard deviation) of cars owned[1] by ABC_1 households and C_2 households and test for a significant difference in the mean values. We could also calculate the proportion of no-car households who are ABC_1's, and a similar proportion for one-car households, and test these proportions for a significant difference. Such tests have the obvious limitation that, being confined to comparing *two* means or proportions, they only involve part of the data. We could however carry out a

[1] The open-ended class, "2 or more" cars, would cause some difficulties, though.

number of similar tests, using different parts of the table, in order to encompass the whole data, e.g. by comparing means of ABC_1 with C_2, ABC_1 with D, and C_2 with D. (In practice one often has to treat tables with many more than three classes in each direction and so a very large number of individual comparison tests would then be required.) But the validity of such a procedure is questionable: if a number of tests are carried out at the same level of significance (5%, say) using different subdivisions of the same table, then there is a greater than 5% chance that one of these will be found significant in conditions where the null hypotheses are all true.[1]

What is preferable in the case of Exhibit 1 is a single overall test. One such test – the **chi-squared** test (pronounced "ki-squared") – provides an overall test of association. Using it we could test whether there is a relationship (association) between car ownership and social class, by employing the null hypothesis that these two variables are independent. Put simply, if we rejected such a null hypothesis we would be asserting that knowledge of social class tells us "something useful" about car ownership.

19.2 INTRODUCTION TO THE CHI-SQUARED TEST

The chi-squared (χ^2) test makes use of data in the form of observed frequencies or "counts". It does not matter whether these observed frequencies arise from a numerical or non-numerical (i.e. attribute) classification scheme,[2] nor whether they are formed into a one-way (e.g. a univariate frequency distribution), two-way (as in Exhibit 1) or many-way table. There is a requirement however that the cells in the table, and hence the categories used in the formation of the table, are mutually exclusive. (This does lead to difficulties in applying the χ^2 test to tables based on newspaper readership – where some persons

---

[1] If one carried out 20 independent tests (all with the null hypothesis true) at the 5% level then we would expect one of these to be significant. In the example under discussion the tests are not independent, but a similar reasoning applies. It is possible, nonetheless, to carry out valid tests within an overall table, but such comparisons have to be planned in advance, not result from arbitrary sub-divisions of the table – the topic itself is beyond the scope of this course. (References 1 and 2).

[2] The fact that the χ^2 test does not distinguish between numerical and non-numerical classification is both an advantage (because it gives the test a wide range of application) and a disadvantage. By ignoring the numerical significance of a numerical classification, the χ^2 test effectively discards information and so weakens the test. For some applications there are consequently more powerful tests (in particular, an overall test of means – CU21).

read more than one newspaper – unless separate categories are formed for each possible combination of readership.)

We use the notation O_i (which we shall later refine) to refer to the observed frequency of the ith cell of our table. (Taking Exhibit 1, for example, and reading from left to right and then from one row to the next, we have: $O_1 = 14$, $O_2 = 26$, $O_3 = 20$, $O_4 = 16$, $O_5 = 54$, $O_6 = 10$, $O_7 = 30$, $O_8 = 20$, $O_9 = 10$. The totals, it should be noted, are *not* part of the original data, but statistics derived from the data.)

The test proceeds by evaluating expected frequencies, E_i, for each cell according to some null hypothesis (the form of which depends on the particular application) and then calculating the chi-squared statistic.[1]

$$\chi^2 = \sum_{\text{all cells}} \frac{(O_i - E_i)^2}{E_i} \qquad (1)$$

The reason for this rather strange statistic is simply that, under the null hypothesis and provided that certain assumptions[2] hold, its sampling distribution (the χ^2 distribution) is well known and tabulated. The properties of this statistic and its sampling distribution are summarised below:

- χ^2 can never be less than zero. This is a consequence of the squared form of the numerators in (1) and the fact that the expected frequencies themselves cannot be negative.
- If, for each cell, the observed and expected frequencies are close, the value of χ^2 will be small; if they are widely different χ^2 will be large. Consequently χ^2 is a relative[3] measure of how different the observed and expected frequencies are. A particularly large value of χ^2 should lead us to reject the null hypothesis – a conclusion that the differences are greater than might reasonably be attributed to chance.[4]

[1] χ^2 is a symbol by itself – its square-root χ has no meaning.

[2] These assumptions are similar to, but in general less stringent than, those required for the valid approximation of the binomial distribution by the normal distribution.

[3] Relative, that is, to the value of the expected frequencies. If, for instance, in the ith cell $O_i = 990$, $E_i = 1{,}000$, the difference of 10 is unremarkable. But if $O_i = 15$, $E_i = 25$, the same difference of 10 is much more significant. The contribution to χ^2 reflects this fact: $(O_i - E_i)^2/E_i = 0.1$ in the first case, 4.0 in the second.

[4] Conversely, a very *small* value of χ^2 might indicate that the correspondence between the observed and expected frequencies is "too good to be true", and thus we might suspect that the data had been faked in some way. This is a more specialised use of chi-squared and is normally only considered in "Goodness of Fit" applications (Section 7).

EXHIBIT 2: Sampling Distributions of χ^2 (with H_0 true)

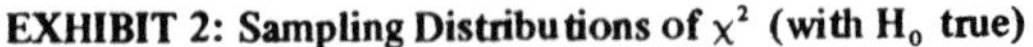

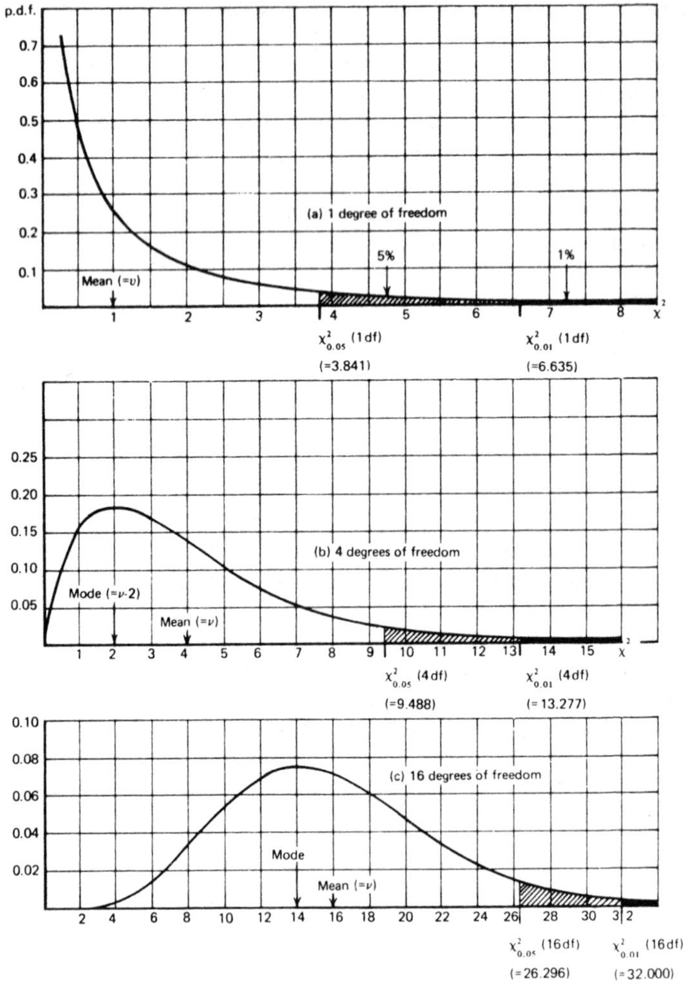

- The expected value of χ^2 (i.e. its mean value over a large number of conceptually identical tests) in conditions where the null hypothesis is true is equal to a quantity known as the number of **degrees of freedom**. This quantity, which is abbreviated by d.f. and given the symbol ν (pronounced "new"), corresponds to the number of independent items of data (frequencies) used in the test. The number of d.f. is closely related to, but somewhat less than, the number of cells in the table to which the test is being applied. Within this Course Unit emphasis is placed upon the determination of the number of d.f. for each application.

- The sampling distribution of χ^2 if the null hypothesis is true (the χ^2 distribution) does not have one single "shape". In fact the χ^2 distribution behaves rather like the Poisson distribution (except that χ^2 is a continuous distribution and Poisson is discrete) in that the distribution is markedly positively skewed for small values of the mean (ν), but becomes more symmetrical as the mean increases. For values of ν greater than 30 (say), the χ^2 distribution is approximately normal – as a consequence of the Central Limit Theorem.

The above properties are conveniently illustrated in Exhibit 2.

19.3 INDEPENDENT TWO-SAMPLE TESTS (2 × 2 TABLES)

Example 1. Two random samples of women were taken – 500 in the London I.T.V. area, and 400 in the Midlands I.T.V. area. It was found that 225 of the former group smoked, and 150 of the latter. Do the proportions of women who smoke differ significantly (at the 5% level) between the two regions, or could the observed differences in proportions be due to sampling errors?

Note we could rephrase the question to read "Is there an association between area and the proportion of women who smoke"? – the two questions being formally equivalent.

Firstly we arrange the data in the form of a 2 × 2 table (i.e. 2 rows and 2 columns) as shown in Exhibit 3, and work out the row and column totals.

EXHIBIT 3: Number of Women

	London I.T.V.	*Midlands I.T.V.*	*Total*
Smoke	225	150	375
Do not smoke	275	250	525
Total	500	400	900

Overall, 375 out of the 900 women smoke: so under the null hypothesis of no difference between the two regions the expected *proportion* of smokers in each should be $\frac{375}{900}$; and the expected proportion of non-smokers, $\frac{525}{900}$. Hence the expected *numbers* of smokers in each region are:

$$\frac{375}{900} \times 500 = 208.3 \text{ in London, and}$$

$$\frac{375}{900} \times 400 = 166.7 \text{ in the Midlands.}$$

Similarly the expected numbers of non-smokers in London and the Midlands are $\frac{525}{900} \times 500 = 291.7$, and $\frac{525}{900} \times 400 = 233.3$ respectively.

We may summarise the original *observed* frequencies and our calculated *expected* frequencies in one table, Exhibit 4 (in which, by convention, the expected frequencies are shown in brackets).

EXHIBIT 4: Number of Women

	London I.T.V.	*Midlands I.T.V.*	*Total*
Smoke	225 (208.3)	150 (166.7)	375
Do not smoke	275 (291.7)	250 (233.3)	525
Total	500	400	900

The calculation of χ^2 is best organised by listing the cells (in any convenient order) as shown in Exhibit 5.

EXHIBIT 5: Calculation of χ^2

i	j	O_{ij}	E_{ij}	$(O_{ij} - E_{ij})$	$(O_{ij} - E_{ij})^2$	$\frac{(O_{ij} - E_{ij})^2}{E_{ij}}$
1	1	225	208.3	16.7	278.89	1.34
2	1	275	291.7	−16.7	278.89	0.96
1	2	150	166.7	−16.7	278.89	1.67
2	2	250	233.3	16.7	278.89	1.20
					χ^2 =	5.17

Since we have a double classification into rows and columns, it is conventional to adopt the usual double-suffix (ij) notation, as in the case of matrices. The first suffix, i, refers to the row, and the second suffix j, to the column. In general:

Row reference $1, 2, \ldots, i, \ldots r$
Column reference $1, 2 \ldots, j, \ldots c$

In our example we have $r = c = 2$.

The number of degrees of freedom of χ^2 can be determined by asking the question, "How many of the O_{ij}'s do we strictly need to know in order to calculate χ^2, in conditions where we already know the row and column totals"?[(1)] It can be seen that if we know just one O_{ij} (say, $O_{11} = 225$), the remaining O_{ij}'s can be obtained by subtraction from the row and column totals, i.e.

$$O_{12} = 375 - 225 = 150$$

$$O_{21} = 500 - 225 = 275$$

$$O_{22} = 400 - 150 = 250$$

Any one of the four observed frequencies, O_{ij}, could have been taken to start with and the remainder obtained by appropriate subtractions. Thus χ^2 has only one degree of freedom for a 2 X 2 table. Quite generally, we have:

ν = number of d.f. = $(r - 1)(c - 1)$, where

r = number of rows (>1), and

c = number of columns (>1).

And so for a 2 X 2 table we have $(2 - 1)(2 - 1) = 1$ d.f.

Returning to the problem, we next compare our χ^2 result with χ^2 tables with 1 d.f. (see Appendix 5 or, alternatively, Exhibit 2a). Since $\chi^2 = 5.17$ and $\chi^2_{0.05}$ (1 d.f.) = 3.841, we reject the null hypothesis and conclude that the proportions of women who smoke in the two samples are significantly different at the 5% level.

It is instructive to rework this example using the two-sample test of differences in proportions. The reader should do this himself in order to demonstrate (numerically) that the two tests are completely equivalent.[(2)]

Finally it is worth noting that the chi-squared test is often applied to data that is percentaged for ease of interpretation and report writing, but the test itself is performed on the actual (unpercentaged) data.

---

(1) The reason for the qualification that "we already know the row and column totals" is as follows. We are basically interested in what changes in the data could be made and yet still lead to the same expected frequencies, E_{ij}. Clearly we can only have the same expected frequencies if the row and column totals remain unchanged.

(2) The reader should find that (rounding errors apart) $Z = \sqrt{\chi^2} = \sqrt{5.17} = 2.27$ and also $z_{0.025} = \sqrt{\chi^2_{0.05}} = \sqrt{3.841} = 1.96$, for 1 d.f. The reason is that χ^2 (ν d.f.) has the same distribution as the *sum of squares* of ν independent standard normal variates. In our case $\nu = 1$, so χ^2 (1 d.f.) $= Z^2$.

Correction for Continuity (Yates' Correction). Resulting from the fact that the observed frequencies can only be whole numbers, the statistic χ^2 is a discrete random variable; and yet the χ^2 *distribution* is continuous. We can improve the accuracy of the χ^2 test by making an allowance for this fact – in much the same way as we did when approximating the binomial distribution by the normal distribution – by using a "continuity correction". The correction is applied by subtracting $\frac{1}{2}$ from the absolute difference between each observed and expected frequency before squaring. That is, we replace $(O_{ij}-E_{ij})^2$ in the χ^2 formula by $(|O_{ij}-E_{ij}|-\frac{1}{2})^2$. With large samples this correction is unnecessary but strictly speaking it should be used. With smaller samples the correction is necessary and must be used to obtain a near correct probability level.

Example 2. In a distribution check 20 grocers out of 60 in Luton were in stock of Royalty cigarettes. In nearby Bedford 18 grocers out of 40 were in stock. Do these sample results differ significantly at the 10% level?

Proceeding in a similar manner to Example 1 we obtain Exhibits 6 and 7.

EXHIBIT 6: Number of Grocers

	Luton	*Bedford*	*Total*
In stock	20 (22.8)	18 (15.2)	38
Not in stock	40 (37.2)	22 (24.8)	62
Total	60	40	100

EXHIBIT 7: Calculation of χ^2 with Continuity Correction

O_{ij}	E_{ij}	$\lvert O_{ij}-E_{ij}\rvert$	$d_{ij} = (\lvert O_{ij}-E_{ij}\rvert-0.5)$	$d_{ij}^2 = (\lvert O_{ij}-E_{ij}\rvert-0.5)^2$	$\frac{d_{ij}^2}{E_{ij}}$
20	22.8	2.8	2.3	5.29	0.23
40	37.2	2.8	2.3	5.29	0.14
18	15.2	2.8	2.3	5.29	0.35
22	24.8	2.8	2.3	5.29	0.21
					$\chi^2 = 0.93$

$\chi^2_{0.10}$ (1 d.f.) = 2.71. Since our calculated value is only 0.93, we accept (the null hypothesis) that the proportion of shops in stock is not significantly different in the two towns. Note that the proportion in stock in Luton is 33.3% and in Bedford 45%: it is too often the case that action is incorrectly taken because of large differences (11.7%) in percentages based on small samples.

Short Cut Computation. If the numbers in Exhibit 6 are represented by letters as follows:

	Sample 1	*Sample 2*	*Total*
In stock	a	b	$(a + b)$
Not in stock	c	d	$(c + d)$
Total	$(a + c)$	$(b + d)$	$n = (a + b + c + d)$

Then the formula

$$\chi^2 = \sum_{\text{all cells}} \frac{(|O_{ij} - E_{ij}| - \frac{1}{2})^2}{E_{ij}}$$

can be simplified to

$$\chi^2 = \frac{n(|ad - bc| - \frac{1}{2}n)^2}{(a + c)(b + d)(a + b)(c + d)} \qquad (2)$$

This formula, although at first sight appearing formidable, is in fact more convenient for computation when a desk calculator is available. It incorporates the continuity correction (*via* the $-\frac{1}{2}n$ in the numerator).

Conditions for Using the Test. It is recommended (Reference 3) that the $\chi^2 (2 \times 2)$ test with a continuity correction can be used when the total sample size is more than 20 and no E_{ij} is less than 5.0; or alternatively, if one E_{ij} is less than 5.0 the total sample size should exceed 40. For sample sizes below 20 (or 40 if an E_{ij} if less than 5.0) an exact test due to R.A. Fisher is available (Reference 2).

19.4 MATCHED SAMPLES (PAIRED COMPARISONS) TESTS

In many studies the opinions of each respondent are sought on two separate occasions, and the results compared. The "before and after" method of studying the effect of advertising in changing opinions on road safety is one example, the study of attitudes to a new bench layout is another. In other types of studies, where it is not possible to administer two different "treatments" to a single respondent, the respondents are matched in as many variables as possible. Examples of this type are: I.Q. studies using identical twins; lung cancer studies using men of the same age, occupational class and area of residence; tests of drugs using patients with similar medical histories.

The previously discussed χ^2 tests cannot be used in such cases since they assume independent samples. A simplified test may however be constructed by careful consideration of the null hypothesis.

Example 3. A film has been made which is designed to improve the attitude of employees and customers to a company. Ninety respondents have their attitude to the company measured. After a suitable interval these same respondents are shown the film and their attitudes remeasured. On each of the two occasions the attitude of each respondent is classified as "favourable" or "unfavourable". The results from the study are shown in Exhibit 8.

EXHIBIT 8: Number of Respondents

		After Film		
		Favourable	*Unfavourable*	*Total*
Before Film	*Favourable*	20(*a*)	15(*b*)	35
	Unfavourable	25(*c*)	30(*d*)	55
				90

Do these results indicate a change in attitudes at the 5% level? From Exhibit 8 we see that 50 persons did not change in their attitudes: 20 being "favourable" both before and after and 30 being "unfavourable" each time. Of the remaining 40 persons 15 were changed from favourable to unfavourable, while 25 went the other way.

Under the null hypothesis, that the film exerted no influence, the number of persons changing their attitudes would be equal – apart from sampling errors. A test for these 40 persons could therefore be based on the binomial distribution with $n = 40$, $\pi = 0.5$. In place of this computationally tedious test we may use a χ^2 test due to McNemar (Reference 2).

The expected number of persons changing attitudes in each direction is, under the null hypothesis, $(15 + 25)/2 = 20$. Hence χ^2 may be calculated as follows:

$$\chi^2 = \frac{(|15-20|-\frac{1}{2})^2}{20} + \frac{(|25-20|-\frac{1}{2})^2}{20},$$

using the continuity correction

$$= \frac{(4.5)^2}{20} + \frac{(4.5)^2}{20}$$

$$= 2.025 \text{ with 1 d.f.}$$

Using the algebraic notation shown alongside the actual figures in Exhibit 8 we may formulate the test as follows. Under the null hypothesis, E(b) = E(c) = $(b + c)/2$. So

$$\chi^2 = \frac{\left(\left|b - \frac{b+c}{2}\right| - \frac{1}{2}\right)^2}{(b+c)/2} + \frac{\left(\left|c - \frac{b+c}{2}\right| - \frac{1}{2}\right)^2}{(b+c)/2},$$

giving

$$\chi^2 = \frac{(|b - c| - 1)^2}{b + c} \qquad (3)$$

McNemar's test, where b and c are the number of respondents who "change direction".

Substituting $b = 15$, $c = 25$, this gives $\chi^2 = 81/40 = 2.025$, as before. Now $\chi^2_{0.05}$ (1 d.f.) = 3.841. As $\chi^2 < \chi^2_{0.05}$, we accept the null hypothesis and state that the changes in attitude observed in the sample could well arise from chance effects – the effect of the film has not been demonstrated.

19.5 FURTHER APPLICATIONS OF THE CHI-SQUARED TEST

Sections 19.6 and 19.7 deal with tests of "association" on so-called "contingency" tables and "goodness of fit" tests, respectively. There are no new concepts in these applications – simply new terminology.

The word "contingency" means "dependency", as in the phrases "A is contingent on B" and "contingency planning". A contingency table is a two-way table where there are more than two categories of respondent, e.g. "heavy smokers", "light smokers" and "non-smokers" and/or where more than two samples are involved, e.g. samples drawn from the London, Midlands, Anglia and Northern I.T.V. regions. Thus the application of χ^2 to contingency tables is little more than an extension of the independent two-sample tests (Section 19.3). The null hypothesis is usually one of "no association", that is, independence between the variables on which the table has been constructed.

The "goodness of fit" applications test the hypothesis that an observed frequency distribution is drawn from a population which follows some theoretical probability distribution.

19.6 CONTINGENCY TABLES AND THE CHI-SQUARED TEST

Example 4. Random samples of grocers shops were taken in Birmingham after stratifying the shopping site into one of three categories – Main Shopping Centres, Local Shopping Centres and Residential Areas. After selection the shops were classified as being Co-operatives, Multiples (10 or more branches) and Independents (1–9 branches). Test at the 1% level if there is any association between the type of shopping area and the type of trading. The sample results are given in Exhibit 9.

EXHIBIT 9: Number of Shops

		Shopping Area			
		Main	*Local*	*Residential*	*Total*
Shop Type	Co-operative	28 (30.0)	18 (15.0)	14 (15.0)	60
	Multiple	84 (50.0)	10 (25.0)	6 (25.0)	100
	Independent	88 (120.0)	72 (60.0)	80 (60.0)	240
	Total	200	100	100	400

The chi-squared statistic is calculated using the same general formula as for the two-sample test, that is:

$$\chi^2 = \sum_{\text{all cells}} \frac{(O - E)^2}{E}.$$

The correction for continuity in contingency tables is complex and is rarely used.

The number of degrees of freedom may be obtained from the general formula $(r - 1)(c - 1)$. In the example we have $(3 - 1)(3 - 1) = 4$ d.f. That is, in conditions where the row and column "marginal" totals are known, we only need to know in addition just 4 of the 9 observed frequencies – the remainder can be deduced by subtraction from the marginal totals.

The expected values for each cell may be derived under the null hypothesis of no association between Shopping Area and Type of Trading. Thus the estimated probability that a shop taken at random is:

(a) from a Main Shopping Area $= \dfrac{200}{400} = \hat{\pi}_{i1}$, say

(b) a Co-operative $= \dfrac{60}{400} = \hat{\pi}_{1j}$

Under the null hypothesis of independence (no association) the joint probability that a shop is both Main Shopping Centre and Co-operative

$$= \frac{200}{400} \times \frac{60}{400} = \hat{\pi}_{11} (= \hat{\pi}_{i1} \cdot \hat{\pi}_{1j})$$

Given that a total sample of 400 (n) is involved the expected number of Main Centre Co-operatives is

$$400 \times \frac{200}{400} \times \frac{60}{400} = \frac{200 \times 60}{400} = 30.0$$

or

$$E_{11} = n \cdot \hat{\pi}_{11}$$

It can be seen that although the expected numbers may be obtained from consideration of the joint probabilities, from a computational point of view the end result can be obtained by multiplying, for each cell, its marginal row and column totals, and dividing by the grand total.

Using the same notation we have:

$$E_{12} = \frac{60 \times 100}{400} = 15.0$$

$$E_{21} = \frac{100 \times 200}{400} = 50.0$$

$$E_{22} = \frac{100 \times 100}{400} = 25.0$$

The remaining expected values may be calculated in the same manner, or by appropriate subtractions. These values have been inserted in Exhibit 9 (in brackets) for the purpose of direct comparison. χ^2 may now be calculated as shown (omitting the (ij) suffices) in Exhibit 10.

EXHIBIT 10

O	E	$(O-E)$	$(O-E)^2$	$(O-E)^2/E$
28	30	−2	4	0.13
84	50	+34	1,156	23.12
88	120	−32	1,024	8.53
18	15	+3	9	0.60
10	25	−15	225	9.00
72	60	+12	144	2.40
14	15	−1	1	0.07
6	25	−19	381	15.24
80	60	+20	400	6.67
400	400	0		χ^2 = 65.76

Now $\chi^2_{0.01}$ (4 d.f.) = 13.28. Our value of χ^2 is considerably greater than this so we reject the null hypothesis and state that the distribution of shop types is associated with the shopping area (as earlier defined).

In the calculation of χ^2 each $(O_i - E_i)^2$ is divided by the expected frequency E_i. For cells that have particularly small E_i's the contribution to χ^2 will be large, possibly leading to a rejection of the null hypothesis. Unfortunately the assumption on which the χ^2 distribution is based is only valid provided that the E_i values are not too small.(1) So to prevent an incorrect conclusion being reached the following rule should be applied: if relatively few expectations (say, less than 1 in 5) are less than 5.0, a minimum expectation of 1.0 is allowable in computing χ^2 (Reference 3). In cases where an occasional expected value is less than 1.0 the row (or column) in which the offending cell lies should be combined (by pairwise addition of expected frequencies, and of observed frequencies) with an adjacent row (or column). This will reduce the number of rows or columns, and hence the number of d.f., of the table. If *a number* of cells have expected values below 5.0 then different computational techniques have to be used (References 1 & 2).

19.7 GOODNESS OF FIT TESTS

These tests are concerned with the comparison of some observed frequency distribution with a theoretical distribution. In the context of this course we are thinking primarily of the binomial, Poisson and normal distributions. The problem is most commonly tackled with a general χ^2 test, but this is not always the best test available. (It can be shown that the most appropriate test depends on the precise form of the hypothesis being tested – Reference 1.) We shall confine ourselves to the problem of determining if our data can be thought to come from a particular form of distribution, with the parameters for that distribution being estimated from our sample.

The general form of the test is the same as is used in significance testing and contingency tables viz:

$$\chi^2 = \sum_i \frac{(O_i - E_i)^2}{E_i}$$

where

O_i is the observed frequency in class i,

E_i is the expected frequency in class i.

However the degrees of freedom are calculated as:

---

(1) This is similar to the requirement, for the binomial distribution, that the expected value $n\pi$ should be greater than 5.0 for a valid normal approximation.

The number of rows (classes)
minus The number of parameters estimated from the sample e.g. one for the Poisson (λ), two for the normal (μ and σ).
minus 1.

Thus when fitting a Poisson distribution with k classes the number of degrees of freedom is $(k - 2)$, while for the normal distribution it would be $(k - 3)$.

Only large values of χ^2 are considered to be significant, and the hypothesis rejected i.e. the conclusion that the fit is a poor one. If the value of χ^2 is too small, implying too good a fit to the model, this can be difficult to interpret but it may be taken that the model is incorrect in some respect, or that the data has been "faked" in some way.

With continuous distributions the data will have to be grouped into classes. The problem of how many classes should be used and where the class boundaries should be does not appear to be critical (Reference 1), However, if the χ^2 approximation is to hold no cells should have an expected frequency of less than 1.0, and there should be fewer than 1 in 5 cells with expected values less than 5.0. Should either of these critieria be violated then each offending class should be combined with an adjacent class or classes until the criteria are met.

In Exhibits 11 and 12 we are mainly concerned with the application

EXHIBIT 11: Fitting and Testing a Normal Distribution

Filled Weight in oz. of Packets Class Limits x_L – x_u	*Observed Frequency* O_i	*Z – Value of* x_u z_u	*Area in the Tail* $Pr(Z > z_u)$	*Normal Prob. in Class*	*Expected Frequency* E_i	$(O_i - E_i)^2$	$\frac{(O_i - E_i)^2}{E_i}$
11.960–11.980	4	−1.64	.0505	.0505	10.10	37.21	3.68
11.980–12.000	28	−0.93	.1762	.1257	25.14	8.15	0.32
12.000–12.020	68	−0.21	.4168	.2406	48.12	395.21	8.21
12.020–12.040	41	+0.50	.3085	.2747	54.94	194.32	3.54
12.040–12.060	29	+1.21	.1131	.1954	39.08	101.61	2.60
12.060–12.080	23	+1.93	.0268	.0863	17.26	32.95	1.91
12.080–12.100 (i.e. over 12.080)	7	"∞"	.0000	.0268	5.36	2.69	0.50
	200				200.00	χ^2 =	20.76

N.B. $\hat{\mu} = \bar{X} = 12.026$ oz. $\hat{\sigma} = s = .028$ oz. (calculated from sample).

of the χ^2 tests – the calculations involved with fitting distributions are given only for completeness sake. Reference should be made to the appropriate Course Units when these are not understood.

Referring to Exhibit 11 we see that no expected frequency is less than 1.0 (no classes need to be combined) and $\chi^2 = 20.76$.

There are $k = 7$ classes: as the mean and standard deviation were estimated from the sample the number of degrees of freedom is $(k - 3) = 4$.

Now $\chi^2_{0.05}$ (4 d.f.) = 9.49. Since χ^2 exceeds $\chi^2_{0.05}$ we reject the hypothesis at the 5% level. Consequently a normal distribution does not provide a suitable model for our sample data.

EXHIBIT 12: Fitting and Testing a Poisson Distribution

Faults Per Vehicle (r)	*No. of Vehicles* O_i	*Poisson Probability*	*Expected Frequency* E_i	$(O_i - E_i)^2$	$\frac{(O_i - E_i)^2}{E}$
0	13	.1108	16.62	13.10	0.79
1	28	.2438	36.57	73.44	2.01
2	63	.2681	40.22	518.93	12.90
3	24	.1966	29.49	30.14	1.02
4	10	.1082	16.23	38.81	2.39
5	8	.0476	7.14	0.74	0.10
6 or more	4	.0249	3.73	0.07	0.02
	150	1.0000	150.00		19.23

N.B. $\lambda = \bar{r} = \frac{330}{150} = 2.2; e^{-2.2} = .1108.$

In Exhibit 12, with $k = 7$ classes and one parameter (λ) being estimated from the sample, the number of degrees of freedom is $(k - 2) = 5$. To test at the 1% level we have $\chi^2_{0.01}$(5 d.f.) = 15.09. Since χ^2 exceeds $\chi^2_{0.01}$ we reject the hypothesis at the 1% level that the Poisson distribution is a suitable model for our data.

In both these "goodness of fit" examples we have rejected hypotheses that we might have expected to hold. We should consequently seek some reason why our hypotheses were rejected. We might, for instance, examine the physical conditions relating to Exhibit 11 to see if there is any reason why the distribution of filled weights appears positively skewed. Similarly, referring to Exhibit 12, we might carry out an investigation to see why "2 faults" occurs so frequently – it could be that one particular kind of fault increases the chances of another.

READING

An account which is highly critical of chi-squared testing, and provides some alternatives, may be found in:
T.H. and R.J. Wonnacott, *Introductory Statistics,* Chapter 17, Wiley, 1972.

REFERENCES

1. M.G. Kendall and A. Stuart, *The Advanced Theory of Statistics,* Vol II Ch. 30. 1961.
2. A.E. Maxwell, *Analysing Qualitative Data,* Ch. 7., 1964.
3. W.G. Cochran, Some Methods for Strengthening the Common Tests, *Biometrics*, 10, pp. 417–451. 1954.
4. B.S. Everitt, *The Analysis of Contingency Tables,* Chs. 1 – 3, 1977.

EXERCISES

1(P) The purpose of this exercise is to generate some data in the form of (2 X 3) contingency tables and apply the chi-squared test to them. The technique used to generate the data is known as simulation, using random number tables.

Independent Case

Consider the following two-way classification.

		Variable 2		
		A	B	C
Variable 1	a	12%	16%	12%
	b	18%	24%	18%

The figures given represent the percentages of a population which would fall into each cell.

Listing the cells (and population %'s) in a single column we obtain:

Cell	*Population %*	*Cumulative Population %*	*Random Number Groups*
aA	12	12	01–12
aB	16	28	13–28
aC	12	40	29–40
bA	18	58	41–58
bB	24	82	59–82
bC	18	100	83–(1)00

Now we can generate a *sample* from this population in the following manner. We use two-digit random numbers to simulate which cell each item of data falls into. Suppose we have "73" as our first random number. Referring to the above table we find that a random number is the range 59–82 inclusive corresponds to cell bB – which gives us the first item of data. Similarly if the next random number is "28" then the second item of data falls into cell aB, and so on. (Note that the random number "00" will be interpreted as 100 for these purposes.)

(i) Show that the population %'s imply that the two variables on which the table has been constructed (leading to classes a and b on the one hand and A, B and C on the other) are statistically independent.
(ii) Explain why using random numbers in the manner indicated amounts to taking a simple random sample from the population.
(iii) Use 50 random numbers (Series C, Standard Data Sheets) to generate 50 items of data and construct the contingency table.
(iv) Forget about the way in which the data was generated. Apply the chi-squared test to the data, explaining carefully the null hypothesis and comment critically on your results.

Non-Independent Case
Now consider another two-way classification:

		Variable 2		
		D	E	F
Variable 1	d	20%	10%	10%
	e	10%	30%	20%

(v) Show that for *this* population the classifying variables are *not* statistically independent.
(vi) Use another 50 random variables to generate a sample of data from this population.
(vii) Apply the chi-squared test to this data, again explaining your null hypothesis and commenting critically on your results.

2(E) Prior to the start of an intense advertising campaign for an existing well-known breakfast cereal in the London I.T.V.

region 1,600 interviews were carried out at random among a sample of housewives. From this survey it was estimated that 20% of the housewives had bought the product in the two weeks before the interview. After 10 weeks of the campaign a different sample of 2,000 interviews were carried out using the same interviewing procedure and it was found that the proportion buying had risen to 26%.

You are asked by the advertising manager of the company to comment on the "significance" of the results. Write a report to the manager showing the calculations you have performed, explaining to him in non-statistical language what is meant by "significance", indicate the reasons for your choice of a particular level of significance (make any assumptions about the background to the situation that you find necessary to answer the question).

HND (Business Studies)

3(E) (a) Pharmaceuticals Ltd. Have developed three new respiratory drugs. They submitted the drugs to clinical trials, the results of which are given in the table below.

Determine whether or not these figures indicate a real difference in the response of the diseases to the drugs, and explain your conclusion.

Use the chi-square (χ^2) distribution as a test of significance with the .01 level of significance table provided.

Drug	*Percentage of patients recovered*	*Percentage of patients not recovered*
A	65	35
B	90	10
C	85	15

Distribution of χ^2

Degrees of freedom	*Probability (.01 level of significance)*
1	6.64
2	9.21
3	11.34
4	13.28

(ACCA 1973)

(b) The Acca Co. Ltd. tests all components built into its products to ensure that they have attained the standard of quality required. There are three checkers, A, B and C,

who undertake this work. The table below shows the numbers of components accepted and rejected by the checkers when they tested three batches of component 1703 recently delivered to the factory. You are required to test the hypothesis that the proportions of components rejected by the three are equal and explain the significance of your conclusion.
Use the chi-square (χ^2) distribution as a test of significance, with the .05 level of significance table provided.

THE ACCA CO. LTD.
Components Quality Test

	Checker A	*Checker B*	*Checker C*	*Total*
Accepted	44	56	50	150
Rejected	16	24	10	50

Distribution of χ^2

Degrees of Freedom	*Probability (.05 level of significance)*
1	3.84
2	5.99
3	7.82
4	9.49

(ACCA 1971)

Authors' note: there are fundamental errors in both these questions. Can you spot them?

4(E) After redecoration of a shop, the suggestion is made that it presents too feminine an appearance, and so discourages male shoppers from entering. To help in the discussion of this problem the opinions of a random sample of customers is taken and the results are tabulated below.
Is there any evidence of difference of opinion between men and women?
State clearly what you test and the level of significance you use.
Is the redecoration an improvement?

	Yes	No difference	No
Men	14	20	15
Women	21	18	10

HND (Business Studies)

5(E) In a nation-wide survey of wage levels, data concerning trades union membership and wage rates were collected for a sample of 100 firms and each aspect was classified as "high" or "low" relative to an arbitrary standard. The results obtained are summarised below:

		Wage Rates	
		"High"	*"Low"*
Union	"High"	37	31
Membership	"Low"	11	21

(i) Is there any evidence of an association in this sample?
(ii) How would you interpret these results in terms of the development of a policy for the encouragement or otherwise of trades unions in your own orgnisation?

(IPM 1974)

6(E) (a) A Credit Control Manager is considering changing the present system of control to reduce the number of "poor" payers. A sample of 90 customers is taken at random, 45 "good" payers and 45 "poor" payers. After the experimental introduction of the proposed scheme to the selected sample the following results are obtained.

		Old Scheme	
		Poor	*Good*
New Scheme	Good	35(*a*)	20(*b*)
	Poor	10(*c*)	25(*d*)

From first principles, but omitting a continuity correction, derive an expression for the value of chi-square (χ^2) for matched samples (McNemar's test) in terms of *a* and *d*. Obtain the values of χ^2 for the given data and comment on the statistical significance of the result at the 5% level.

(b) As part of the investigation by the Credit Control manager the following classification of firms by their credit rating was derived.

		Number of Firms		
Size of Firm		*Large*	*Medium*	*Small*
Credit Rating	Good	28	60	57
	Poor	12	40	53

Test at the 5% level if there is any association between the size of firm and their Credit Rating.

BA (Business Studies)

7(E) Two factories using materials purchased from the same supplier and closely controlled to an agreed specification produce output for a given period classified into three quality grades as follows:

	Output in tons			
Quality grade:	*A*	*B*	*C*	*Total*
Factory:				
X	42	13	33	88
Y	20	8	25	53
Total	62	21	58	141

(a) Do these output figures show a significant difference at the 5% level?

(b) What hypothesis have you tested?

(ICMA 1971)

8(E) What theoretical assumptions underlie the use of the Poisson distribution?

Records are kept of the issues of a certain machinery spare from the stores. From the records, extending over the past two years (100 weeks), a summary of the pattern of issues was compiled as follows:

No. of issues during week	*No. of weeks (frequency)*
0	39
1	32
2	19
3	10
4 or more	0

Find the mean number of weekly issues. Using this mean, fit a Poisson distribution to this data.

Carry out a test for the "goodness of fit" of the distribution, and comment critically on your findings.

(DMS 1975)

9(E) (i) The number of breakdowns that have occurred during the last 100 shifts is as follows:

Number of breakdowns per shift	0	1	2	3	4	5
Frequency:						
Expected number of shifts	14	27	27	18	9	5
Actual number of shifts	10	23	25	22	10	10

Show whether the manager is justified in his claim that the difference between the number of actual and expected breakdowns is due to chance. It has been customary to use a significance level of 0.05.

(ICMA 1973)

(ii) (a) What does the χ^2 test test?
(b) The number of rejects in six batches of equal size were:

Batch	*Number of rejects*
A	270
B	308
C	290
D	312
E	300
F	320

Test the hypothesis that the difference between them is due to chance using a level of significance of 0.05.

(ICMA 1973)

10(E)* A seed merchant operates a mail order business. In the third week of January the company received 1,445 orders, each order asking for a number of different seed types. The following data prepared by the company statistician gives N, the number of orders received asking for R *or more* seed types. (i.e. the cumulative frequency distribution.)

R	1	2	5	10	20	50
N	1445	1312	1002	631	251	16

The company statistician also "fitted" the following empirical model to the data: $\text{Log}_{10} N = 3.20 - 0.04 R$.

(i) Using the empirical model calculate the expected value of N for each given value of R.
(ii) Construct an observed *frequency distribution* and compare with the expected frequency distribution.
(iii) Calculate an approximate value of χ^2 (chi-squared) to test the goodness of fit of the empirical model.
(iv) What conclusions do you draw from the value of χ^2 you have obtained?

BA (Business Studies)

SOLUTIONS TO EVEN NUMBERED EXERCISES

2. This question may be answered using a large sample test of differences in proportions (CU18) or by a chi-squared test as follows:

	Sample 1	*Sample 2*	*Total*
Had bought	320 (373.3)	520 (466.7)	840
Had not bought	1,280 (1,226.7)	1,480 (1,533.3)	2,760
Total	1,600	2,000	3,600

O	*E*	*(O–E)*	*(O–E)²*	*(O–E)²/E*
320	373.3	–53.3	2,840.89	7.610
520	466.7	+53.3	2,840.89	6.087
1,280	1,226.7	+53.3	2,840.89	2.316
1,480	1,533.3	–53.3	2,840.89	1.853
			X^2 =	17.866

N.B. (i) With such large numbers the continuity correction may be ignored.

(ii) The result may be obtained from the "short-cut" Formula (2) in the text.

(iii) There is $(r - 1)(c - 1) = 1$ degree of freedom. With

$$\chi^2_{0.05} \text{ (1 d.f.)} = 3.841 \text{ and}$$

$$\chi^2_{0.01} \text{ (1 d.f.)} = 6.635,$$

the null hypothesis is rejected.

4.

	Y	*ND*	*N*	*Total*
Men	14 (17.5)	20 (19.0)	15 (12.5)	49
Women	21 (17.5)	18 (19.0)	10 (12.5)	49
Totals	35	38	25	98

O	*E*	*(O–E)*	*(O–E)²*	*(O–E)²/E*
14	17.5	–3.5	12.25	0.700
21	17.5	3.5	12.25	0.700
20	19.0	1.0	1.00	0.052
18	19.0	–1.0	1.00	0.052
15	12.5	2.5	6.25	0.500
10	12.5	–2.5	6.25	0.500
			χ^2 =	2.504

$\nu = (r - 1)(c - 1) = (2 - 1)(3 - 1) = 2$ d.f.

$\chi^2_{0.05} = 5.991$ and hence the null hypothesis is accepted.

6. (a) Under the null hypothesis E(a) = E(d) = $(a+d)/2$. So

$$\chi^2 = \frac{\left(a - \frac{(a+d)}{2}\right)}{\frac{a+d}{2}} + \frac{\left(d - \frac{(a+d)}{2}\right)}{\frac{a+d}{2}},$$

which simplifies to

$$\chi^2 = \frac{(a-d)^2}{a+d}$$

Substituting $a = 35, d = 25$ we obtain

$$\chi^2 = \frac{(35-25)^2}{35+25} = 1.67.$$

As $\chi^2_{0.05}$ (1 d.f.) = 3.841 we accept the null hypothesis

6. (b)

	L	*M*	*S*	*Total*
G	28 (23.2)	60 (58)	57 (63.8)	145
P	12 (16.8)	40 (42)	53 (46.2)	105
	40	100	110	250

O	*E*	*(O–E)*	*(O–E)*^{2}	*(O–E)*^{2}*/E*
28	23.2	4.8	23.04	0.993
12	16.8	–4.8	23.04	1.371
60	58.0	2.0	4.00	0.069
40	42.0	–2.0	4.00	0.095
57	63.8	–6.8	46.24	0.725
53	46.2	6.8	46.24	1.001
				χ^2 = 4.254

We have $(r-1)(c-1) = (2-1)(3-1) = 2$ d.f.
$\chi^2_{0.05}$(2 d.f.) = 5.991 and hence the null hypothesis is accepted.

8.

x_i	$f_i(=O_i)$	$x_i f_i$	$Pr(x_i \mid \lambda = 1.0)$	E_i	$(O_i - E_i)$	$(O_i - E_i)^2/E_i$
0	39	0	0.3679	36.8	+2.2	0.135
1	32	32	0.3679	36.8	–4.8	0.626
2	19	38	0.1839	18.4	+0.6	0.020
3	10	30	0.0613	6.1	+3.9	2.493
4+	0	0	0.0190	1.9	–1.9	1.900
						χ^2 = 5.174

$\bar{x} = \hat{\lambda} = \sum x_i f_i / \sum f_i = 100/100 = 1.0$

ν = number of rows − number of parameters estimated − 1

= 5 − 1 − 1

= 3 (d.f.).

So $\chi^2_{0.05}$ (3 d.f.) = 7.815; the null hypothesis is accepted.

10. (i)

R	1	2	5	10	20	50
N	1,445	1,312	1,002	631	252	16
$Y = (3.20-0.04R)$	3.16	3.12	3.00	2.80	2.40	1.20
Antilog$_{10}$ (Y) (= Expected N)	1,445.4	1,318.2	1,000	630.9	251.2	15.8

The observed and expected frequency distributions are then obtained by subtraction of adjacent N values in the above *cumulative* frequency distribution.

(ii) Number of Seed Types.

Class Limits	*Observed Frequency (O)*	*Expected Frequency (E)*	*$(E-O)^2/E$*
1	133	127.2	0.26
2–4	310	318.2	0.21
5–9	371	369.1	0.01
10–19	380	379.7	0.00
20–49	235	235.4	0.00
50 and over	16	15.8	0.00
	1,445	1,445.4	χ^2 = 0.48

We see that the observed and expected frequencies are very close. Now

ν = number of rows − number of parameters estimated − 1

= 6 − 2 − 1

= 3 d.f.

$\chi^2_{0.05}$ (3 d.f.) = 7.82 and

$\chi^2_{0.95}$ (3 d.f.) = 0.352 (i.e. a *left-hand* tail test).

It can be seen that the null hypothesis is accepted on both a right-hand fail test (poor fit) and a left-hand tail test (too good a fit).

In questions 2, 4, 6 and 10 we have used arbitrary but conventional significance levels since they were not given in the questions. The reader may well choose with good reason to use other levels.

COURSE UNIT 20 – SMALL SAMPLE TESTS – THE *t*-DISTRIBUTION

20.1 INTRODUCTION

In our earlier tests on sample means (CU18) we made use of the normal deviate Z for testing purposes. In the single sample test[1], for example, we calculated

$$Z = \frac{\bar{X} - \mu}{\sigma/\sqrt{n}}.$$

When using this formula we assumed that either σ^2, the population variance, was known or (more commonly) that it could be approximated, to a sufficient degree of accuracy, by the sample variance, s^2. Provided that the sample size is large ($n \geqslant 30$) this latter assumption holds: consequently the statistic

$$\frac{\bar{X} - \mu}{s/\sqrt{n}}$$

is still *approximately* normally distributed with zero mean and unit variance, and the standard normal distribution test may be applied. With small samples ($n < 30$), however, this approximation is no longer valid – the normal distribution is then inappropriate. There are two basic problems. Firstly s^2 is a biased estimator of σ^2 (CU5) and hence we should correct for this by replacing s by

$$\hat{\sigma} = s \cdot \sqrt{\frac{n}{n-1}}$$

in our test statistic to give

$$t = \frac{\bar{X} - \mu}{\hat{\sigma}/\sqrt{n}} = \frac{\bar{X} - \mu}{s/\sqrt{n-1}} \qquad (1)$$

(We use the symbol t for the test statistic for small samples, rather than Z, in order to emphasise that it is not normally distributed.) The second

---

[1] Similarly for an independent two-sample test of means we evaluated the normal deviate

$$Z = \frac{(\bar{X}_1 - \bar{X}_2) - (\mu_1 - \mu_2)}{\sqrt{\frac{\sigma_1^2}{n_1} + \frac{\sigma_2^2}{n_2}}}$$

problem is that s will itself vary from sample to sample, and this results in the variance of t being greater than unity.

Provided that the *measurement quantity*, X, is itself normally distributed (and only if this is the case) the sampling distribution of t is called, simply, the t-distribution, tables of which are to be found in most statistical texts.[(1)] An important feature of this distribution is that it does not depend on any *unknown* population parameters: in fact it has just one parameter ν (the number of degrees of freedom) which is *exactly* related to the sample size (n) involved in the test.[(2)]

The p.d.f. of t, $f(t)$, is given by

$$f(t) = c_\nu \left(1 + \frac{t^2}{\nu}\right)^{-\left(\frac{\nu+1}{2}\right)},$$

where c_ν is a normalising constant (ensuring that the area under the graph is unity) depending only on ν. One example of the p.d.f. (for ν = 6) is shown in Exhibit 1 from which it can be seen that the shape is *roughly* the same as the normal distribution.

EXHIBIT 1: Comparison of the t-distribution (ν = 6) and the Standard Normal Distribution

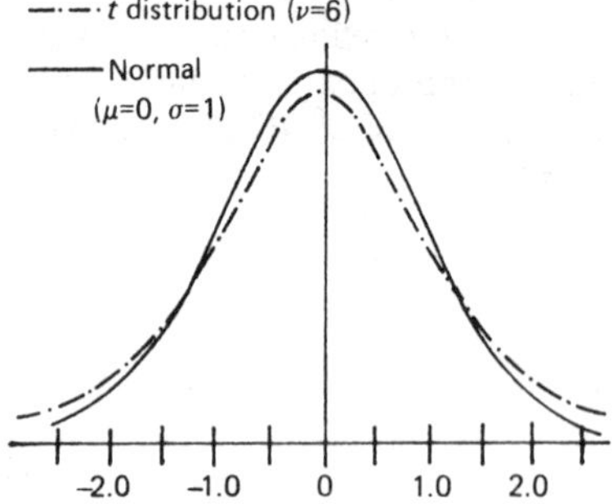

A few important properties of the t-distribution may be summarised as follows:

---

(1) The distribution and tests are commonly called "Student's t", Student being the pseudonym under which W.S. Gossett first published details of the distribution.

(2) Readers who have completed CU19 will not fail to notice the similarity in description to the χ^2 distribution. In fact the χ^2 distribution and the t-distribution have a close theoretical relationship.

- The distribution is symmetrical about $t = 0$, and hence the expected value of t is zero.
- The variance of t is given by $\nu/(\nu - 2)$ for $\nu > 2$, which is always greater than unity, although only marginally so for large values of ν. For very large ν the variance is effectively unity and the p.d.f. is then closely approximated by the standard normal distribution (for an infinite ν the two distributions are identical).
- The t values which result in a (small) given "area in the tail" of the distribution are *larger* than the corresponding z values, the discrepancy being greater for small values of ν. For example $z_{0.025} = 1.96$, while $t_{0.025}$ is 2.086 for $\nu = 20$, 2.228 for $\nu = 10$ and 2.571 for $\nu = 5$.

It cannot be emphasised too strongly that the t-distribution (and the tests based on it) strictly applies only when employing simple random sampling from a normally distributed population. It is the responsibility of the researcher to evaluate the effect on probability levels when using t-tests if these assumptions do not hold. But when these assumptions are correct the t-tests to be described are *exact*; and, in fact, the large sample tests of CU18 (involving *estimated* variances) are approximations to these exact tests.

20.2 ONE-SAMPLE TESTS

These are similar to the corresponding large sample tests and, as such, they can be either one or two-sided. The number of degrees of freedom of a one-sample test $\nu = n - 1$, where n is the sample size.(1)

Example. The standard time allowed to first-stage forge a racing car crankshaft is 14 mins 40 secs. An order for six crankshafts has been received and the times taken for forging are shown in Exhibit 2.

In order to demonstrate the close relationship between confidence intervals and significance tests we shall use this example (a) to place 95% confidence limits for the population mean forging time, and (b) to test whether the observed mean forging time differs significantly from the allowed 14 mins 40 secs at the 5% level. For convenience we work in terms of the *coded* values, Y (= X – 14 mins 40 secs), and make the assumption, as is required for the t-distribution, that the values X

(1) As in CU19 we interpret ν as the number of independent observations on which the test is based. In the case of the $n = 3$ observations (10, 6, 11) the value of $\bar{X}$ is $27/3 = 9$. But knowing this value only *two* (= $n - 1$) of the observations are now independent because the third observation can be *calculated* from these two and the known value of $\bar{X}$. Hence the knowledge of $\bar{X}$ effectively reduces the number of degrees of freedom in a sample by one.

EXHIBIT 2: Forging Times for Six Crankshafts

Forging Times (X)		*Y*	*Y^2*
Mins.	*Secs.*	*(= X – 14 mins. 40 secs.)*	
14	45	+5	25
14	47	+7	49
14	41	+1	1
14	44	+4	16
14	40	0	0
14	38	–2	4
		$\Sigma Y = 15$	$\Sigma Y^2 = 95$

(and Y) are randomly "selected" from a normal distribution. Now

$$\bar{Y} = \sum Y/n = 15/6 = 2.5$$

and

$$s^2 = \frac{1}{n}\sum Y^2 - \bar{Y}^2 = 95/6 - (2.5)^2 = 9.583.$$

We could next correct s^2 for bias, giving $\hat{\sigma}^2 = ns^2/(n-1)$, and then evaluate $\hat{\sigma}^2/n$; but it is simpler to combine these two steps into one since $\hat{\sigma}^2/n = s^2/(n-1)$. So

$$\hat{\sigma}^2/n = s^2/(n-1) = 9.583/5 = 1.917,$$

and the

$$\text{SE of } \bar{Y} = \sqrt{\hat{\sigma}^2/n} = \hat{\sigma}/\sqrt{n} = \sqrt{1.917} = 1.384$$

The number of d.f., $\nu = n - 1 = 5$, and having calculated $\bar{Y}$ and its SE we can now proceed with the confidence interval (a) and significance test (b).

(a) For a 95% CI we require 2.5% (= 0.025) in *each* tail of the distribution. From tables, with $\nu = 5$, we obtain $t_{0.025} = 2.571$ (compare with $z = 1.96$ for the normal distribution).
A 95% CI for the true mean, μ_Y, is given by

$$\bar{Y} \pm t_{0.025}(\text{SE of } \bar{Y})$$
$$= 2.5 \pm 2.571 \times 1.384$$
$$= 2.5 \pm 3.56$$
$$(\text{i.e. } -1.06 \text{ to } +6.06)$$

This interval includes zero (i.e. actually 14 mins 40 secs) and so our forging times could well be attributable to sampling errors about the population mean forging time of 14 mins 40 secs.

(b) Formally we have:

$$H_0 : \mu_Y = 0$$

$$H_1 : \mu_Y \neq 0 \quad \text{(hence a 2-sided test)}$$

$$\alpha = 0.05, \alpha/2 = 0.025, t_{\alpha/2}(5 \text{ d.f.}) = 2.571,$$

as before. Our observed value of

$$t = \frac{\bar{Y} - \mu_Y}{\hat{\sigma}/\sqrt{n}} = \frac{2.5 - 0}{1.384} = 1.806$$

Now $1.806 < t_{\alpha/2}$, and thus we accept (the null hypothesis) that the allowed time of 14 mins 40 secs is correct -- the same conclusion as was obtained in (a).

EXHIBIT 3: A 2-Sided *t*-Test

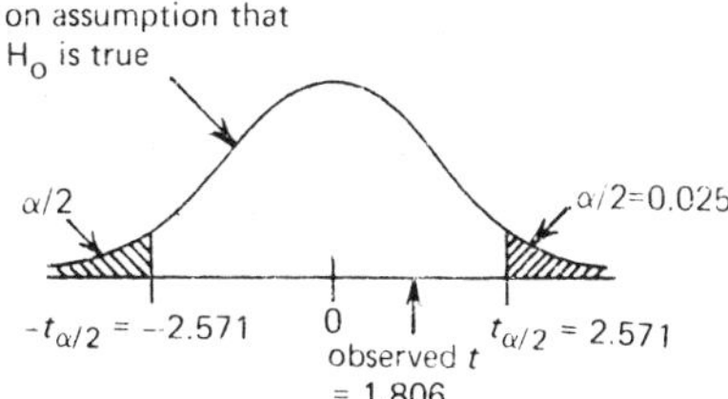

20.3 INDEPENDENT TWO-SAMPLE TESTS

As with large sample tests we can compare two independent sample means. The small sample t-test relies on the following two assumptions:

- each sample is randomly selected from a *normal* population.
- the two samples come from populations with *equal* variances.

Neither of these assumptions were required for the corresponding large sample tests (Section 18.3). When the population variances are unknown, but assumed equal, the t-test may be employed provided that we obtain a *pooled* estimate of the population variance $\hat{\sigma}^2$, given by[1]

---

[1] Alternatively, for raw data,

$$\hat{\sigma}^2 = \frac{\Sigma X_1^2 - n_1 \bar{X}_1^2 + \Sigma X_2^2 - n_2 \bar{X}_2^2}{n_1 + n_2 - 2}$$

$$\hat{\sigma}^2 = \frac{n_1 s_1^2 + n_2 s_2^2}{n_1 + n_2 - 2}, \tag{2}$$

where

$$s_1^2 = \frac{1}{n_1} \sum^{n_1} (X_1 - \bar{X}_1)^2 \text{ and } s_2^2 = \frac{1}{n_2} \sum^{n_2} (X_2 - \bar{X}_2)^2,$$

the variances of samples 1 and 2 respectively.

Before proceeding with an example of the test we pause to examine this *pooling* process in a little more detail.

Pooling of Variances. Formula (2) may be rewritten as

$$\hat{\sigma}^2 = \frac{\sum^{n_1} (X_1 - \bar{X}_1)^2 + \sum^{n_2} (X_2 - \bar{X}_2)^2}{n_1 + n_2 - 2} = \frac{\text{Total Sum of Squares}}{\text{Total Degrees of Freedom}}$$

The phrase "sum of squares" is a commonly used piece of statistical shorthand for the *sum of squared deviations from the sample mean.* The denominator is simply the number of degrees of freedom for sample 1 ($= n_1 - 1$) added to the number for sample 2 ($= n_2 - 1$). This technique of dividing the Total Sum of Squares by the Total Degrees of Freedom in order to estimate a (common) population variance from a number of independent samples is one which is widely used in more advanced work – we shall employ it again in CU21. Now suppose, to take a particular example, we have equal sized samples ($n_1 = n_2 = n$): Formula (2) can then be written as

$$\hat{\sigma}^2 = \frac{ns_1^2 + ns_2^2}{2n - 2} = \frac{1}{2}\left\{\frac{ns_1^2}{n-1} + \frac{ns_2^2}{n-1}\right\} = \frac{1}{2}\left\{\hat{\sigma}_1^2 + \hat{\sigma}_2^2\right\};$$

that is, the simple average of the two independent estimates.

Example of a Two Sample Test. A sample of 10 key manual workers and 8 clerical workers were given an expensive anti-cold treatment at the beginning of the winter. The number of days lost through colds during the winter by the two groups is summarised as follows:

Manual	$\bar{X}_1 = 8.0$	$s_1 = 2.4$	$(n_1 = 10)$
Clerical	$\bar{X}_2 = 12.0$	$s_2 = 2.6$	$(n_2 = 8)$

Is the difference between the groups significant at the 2% level? First, we obtain the pooled variance estimate.

$$\hat{\sigma}^2 = \frac{n_1 s_1^2 + n_2 s_2^2}{n_1 + n_2 - 2} = \frac{10(2.4)^2 + 8(2.6)^2}{10 + 8 - 2} = 6.98$$

Next the SE of the difference between two independent means is calculated as

$$\sqrt{\frac{\hat{\sigma}^2}{n_1} + \frac{\hat{\sigma}^2}{n_2}} = \sqrt{\hat{\sigma}^2\left(\frac{1}{n_1} + \frac{1}{n_2}\right)} = \sqrt{6.98\left(\frac{1}{10} + \frac{1}{8}\right)} = \sqrt{1.571} = 1.253$$

(Note that the pooled variance has been used for *both* samples[1].) The null hypothesis is,

$$H_0 : \mu_2 = \mu_1 \text{ or } \mu_2 - \mu_1 = 0$$

and the alternative is

$$H_1 : \mu_2 \neq \mu_1 \text{ or } \mu_2 - \mu_1 \neq 0 \quad \text{(hence a 2-sided test)}$$

So $\alpha = 0.02$, $\alpha/2 = 0.01$ and $t_{\alpha/2} = t_{0.01}$ (with $10 + 8 - 2 = 16$ d.f.) = 2.583. Our observed value of t is evaluated as

$$t = \frac{(\bar{X}_2 - \bar{X}_1) - (\mu_2 - \mu_1)}{\sqrt{\hat{\sigma}^2\left(\frac{1}{n_1} + \frac{1}{n_2}\right)}} = \frac{(12.0 - 8.0) - 0}{1.253} = 3.192$$

(It can be seen that the test procedure corresponds to the large sample test but differences occur in the SE estimation and the use of the t-distribution with $n_1 + n_2 - 2$ d.f.) Now the observed value of t = $3.192 > t_{0.01}$ (= 2.583) and hence we reject the null hypothesis and conclude that the difference between the groups is significant at the 2% level.

20.4 MATCHED OR PAIRED SAMPLES

In CU19 we considered a χ^2 test (McNemar's) designed to detect significant differences between matched or paired samples. That test applied to non-numeric (attribute) data. Here we consider the corresponding

---

(1) Note too that this procedure is similar to the pooling of proportions in the large sample test of difference between proportions (CU18). We obtained

$$\hat{\pi} = \frac{n_1 P_1 + n_2 P_2}{n_1 + n_2},$$

and used this value to work out the SE of difference in proportions as

$$\sqrt{\hat{\pi}(1 - \hat{\pi})\left(\frac{1}{n_1} + \frac{1}{n_2}\right)}$$

t-test for paired comparisons (especially those involving two different treatments applied to the same objects or persons) in which the results can be quantified by some *numerical* score. We treat one problem in a little detail in order to illustrate some elementary aspects of statistical design.

Suppose we wished to test whether alcoholic drink affected the ability of drivers to brake in time to avoid an accident. As a "live" test seems rather hazardous we might employ a road test "simulator" machine on which braking distance can be measured. Thinking about the reasons why braking distance varies from one driver to another or from one occasion to another we might identify the following factors:

(a) Innate difference between drivers (experience, age, alertness)
(b) Differences between different occasions for the same driver (statistical fluctuations)
(c) Experience of using the simulator
(d) Quantity of alcohol consumed.
(e) Effect of a given quantity of alcohol

Clearly if we wish to test whether alcohol affects braking distance we should set about attempting to reduce the variability in braking distance brought about by the other factors. We could for instance choose experienced drivers of the same age (even though this may limit the validity of our conclusions to such drivers), which would reduce the variability due to (a). We could give all drivers sufficient experience on the simulator so that they are thoroughly familar with it and administer a fixed amount of alcohol to randomly chosen drivers a fixed time before the test. Even with such careful planning we would be unlikely to detect a significant difference from two *independent* samples (one group having drunk the alcohol, the other not) because of the remaining variation due to (b) and (e), unless the sample sizes are very large. An alternative approach is to carry out a "before and after" (drinking) test on the *same* group of drivers, calculating for *each* the increase in braking distance (D) between the two tests. Not only is this procedure more effective in eliminating between-driver variation (a), it has the added advantage of reducing the test to that of a single sample of differences. Now the example.

Example. Eight experienced drivers are taught to use a road test simulator machine. After they are satisfied they can use the machine properly an accident is simulated and the braking distance measured. Each driver is then given two double whiskies and a half-hour later repeats the test. The results are shown in Exhibit 4.

EXHIBIT 4: Braking Distance (feet)

Driver	*Before*, X_1	*After*, X_2	$D = X_2 - X_1$	D^2
A	45	52	+7	49
B	62	58	−4	16
C	38	44	+6	36
D	50	54	+4	16
E	44	48	+4	16
F	47	50	+3	9
G	42	46	+4	16
H	45	49	+4	16
			$\Sigma D = +28$	$\Sigma D^2 = 174$

Having calculated the D values we may treat these as a single sample (i.e. ignoring the X_1 and X_2 values from which they were obtained). Our null hypothesis that drinking does not affect braking distance can then be expressed as

$$H_0 : \mu_D = 0,$$

and the alternative hypothesis is that drinking *increases* braking distances, $H_1 : \mu_D > 0$ (hence a one-sided test). Now

$$\bar{D} = \sum D/n = 28/8 = 3.5$$

$$s_D^2 = \sum D^2/n - \bar{D}^2 = 174/8 - (3.5)^2 = 9.5.$$

The SE of $\bar{D} = \sqrt{s_D^2/(n-1)} = \sqrt{9.5/7} = \sqrt{1.357} = 1.165$

Hence our observed value of

$$t = \frac{3.5 - 0}{1.165} = 3.004$$

Assuming that we wish to carry out the test at the 1% level we compare our result with $t_{0.01}$ with $(n - 1) = 7$ d.f., which from tables is 2.998. Now we obtained $t = 3.004 > t_{0.01}$ (7 d.f.), and so we reject the null hypothesis and conclude that the alcohol increases the braking distance (see Exhibit 5).

EXHIBIT 5: A One-Sided t-Test

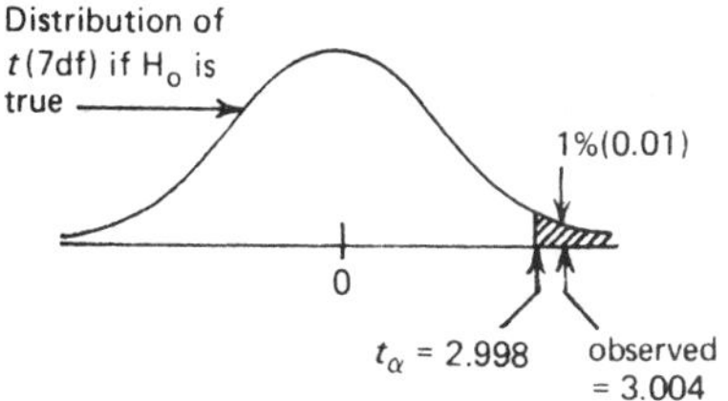

Had we mistakenly carried out an independent two-sample test on this data this would have let to an *acceptance* of the null hypothesis. (For a comparison of the independent two-sample and paired comparison tests see Reference.) The assumption underlying the paired comparison test is that the differences, D, are independent random variables from the normal distribution, but the restriction (as in the independent two-sample test) that the X_1 and X_2 values have equal variance no longer applies.

20.5 TEST ON REGRESSION (AND CORRELATION) COEFFICIENTS

Regression (and correlation) analysis is often carried out with small samples: provided that certain assumptions hold we may use a t-test to examine the significance of the coefficients.

Taking regression first, to ensure the validity of the t-tests we need to assume the *standard regression model*[(1)] (CU14) and that the Y-errors about the true regression line are independently *normally* distributed. We may then test the regression coefficients a and b in the least-squares regression line, $\hat{Y} = a + bX$, of Y on X. If b, in particular, is found to be "significant" we mean in fact that b is *significantly different from zero*. If b is *not* significantly different from zero then the regression line is of no use for predictive purposes.

So our null hypothesis is $H_0: \beta = 0$ (i.e. the slope of the *population* regression line is zero). Now it may be shown that, under the stated assumptions, the value of b is normally distributed about the true value β (= 0, under H_0).

Consequently the statistic

$$t = \frac{b - \beta}{\text{SE of } b} = \frac{b - 0}{\text{SE of } b} \quad (\text{if } H_0 \text{ is true}) \qquad (3)$$

will have a t-distribution with $(n - 2)$ degrees of freedom. (The subtraction of 2 from the sample size results from the fact that *two* statistics $\bar{X}$

(1) In short the standard regression model assumes:

- the X_i (predictor) values can be measured without inaccuracy
- the Y-errors $(Y_i - \hat{Y}_i)$ about the true regression line are randomly drawn from the *same* distribution.

and $\bar{Y}$ are needed to evaluate b.) Using results from CU14 Formula (3) may be written more conveniently as[1]

$$t = \sqrt{\frac{r^2}{1 - r^2}(n - 2)}, \qquad (4)$$

where r is the *product moment correlation coefficient*

So to test the *regression* coefficient b, we use the *correlation* coefficient r. We should not be too surprised by this – we have already seen (CU14) that r^2 provides a measure of how well the least-square line fits the data.

Example. We use the bivariate data from the mail-order example (CU14) reproduced here as Exhibit 6.

EXHIBIT 6: Data from Mail-Order Problem

X (No. of items)	*Y (Packing Time in Mins.)*	*X (No. of items)*	*Y (Packing Time in Mins.)*
8	6	6	4
11	8	12	9
8	8	9	6
11	9	6	8
4	5	5	7

[1] This is not too difficult to prove:

From CU14 we have

$$\text{Var}(b) = \frac{\sigma_e^2}{\Sigma(X_i - \bar{X})^2} = \frac{\sigma_e^2}{ns_X^2}$$

But

$$\hat{\sigma}_e^2 = \frac{1}{n-2}\Sigma(Y_i - \hat{Y}_i)^2 = \frac{(1 - r^2)}{n-2}\Sigma(Y_i - \bar{Y})^2 = \left(\frac{1 - r^2}{n - 2}\right)\cdot ns_Y^2$$

So

$$\hat{\sigma}_b = \left(\frac{1 - r^2}{n - 2}\right)\cdot\frac{s_Y^2}{s_X^2},$$

but b itself is given by

$$b = \frac{s_{XY}}{s_X^2}$$

Substituting into (3) gives

$$t = \frac{s_{XY}}{s_X^2}\sqrt{\frac{n-2}{1-r^2}}\cdot\frac{s_X}{s_Y} = \sqrt{\frac{r^2}{1-r^2}(n-2)}.$$

The data shown led to a regression line (of Y on X) of $\widehat{Y} = 3.704 + 0.412X$; so $b = 0.412$. Also the value of r^2 was calculated in CU14 as 0.444. So

$$t = \sqrt{\frac{r^2}{1 - r^2}(n - 2)} = \sqrt{\frac{0.444}{0.556}(10 - 2)} = \sqrt{6.388} = 2.528.$$

Testing against $H_1 : \beta > 0$ (a one-sided test) at the 5% level, we have $t_{0.05}$ (8 d.f.) = 1.860. Hence b is significantly greater than zero at the 5% level.

In principle we could use the t-distribution to test for β being *any* fixed value (rather than zero, as in this example). Similarly we could test for any fixed value of α, the intercept on the y-axis. But such tests lack the computational simplicity of the test we have carried out.(1) Finally, it is interesting to note that the corresponding test on the correlation coefficient r involves precisely the same test-statistic – that of Formula (4) – as for the test of b: in fact by testing b we are simultaneously testing r.(2)

It may have appeared strange to the reader that t-tests can be applied to such diverse problems as those involving means from two independent samples and the regression coefficient. In fact, however, there is a simple link between the two. Referring back to the anti-cold treatment problem (Section 20.3), we might reformulate it as a problem in *regression* as follows. Let us "code" each result from a manual worker using $W = 1$, and from a clerical worker using $W = 2$. Exhibit 7 shows a scatter diagram of the results on which the regression line of X on W has been drawn. The regression line will obviously pass through the individual sample means and so has a slope

$$b = \frac{\bar{X}_2 - \bar{X}_1}{2 - 1} = \bar{X}_2 - \bar{X}_1.$$

(1)To test for $\beta = \beta_0$ we use $t = (b - \beta_0)/(\text{SE of } b)$, where the SE of b is given (from CU16) as

$$\frac{s_Y}{s_X}\sqrt{\frac{1 - r^2}{n - 2}}.$$

(2)In fact b and r are linearly related by

$$b = \frac{s_Y}{s_X} \cdot r,$$

so if one is significant the other will be too.

EXHIBIT 7

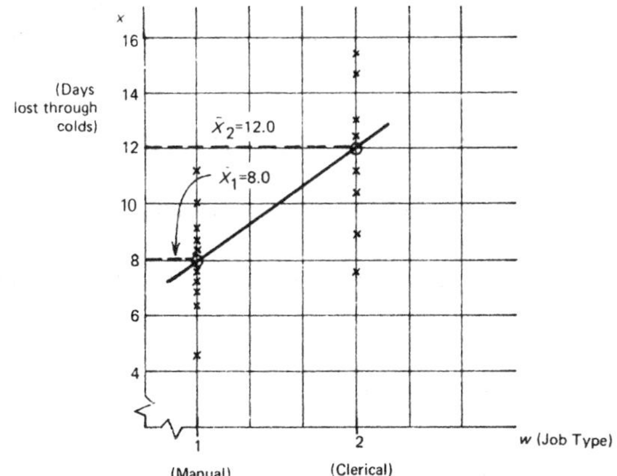

Consequently the slope of the regression line is formally identical to the difference between the two sample means: as the sample means were found to be significantly different, so would the slope be found significant.

READING

F.E. Croxton, D.J. Cowden & B.W. Bolch *Practical Business Statistics* 4th Ed. Prentice-Hall, Chapter 11. (But note that the authors use s in place of our $\hat{\sigma}$.)

REFERENCE

G.W. Snedecor and W.G. Cochran, *Statistical Methods* (6th Ed.), Chapter 2, Iowa State.

EXERCISES

1(P) Take a ruler and examine the cm. scale. Try and fix in your mind the length 10 cm. (but *without* using any aids like comparing it with the length of a finger). Put the ruler out of sight.

(i) Cut up five lengths of string (or strips of paper) to an estimated length of 10 cm. As each one is cut put it aside so that you cannot compare one with another. When completed, measure each length to the nearest 0.1 cm. and then carry out an appropriate single sample t-test on your sample at the 5% level.

(ii) Repeat part (i).

(iii) Carry out a two-sample t-test on your samples from parts (i) and (ii) at the 5% level.

(iv) Comment on your results and state in each case, both statistically and in more practical terms, what conclusions you reach. Now consider all possible results from these three tests.

	(i)	*(ii)*	*(iii)*
I	Sig	Sig	Sig
II	Sig	Sig	NS
III	Sig	NS	Sig
IV	NS	Sig	Sig
V	Sig	NS	NS
VI	NS	Sig	NS
VII	NS	NS	Sig
VIII	NS	NS	NS

What combination of results would you expect many people to achieve, and why? Are any combinations especially unlikely or impossible? Explain.

2(E) A pharmaceutical manufacturer claims that its slimming aids will achieve an average weight reduction of 5 lb in the first week of use, provided that the instructions are followed strictly. A chemist selling the product enlisted the help of the ten women who bought the product on a particular day to test this claim. They reported back as follows:

Women	*A*	*B*	*C*	*D*	*E*	*F*	*G*	*H*	*I*	*J*
Weight Loss (lb) in First Week of Diet	3	8	2	5	0	−1	2	3	0	4

(i) On the assumption that the instructions have been followed strictly, carry out an appropriate test at the 5% level to determine whether these results are consistent with the manufacturer's claim.

(ii) Calculate 95% confidence limits for the true average weight loss.

HND (Business Studies)

3(E) (a) What do you understand by the t-distribution?

(b) From a random sample of 9 machine components an average life of fifteen months is expected with a standard deviation of two months.

(i) Calculate the limits between which the actual average life of the components could vary, based on this sample, at 95% and 99% confidence limits.

(ii) Considering that the components are expected to be of high quality, what comments would you make on the results?

(ICMA 1974)

4(E) Fifteen applicants for a post were divided at random in two groups containing 8 and 7 persons. The two groups were interviewed by two different Personnel Officers. The two interviewers gave marks from 1 to 10 for the appearance of each of the applicants. Is there any evidence from the marks below that Interviewer A has different standards from Interviewer B. Carry out the appropriate test at the 1% level.

Interviewer A	*Interviewer B*
8	7
7	4
6	6
9	8
7	5
5	6
8	7
9	

(DMS)

5(E) A trade union asserts that workers on one production line (Line X) are being underpaid in relation to workers on another production line (Line Y). The average weekly earnings (£) on the two lines are denoted $x_1\ x_2, \ldots, x_{10}$ and $y_1, y_2, \ldots, y_{17}$ respectively. Assuming that both sets of measurements can be regarded as random samples from normal distributions with the same variance and that

$$\sum_{i=1}^{10} (x_i - \bar{x})^2 = 0.92, \quad \sum_{i=1}^{17} (y_i - \bar{y})^2 = 0.52, \bar{x} = 31.32 \text{ and}$$

$\bar{y} = 31.14$,

(a) is there evidence of any significant difference in wages between the two production lines?
(b) is there any evidence that workers on line X are paid *less* than workers on line Y?
(c) which of (a) and (b) would you consider to be the more appropriate test in the circumstances and why?

(IPM 1972)

6(E) The sales of tinned and bottled pears were compared in 8 self-service stores over a period of two weeks. Carry out the appropriate test to determine if the average difference obtained is statistically significant at the .05 (5%) level.

	Sales of Pears	
Store	*Tinned*	*Bottled*
A	230	245
B	180	185
C	316	385
D	84	107
E	41	69
F	85	104
G	183	202
H	217	193

Describe how you would modify your test if 8 stores had sold tinned pears and a *different* 8 stores had sold bottled pears.

BA (Business Studies)

7(U) Refer to Exercise 8, CU14 (Linear Regression). Test whether the regression coefficient b in $\widehat{Y} = a + bX$ is significant at the 1% level.

8(E) "Babycare" is a large chain of shops selling "mother and child" equipment. Twelve of the shops were selected at random for the purpose of a promotional test. In a particular week (week 1) each of the shops was asked to keep a record of the number of high-chairs sold. In the following week (2) each customer buying a high-chair was given a voucher for £1 to spend as she liked on other equipment. (The offer was announced by a poster placed in the window of the selected shops for the duration of week 2.) Again, in week 2, the shops were asked to keep a record of the sales of high-chairs. The results were as follows (in random order):

Shop Code	*Week 1 Sales*	*Week 2 Sales*
A	8	11
B	4	6
C	5	*
D	7	7
E	9	8
F	*	*
G	6	*
H	10	13
I	*	10
J	*	5
K	8	13
L	*	*

The asterisks in the above table indicate that the sales figures have been lost or destroyed.

(i) By making an *appropriate selection* of data from the table, conduct a *paired comparison* test to establish whether sales had increased significantly, in the week of the promotion, at the 5% level.

(ii) Make an *appropriate selection* of data from the table that would enable a valid *independent* two-sample test to be carried out and explain your reasoning carefully. (Do NOT actually carry this test out.) What would you conclude if this test were to prove non-significant?

(DMS)

SOLUTIONS TO EVEN NUMBERED EXERCISES

2. Let us denote weight loss by the variable X. Then

$$\bar{X} = \sum X/n = 26/10 = 2.6$$

$$s^2 = \sum X^2/n - \bar{X}^2 = 132/10 - (2.6)^2 = 6.44$$

So

$$\frac{\hat{\sigma}^2}{n} = \frac{s^2}{n-1} = \frac{6.44}{9} = 0.7155.$$

Consequently the SE of $\bar{X} = \sqrt{0.7155} = 0.846$

(i) Our test is $H_0 : \mu = 5$ against $H_1 : \mu \neq 5$ (a two-sided test). $\alpha = 0.05$, $\alpha/2 = 0.025$ and $t_{0.025}$ (9 d.f.) = 2.262. From our sample

$$t = \frac{\bar{X} - \mu}{(\text{SE of } \bar{X})} = \frac{2.6 - 5.0}{0.846} = -2.837$$

Ignoring the minus sign (because it is a two-sided test), we find $2.837 > t_{0.025}$ (9 d.f.) and so we reject the null hypothesis and conclude that the manufacturer's claim is false.

(ii) $\mu = \bar{X} \pm t_{0.025}$ (SE of $\bar{X}$) $= 2.60 \pm 2.262 \times 0.846 = 2.60 \pm 1.91$.

4. Our test is $H_0 : \mu_A - \mu_B = 0$ against $H_1 : \mu_A - \mu_B \neq 0$ (two-sided test)

$$\bar{X}_A = 59/8 = 7.375, s_A^2 = 449/8 - (7.375)^2 = 1.734$$

$$\bar{X}_B = 43/7 = 6.143, s_B^2 = 275/7 - (6.143)^2 = 1.549$$

$$\hat{\sigma}^2 = \frac{8 \times 1.734 + 7 \times 1.549}{8 + 7 - 2} = \frac{24.715}{13} = 1.901$$

So

$$t = \frac{7.375 - 6.143}{\sqrt{1.901\left(\frac{1}{8} + \frac{1}{7}\right)}} = \frac{1.232}{0.7136} = 1.727$$

Now $\alpha = 0.01$, so $t_{\alpha/2} = t_{0.005}$ $(8 + 7 - 2 = 13$ d.f.$) = 3.012$ which exceeds out calculated value. We accept the null hypothesis and conclude that there is no evidence of a difference in marking standards.

6. This requires a paired comparison test, conducted on the *single* sample of differences (D) between tinned and bottled sales.
We find $\bar{D} = 154/8 = 19.25, s_D^2 = 7622/8 - (19.25)^2 = 582.19$. So

$$\frac{\hat{\sigma}^2}{n} = \frac{s^2}{n - 1} = \frac{582.19}{7} = 83.170,$$

and the

$$\text{SE of } \bar{D} = \sqrt{83.170} = 9.120$$

From the sample,

$$t = \frac{19.25 - 0}{9.120} = 2.111 \text{ under } H_0 : \mu_D = 0.$$

As $H_1 : \mu_D \neq 0$ we require a two sided test at the 5% level.
$t_{0.025}$ (7 d.f.) $= 2.365$ which exceeds our sample result. Hence we accept the null hypothesis that there is no significant difference in sales.

8. (i) Choosing only those shops where sales had been recorded in both weeks, we have:

	Wk 1	*Wk 2*	*D* (= *Wk 2 – Wk 1*)	D^2
A	8	11	3	9
B	4	6	2	4
D	7	7	0	0
E	9	8	−1	1
H	10	13	3	9
K	8	13	5	25
		Totals ΣD =	12	ΣD^2 = 48

$$n = 6; \bar{D} = 12/6 = 2; s_D^2 = 48/6 - 2^2 = 4$$

$$\frac{\hat{\sigma}^2}{n} = \frac{s^2}{n-1} = \frac{4}{5} = 0.80$$

Hence

$$\text{SE of } \bar{D} = \sqrt{0.80} = 0.894$$

We compute

$$t = \frac{2-0}{0.894} = 2.24,$$

on the basis of $H_0 : \mu_D = 0$ and against $H_1 : \mu_D > 0$ (a one-sided test).

Now $t_{0.05}$ (5 d.f.) = 2.015 and so we conclude that the increase in sales is significant.

(ii) Shops C and G should be included in Wk 1 sample (as Wk 2 not recorded).

Shops I and J should be included in Wk 2 sample (as Wk 1 not recorded).

Other shops are then allocated systematically to the two samples choosing sales figure for Wk 1 for Wk 1 sample, and Wk 2 for Wk 2 sample.

For example

Wk 1 Sample		*Wk 2 Sample*	
C	5	I	10
G	6	J	5
A	8	B	6
D	7	E	8
H	10	K	13

If the test were non-significant we should adopt a verdict of "not proven" – a larger sample may be required to overcome the between-shops variation.

COURSE UNIT 21 – THE *F*-DISTRIBUTION AND AN OVERALL TEST OF MEANS

21.1 INTRODUCTION

In the preliminaries to CU19 (Chi-squared tests) we discussed the problem of conducting a significance test to compare *three* distinct samples. We concluded that this *could* be achieved by carrying out three separate tests, each comparing the two sample means from a *pair* of the samples; but by doing so we should be unsure of the level of significance of the test overall.

In this Course Unit we develop an improved procedure, a *single* test to compare the sample means from any number, k, of independent samples. Using it we might investigate whether the average weekly expenditure on bread differs between four social classes, whether there is a significant difference between the mean number of days lost through sickness of five groups of workers, or if there is any real difference in quality between the components from six suppliers.

Formally, if we obtain the sample means, $\bar{X}_1, \bar{X}_2, \bar{X}_3 \ldots, \bar{X}_k$ of k independent samples, we test the null hypothesis

$$H_0 : \mu_1 = \mu_2 = \mu_3 = \ldots = \mu_k$$

against the alternative

$$H_1 : \mu_1, \mu_2, \mu_3 \ldots \text{ and } \mu_k \text{ are not } \textit{all} \text{ equal.}$$

Note that the alternative hypothesis does *not* state that *all* the population means are different, simply that *at least one* is different from the rest. This is the only alternative hypothesis available for the test we shall develop and, rather oddly as we shall see, the test is always conducted in a one-sided fashion. When $k = 2$ the test is directly equivalent to a *two-sided* independent two-sample t-test.

The testing procedure is not quite so straight forward as for the other tests we have encountered, so we must first develop some "tools for the job": the main one is the *Variance Ratio* test and we degress to consider this separately before returning to the test on several means.

21.2 THE VARIANCE RATIO (*F*) TEST

If we obtain two *independent* unbiased estimates of the variance of a normal population (from two separate samples, for example) we may calculate the variance ratio

$$F = \frac{\text{Estimate 1 of Population Variance}}{\text{Estimate 2 of Population Variance}} = \frac{\hat{\sigma}_1^2}{\hat{\sigma}_2^2}$$

Now this F ratio will vary between one pair of estimates and another, sometimes being greater, sometimes less, than its *expected* value of unity. Its sampling distribution is called, simply, the F-distribution.

Now consider taking a sample from each of two *separate* normal populations – whose variances are *hypothesised* as being equal – and obtaining the F ratio of the variance estimates calculated from the two samples. We may then test the hypothesis of equal variance by comparing this F ratio with the tabulated percentage points of the F-distribution.

Example. A company developing an electric town-car has to decide which of two manufacturers' batteries should be used. The company buyer visits both manufacturers and selects at random 25 batteries of a chosen type from manufacturer 1 and 15 equivalent batteries from manufacturer 2, for the purpose of extended testing. The results of the tests, which measure the effective working life of the batteries in hundreds of hours, are shown below.

	Sample Mean ('00 hours)	*Standard Deviation ('00 hours)*	*Sample Size*
Manufacturer 1	$\bar{X}_1 = 37.2$	$s_1 = 6.2$	$n_1 = 25$
Manufacturer 2	$\bar{X}_2 = 36.9$	$s_2 = 4.4$	$n_2 = 15$

Now the sample means are virtually identical – with the sample sizes employed they would not be found significantly different – and so there is little to choose between the batteries on the basis of average life. The standard deviations, on the other hand, are widely different. If it can be established that manufacturer 2's batteries have a significantly smaller standard deviation (or variance) than those of manufacturer 1, the company would do well to use the former as they would give a more *consistent* working life. This can be checked with an F-test (on the assumption of normally distributed effective working life of both batteries).

Our null hypothesis is

$$H_0 : \sigma_1^2 = \sigma_2^2$$

with the alternative

$$H_1 : \sigma_1^2 > \sigma_2^2$$

To calculate the F-ratio we must first obtain *unbiased* estimates of the population variances:

$$\hat{\sigma}_1^2 = \frac{n_1}{n_1 - 1} \cdot s_1^2 = \frac{25}{24}(6.2)^2 = 40.04,$$

and

$$\hat{\sigma}_2^2 = \frac{n_2}{n_2 - 1} \cdot s_1^2 = \frac{15}{14}(4.4)^2 = 20.74$$

so

$$F = \frac{\hat{\sigma}_1^2}{\hat{\sigma}_2^2} = \frac{40.04}{20.74} = 1.93$$

Testing at a 5% level, from F-distribution tables (to be discussed shortly) we find the appropriate $F_{0.05} = 2.35$. So, somewhat surprisingly perhaps, we accept the null hypothesis and conclude that there is no significant difference between the variances (and thus standard deviations) of the lives of the two batteries.

Using Tables of the F-Distribution. An examination of the F-distribution tables shows that the number of degrees of freedom for Sample 1 (ν_1) are listed along the top of the table, while those for Sample 2 (ν_2) are listed down the side. Some differences will be found in the various sets of tables available for use: in some cases there will be found more than one entry for each combination (ν_1, ν_2) of degrees of freedom, corresponding to various levels of significance (α = 0.05, 0.025, 0.01, 0.001 etc); in others, where only one entry is given, the whole table relates to one particular level of significance. The reader should study carefully the instructions given at the head of a particular set of tables before use.

This complication arises (in comparison with t and χ^2 tables) because the F-distribution has two parameters (ν_1 and ν_2) and, as with the binomial distribution, this makes tabulation cumbersome. The distribution itself has a shape which varies with ν_1 and ν_2, but generally – for most values of ν_1 and ν_2 – it can be thought of as being positively skewed as shown in Exhibit 1.

The two variance estimates being tested will be held to be significantly different if the F-ratio is either significantly greater than 1.0 (i.e. in the right-hand tail of the distribution) or significantly less than 1.0 (close to the origin) – depending on which estimate is used for the numerator of the F-ratio. But in order to save space in printing the tables the percentage points are only quoted for the *right-hand* tail of the distribution (a glance at the tables will show that all values given are greater than 1.0). Consequently we need to adopt the convention, when

EXHIBIT 1: P.d.f. of F-Distribution with (ν_1, ν_2) Degrees of Freedom

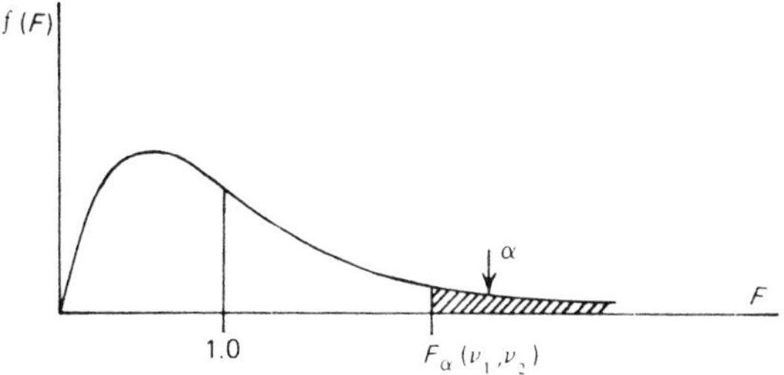

carrying out a sensible[(1)] test, that F is calculated as:

$$F = \frac{\textit{Larger} \text{ Estimate of Variance}}{\textit{Smaller} \text{ Estimate of Variance}},$$

a convention we adopted in our example. Using this convention ν_1 is the number of degrees of freedom of the *larger* estimate, ν_2 is that of the *smaller* estimate. Where the estimates come from two independent samples $\nu_1 = n_1 - 1$ and $\nu_2 = n_2 - 1$, as for the t-tests: in our example $\nu_1 = 25 - 1 = 24$ and $\nu_2 = 15 - 1 = 14$.

As with other significance tests, the F-test may be one-sided or two-sided. In our example we tested against the alternative

$$H_1 : \sigma_1^2 > \sigma_2^2 \text{ (a one-sided test)}$$

and obtained $F_\alpha(\nu_1, \nu_2) = F_{0.05}(24, 14) = 2.35$. Had we wished to test against

$$H_1 : \sigma_1^2 \neq \sigma_2^2 \text{ (a two-sided test)}$$

we would have obtained $F_{\alpha/2}(\nu_1, \nu_2) = F_{0.025}(24, 14) = 2.79$.

Uses of the F-Test. The test has three main uses:

- where we are interested in comparing two population variances *for their own sake,* because they provide a measure of *reliability* or *consistency* (as in our example).
- as a qualifying test prior to carrying out an independent two-sample t-test of means (CU20). That test assumed equal variances of the two populations and consequently this assumption may first be tested to ensure the validity of the subsequent t-test. However,

---

(1) A sensible test is one where the alternative hypothesis could be found true. If, for example, we tested against $H_1 : \sigma_1^2 > \sigma_2^2$ but yet $\hat{\sigma}_1^2 < \hat{\sigma}_2^2$, this would not be a sensible test – the null hypothesis is accepted automatically.

some practitioners do not recommend this practice since the t-test then becomes conditional on the F-test and the overall significance level is thereby changed.

as an essential tool in the Analysis of Variance. The central subject matter of this Course Unit is a simultaneous test on several sample means and this is one of many techniques which are collectively known as the Analysis of Variance (ANOVAR). This is by far the most important application of the F-test and it explains why the F-distribution tables are published with many more values for ν_2 than for ν_1.

21.3 TESTING SEVERAL MEANS

Returning to our original problem of a simultaneous test on several sample means, an example will now be given to illustrate the general principles involved. In Section 21.4 we shall rework the same problem using a more convenient computational method.

Example. A typing pool employs twenty typists. At present manual typewriters are used, but it is the company's intention to change over to electric machines. Four electric typewriters (machines A, B, C and D) have been supplied on trial by various manufacturers. The twenty typists are allocated at random into four groups, each of five typists – each group to be concerned with the evaluation of one of the machines. On the day of the test each typist in a group uses the trial machine for part of her work and completes a questionnaire on it. In the final question the typist gives a "score out of 10" to the machine. The results of these evaluations are given in Exhibit 2. The question arises: Is there any real difference in the performance of the four machines, or do the different scores simply reflect the individual whims of the various typists?

EXHIBIT 2: Scores (X) Out of 10

A	2	6	4	1	2	$\therefore \bar{X}_A = 3$
B	7	4	6	4	4	$\therefore \bar{X}_B = 5$
C	6	6	5	5	8	$\therefore \bar{X}_C = 6$
D	4	7	7	5	7	$\therefore \bar{X}_D = 6$

The situation is one where we have 4 samples – corresponding to the four machines – each containing 5 observations. (In general we might be concerned with k samples each of n observations.) Furthermore the samples are independent because each observation arises from a different typist. If instead we were comparing just *two* machines we could consequently consider applying an independent two-sample *t*-test to the data). The values of the four sample means are given alongside Exhibit 2 – on the face of it, it seems that machine A is a rather inferior machine and machines C and D are the best.

Firstly, let us consider the assumptions we need to make in order to ensure the validity of the test we are about to develop. We assume that:

Sample A is "taken from" a normal distribution with mean μ_A and variance σ^2

Sample B is "taken from" a normal distribution with mean μ_B and variance σ^2

Sample C is "taken from" a normal distribution with mean μ_C and variance σ^2

Sample D is "taken from" a normal distribution with mean μ_D and variance σ^2

It is particularly important to note the assumption of normality (which is required for the *F*-test we shall use later) and that of a *common* population variance, σ^2, for all samples. These are precisely the same assumptions that were made for the independent two-sample *t*-test.

Our null hypothesis is that the population means are all equal:

$$H_0 : \mu_A = \mu_B = \mu_C = \mu_D,$$

and the alternative hypothesis is that they are not *all* equal, i.e. at least one population mean is different from the others.

Let us now use the data to estimate the common variance, σ^2, in *three* separate ways.[(1)]

The purpose of this exercise is not to demonstrate that such ingenuity is possible but to provide a justification for the significance test. As a preliminary, we notice that the *overall* mean of all 20 observations is 5 (as the sum of all 20 observations is 100). This overall mean could alternatively be evaluated by finding the mean value of the four individual sample means (3, 5, 6 and 6) and for this reason it is denoted by $\bar{\bar{X}}$ – the double-bar indicating a process of double averaging.

Method 1. If H_0 is true then the division of the data into 4 samples is quite artificial – all 20 items can be regarded as a *single* sample from a

---

[(1)] This is for demonstration purposes only In practice we would use the simplified method given in Section 21.4.

normal distribution with mean μ (= $\mu_A = \mu_B = \mu_C = \mu_D$) and variance σ^2. Consequently we may estimate σ^2 by simply dividing the "sum of squared deviations from the overall mean, $\bar{\bar{X}}$" by the number of degrees of freedom (See CU20).

Now the sum of squares =

$$(2-5)^2 + (6-5)^2 + (4-5)^2 + (1-5)^2 + (2-5)^2$$
$$+ (7-5)^2 + (4-5)^2 + (6-5)^2 + (4-5)^2 + (4-5)^2$$
$$+ (6-5)^2 + (6-5)^2 + (5-5)^2 + (5-5)^2 + (8-5)^2$$
$$+ (4-5)^2 + (7-5)^2 + (7-5)^2 + (5-5)^2 + (7-5)^2$$
$$= 68$$

The number of degrees of freedom for this method is $20 - 1 = 19$, as for any simple variance estimate based on 20 observations. Hence our (unbiased) estimate of σ^2,

$$\hat{\sigma}^2 = \frac{68}{19} \qquad \text{(Estimate 1)}$$
$$= 3.58$$

Method 2. If H_0 is true then the variation in sample means from one sample (machine) to another is simply a reflection of the common population variance – if σ^2 were large we would expect high variability between the sample means; if σ^2 were small we would expect a lower variability. We can use this fact to estimate σ^2. Firstly we estimate the *variance of sample means,* $\sigma^2_{\bar{X}}$, from the four sample means 3, 5, 6 and 6 by dividing their "sum of squared deviations from *their* mean, $\bar{\bar{X}}$" by the number of degrees of freedom.

$$\text{Sum of Squares} = (3-5)^2 + (5-5)^2 + (6-5)^2 + (6-5)^2$$
$$= 6$$

The number of degrees of freedom is 3 (one less than the number of sample means). Consequently

$$\hat{\sigma}^2_{\bar{X}} = \frac{6}{3}$$

But we know that $\sigma^2_{\bar{X}} = \sigma^2/n$, where n is the number of observations (= 5, in this case) used to calculate each sample mean (CU15). So

$$\hat{\sigma}^2 = n\,\hat{\sigma}^2_{\bar{X}} = 5 \times \frac{6}{3} = \frac{30}{3} \qquad \text{(Estimate 2)}$$
$$= 10$$

Method 3. Whether H_0 is true *or not*, we can still obtain an estimate of σ^2 by pooling the sums of squares from the four samples, taken separately, in much the same way as we did for the independent two-sample t-test.

Sum of Squares (A)

$$= (2-3)^2 + (6-3)^2 + (4-3)^2 + (1-3)^2 + (2-3)^2 = 16$$

Sum of Squares (B)

$$= (7-5)^2 + (4-5)^2 + (6-5)^2 + (4-5)^2 + (4-5)^2 = 5$$

Sum of Squares (C)

$$= (6-6)^2 + (6-6)^2 + (5-6)^2 + (5-6)^2 + (8-6)^2 = 6$$

Sum of Squares (D)

$$= (4-6)^2 + (7-6)^2 + (7-6)^2 + (5-6)^2 + (7-6)^2 = 8$$

Total = "Pooled" Sum of Squares = 38

Notice that in forming the sum of squares for each sample the appropriate sample means have been used in each case.

Now the number of d.f. in this particular case = 20 − 4 = 16. The "−4" arises because we have used all 4 sample means in arriving at the pooled sum of squares.[1] Consequently

$$\hat{\sigma}^2 = \frac{38}{16} \qquad \text{(Estimate 3)}$$

$$= 2.375$$

Let us now stand back and see what insights these three methods of estimating σ^2 provide. If H_0 is true, and *only* if it is true, the differences between the values of the three estimates (3.58, 10 and 2.375 in our example) must be the result of sampling fluctuations alone. Furthermore Estimate 2 (= 10) and Estimate 3 (= 2.375) are *independent* estimates of σ^2, because Estimate 2 used the variation "between samples" whereas Estimate 3 used the variation "within samples" to estimate σ^2. (Estimate 1 used *both* sources of variation). Consequently the statistic

$$F = \frac{\text{Estimate 2}}{\text{Estimate 3}} = \frac{\text{Between Samples Variance Estimate}}{\text{Within Samples Variance Estimate}}$$

$$= \frac{10}{2.375} = 4.21$$

---

[1] Put another way, *each* machine sum of squares has $n - 1 = 5 - 1 = 4$ d.f. We have 4 independent samples and so the pooled sum of squares has $4 \times 4 = 16$ d.f.

can be used to test for acceptance of the null hypothesis by reference to the F-distribution with 3 and 16 d.f. Notice that we are testing H_0 with a *variance* ratio despite the fact that H_0 is a statement about population *means*.

The next issue to be resolved is whether we should conduct a one-sided or two-sided test. At first sight (though wrongly) we might conclude that as the alternative hypothesis states that the population means are not all equal (as distinct from saying that some are greater or smaller than the others) a two-sided test is appropriate. This is, in fact, not the case as the following argument shows.

Let us consider some artificially generated data where H_0 is clearly false, as in Exhibit 3.

EXHIBIT 3: Scores (X) Out of 10 with H_0 Obviously False

A	3	3	3	3	3	$\bar{X}_A = 3$
B	5	5	5	5	5	$\bar{X}_B = 5$
C	6	6	6	6	6	$\bar{X}_C = 6$
D	6	6	6	6	6	$\bar{X}_D = 6$

This data has been chosen partly because we would immediately conclude that the population means are unequal (H_0 false) and partly because it has the same sample means as our data in Exhibit 2. What are the values of Estimate 2 and Estimate 3 in this case? Estimate 2 will be the same as for the data in Exhibit 2 (= 10) because the sample means are identical but Estimate 3 will be zero as there is no variation *within* the samples. Consequently the F value for Exhibit 3 will be $10/0 = \infty$. This demonstrates that in cases where H_0 is false (and consequently the alternative hypothesis is true) the value of F will be large. Our test, then, is to establish whether the observed value of F is significantly *greater than* unity (the expected value), and hence a one-tailed test is appropriate.

Conducting the test at the 5% level of significance, we have:

$$F_{0.05}\ (3, 16) = 3.24$$

As F (= 4.21) $> F_{0.05}$ (3,16), we reject the null hypothesis. Thus our intuitive feeling that the machines are *really* different in performance has been established – the typists certainly vary a great deal in their opinions but not so much as to lead us to doubt that the machines are different.

Before moving on to the simplified method of computation, let us look once again at our three estimates of σ^2 expressed in the form of fractions rather than their decimal equivalents. We find the following results.

Numerator of Estimate 1 (= 68)	=	Numerator of Estimate 2 (= 30)	+	Numerator of Estimate 3 (= 38)
Denominator of Estimate 1 (= 19)	=	Denominator of Estimate 2 (= 3)	+	Denominator of Estimate 3 (= 16)

These rather startling results are no arithmetic coincidence. They can be proved quite generally for any set of data involving k independent samples each of n observations.[1] we say that the total Sums of Squares (and Degrees of Freedom) have been "partitioned" into two components – the "between samples" component and the "within samples" (or "residual") component. (In more complex situations we can partition the total sum of squares and degrees of freedom into three or more independent components, corresponding to different sources of variability.) One practical benefit of this property in our case is as follows. It is easy to simplify the computation of the Sums of Squares for Estimate 1 and Estimate 2 – but not for Estimate 3. Consequently the Sums of Squares for Estimate 3 is more conveniently obtained by an appropriate subtraction.

---

[1]These results can be proved directly from the algebraic equivalents of the three estimates. In the case of k samples each of n observations we can denote the jth observation in the ith sample by X_{ij}. We also use $\bar{X}_i$ to denote the ith sample mean and, of course, $\bar{\bar{X}}$ for the overall mean. A careful examination of the calculations involved in producing our three estimates will show that they can be represented, in general, by:

Estimate 1 $$\hat{\sigma}^2 = \frac{\sum_{i,j}(X_{ij} - \bar{\bar{X}})^2}{nk - 1}$$

Estimate 2 $$\hat{\sigma}^2 = \frac{n\sum_{i}(\bar{X}_i - \bar{\bar{X}})^2}{k - 1}$$

Estimate 3 $$\hat{\sigma}^2 = \frac{\sum_{ij}(X_{ij} - \bar{X}_i)^2}{k(n - 1)}$$

It is trivial to show that the quoted result for the denominators, as $nk - 1 = k - 1 + k(n - 1)$. The quoted result for the numerators is a tedious but otherwise straight forward exercise in summations.

21.4 A CONVENIENT COMPUTATIONAL PROCEDURE FOR TESTING SEVERAL MEANS

We now repeat the test carried out in the last section in a much more convenient fashion. The method to be described is quite general and may even be applied to sample means from k *unequal*-sized samples (n_1, $n_2, \ldots n_k$).

We first complete the table of data by forming the sample totals (T_i), and the Grand Total (T) of all observations (Exhibit 4).

EXHIBIT 4: Scores out of 10 (X_{ij})

j =	*1*	*2*	*3*	*4*	*5*	*Totals* (T_i)
i = 1 (A)	2	6	4	1	2	15
2 (B)	7	4	6	4	4	25
3 (C)	6	6	5	5	8	30
4 (D)	4	7	7	5	7	30
(= k)						
					Grand Total, T =	100

The following calculations are then performed.

Correction Factor,

$$T^2/N = (100)^2/20 = 500$$

where N is the total number of observations.

Total Sum of Squares,

$$\sum X_{ij}^2 - T^2/N$$

$$= (2^2 + 6^2 + 4^2 + 1^2 + 2^2 + 7^2 + 4^2 + \ldots 5^2 + 7^2) - 500$$

$$= 568 - 500 = 68$$

(The notation ΣX_{ij}^2 means square each observation in each sample and add together).

Between-Samples Sum of Squares,

$$\sum \frac{T_i^2}{n_i} - \frac{T^2}{N},$$

where n_i is the number of observations in the ith sample

$$= \frac{15^2}{5} + \frac{25^2}{5} + \frac{30^2}{5} + \frac{30^2}{5} - 500$$

$$= 530 - 500 = 30$$

Using these results we next set out an Analysis of Variance (ANOVAR) table as shown in Exhibit 5. The "Between Samples" and "Total" rows are entered first; the entries for the Sum of Squares and degrees of freedom for the "Within Samples" row are then obtained by subtraction. Finally the F-ratio is obtained (as in Section 21.3) as

$$F = \frac{\text{Between Samples Variance Estimate}}{\text{Within Samples Variance Estimate}}$$

EXHIBIT 5: ANOVAR Table

Source of Variation	*Sum of Squares*	*d.f.*	*Variance Estimate*	*F-ratio*
Between Samples	30	$(k-1) = 3$	$30/3 = 10$ (B)	$B/W = 4.21$
Within Samples (Residual)	38	$(N-k) = 16$	$38/16 = 2.375 (W)$	–
Total	68	$N-1 = 19$	–	–

The test (or the equality of sample means) is conducted at the α-level by comparing our F-ratio with the critical value,

$$F_{\alpha}(k-1, N-k) = F_{0.05}(3, 16) = 3.24$$

It can be seen that all the figures in this method of calculation correspond exactly to those in Section 21.3: the only difference is that this method, with practice, is much quicker. The reader will have noticed that it makes use of sample totals. (Despite the fact that we are testing sample means, we never actually calculate them!) The reasons for this are to be found in the algebraic derivation, the details of which do not concern us.

The technique we have used is a *one-way* Analysis of Variance – we have been concerned to test one set of differences, those between the typewriters. The way we conducted the experiment reflected this fact by having each typist evaluate just *one* machine, and by randomising the allocation of typists to the four machines. We could have conducted the experiment differently: we could have had a number of typists each evaluating all four machines, in which case a *two-way* Analysis of Variance would have been appropriate. Our ANOVAR table would then have three rows (apart from the totals) – a between-machines row, a between-typists row and a residual row. The analysis required for this situation follows closely our one-way analysis, although in general it requires careful consideration of the model chosen to represent the variability in order that valid tests may be carried out. Such matters are best left to a more advanced course.

READING

Some idea of the scope of the Analysis of Variance may be obtained from:
M.J. Moroney, *Facts from Figures*, Penguin Books, Chapter 19.

REFERENCE

G.W. Snedecor & W.G. Cochran *Statistical Methods 6th Ed.* Iowa State Chapter 10.

EXERCISES

1(P) Refer to Exercise 1, CU20. Make use of the samples (lengths of string cut to an estimated length of 10 cm) obtained by a small number (3 or 4) of colleagues. In each case use the *second* sample taken in part (ii). Test for the equality of means, setting out your results in an ANOVAR table in the manner shown in Section 21.4. State the assumptions made. (You may have to complete Exercise 2 before attempting the test.)

2(U) Test for equality of the following variances (assumed independent and from normal populations) against the stated alternatives at the stated levels of significance.

(i) $\hat{\sigma}_1^2 = 14.0(n_1 = 6)$, $\hat{\sigma}_2^2 = 3.2(n_2 = 10)$; $H_1 : \sigma_1^2 > \sigma_2^2$; $\alpha = 0.05$.
(ii) $\hat{\sigma}_1^2 = 23.2(n_1 = 11)$, $\hat{\sigma}_2^2 = 9.3(n_2 = 25)$; $H_1 : \sigma_1^2 \neq \sigma_2^2$; $\alpha = 0.05$.
(iii) $\hat{\sigma}_1^2 = 14.7(n_1 = 8)$, $\hat{\sigma}_2^2 = 30.7(n_2 = 11)$; $H_1 : \sigma_2^2 > \sigma_1^2$; $\alpha = 0.05$.
(iv) $\hat{\sigma}_1^2 = 101(n_1 = 25)$, $\sigma_2^2 = 200(n_2 = \infty)$; $H_1 : \sigma_1^2 < \sigma_2^2$; $\alpha = 0.01$.
(v) $\hat{\sigma}_1^2 = 38(n_1 = 10)$, $\hat{\sigma}_2^2 = 36(n_2 = 12)$; $H_1 : \sigma_1^2 < \sigma_2^2$; $\alpha = 0.05$.
(vi) $s_1^2 = 140(n_1 = 25)$, $s_2^2 = 200(n_2 = 13)$; $H_1 : \sigma_1^2 \neq \sigma_2^2$; $\alpha = 0.05$.

3(E) In a social survey the distribution of the number of cars owned by a household was obtained for random samples of households in four social classes. Outline the techniques involved with carrying out:

(i) overall test of association
(ii) an overall test of means.

Distinguish between the hypotheses being tested and the conclusions that could be drawn from a significant result in each case.

(DMS)

4(U) Refer to Exercise 4, CU20. Conduct the *same* test (for equality of means) using the one-way analysis of variance technique and hence show that

$$t = \sqrt{F} \quad \text{for } two \text{ samples,}$$

and $t_{\alpha/2} = \sqrt{F_\alpha}$, where the numerator of the F-ratio has *one* degree of freedom and the denominator, the same number $(n_1 + n_2 - 2)$ as t.

5(E) As part of an experiment, to determine the effectiveness of a teaching machine programme on business statistics, five "A" level mathematics students were selected from each of four grammar schools. Each student was given a standard test to determine his "mathematical retention" ability. The marks are given below (out of 60 marks maximum).

School

A	*B*	*C*	*D*
42	44	41	43
48	53	50	50
51	50	47	54
43	49	46	51
47	45	42	46

Use the Analysis of Variance technique at the 5% level to determine if the mean scores of the schools are significantly different. (Hint: subtract a suitable constant from all values.)

BA (Business Studies)

6(E) Sixteen similar towns were selected for a test marketing operation, introducing a new product via local newspaper advertising. Four different advertising themes were selected. Four towns were allocated at random to each theme. After a period of 8 weeks the total sales per 100 households were as follows:

Sales per 100 Households
Advertising Theme

A	*B*	*C*	*D*
6.1	14.6	11.5	13.4
13.8	15.7	16.0	20.2
8.7	11.8	9.0	12.9
12.0	16.5	13.3	12.5
40.6	58.6	49.8	59.0

Using a significance level of .05 (5%), test whether the sample means differ significantly. (Hint: a constant amount may be subtracted from each entry.)

(DMS)

7(E) Five samples were taken from 4 successive deliveries of steel rod for stressing concrete. The tensile strength of the samples were obtained, and the coded (i.e. disguised) results are given below in lbs. per square inch. Five laboratory assistants were involved in the test.

	Delivery			
Assistant	*I*	*II*	*III*	*IV*
A	8	7	9	6
B	7	6	8	5
C	9	8	10	7
D	6	5	7	6
E	8	7	9	4

(a) Carry out the Analysis of Variance to determine if the 4 sample means are significantly different at the 1% level.
(b) Is there any evidence to suggest that variation exists among the five *assistants* (use 5% level).
(c) Write a short note explaining the implication of the results in non statistical language.

BA (Business Studies)

N.B. This exercise should properly be tackled using a two-way analysis of variance. It is suggested that two one-way analyses be carried out.

SOLUTIONS TO EVEN NUMBERED EXERCISES

2. (i) $F = 4.375$, $F_{0.05}(5, 9) = 3.48$. Significant, H_0 rejected.
(ii) $F = 2.49$, $F_{0.025}(10, 24) = 2.64$. Not significant, H_0 accepted.
(iii) $F = 2.09$, $F_{0.05}(10, 7) = 3.64$. Not significant, H_0 accepted.
(iv) $F = 1.98$, $F_{0.05}(\infty, 24) = 1.73$. Significant, H_0 rejected.
(v) Not a sensible test. H_0 accepted automatically.
(vi) $\hat{\sigma}_1^2 = \frac{25}{24} \cdot 140 = 145.8$, $\hat{\sigma}_2^2 = \frac{13}{12} \cdot 200 = 216\ 7.$

So $F = 216.7/145.8 = 1.486$, $F_{0.025}(12, 24) = 2.54$. Not significant.

4.

A	8	7	6	9	7	5	8	9	$T_A =$	59
B	7	4	6	8	5	6	7		$T_B =$	43
									$T =$	102

N (Total number of observations) = 8 + 7 = 15

So Correction Factor $T^2/N = (102)^2/15 = 693.6$

Total Sum of Squares $= \sum X_{ij}^2 - T^2/N$

$= 724 - 693.6 = 30.4$

Between-Samples Sum of Squares $= \sum \frac{T_i^2}{n_i} - \frac{T^2}{N}$

$= \frac{59^2}{8} + \frac{43^2}{7} - 693.6 = 5.668$

Source of Variation	*Sum of Squares*	*d.f.*	*Variance Estimate*	*F-ratio*
Between-Samples	5.668	1	5.668	2.980
Within-Samples	24.732	13	1.902	–
Total	30.4	14	–	–

$F_{0.01}$ (1,13) = 9.07. Hence the result is not significant.

6. Subtracting 12 from each value, we obtain:

A	*B*	*C*	*D*	
−5.9	2.6	−0.5	1.4	
1.8	3.7	4.0	8.2	
−3.3	−0.2	−3.0	0.9	
0	4.5	1.3	0.5	
−7.4	10.6	1.8	11.0	Totals (T_i)

So $T = 16.0$, $X_{ij}^2 = 186.88$, $N = 16$.

Correction Factor

$T^2/N = 256/16 = 16.0$

Total Sum of Squares

$\sum X_{ij}^2 - T^2/N = 186.88 - 16.0 = 170.88$

Between-Samples Sum of Squares

$\sum \frac{T_i^2}{n_i} - \frac{T^2}{N} = 291.36/4 - 16.0 = 56.84$

ANOVAR Table

Source of Variation	*Sum of Squares*	*d.f.*	*Variance Estimate*	*F-ratio*
Between Themes	56.84	3	18.95	1.99
Within Themes	114.04	12	9.50	–
Total	170.88	15		

$F_{0.05}$ (3, 12) = 3.49 and so the differences are not significant.

COURSE UNIT 22 – NON-PARAMETRIC STATISTICS AND DISTRIBUTION-FREE TESTS

22.1 INTRODUCTION

In earlier Course Units we have discussed two classes of statistical significance tests:

- those where large samples are taken and we justifiably assume that the test statistic has a *normal* sampling distribution (as in CU18).
- those where smaller samples are involved, but we specify that the population from which the sample is drawn is itself normally distributed (as for the *t*- and *F*-tests of CU20 and CU21).

This obviously leaves a gap in our procedures – namely those occasions where we are forced to use small samples and yet cannot rely on the populations being normally distributed; indeed it is often the case that we may know for certain that a population is skewed or is non-normal in some other respect.

Before embarking on a series of new test procedures to remedy this deficiency let us first establish that they would meet a practical need. Let us consider some problems facing a psychologist working in the field of management selection – trying to ensure that persons of suitable ability and personality are recruited for vacant jobs. Many such psychologists use questionaires to measure intelligence and evaluate personality. In some general ability tests the questionnaires have been widely tested with large numbers of persons and the questions (or marks per question) amended so that the population distribution is normal in shape, in which case the psychologist is happy to use *t*- and *F*-tests on small samples of applicants. (Also, in larger scale studies, the Market Research Officer is also quite satisfied that it is valid to compare the mean scores on a verbal rating scale (SU1b) even though the distribution of answers are commonly markedly skewed, since the sampling distribution of the means can be assumed to be normally distributed.)

But now suppose a psychologist has been called in to help decide whether an accountant or a statistician would, in principle, be the best Head of a newly created Management Services Department in a company which has ten employees of each type. Here the psychologist is faced with creating his own measurement techniques but it is not practicable to validate the tests with large samples in order to obtain a normal population of scores on each test devised. Thus whilst he will be able to give a "score" on a personality factor he is not prepared to assume an underlying normal distribution for this score. Further, it is unlikely that the score is anything other than an indicator of a person's position on the measuring scale relative to other job candidates.

Much of the impetus for the search for new statistical procedures has come from the social scientist faced with problems similar to that outlined above. However, before new procedures were devised a number of research studies were conducted into the effect of non-normality on the standard tests when the sample was small. A broad assessment of their findings is that tests of hypotheses about population means (e.g. t-tests) are relatively insensitive to departures from normality; these tests are said to be "robust". However, tests involving population variances, such as F-tests, are more sensitive; the probability levels arising from such tests on small samples drawn from non-normal populations are liable to be substantially wrong.

Significance tests which do *not* rely on the assumption of a normal population (or, in fact, any other specified distribution) are known as *non-parametric* or *distribution-free* tests – these two terms are often used interchangeably, but they actually have quite different meanings. A non-parametric test is one where the hypothesis being tested does not involve population parameters, unlike the standard z-, t- and F-tests treated in earlier Course Units. Thus the "goodness of fit" chi-squared test of CU19 is a non-parametric test since the hypothesis concerns the *form* of the population distribution, not the specific value of a parameter. The term "distribution-free", on the other hand, describes a class of tests which remain valid for a wide range of population distributions.[(1)] Hence distribution-free tests may be applied to both parametric and non-parametric problems: later we shall discuss a test about a population median – which is parametric – and another test concerning randomness in a time series which is non-parametric. In summary we may regard problems as being either parametric or non-parametric and they may be tackled by methods which may, or may not, be distribution-free.

There are a very large number of distribution-free tests available for use, and more seem to appear with every new issue of the statistical research journals; here we introduce several of the most widely known and used tests. In general they tend to use ranking procedures to a greater extent than we have met previously; also the use of signs (+ or –) of data without their numerical values is quite common.

From our earlier discussion it should be clear that the real value of

---

[(1)]The validity of distribution-free tests does however require that the measurement quantities or "scores" are *in principle* continuous variables, even if the scores are quoted as whole numbers – the assumption of an *underlying* continuous distribution.

distribution-free methods lies in small sample applications where the normality of the population is suspect. Unfortunately many such tests require the use of specially constructed satistical tables. These are readily available (see *Siegel*, for instance) but since a familiarity with their use contributes little to an understanding of the tests they are not reproduced here. The methods may also be applied to large samples, and in some cases we illustrate the tests using the large sample technique.

We shall take examples from the various classes of test available, namely:

- randomisation tests – independent and paired samples
- sign test – paired samples
- ranking tests – independent and paired samples
- runs test – one sample randomness

We point out here that further methods covering more than two samples are available i.e. non-parametric versions of the analysis of variance, but these are beyond the scope of this Course Unit. Also, it should not be too late to remind the reader that *any* test of significance assumes that the observations are *randomly* selected from the population.

22.2 RANDOMISATION TESTS

These tests are presented first since they represent perhaps the most radical departure from the parametric tests so far studied, and yet are dependent upon the *same* concepts as all statistical significance tests.

The parametric tests for both large and small samples depend upon the concept of the sampling distribution of the statistic being known in advance of the test – a normal sampling distribution for large samples, a t-distribution for tests on the mean of a small sample from a normal population. In contrast, randomisation tests are based on the sampling distribution of the sample results *after* the sample results are available. Essentially the procedure is to obtain every possible value of the sample statistic by permutating the sample results (the detailed procedure is illustrated below) and hence obtain a sampling distribution that is based entirely on the sample results themselves. The actual value of the statistic is then compared with the derived sampling distribution to determine if it lies in the tail of the distribution and hence its significance determined in the usual probability sense. Note that no reference is being made to the population from which the sample came, excepting the usual requirement of an underlying continuous distribution (which hereafter we shall not keep repeating), and that the measurement should be on at least an interval scale.(1)

The drawback to these tests is that they can involve a large amount of computation; however, the principles may be used in a wide variety of situations so that the student or practitionner who has unlimited access to a computer could conduct all his significance testing (e.g. of a correlation coefficient) using the same basic idea. See Exercise 2(ii).

The Randomised Matched Samples Test. We now return to the psychologist's problem of comparing 10 accountants with 10 statisticians. Suppose he is able to "pair up" 8 accountants with 8 statisticians by using some combination of critieria such as age, I.Q. and motivation (the other two employees of each type cannot be paired, it is supposed, because one accountant is much older than all the statisticians and one statistician is so poorly motivated). Now consider testing whether there is any significant difference in the salaries earned between the two groups of employees. Subtracting the statistician's salary from the accountant's for each pair the psychologist might arrive at the following set of results, where d_i denotes the difference (in £'s per month) of the *i*th pair:

$$d_i\colon -15, -5, -69, -23, -28, -19, -19, +24 \text{ and } \sum d_i = -154.$$

The reasoning behind the randomisation test is that under the null hypothesis that the population medians are equal each of the eight differences d_i is equally likely to be plus or minus in sign (if any pair has a zero difference then it is omitted from the argument).

Hence with 8 pairs there are $2^8 = 256$ equally likely outcomes for Σd_i. Each of the 256 possible values of Σd_i are then calculated and ranked in ascending order. With the above data we can see that the smallest value of Σd_i (when all the signs are negative) is -202 and the largest value of Σd_i (when all the signs are positive is $+202$. The second smallest values occurs when all but the second observation are negative, i.e. when $d_2 = +5$, $\Sigma d_i = -192$; the third smallest Σd_i occurs when $d_1 = +15$ and the remaining signs are negative, and so on. By this procedure we can obtain the distribution of sample results under H_0. As usual in significance testing we compare the actual sample result with the tail (or tails) of this distribution.

Suppose we wished to conduct a one-sided test using a level of significance of $\alpha = 0.05$. The critical value of the test is formally given by the $256\alpha = 256 \times 0.05 = 12.8$*th* extreme value of the distribution. In

---

(1) By "interval scale" we mean a measurement scale where, although the zero value and the units of measurement may be arbitrarily fixed, equal *distances* on the scale have the same importance.

practice this means that the sample result of −154 is considered significant if it is one of the 12 extreme values in the (left-hand) tail of the distribution. Similarly, for a two-sided test the critical values are given by $256\alpha/2 = 6.4$. Hence our $\Sigma d_i = -154$ would be significant if it were one of the 6 extreme values in *either* tail of the distribution.

It can be seen that this test requires no assumption about the basic distribution of differences or sampling from a normal population; it is then a most acceptable alternative to the paired t-test of CU20.

The Randomised Independent Two-Samples Test. In this test we denote the two sample sizes by n_1 and n_2 for samples 1 and 2 respectively. The procedure this time is to rank in order the combined $(n_1 + n_2)$ observations, carefully noting to which sample (1 or 2) each observation belongs. The total "score" for each sample is obtained by summing the ranks of its members.

Under the null hypothesis (equal medians) it is a matter of chance where each observation lies from position (rank) 1 to position $(n_1 + n_2)$. Now the total number of ways of arranging $(n_1 + n_2)$ objects in a straight line, n_1 of one type and n_2 of another, is given by $(n_1 + n_2)!/(n_1!n_2!)$ (CU9). Using a significance level of α, the rejection region for H_0 is given by the extreme $\alpha \cdot (n_1 + n_2)!/(n_1!n_2!)$ scores. As in the previous example the test may be either one-sided or two-sided by considering the extreme scores in one tail or both tails respectively.

The use of this test does not require assumptions of normal distributions or homogeneity of variance as is required by the parametric t-test for independent samples.

If our psychologist had treated the ten accountants and ten statisticians as two independent random samples (rather than forming 8 pairs from them as in the previous test) he may have obtained the following 20 monthly salaries in ascending order of £'s per month:

280,	285,	300,	319,	332,	348,	351,	356,	363,	365
A(1)	S(2)	A(3)	S(4)	A(5)	A(6)	S(7)	A(8)	S(9)	A(10)
368,	378,	384,	386,	390,	400,	408,	420,	450,	520
A(11)	S(12)	S(13)	A(14)	A(15)	S(16)	S(17)	S(18)	S(19)	A(20)

The letter A denotes that the salary relates to an accountant, and S to a statistician; the bracketed number indicates the rank (score).

Let R_a and R_s represent the sum of the ranks of the 10 accountants and 10 statisticians respectively, then

$$R_a = 1 + 3 + 5 + 6 + 8 + 10 + 11 + 14 + 15 + 20 = 93,$$

$$R_s = 2 + 4 + 7 + 9 + 12 + 13 + 16 + 17 + 18 + 19 = 117.$$

We now have to determine if R_a is significantly different from R_s. Now the total number of combinations of 10 objects from 20 objects is given by $N = 20!/(10!10!) = 184{,}756$. Thus the sampling distribution of R_a consists of 184,756 members, the extreme values of which are the sums of the numbers 1 to 10 = 55 and the numbers 11 to 20 = 155. We do not propose to list the remaining 184,754 members here! With smaller sample sizes than above it is more feasible to work out the full sampling distribution. For larger sample sizes, however, it has been shown that we may approximate the procedure by conducting an independent two-sample t-test (having $n_1 + n_2 - 2$ degrees of freedom) on the *ranks* (rather than the original values) of the two samples – a strange result at first sight since R_a and R_s, for example, are clearly negatively correlated. The test, in this form, remains distribution-free.

22.3 THE SIGN TEST

This test may be used as another alternative to the matched/paired comparisons t-test of CU20. As its name implies it uses only the sign (+ or –) of the difference between each matched pair and not the numerical value of the difference. Returning to the example in Section 22.2 – the randomised paired samples test – we observed that the eight pairs gave the differences d_i: –15, –5, –69, –23, –28, –19, –19, +24. If we now omit the numerical part of the differences we are left with the eight signs

–, –, –, –, –, –, –, +.

Under the null hypothesis of no difference in medians of the two populations from which the samples are drawn the expected number of plusses is equal to the expected number of minuses. Using X_a and X_s to denote the salaries of the ith pair of accountant/statistician we have in probability terms $\Pr(X_a > X_s) = \Pr(X_s < X_a) = \frac{1}{2}$ for the n pairs of values. In the above example with $n = 8$ the sampling distribution of the number of plusses is a binomial distribution with $\pi = (1 - \pi) = \frac{1}{2}$, shown in Exhibit 1.

The test may be restated in terms of the binomial distribution as

$H_0 : \pi = \frac{1}{2}$

$H_1 : \pi \neq \frac{1}{2}$ for a two sided test.

From the binomial table we can deduce that the probability of obtaining one or less plus (or minus) signs = (.0351 + .0351) = .0702. The result is therefore statistically significant at the .07(02) or 7% level. Similarly for a one-sided test alternative we have

EXHIBIT 1: Binomial Distribution with $n = 8, \pi = \frac{1}{2}$

x	$p(x)$
0	.0039
1	.0312
2	.1094
3	.2188
4	.2734
5	.2188
6	.1094
7	.0312
8	.0039

$$H_1 : \pi < \tfrac{1}{2}$$

in which case we would have a significant result at the .0351 or 3.5% level.

It can be seen that this test is comparatively simple to carry out in practice for small samples provided binomial tables for the appropriate n are available. However, we are discarding information when we omit the numerical value of the differences – a difference of £1 is given the same weight in the calculations as a difference of £100. Any such loss of information reflects upon the performance of the test as is explained in Section 22.6. Nonetheless we have gained in generality of application of the test compared with the randomised test. We could, for example, perform the test if we only knew which member of the pair had the higher salary and not (for reasons of confidentiality) the precise values of the two salaries. Again, the test may be used where we cannot obtain a score at all, but where we can nontheless rank the two objects.

To compensate for the effect of losing information by omitting the numerical score it is desirable (with the cost constraints of the investigation) to take as large a sample as possible. When large samples are used we may also use the normal approximation to the binomial distribution (or the chi-squared test) to evaluate the significance of our test results.

Example. 100 samples of turf are treated in two halves by two brands of lawn conditioner (P and L); after 6 weeks the "appearance" of each of the 100 samples is evaluated by eye, and the two halves ranked in order. Suppose conditioner P is ranked first 73 times and Conditioner L ranked first 27 times. Using a two-sided test is this result significant at the 5% level? Using the normal curve test we have (with continuity correction)

$$n = 100 \qquad \pi = 0.5 \qquad X = 73 \qquad Z = \frac{73 - 0.5 - 50}{\sqrt{100 \times 0.5 \times 0.5}} = 4.5$$

The result is significant at the .05 level since $z_{\alpha/2} = 1.96$.

22.4 RANKING TESTS

As we have already noted the use of ranks, in place of the original data, is a common device for constructing distribution-free tests. (Before proceeding the reader is advised to reread Section 13.6 on the treatment of *tied* ranks.)

The Wilcoxon Signed Ranks Test. In the Sign Test of Section 22.3 we omitted the numerical value of the differences, and noted that we were losing information. This present test retains these values and in the sense of Section 22.6 is a "better" test. We shall here give only the large sample ($n \geqslant 30$) version of the test, but for illustrative purposes apply it to the small sample situation already analysed by the Sign Test.

The eight pairs of differences d_i are −15, −5, −69, −23, −28, −19, −19, +24. We first rank the differences in order of size ignoring their signs viz 5, 15, 19, 19, 23, 24, 28, 69. We then replace the signs with the differences in their new rank order, and assign the rank scores:

d_i	−5,	−15,	−19,	−19,	−23,	+24,	−28,	−69
rank	1	2	$3\frac{1}{2}$	$3\frac{1}{2}$	5	6	7	8

The data is now in the form required for the test, and the remaining steps are as follows:
Calculate either the sum of the ranks for the "plus" differences (T_+), or for the minus differences (T_-). So

$$T_+ = 6$$

or

$$T_- = 1 + 2 + 3\tfrac{1}{2} + 3\tfrac{1}{2} + 5 + 7 + 8 = 30$$

In our case T_+ is obviously easier to calculate and so for the remainder of this example we use this value. Under the null hypothesis it can be shown that the expected value of T (i.e. of both T_+ and T_-) is given by

$$\mu_T = (T_+ + T_-)/2 = n(n+1)/4$$

and the standard deviation of T by

$$\sigma_T = \sqrt{n(n+1)(2n+1)/24}$$

For large values of n we may use the normal deviate

$$Z = \frac{T_+ - \mu_T}{\sigma_T} = \frac{T_+ - n(n+1)/4}{\sqrt{n(n+1)(2n+1)/24}}.$$

In our example $T_+ = 6, n = 8$, so

$$Z = \frac{6 - 8 \times 9/4}{\sqrt{\dfrac{8 \times 9 \times 17}{24}}} = \frac{-12}{\sqrt{51}} = -1.68$$

Based on a two-sided test at the 5% level, this is not a statistically significant result ($z_{\alpha/2} = 1.96$). From normal curve tables with $Z = 1.68$ the actual, two sided, probability level is $0.0465 \times 2 = 0.093$ or 9.3% – a result in line with that previously obtained in Section 22.3, even though we have (incorrectly) used a large sample method.

The Mann-Whitney Independent Samples Test. This test may be used as an alternative to the independent samples randomisation test, and for large samples it is computationally simpler. It can be shown that the test is a generalisation of another test first proposed by Wilcoxon. The test procedure starts off in the same manner as for the independent samples randomisation test: the scores from the two samples are combined and ranked noting to which sample each rank score belongs. The procedure for small samples differs from that of large samples from this point onwards.(1) We illustrate only the large sample method, requiring the use of normal curve tables, and by "large" we mean that the larger of the two samples contains a minimum of 20 observations. In the example below we also show how to handle the problem that arises with "tied" ranks.

Example. A sample of 22 sweet shop owner/managers with shops that also sold ice creams were compared with 14 others who refused to sell ice cream in their shops. All 36 owner/managers completed a questionnaire designed to measure their attitude towards children. From a "battery" of questions a score was derived that indicated their attitude on a favourable-unfavourable scale. The scores were as follows:

Without ice cream (N)	10, 14, 15, 15, 15, 19, 19, 21, 22, 28, 29, 37, 41, 45.
With ice cream (I)	11, 13, 14, 15, 15, 17, 18, 19, 20, 25, 25, 27, 31, 31, 35, 36, 38, 40, 42, 43, 44, 46.

The combined sample with the designatory letters N and I are

(1) For small sample test see *Siegel* (Reference).

Score	10,	11,	13,	14,	14,	15,	15,	15,	15,	15,	17,	18,	19,	19,
N/I	N	I	I	N	I	N	N	I	N	I	I	I	N	N
Rank	1	2	3	$4\frac{1}{2}$	$4\frac{1}{2}$	8	8	8	8	8	11	12	14	14

Score	19,	20,	21,	22,	23,	25,	27,	28,	29,	31,	31,	35,	36,	37,
N/I	I	I	N	N	I	I	I	N	N	I	I	I	I	N
Rank	14	16	17	18	19	20	21	22	23	$24\frac{1}{2}$	$24\frac{1}{2}$	26	27	28

Score	38,	40,	41,	42,	43,	44,	45,	46
N/I	I	I	N	I	I	I	N	I
Rank	29	30	31	32	33	34	35	36

The steps in the calculations are as follows:

1. Obtain the sum of the ranks of either of the groups. Since the "N" group has fewer members it is simpler to sum that group giving,

$$R_N = 1 + 4\tfrac{1}{2} + 8 + 8 + 8 + 14 + 14 + 17 + 18 + 22 + 23 + 28 + 31 + 35 = 231.5$$

2. Compute

$$U = n_N \cdot n_I + \frac{n_N(n_N + 1)}{2} - R_N$$

$$= 14 \times 22 + \frac{14(14 + 1)}{2} - 231.5$$

$$= 181.5$$

(Under the null hypothesis that the two populations have the same *median* the expected value of U is $n_N \cdot n_I/2 = 154$.)

3. Compute the variance of U;

$$\sigma_U^2 = \frac{n_N \cdot n_I}{n(n-1)} \left(\frac{n^3 - n}{12} - T \right)$$

where $n = n_N + n_I = 36$, the total numbers of observations, and

$$T = \sum \frac{t^3 - t}{12},$$

where t is the number of ties for a rank and the summation is carried out over all ranks with ties. Since there are two scores of $4\frac{1}{2}$, five of 8, three of 14 and two of $24\frac{1}{2}$ we have

$$T = \frac{2^3 - 2}{12} + \frac{5^3 - 5}{12} + \frac{3^3 - 3}{12} + \frac{2^3 - 2}{12}$$
$$= 0.5 + 10.0 + 2.0 + 0.5$$
$$= 13.0$$

and hence

$$\sigma_U^2 = \frac{14 \times 22}{36 \times 35}\left(\frac{36^3 - 36}{12} - 13.0\right)$$

$$= 946.49$$

4. Compute the normal deviate

$$Z = \frac{U - n_N \cdot n_I/2}{\sigma_U} = \frac{181.5 - 154}{\sqrt{946.49}}$$

$$= 0.89$$

If we chose to test the null hypothesis with a two-sided test at the 5% significance level then $z_{\alpha/2} = 1.96$ and we would accept the null hypothesis that the samples come from populations which have the same median.

This test is among the most suitable alternatives for the parametric t-test; it requires only ranked observations and does not rely on the assumptions of normal distributions with equal variances.

22.5 RUNS TESTS AND TRENDS

In this section we illustrate by the use of two examples a larger number of test procedures in which the sample results are in sequence (or may be put into sequence). Typical applications are to time series analysis of sales patterns throughout the year, learning curves for new training methods or perhaps experiments carried out in random order but thereafter placed in a sequence determined by some controlling factor. We distinguish between two types of test; those based on runs (order or sequence) and tests for trends.

A Runs Test. For our purposes a run may be defined as a sequence of identical symbols occuring in a longer sequence containing other symbols.

If our psychologist had taken the eight "signs" of salary difference (Section 22.3),

$$- \quad - \quad - \quad - \quad - \quad - \quad - \quad +,$$

and placed them in some relevant sequence, for instance in increasing order of the average age of the pairs of accountants/statisticians to which they relate, they may have appeared as

$$- \quad - \quad - \quad + \quad - \quad - \quad - \quad -$$

In this example, by our definition, we have 3 runs, viz:

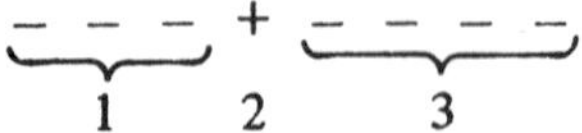

We may test the hypothesis that this is a random sequence against the alternative that there is some order (non-randomness) or trend. The test involves counting the number of runs – as we have done above – and comparing the result with the expected number of runs under the null hypothesis. As in the previous examples, if we have a large sample, the normal distribution provides an adequate approximate test.

Example. A production control manager keeps a record of the sequence of breakdowns (major stoppages) on a production line, 0 indicating a day in which no such breakdown occurs and B a day when a breakdown does occur. (We assume that two breakdowns cannot occur on a single day since the remainder of the day is taken up in repairing the fault.) Over a period of 35 days the sequence shown below was obtained: the manager wishes to test if this is a random sequence and he considers a significance level of $\alpha = .05$ to be appropriate.

0 0 0	B	0 0	B B	0 0 0 0 0	B	0	B	0 0	B	0	B
1	2	3	4	5	6	7	8	9	10	11	12

0	B	0 0 0 0 0	B B	0 0 0 0 0
13	14	15	16	17

We can see that 17 runs have occurred. If this is significantly smaller than the expected number of runs under the null hypothesis then we may suspect either a time trend or some clustering of breakdowns due to lack of independence. If it is significantly greater, we may suspect some systematic time (or quantity) cyclic effect is influencing the process.

Our analysis proceeds as follows:

Let n_1 = the number of days with no breakdown: $n_1 = 25$
Let n_2 = the number of days with a breakdown; $n_2 = 10$
Let $n = n_1 + n_2$; $n = 35$
Let r = the number of runs; $r = 17$

Denoting the expected number of runs under the null hypothesis by R, then it can be shown that $R = (2n_1n_2 + n)/n = (2 \times 25 \times 10 + 35)/35 = 15.29$. It can also be shown that the standard deviation σ_r is given by

$$\sigma_r = \sqrt{\frac{2n_1n_2(2n_1n_2 - n)}{n^2(n-1)}}$$

$$= \sqrt{\frac{2 \times 25 \times 10(2 \times 25 \times 10 - 35)}{35^2 \times 34}} = 2.36$$

The normal deviate

$$Z = \frac{r - R}{\sigma_r} = \frac{17 - 15.29}{2.36} = +0.72$$

As phrased above this is a two sided test with $\alpha/2 = .025$, $z_{\alpha/2} = 1.96$. Hence we would accept the null hypothesis that the series is a random one. By specifying that we were expecting a trend in a particular direction as our alternative hypothesis we would have performed a one-sided test in the usual manner.

A Test for Trend. This test may be used in a variety of applications in which the probability of "success" is known.

Let π by the probability of success in a series of n trials
Let the n trials take the numbers $1, 2, 3, \ldots, n$.
Let s = the sum of the numbers of the trials in which a success is obtained.

Denoting the expected value of s by S, it can be shown that

$$S = n(n + 1) \cdot \pi/2$$

and the standard deviation $\sigma_s = \sqrt{n(n + 1)(2n + 1)\pi(1 - \pi)/6}$.

In CU17 we discussed the problem of discrimination testing of two meat paste sandwiches for market research purposes. Let us now suppose that we wish to test if a respondent is subject to a learning effect, i.e. to test whether, in a long series of trials, a housewife tends to obtain the correct answer the more experienced she becomes in taking part in the experiment. (Similar learning problems occur in industrial training and experimental behaviour tests with animals.) On this occasion a housewife is asked to identify which of three sandwiches ($H_0: \pi = \frac{1}{3}$) contains her usual brand of meat paste. A sequence of 30 trials is carried out, where 0 indicates an incorrect identification and C a correct identification of the usual brand. The sequence obtained is

0	0	0	C	0	C	C	0	0	C	C	0	0	C	0
1	2	3	4	5	6	7	8	9	10	11	12	13	14	15

0	0	C	C	0	0	C	C	0	0	C	0	C	C	C
16	17	18	19	20	21	22	23	24	25	26	27	28	29	30

So

$$s = 4 + 6 + 7 + 10 + 11 + 14 + 18 + 19 + 22 + 23 + 26 + 28 + 29 + 30 = 247,$$

whilst

$$S = 30 \times 31 \times \tfrac{1}{3}/2 = 155$$

and

$$\sigma_s = \sqrt{\frac{30 \times 31 \times 61 \cdot \frac{1}{3} \times \frac{2}{3}}{6}} = \sqrt{2{,}101} = 45.94$$

The normal deviate

$$Z = \frac{s - S}{\sigma_s} = \frac{247 - 155}{45.94} = 2.00.$$

We would in this instance specify a one-sided test as being appropriate and the above result is significant at the 5% level. We conclude that the housewife has learned to discriminate with experience.

22.6 CONCLUSIONS

In summary we may state the *advantages* of distribution-free tests as follows:

- They are more *robust* than their classical counterparts; that is, they do not require as many assumptions to be made about the populations from which the samples are drawn in order to ensure their validity. In particular, probability statements obtained from such tests are exact, regardless of the shape of the population distribution(s).
- They have a wider range of application in that they may be applied to data which is inherently in the form of ranks, or simple categories of the "better or worse" variety.
- In some cases they are computationally simpler.

What then of their disadvantages? It is often stated that distribution-free tests are less powerful – less able to reject a null hypothesis which is actually false, and thus should be rejected (CU17). Thus in Exhibit 2 the power curve A might correspond to a particular matched sample t-test – the most powerful – while B is for the corresponding Wilcoxon signed ranks test and C for the Sign test (all tests having the same level of significance, α). The power of a test is generally related to the "amount of information" it uses: the t-test uses all numerical information, the Wilcoxon test uses only signed ranks and the Sign test only signs.

EXHIBIT 2: Power Curves for One-Sided Significance Tests

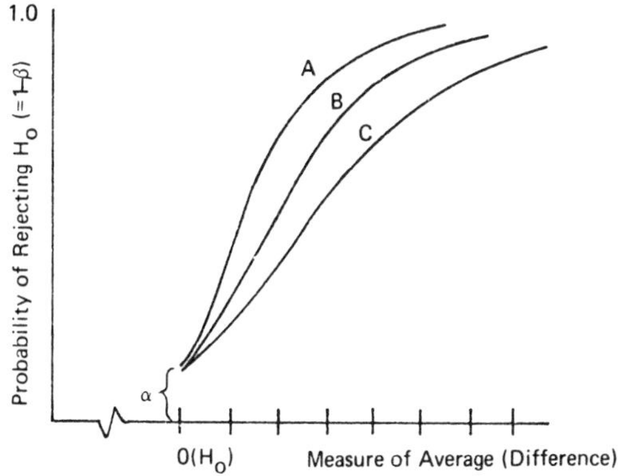

But such arguments are rather superficial. The *t*-test is only clearly superior if it is valid – where the normality assumptions are satisfied. The Sign test is only "throwing away information" if we can rely on the scale of measurement used being meaningful, and not giving a completely spurious air of precision. A useful comparison by *Wonnacott and Wonnacott* (see reading recommendations) shows that under certain conditions distribution-free tests can be more powerful than their classical counterparts.

Which test then should be used in a given situation? We shall content ourselves by generalising that if the assumptions of normality are satisfied then the parametric tests, t and F, are the most suitable. If it can be shown that the assumptions are grossly not satisfied then a distribution-free alternative should be used. If the amount of non-normality is "moderate" then either test may be used.

READING

T.H. & R.J. Wonnacott *Introductory Statistics* 2nd Ed. Wiley. Chapter 16.

REFERENCE

Sidney Siegel *Nonparametric Statistics for the Behavioural Sciences* McGraw-Hill.

EXERCISES

1(P) (a) Take a pack of ordinary playing cards and divide into two packs of 26 red and 26 black cards. Select 10 pairs of cards i.e. one card from each pack, ten times. Take the red card score away from the black card score for each pair so that you have ten "differences", each with a + or – sign or a zero difference. (For these purposes count the picture cards as having scores of 11, 12 and 13 for Jack, Queen and King.)

(b) Discard any zeros and use a table of the binomial distribution with the appropriate n and $\pi = 0.5$ to carry out a Sign test.

(c) Carry out the parametric t-test on the 10 pairs of differences. Comment on the validity of the test.

(d) Repeat (a) five times, then combine the 50 signs and carry out a Sign test using the normal approximation to the binomial distribution to evaluate the result.

(e) Using a calculator carry out a large sample z-test on the 50 pairs of differences from (d). Comment on the validity of the test.

(f) Say precisely what you have been testing in each case and comment on the results obtained.

2(U) *Randomised Tests*

(i) Referring to the Randomised Matched Samples Test of Section 2 derive 15 values in each tail of the sampling distribution. Hence determine whether $\Sigma d_i = -154$ is significant on a one-sided test with a significance level of 0.05.

(ii) Outline a procedure for carrying out a randomised test of a correlation coefficient, r.

3(E) (a) Explain what is meant by the term "statistical test"?

(b) What are the differences between parametric and non-parametric tests and what are the advantages and disadvantages of each type? (Your answer should be illustrated by one specific problem for which both types of test are appropriate.)

(IPM June 1974)

4(U) Look critically at the exercises in CU20 – the t-test. Taking Exercises 2, 4, 6 and 8:

(a) Is the normality assumption justified for the underlying distribution?

(b) Which tests in this Course Unit would you use in place of the parametric t-test?

(c) Perform a suitable test in those cases where it is computationally feasible to do so. Compare the resulting significance levels obtained. (You may care to use the small sample methods and tables given in *Siegel.*)

5(E) A new wage agreement involving a major re-grading exercise and the introduction of bonus payments is being negotiated. A random sample of employment and production records was selected and two sets of calculations were made of weekly remunerations

(a) under the existing system, plus 20% to cover the rise in the cost of living, and

(b) under the proposed new system.

The results obtained (£ per week) were as follows:

Subject	1	2	3	4	5	6	7	8	9	10	11	12	13
Old (plus 20%)	56	43	59	62	38	49	53	37	71	53	47	39	37
New	67	58	58	75	47	51	52	49	75	59	56	41	42

Subject	14	15	16	17	18	19	20	21	22	23	24	25
Old (plus 20%)	68	27	68	75	42	53	61	56	58	35	46	37
New	65	31	72	84	45	54	65	61	57	39	49	39

By the use of a Sign Test, or otherwise, determine whether the new agreement will lead, on average, to an increase of wages of more than 20%.

(IPM June 1975)

6(E) A staff promotion board ranked 10 candidates in terms of their managerial potential and the results obtained were summarised in terms of their educational background ("public school (P)" or "other (O)") as follows:

Rank	1	2	3	4	5	6	7	8	9	10
Education	P	O	O	P	O	O	O	P	O	O

How would you assess these results to threw light upon the proposition that "public school education is a good training for management" using *either*

(i) a "randomisation" test (detailed calculations are *not* required), or

(ii) a Mann-Whitney (rank sum) test.

To what limitations would any conclusion concerning the proposition based on data of this kind be subject?

(IPM June 1973)

SOLUTION TO EXERCISE 2

2. (i) The sampling distribution tails are:

−202	−192	−172	−164	−164	−162	−156	−154	−154	−154
−146	−146	−144	−136	−134	....				
				....	134	136	144	146	146
154	154	154	156	162	164	164	172	192	202

To be significant at the $\alpha = 0.05$ level (one-sided left-hand test) the sample result should be less than the 12th extreme value, i.e. −146. As our $\Sigma d_i = -154$, the result is significant.

(ii) The procedure is:

1. Calculate r from the bivariate data (CU13)
2. Holding the X values constant, write down all possible permutations of the Y values. For a bivariate sample of n observations there will be $n!$ permutations (one of which is the original data).
3. Evaluate r for X and Y for each permutation.
4. Arrange the values of r from 1 and 3 in increasing order.
5. To test at the α-level (two-sided test) determine the $n!\alpha/2$ extreme values in each tail.
6. Compare the original value of r from 1 with the two critical values of r from 5. If it falls outside these values the result is significant.

SEMINAR UNITS

SEMINAR UNIT 1a – THE COLLECTION OF DATA

1a.1 INTRODUCTION

Many statistical text books convey the impression that data is readily available for analysis. In fact, of course, someone at some stage has to make a conscious effort to collect it.

In the business world data is collected while carrying out investigations designed to help solve problems or make decisions. A financial audit, a quality control check and a market research survey are all readily recognisable as investigations conducted for a specific purpose. But even the more mundane activities associated with the routine collection of sales figures, labour turnover, weekly wage bills and stock levels can be considered as part of an investigation with the broader purpose of ensuring that the company is functioning smoothly and efficiently. Statistics compiled from these sources, correctly presented and interpreted, provide the basis for effective management control over the activities of the various departments.

A course in Statistics, with any emphasis on techniques, has to be presented in a rather piecemeal fashion. So before considering various aspects of data collection it seems worthwhile to examine briefly just one example of an overall investigation in which data collection plays a major part. This investigation (see Exhibit 1) has as its purpose the control of a sales force.

EXHIBIT 1: Nature of a Statistical Enquiry

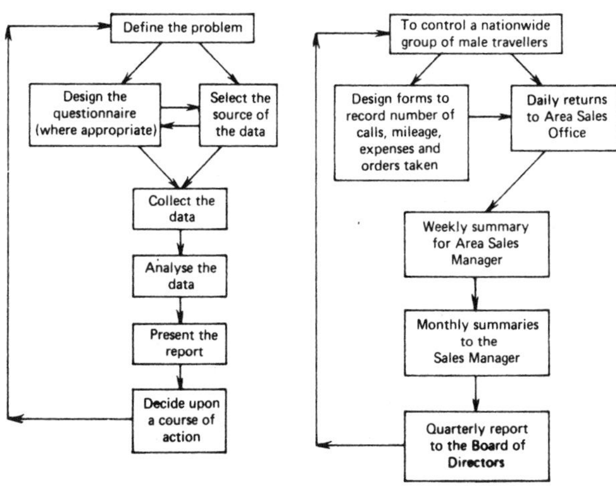

In Exhibit 1 we can see how the daily figures compiled by a company salesman are used first of all as a check to see that the man is carrying out correctly his function of calling on potential buyers. From the daily returns weekly summaries can be produced for each salesman by his immediate superior, the Area Sales Manager, listing such items as the number of calls made (and orders taken), car mileage and expenses. The Area Sales Manager can then compare these statistics for each man under his control to see if any salesman needs extra help or training; and if one salesman seems outstanding, to consider whether his approach can be taught to the others. The Area Sales Manager will then summarise all the weekly reports into his monthly report to the Sales Manager, who in turn can compare and contrast the reports received from different areas, taking any action he feels appropriate. From the monthly summaries a quarterly report may be written setting out briefly the achievements of the field force for the benefit of the Directors.

It can be seen that the responsibility for selling the company's products is delegated by the Directors, in stages, to the salesmen who may be hundreds of miles away. Effective management control over this situation is achieved by "inspection" of various statistics – some detailed, others less so – at each control stage. The final level of control is exercised in the boardroom with the aid of a few essential figures on a quarterly report.

1a.2 CLASSIFICATION AND DEFINITION

While it should be self-evident that the purpose of an enquiry must be defined before any data collection takes place (otherwise the *wrong* data may be collected), it may not be so apparent that a carefully considered classification scheme for the data must also be prepared beforehand. The reason for this is to ensure that all those involved in the enquiry will record the same objects under the same headings. Consider, for example, conducting a motorway traffic survey, the purpose of which is to determine the pattern of road usage. It might be tempting to classify vehicles into the groups (a) cars, (b) vans, (c) lorries, (d) coaches, (e) motorcycles and (f) others. But imagine being stationed on a bridge spanning the road and trying to decide where to record such items as taxis (cars or others?), vans fitted with extra seats (cars or vans?), oil tankers (lorries or others?) etc. If we were responsible for 50 persons recording the information (using this classification scheme) at different periods during the month, we would find the final result of limited value, because of different interpretations as to how the various vehicles should be classified amongst those collecting the data.

In deciding upon a system of classification it is necessary to strike a balance between having too few classes, in which case objects which are not really similar have to be grouped together for convenience, and too many classes, with only a few items in each class. For example, in a household paint survey it may be possible to classify 95% of paint used as belonging to one of 12 colours or shades, while the remaining 5% of paints may require a further 25 classifications. For some purposes it would be quite sufficient to have a separate class for each of the 12 main colours and an "other" class for the rest.

Exact definitions must be provided for a classification scheme. For the motorway survey notes should be added to clarify the *principles* underlying the classification scheme and examples should be given on the classification of "unusual" vehicles. In the paint survey definitions could be given by means of colour charts. If the definitions are poor in that they are vague or incomplete no amount of refined analysis, carried out at a later stage, can compensate.

Should two or more investigations be conducted using different definitions, then the results may not be comparable. Thus if one of two market research organisations defines a shop as being "large" if it employs 10 or more assistants, while the other defines a "large" shop as having 5 or more assistants, results from the surveys cannot easily be compared. Should yet another organisation use an entirely different system, e.g. by defining a large shop as having an annual turnover of £100,000 or more, then comparison becomes almost impossible.

It cannot be emphasised too strongly that unless the data is collected in a correct manner then the time spent on the subsequent calculations may be wasted: and, what is more important, any conclusions reached may be in error, and action based on these conclusions incorrect.

We next consider what method should be used to collect the data. The related topics of questionnaire design, the sample size, and the process for selecting respondents are covered elsewhere (in SU1b, CU15 and SU5 respectively).

1a.3 PRIMARY AND SECONDARY DATA

Before considering whether to instigate a data collection exercise at all it is wise to ascertain whether data which could serve the purpose of the current enquiry is already available, either within the organisation or in a readily accessible form elsewhere. Thus when starting a Market Research department in a company much data, in particular sales data (by region, wholesaler or retailer, by size or by type of product), may already be filed away in the Accounts or Invoice department, where it has been used to bill the customer. It may also be possible for the Market Research department to estimate the company's share of the total market by comparing the company's sales statistics with those published by the government in relation to production and imports and exports of the same type of product.

When data is used for the purpose for which it was originally collected it is known as **primary data**; when it is used for any other purpose subsequently, it is termed **secondary data.** In the above example the invoice data as used by the Accounts department is primary data: so too are the statistics of production, imports and exports collected by the government when used to guide national economic policy. When the Market Research department uses such data, appropriately summarised, for another purpose – to help in the taking of decisions concerned with advertising and sales policies – it is secondary data. Similarly if a company Buyer obtains quotations for the price, delivery date and performance of a new piece of equipment from a number of suppliers with a view to purchase, then the data as used by the Buyer is primary data. Should this data later be used by the Budgetary Control department to estimate price increases of machinery over the past year, then the data is secondary.

The above distinctions are important simply because the danger of misinterpretation is greater for secondary data. A user of secondary data may be faced with the following difficulties:

- The coverage of the original enquiry may not have been the same as that required, e.g. a survey of house building may have excluded council built dwellings.
- The information may be out of date, or may relate to a different period of the year to that required. Intervening changes in price, taxation, advertising or season can and do change people's opinions and buying habits.
- The exact definitions used may not be known, or may simply be different from those desired, e.g. a company which wishes to estimate its share of the "fertilizer" market will find that the government statistics include lime under "fertilizers".
- The sample size may have been too small for reliable results, or the method of selecting the sample a poor one. (SU5).
- The wording of the questions may have been poor, possibly biasing the results (see Questionnaire Design, SU1b).
- No control is possible over the quality of the collecting procedure, e.g. by seeing that measurements were accurate, questions were properly asked and calculations accurate.

If secondary data is used time must be spent to determine whether any of the above mentioned shortcomings apply, and if so to estimate their likely effects on the accuracy of the data for the purposes of the current enquiry. However, the advantage of secondary data, when available and appropriate, is that a great deal of time and money may be saved by not having to collect the data oneself. Indeed in many cases, for example with import-export statistics, it may be impossible for a private individual or company to collect the data which can only be obtained by the government using its legal powers and by giving the assurance that the information will be kept confidential. One can, for example, obtain from the Census of Distribution the sales through grocery outlets, but the data is published for groups of shops so that the sales from any individual outlet cannot be deduced.

1a.4 THE NATURE OF BIAS

It is well known that results from sample surveys, however well designed, are subject to error. (The reason for this is that it is impossible to ensure that the sample is completely representative of the population from which it is drawn.) Such errors are to some extent unavoidable and are known as sampling errors. There are, however, other types of error (non-sampling error) which result from the procedures or questions used in the survey and tend to make the results obtained differ consistently from the "true" results. These errors, called **biases**, can be avoided by careful conduct of the survey.

Bias may arise at any of the following stages of a statistical survey:

- in the compilation of a **sampling frame**,[(1)] by omitting some members of the population e.g. by leaving out sparsely populated areas or new customers for whom a permanent invoice record is not yet available,
- in the selection of the respondents, by not giving every person (company or object) an equal chance of being selected,
- in the wording of questions, e.g. questions that lead the respondent to give a certain answer,
- by the interviewers imposing their views on the respondents,
- in the coding of answers consistently into the wrong classes,
- in the methods of calculation used,
- by interpreting the data in terms of preconceived ideas.

(1) A list of all members of the population from which the sample is selected.

In fact it is precisely by using one or more of these devices that statistical methods may be used to "prove anything" if one sets out to do so. Great care, therefore, must be taken at all stages to ensure that nothing in the questionnaire, the sampling procedure and instructions to the interviewer, will influence or bias the results.

An example in Great Britain of the effect of bias can be seen quite frequently in the daily papers that publish results of public opinion polls. The major organisations carrying out the polls use different methods of sampling the population and different wording of the questions. The results from their polls (all of which attempt to measure the voting intentions of the electorate) often differ by several percentage points – more than can be reasonably explained by pure sampling errors.

1.5 METHODS OF OBTAINING DATA

In many situations within a company the appropriate method of collecting data may well be "obvious" – time sheets, wage slips, stores requisitions, dispatch invoices etc. In cases which are mainly concerned with the efficient day-to-day running of a firm the problem is one of deciding upon the most suitable office procedure – should each person "sign-in", punch a card, or should only late-comers? Should the time clock be at the factory gate or at the workshop entrance? The firm will decide which is the most appropriate method in the light of its own particular circumstances, aiming at obtaining the essential data sufficiently accurately with minimum cost.

In other, non-routine, situations the choice of method for data collection is much wider. In such a case it has to be determined which of the various methods discussed below is the most efficient – bearing in mind the degree of accuracy required, the facilities and finance available. These methods may be used within the company, on other companies, or on the general public.

The Postal Questionnaire. The principal *advantages* are:

- The apparent low cost compared with other methods although the cost per *useful* answer may well be high.
- No need for a closely grouped sample as in personal interviews, since the Post Office is acting as a field force.
- There is no interviewer bias.
- A considered reply can be given – the respondent has time to consult any necessary documents.

The principal *disadvantages* are:

- The whole questionnaire can be read before answering (and in some circumstances this is undesirable).
- Spontaneous answers cannot be collected. Only simple questions and instructions can be given.
- The wrong person may complete the form.
- Other persons' opinions may be given e.g. by a wife consulting her husband.
- No control is possible over the speed of the reply.
- Possibly a poor "response rate" (a low percentage of replies) will be obtained.

The fact that only simple questions can be asked and the possibility of a poor response rate are the most serious disadvantages and are the reasons why comparatively little use is made of the postal questionnaire among the general public in Great Britain. Only simple questions can be asked because there is nobody available to help the respondent if they do not understand the question. The respondent may supply the wrong answer or not bother to answer at all. If a poor response rate is obtained only those that are interested in the subject may reply and these may not reflect general opinion. The postal questionnaire has been used successfully on a number of topics by the Social Survey Unit, and in the U.S.A. there are a number of market research companies who specialise in this technique. Postal surveys seem to be more successful in industrial market research and government enquiries, e.g. income tax, when a reply is compulsory by law.

Telephone Interviewing. The main *advantages* are:

- It is cheaper than personal interviews but tends to be dearer on average than postal questionnaires.
- It can be carried out relatively quickly.
- Help can be given if the person does not understand the question as worded.
- The telephone can be used in conjunction with other survey methods, e.g. for encouraging replies to postal surveys or making appointments for personal interviews.
- Spontaneous answers can be obtained.

The main *disadvantages* are:

- In most countries not everybody owns a telephone: therefore, a survey carried out among telephone owners would be biased towards the upper social classes of the community. However, in the

United States this method is more widely used than in Great Britain because of the greater coverage of the population by the telephone; but the telephone can be used in industrial market research anywhere since businesses are invariably on the telephone.

- It is easy to refuse to be interviewed on the telephone simply by replacing the receiver. The response rate tends to be higher than postal surveys but not as high as when personal interviews are used.
- As in the postal questionnaire, it is not possible to check the characteristics of the person who is replying, particularly with regard to age and social class.
- The questionnaire cannot be too long or too involved.

The Personal Interview. In market research this is by far the most commonly used way of collecting information from the general public. Its main *advantages* are:

- That a trained person may assess the person being interviewed in terms of age and social class and area of residence, and even sometimes assess the accuracy of the information given (e.g. by checking the pantry to see if certain goods are really there).
- Help can be given to those respondents who are unable to understand the questions, although great care has to be taken that the interviewer's own feelings do not enter into the wording of the question and so influence the answers of the respondents.
- A well-trained interviewer can persuade a person to give an interview who might otherwise have refused on a postal or telephone enquiry, so that a higher response rate, giving a more representative cross-section of views, is obtained.
- A great deal more information can be collected than is possible by the previous methods. Interviews of three quarters of an hour are commonplace, and a great deal of information can be gathered in this time.

Its main *disadvantages* are:

- It is far more expensive than either of the other methods because interviewers have to be recruited, trained and paid a suitable salary and expenses.
- The interviewer may consciously or unconsciously bias the answers to the question, in spite of being trained not to do so.
- Persons may not like to give confidential or embarassing information at a face-to-face interview.
- In general, people may tend to give information that they feel will impress the interviewer, and show themselves in a better light, e.g. by claiming to read "quality" newspapers and journals.

- There is also the possibility that the interviewers will cheat by not carrying out the interviews or carrying out only parts of them. All reputable organisations carry out quality control checks to lessen the chances of this happening.
- Some types of people are more difficult to locate and interview than others, e.g. travellers. While this may not be important in some surveys, it will be on others, such as car surveys. One particular problem is that of the working housewife who is not at home during the day: hence special arrangements have to be made to carry out interviews in the evenings and at weekends.

Observation. This may be carried out by trained observers, cameras, or closed circuit television. Observation may be used in widely different fields; for example, the anthropologist who goes to live in a primitive society, or the social worker who becomes a factory worker, to learn the habits and customs of the community they are observing. Observation may also be used in "before and after" studies, e.g. by observing the "traffic" flow in a supermarket before and after making changes in the store layout. In industry many Work Study techniques are based upon observing individuals or groups of workers to establish the system of movements they employ with a view to eliminating wasteful effort. If insufficient trained observers are available, or the movements complicated, cameras may be used so that a detailed analysis can be carried out by running the film through a number of times. Quality control checks and the branch of market research known as retail audits may also be regarded as observation techniques.

The *advantages* of the observational technique are:

- The actual actions or habits of persons are observed, not what the persons say they would do when questioned. It is interesting to note that in one study only 40% of families who stated they were going to buy a new car had actually bought one when called upon a year later.
- In some cases it is undesirable for people to know an experiment or change is to be made, or is taking place since their actions would change and the experiment would be spoiled. In one case experiments showed that production improved even though working conditions were worsened, because the operators knew they were taking part in a series of experiments.

The main *disadvantages* are:

- In the sociological example mentioned above the results of the observations depend on the skill and impartiality of the observer.
- It is often difficult in practice to obtain a truly random sample of persons or events.

- It is difficult to predict future behaviour on pure observation.
- It is not possible to observe actions which took place before the study was contemplated.
- Opinions and attitudes cannot usually be obtained by observation.
- In marketing, the frequency of a person's purchase cannot be obtained by pure observation. Nor can such forms of behaviour as church-going, smoking and crossing roads, except by employing a continuous and lengthy (and hence detectable) period of observation.

Reports and Published Statistics. Much information published nowadays by international organisations such as the United Nations Organisation can be useful in overseas marketing. Most governments publish statistics of population, trade, production etc. Reports on specialised topics including scientific research are published by governments, trade organisations, trade unions, universities, professional and scientific organisations and local authorities. Finally mention should be made of reports which are circulated within firms giving summaries of the work carried out in various departments.

The advantages and disadvantages of using these statistics have already been given in detail under the heading of secondary data. A final warning is given here against assuming that published data is correct simply because of the reputation of the organisation issuing the figures. The authors have met mistakes in the Registrar General's Annual Population Estimates and have also had the task of "reconciling" information that has come from two or more sources (each apparently impeccable) where the findings have differed.

REFERENCES

1. W.J. Reichmann, *The Uses and Abuses of Statistics*, Penguin.
2. C.A. Moser & G. Kalton, *Survey Methods in Social Investigation*, Heinemann.

EXERCISES

1(P) Re-read Section 1a.2 (Classification and Definition), then tackle for yourself the problem of defining and classifying vehicles using a motorway.

2(U) Briefly discuss whether you would use:

(a) a postal questionnaire,
(b) a telephone interview,
(c) a personal interview,
(d) an observational study for the following enquiries:
 (i) traffic census
 (ii) sales of a baby-food product
 (iii) attitudes towards capital punishment.

3(E) Outline the main steps in a statistical enquiry indicating some of the problems that may be involved.

(ICMA)

4(E) Discuss the advantages and disadvantages of possible methods of collecting information on workers attitudes and opinions in a multi-factory organisation from a methodological point of view: personal interviews, postal questionnaires, etc.

(DMS)

5(E) (a) What do you consider to be the advantages and disadvantages of using postal questionnaires to collect statistical information from business organisations?
(b) List five important points to be borne in mind when designing a questionnaire to be posted to firms in order to collect statistical data.

(ACCA, June 1973)

6(E) What is the difference between primary and secondary data? Why is it important that statisticians should make a distinction in their use of these categories of data?

(ACCA, June 1971)

SEMINAR UNIT 1b – THE DESIGN OF QUESTIONNAIRES

1b.1 INTRODUCTION

It is not often appreciated by the layman that the results of a statistical enquiry will be affected by the wording of any questions asked. That is, if two enquiries are carried out under identical circumstances, the only difference being in the wording used in the questionnaires, the results themselves may well differ.

It has already been noted in SU1a that with postal enquiries the questions have to be kept very simple, while in the personal interview it is possible to ask more complicated questions. It follows that the precise wording of questions must be deferred until the form of the enquiry has been determined and until it is known whether the questions are to be asked by post, telephone or by a personal interview.

1b.2 LAYOUT OF THE QUESTIONNAIRE

Clarity, and the ease with which a questionnaire can be completed, must be the first consideration in the design of the layout. The person completing the questionnaire must be able to see at a glance where an answer goes and what form that answer can take. With questionnaires to be completed by interviewers, it should be made easy to ring the "codes" provided when working under difficult conditions, such as when standing on a doorstep on a wet winter's evening.

Example 1.

What is the month and year of the first registration of your car?	MONTH	YEAR

Example 2.

Did you go abroad for your main holiday last year?	YES/NO (delete inappropriate answer)

Example 3. (for a personal interview)

How many cigarettes do you usually smoke per day?	None	1
	1–10	2
	11–20	3
	21–30	4
	31–40	5
	41 or more	6

In the design of the layout we must also attend to the needs of the coder, that is, the person who converts written answers into numerical codes that are later to be punched onto cards used for analysis. Space must be left for them to write in the codes clearly and in sequence. This is important so that the punch operator can work quickly and accurately.

1b.3 CONSTRUCTION AND WORDING OF THE QUESTIONNAIRE

Ease of completing a questionnaire is one thing: willingness to do so is quite another. Every possible encouragement should therefore be given to respondents to co-operate fully in the enquiry and to answer the questions truthfully. It is only by paying careful attention to the construction of the questionnaire and the wording of the questions that this can be achieved.

The questionnaire should not be too long for its intended purpose. The length of time that a person is prepared to answer questions will depend primarily on where the questionnaire is completed (in the home, office or in the street), the degree of interest of the subject matter to the respondent, and the time available to the respondent at that particular moment. A host of other considerations such as the respondent's (and interviewer's) personalities and chance interruptions will also affect the situation. The result of an interview that is too long for a particular situation is that the respondent may refuse to answer the later questions or may give hurried or deliberately misleading answers in order to terminate the interview. Incidentally, in postal surveys it has been found that a long questionnaire tends to increase the length of time taken to reply (as well as affecting the response rate).

The questions should be in some logical sequence. It will help the respondent's memory if he can be taken back, step by step, through thoughts and actions which may have taken place several days or months ago. Since many questionnaires depend heavily for their accuracy on the respondent's memory, every possible aid should be used.

Respondents should not be asked to carry out calculations: for example, weekly paid workers should not be asked their annual salary. In an interview, cards can be supplied to the interviewer to show the respondent, and the answer given easily coded on the spot, as shown in Exhibit 1 on the following page.
This device may also elicit an answer from those who would hesitate to give a more precise answer to a personal question such as earnings; often a high degree of accuracy is not required.

Hypothetical questions should be avoided on factual surveys. Ques-

EXHIBIT 1

Weekly Earnings	*Code*	*Annual Earnings*
Under £20	0	Under £1,000
£20–£30	1	£1,000–£1,500
£30–£40	2	£1,500–£2,000
£40–£60	3	£2,000–£3,000
£60 or more	4	£3,000 or more

tions such as "If a canteen were built would you use it?" may obtain an enthusiastic response, and yet when built the canteen is little used. People cannot easily predict their actions in situations which have never arisen. However, this type of question may usefully be employed in psychological research when the object of the question is not "factual" in the sense of the above question.

Some Rules for the Wording of Questions. A number of general principles can be given, but these are little more than a list of pitfalls to be avoided. The actual questions used will of course depend upon the designer's previous experience, knowledge of the subject matter, and the type of respondent that is to be questioned.

- Questions should not be *ambiguous*. Consider: "Did you change your car tyres to cope with the winter weather?" A "no" to this question could mean "No, I didn't change my tyres for that reason" or simply, "No, I didn't change my tyres".
- Questions should not contain words of *vague meaning*. Consider: "Do you buy your coal in large quantities?" A "large" quantity to a person living alone may be 5 cwt.; to a family it may mean 30 cwt. Other words falling into this category are "skilled", "expensive", "frequently", etc.
- Questions should contain only *simple words*. Thus, "What is the total income after tax of all members of the family?" is to be preferred to "What is the aggregate net income of the household?" Also the use of technical jargon or slang should be eschewed.
- *Leading questions* should be avoided, such as "Do you agree that the church should stay out of politics?" This type of question would "lead" the respondent to say yes; partly because it is easier to agree than disagree during an interview, and partly because of the unpleasant overtones suggested by the words "stay out of".
- The use of *emotive words* should be avoided. "Do you feel the working classes are fairly treated by the bosses?" is an exaggerated example that brings to mind the conflicts of the early days of trade

unionism and would elicit a different pattern of answers from the question, "Do you feel that employees are fairly treated by the employers?"

- Questions should have a *specific answer*. The question: "Do you like your new car?" would usually provide too vague a set of answers for most purposes from a simple "yes" to a full page reply. Such a question should be broken down into sub-questions, each concerned with a particular aspect of the car's performance.
- Questions may sometimes by asked successfully on *personal matters* and attitudes by putting the question into the third person. Using an earlier example, if a question were asked, "If a canteen were built do you think many people would use it?" it might obtain a response closer to the respondent's true feelings on the matter, than if it were phrased in a directly personal way.

Rating Scales: A Further Development. In market and social research a number of techniques, collectively known as rating scales, are used to evaluate attitudes, opinions and beliefs and in particular the "strength of these attitudes. In the case of a housewife who had just tasted three soups, the simple question "Did you like soup K? YES/NO" would give no indication of the degree to which she liked or disliked it. In order to estimate more accurately the strength of her reaction to the flavour, the question could be rephrased, asking her to study the card shown in Exhibit 2 and to decide which statement came closest to her own opinion of each soup.

EXHIBIT 2: Rating Scale Card

I liked it very much
I quite liked it
I neither liked nor disliked it
I disliked it slightly
I did not like it at all

The principle behind the use of rating scales is to enable the analyst to decide how much weight should be attached to the answer (which is simply not possible with a YES/NO type question). Commonly, scores are given to these scales ranging from +2 for those who "liked it very much", through 0 for those who "neither liked nor disliked it", to –2 for those who who "did not like it at all". Such scores from a sample of respondents can be formed into a frequency distribution and depicted as a histogram (CU2). Sample statistics, such as the mean, standard deviation and correlation coefficient, can be calculated from them (CU's 4, 5 and 13, respectively) and the mean scores tested for significant differences (CU18).

Another form of scale in common use is known as the *Semantic Differential*. This usually takes the form of a 5 point or 7 point scale and is shown to the respondent in the form of a line of boxes with suitable words or phrases at each end, as shown.

Weak								Strong

The respondent then indicates which "box" on the scale corresponds most accurately to his opinion (e.g., of the flavour of cheese or his view of company management). Sometimes the individual boxes are labelled with words or statements of graded opinion in order to obtain a more structured scale, viz:

	Extremely	Very	Fairly	Neither/ Nor	Fairly	Very	Extremely	
Weak								Strong

The use of all such scales is open to a number of objections, the main one being that we do not all place the same interpretation on the descriptive words used. The simple scoring system of +3 to −3, if used, does assume "equal intervals" of the points on the scale in much the same way as length or weight is measured: but is it reasonable to say that "extremely" is three times as strong as "fairly", as the scores of +3 and +1 imply?

Yet another, and perhaps simpler, method is to ask the respondent "How many marks out of 10 do you give the product for flavour?" on the assumption that (in Great Britain) respondents are familiar with the marking system from their school days. Finally, when several objects (brands, colours, people) are being evaluated we may simply ask the respondent to put them in order (rank) of preference. Whilst this does not assume any underlying scale it does assume "unidimensionality", that is only *one* scale is being used. If we are ranking in order several applicants for a job our concept of "suitability" implies a subjective weighting of such characteristics as "ability", "appearance" and "motivation", each of which could be assessed on a separate scale.

REFERENCE

R.M. Worcester & J. Downham (Eds), *Consumer Market Research Handbook,* (Chapter on questionnaire design by J. Morton-Williams), McGraw-Hill, 1978.

EXERCISE

Write a questionnaire for one of the enquiry topics in SU1a.

SEMINAR UNIT 2 – INDEX NUMBERS

2.1 INTRODUCTION

The simplest way of making a comparison between two similarly measured quantities is to express one as a ratio or percentage of the other. Because it is readily understood the percentage method is more frequently adopted: where the percentage sign (%) is omitted the resulting measure of comparison is called an *index number* (this is its simplest form). Thus, if we were told that the index number corresponding to a measured quantity is 140 this is to be interpreted as a 40% *increase* over some similarly measured reference figure (referred to as the **base**). In economic statistics in particular, comparisons are frequently made of prices and volumes (among other economic variables) between one year and another or between different countries or industries. As it is vitally important to identify the base used, a special convention is adopted, of which we quote two examples:

(1970 = 100), for a comparison over time, and
(U.K. = 100), for a comparison between countries.

Now it is rare that one is able to make an effective comparison from two *directly* measured quantities. It is usually the case that the comparison has to be made between figures which are themselves *averages*, and, because of the nature of the comparisons made, *weighted*[(1)] averages are often used in preference to simple arithmetic averages.

To take an example quite outside the field of economic measurement, there exist a number of standard tests given to school children each of which measures a particular aspect of intelligence (verbal reasoning, spatial ability etc.). The scores obtained by a particular child in these tests are combined by forming a weighted average. The result, expressed in the form of an index number (the base being the result that would be obtained by an "average" child of the same age), is called the *Intelligence Quotient* (IQ) which gives a measure of "overall" intelligence.

[(1)] The three values 6.2, 5.0 and 6.8 have the simple arithmetic average (mean) of $(6.2 + 5.0 + 6.8)/3 = 6.0$. Implicit in this calculation is the assumption that all three values are equally important. But if we were to assign "weights" to the values (reflecting their relative importance) in the ratio 1:2:4 respectively, we could calculate a weighted average as $(1 \times 6.2 + 2 \times 5.0 + 4 \times 6.8)/(1 + 2 + 4) = 6.2$. Here we have divided the "weighted sum of the values" by the "sum of the weights". By comparison it can be seen that the arithmetic average is simply a weighted average with weights of unity attached to each value.

Index numbers are most frequently constructed for the purpose of monitoring changes in price or volume over a period of time. Of the many official index numbers published,[(1)] the Index of Retail Prices (IRP) and the Export Volume Index (EVI) are commonly used and quoted. The Financial Times Index of Share Prices is an example of a well-respected index number published outside the government sector.

Index numbers have become so fashionable that some companies even construct their own index numbers for measuring such quantities as productivity, quality of goods and attitudes of consumers. But there is no doubt that in terms of practical usefulness price indices are the most important. In the remainder of this Seminar Unit we shall centre our discussion on the best known of these, the abovementioned IRP (the so-called "Cost of Living" index).

2.2 DEVELOPMENT OF A PRICE INDEX

The Index of Retail Prices effectively compares the cost of buying a shopping basket of goods at one point in time with the cost of buying the same goods at subsequent points in time. The present method of calculating the IRP is quite complex, so we will approach it by first considering some simpler formulations.

Simple Price Index. By "simple" we mean relating to one single commodity. Consider, for example, the following prices for a "standard" loaf of bread:

Year	*Price*
1972 (base year)	10p = P_0
1973	12p = P_1
1974	14p = P_2

To obtain the relative changes in price we divide the price in the subsequent years by the price in the base year to arrive at **price relatives.**

Year	*Price Relatives*
1972	P_0/P_0 = 10/10 = 1.00
1973	P_1/P_0 = 12/10 = 1.20
1974	P_2/P_0 = 14/10 = 1.40

(1) British readers will find a whole range of index numbers listed in "Monthly Digest of Statistics", published by HMSO.

The simple price index is obtained by multiplying the price relatives by 100.

Year	*Price Index*
1972	100
1973	120
1974	140

Simple Aggregate Price Index. If we proceed from a single commodity to a basket of goods we are faced with the problem of how to combine changes in the price of the individual goods. We set out by specifying the quantity of each item in the basket as shown in Exhibit 1.

EXHIBIT 1

Goods	*Quantity*	*1972 Price (p)* P_0	*1975 Price (p)* P_n
Bread	Loaf	10	15
Beef	$1\frac{1}{2}$ lbs.	120	150
Milk	4 pints	20	22
Sugar	2 lbs.	10	28
Potatoes	10 lbs.	40	70
	Totals	200	285

So a basket of goods which cost 200p in 1972 costs 285p in 1975. A price index may be calculated for 1975 as

$$\text{Price Index 1975 } (1972 = 100) = \frac{285}{200} \times \frac{100}{1} = 142.5$$

This may be interpreted as a 42.5% increase in prices over the base year. The basic fault with a price index calculated in this way is that the value obtained depends not only upon the foods selected for the basket but the precise quantities chosen for each. If, for example, we chose to have only $\frac{3}{4}$ lb. of beef in our basket, then the totals for 1972 and 1975 would be 140p and 210p respectively. The price index would then be:

$$\text{Price Index 1975 } (1972 = 100) = \frac{210}{140} \times \frac{100}{1} = 150$$

This indicates a price rise of 50% as compared with 42.5% when we had $1\frac{1}{2}$ lbs. of beef in the basket.

This type of index, called a "Simple Aggregative Price Index", may be expressed in mathematical symbols as:

$$\text{P.I. Year } n \text{ (Year 0 = 100)} = \frac{\sum_{i=1}^{k} P_{in}}{\sum_{i=1}^{k} P_{i0}} \times \frac{100}{1}$$

The notation $\Sigma_{i=1}^{k} P_{in}$ indicates that we are adding the prices of the k goods in the basket in year n. For the remainder of this Seminar Unit we omit the suffix i without introducing any ambiguity. Hence P_n will indicate the price of an item in year n, while ΣP_n will indicate the summation of the prices of k goods in year n. Similarly P_0 indicates the price of a good in the base year and ΣP_0 the summation over the same k items in that year.

Average Price Relative Index. This index is an attempt to eliminate the effect of the quantity bought of each of the goods. Firstly we obtain the price relatives of each of the goods separately (Exhibit 2): we see immediately that some goods (e.g. sugar) have risen in price by a greater percentage than others. (Note that the price relative for beef is 1.25 whether we buy 1½ lbs. or ¾ lb. since the ratio of the prices of an item is independent of the quantity bought.) The price index is then obtained from an *average* of these five price relatives.

EXHIBIT 2: Average Price Relative Index Number

Item	P_0 *(1972)*	P_n*(1975)*	(P_n/P_0)
Bread	10	15	1.50
Beef	120	150	1.25
Milk	20	22	1.10
Sugar	10	28	2.80
Potatoes	40	70	1.75
		Total	8.40

$$\text{Price Index 1975 (1972 = 100)} = \frac{8.40}{5} \times \frac{100}{1} = 168.0$$

In our abbreviated mathematical notation this index may be written as:

$$\text{P.I. Year } n \text{ (Year 0 = 100)} = \frac{\Sigma(P_n/P_0)}{k} \times 100$$

In this formulation we are implicitly assuming that all the goods have equal importance: but we all know that we do not spend *equal amounts of money* on bread and beef, milk and potatoes. (It is this fact that explains the vast discrepancy between the value of this index, 168.0, and that of the Simple Aggregate Price Index, 142.5.) In practice the Average Price Relative Index is rarely used for just these reasons – we have introduced it simply as a convenient way to lead into another, much more realistic, formulation.

Weighted Price Relative Index. A measure of the *importance* of each of the goods in our basket, as far as changes in price are concerned, is the amount of money spent on each. It is most convenient to take the amounts spent on the goods[(1)] in the *base year* and use these figures as *weights* in forming a weighted average of price relatives. For the purpose of the calculations it does not matter whether we use absolute values (actual amounts spent in £p) for the weights or their relative values (obtained by scaling the weights to add up to some convenient number). In Exhibit 3 we show some relative weights, w, and the calculation of the price index in the form of a weighted average.

EXHIBIT 3: A Weighted Price Relative Index

Good	*Weight (w)*	P_n/P_0	$w \times (P_n/P_0)$
Bread	16	1.50	24.00
Beef	17	1.25	21.25
Milk	23	1.10	25.30
Sugar	5	2.80	14.00
Potatoes	9	1.75	15.75
Totals	70		100.30

$$\text{Price Index 1975 } (1972 = 100) = \frac{100.30}{70} \times 100 = 143.3$$

The weights used in Exhibit 3 are typical of those in the Food Section of the Index of Retail Prices in which the weights for all items covered by the Index are scaled to add to 1,000. Hence we may interpret the weights used in this example to mean that, out of every £1,000 spent on items covered by the Index, £16 are spent on bread, £5 on sugar, etc.

If we use weights which are related to *base year* expenditure then, for a particular item, we may write:

---

(1) In the Index of Retail Prices this data is collected by means of the annual Family Expenditure Surveys, corrected by the use of other data using, for example, Customs and Excise records for goods (especially alcohol and tobacco) which are found to be incorrectly recorded in the survey.

weight(w) = money spent on this item
= price × quantity bought (in year 0)
= P_0Q_0.

When we multiply the weights applied to each item by its price relative, as in the last column of Exhibit 3, we obtain $w \times (P_n/P_0) = (P_0Q_0) \times (P_n/P_0) = P_nQ_0$. Summing over all goods leads to ΣP_nQ_0, and by dividing by the sum of the weights ΣP_0Q_0, we arrive at the following mathematical equivalent of our calculation.(1)

$$\text{Laspeyre's Price Index for year } n \text{ (Year 0 = 100)} = \frac{\sum P_nQ_0}{\sum P_0Q_0} \times 100 \qquad (1)$$

The name "Laspeyre" associated with this formulation of the index may be taken to mean "base year weighted". Laspeyre's method was used for the U.K. Index of Retail Prices until 1962. Unfortunately it assumes that if prices have risen between year 0 and year n we would still purchase the same quantities. This may be true of some goods but not all. Over a large range of goods one would expect the quantities bought to fall: people will either use less or substitute a cheaper product if one exists. This leads to the above form of index *overestimating* price increases in a period of rising prices. Since rising prices are usually regarded as being "bad" economically it is even more unfortunate when the measuring instrument makes the situation appear worse than it really is.

An alternative method is to use current period quantities, Q_n, in forming the weights. This is known as the Paasche method and may be expressed as

$$\text{Paasche Price Index} = \frac{\Sigma P_nQ_n}{\Sigma P_0Q_n} \times 100 \qquad (2)$$

This index suffers from the reverse effect; that is, it assumes that if in the base year prices were lower we would have bought the same quantities as in the current year. Hence price rises will tend to be *underestimated*. This index has two further drawbacks. Firstly it is difficult (or

(1) This formulation is in a suitable form for calculation only in cases where we have the quantities Q_0 as "raw data" rather than the weights, w. It should also be noted that, expressed in this way, the numerator ΣP_nQ_0 is the total amount of money spent on the goods in year n, and the denominator, the amount spent in the base year. Hence if the calculations are performed using quantities Q_0 rather than weights the index is sometimes called a "weighted aggregate" despite the fact that it is formally identical to the "weighted price relative".

expensive) to obtain current year weights in time to use them in the index: it may take several weeks, if not months, to collect, process and analyse the data and so delay the calculation of the index. Secondly, with the weights changing every period, it is difficult to make comparisons between time periods, since changes in the index reflect both changes in prices and changes in weights.

The over- and underestimation of price changes when using base or current quantities as weights has led to the concept of an "Ideal" index number, two of which are

(a) the Irving Fisher Index $= 100 \sqrt{\dfrac{\sum P_n Q_0}{\sum P_0 Q_0} \times \dfrac{\sum P_n Q_n}{\sum P_0 Q_n}}$

which is the geometric mean of the Laspeyre and Paasche numbers, and

(b) the Marshall-Edgeworth Index $= \dfrac{\sum P_n \dfrac{(Q_n + Q_0)}{2}}{\sum P_0 \dfrac{(Q_n + Q_0)}{2}} \times 100$

which uses the arithmetic mean of the base year and the current year quantities in forming the weights.

Although correcting for bias, these indices are still open to the same criticisms made of the Paasche index in obtaining and interpreting index numbers with current weights.

The weighting system used in the earlier (1947–61) published Price Index Numbers used to follow Laspeyre's method using base period quantities. The present index, however, is more like that of the Paasche method. The weights are revised every year based on data collected in the past three years. The present index, therefore, tends to fall between the two methods, attempting to overcome the difficulty of obtaining exact current period weights by using weights which are at least more up-to-date than static base-period weights.

To summarise then, the idea behind the present index is one of calculating the cost of buying a large representative basket of goods in January of each year, and then finding the cost of the same basket each month for the rest of the year. The basket is changed (if necessary) every January to make sure that it is still representative of the goods (and services) that are being bought.

Turning very briefly to *volume* index numbers, for each formulation of a price index the equivalent volume index may be expressed mathematically by replacing quantities (Q) by prices (P) in the formulation, and vice versa. The rationale behind this is that we may first obtain *quantity* relatives (Q_n/Q_0) and weight these by the relevant amounts of money spent on each good. So for a base year weighted volume index we use weights P_0Q_0, leading to a Laspeyre volume index:

$$\text{Laspeyre's Volume Index for year } n \text{ (year 0 = 100)} = \frac{\sum P_0 Q_n}{\sum P_0 Q_0} \times 100 \qquad (3)$$

2.3 USES OF PRICE INDEX NUMBERS

Index numbers have two main areas of use: those concerned with analysing and understanding the past, and those concerned with controlling the present and influencing the future. But it would be unfair to draw too formal a distinction between these two areas of use since an understanding of the past often provides a guide for future provisions.

Once again, our main focus will be on the use of price index numbers, but the reader should find no difficulty in envisaging uses to which other index numbers could be put.

Research. The essential problem in research is that of making a just comparison of economic conditions over a period of time and/or between a number of countries or industries. And in particular the need is to compare quantities expressed in monetary terms: wage rates, the value of exports, government spending on social services, etc. As a price index calculated on a yearly basis provides a measure of changing prices, it can consequently be used as a means to remove the effects of such price changes, in order that the comparisons be made in what may be called "real" terms. So while it may be true that the average level of wages in £p has steadily increased over a period of years, this means little unless the increase is evaluated against similar increases in prices over the same period. The concept of "real wages" is the quantity of goods which can be bought with the wages at each point in time. (If prices rise faster than wages, then although the number of £p in the pay packet has risen, the wage-earner is worse off.)

Consider the example in Exhibit 4 where the value of exports has apparently risen over the years. Fortunately, a relevant price index exists for the commodity under study. Note the change in interpretation of the export figures when the value is "deflated" by dividing by the Price Index – an apparent growth in exports is really a decline when the change in money values is taken into account.

EXHIBIT 4: Real Value of Exports

Year	*Value of Exports (£m)*	*Wholesale Price Index (1964 = 100)*	*Value of Exports at Constant (1964) Prices (£m)*
1964	200	100	200
1965	205	105	$195 = \frac{205}{105} \times 100$
1966	210	110	$191 = \frac{210}{100} \times 100$
1967	220	115	191 etc.
1968	224	120	187
1969	230	125	184

In the study of the relationships between economic variables (econometrics) it is particularly important to eliminate the effect of price changes in order to avoid "spurious correlations". Without correcting for price changes we might be led to believe, for instance, that because the values of two economic variables (say wage rates and the value of exports) rose steadily over a period of time, one was the *cause* of the other: quite a different picture might emerge after deflating the values of the economic variables. The methods used in the examination of these relationships are based on correlation and regression techniques (CU13 and 14). If relationships can be established then the likely effect of changes in policy can be determined beforehand and some (if not all) unfortunate policies avoided.

Government and Trade Union Decision Making. If a government has a reliable measure of price changes then it is in a better position to make decisions which will affect prices by way of taxation, subsidies, etc. Moreover it is in a position to estimate price rises *before* legislation is brought in. (In recent years the Chancellor has stated the likely effect on the Index of Retail Prices of Budget changes.) Thus large increases in taxation on goods which have a small weight in the index (e.g. luxury items) will not raise the index as much as smaller increases in other items with higher weights (e.g. food). Subsidies may be introduced on some staple items specifically in order to balance out price increases in other less necessary goods. The cynic will quickly realise that increased charges in items not covered by the index (e.g. National Insurance contributions) will not change the index at all.

Price Index Numbers may be constructed to compare the cost of living in different parts of a country or between different sectors of the community. In Britain, government and local government employees receive a "London Allowance" for the higher costs incurred in living and working in that area. A price index could obviously be used to assess the appropriate amount of such an allowance. Other examples of financial assistance, such as state pensions and Family Allowances, could be similarly assessed and kept up to date by means of an appropriate price index. However, the difficulty of using for such purposes a price index based on the general population is that special groups receiving government benefits (retired persons and families with several children) may have quite different needs and spending patterns: such differences are ignored when a price index with "average" weights is used.

Trade Unions find ready use for the IRP in wage negotiations. (Indeed, this is one of the reasons for the Chancellor's pre-occupation with the index, in that wage increases themselves tend to raise prices, so leading to the wages-prices spiral of inflation.) Union negotiators and others depend upon the index to ensure that the benefit to their members of previous negotiations is not eroded away over time due to price rises, and as one guide as to the size of future increases necessary to maintain living standards. But there is some political controversy as to whether wage rates should keep in step with past price increases or anticipate future ones – it is maintained by some that the latter approach makes inflation a self-fulfilling prophecy.

"Automatic" Uses: Index Linking. In tenders for long-term contracts (e.g. dams, roads, continuous market research) there is the possibility that one of the parties to the contract may make a large profit or loss due to factors outside the control of either party, as, for example, an increase in national wage rates or a fall in world raw material prices. One method of allowing for such contingencies is to permit the charges to vary each year in proportion to a relevant index. If a large percentage of costs is due to the labour content – as in a market research contract – an index of wage rates or retail prices may be relevant; where raw materials form a substantial proportion of costs (e.g. cement, copper), an index based on world commodity prices may be appropriate.

This same idea has more recently been adopted by governments to protect particular sections of the population against the worst effects of inflation. One example is the "index-linking" of investments, where savings bonds issued by the government have their "face value" re-valued in line with increases in the IRP. The basic objection to official index-linking is that it reduces the number of areas over which government can exert direct control – a philosophy of correcting for (and hence accepting) inflation rather than controlling it.

2.4 PROBLEMS IN THE CONSTRUCTION OF INDEX NUMBERS

A number of common problems arise in the construction of all index numbers: we next consider some such problems and how they are tackled in the cases of the Export Volume Index (EVI) and the Index of Retail Prices (IRP).

The Selection of Items for Inclusion in the Index. It is obviously too laborious a process to monitor changes in all the items under consideration. Instead, it is necessary to select a sample of items in such a way that changes which occur to these items will reflect as accurately as possible changes occurring overall. To do this a list of all possible items is first compiled; they are then grouped into "sections" and a few items selected from each section as being representative. In the construction of the EVI, for example, the complete list contained all exported items such as fertilizers, cars and tennis rackets. Fertilizers were then grouped with other chemicals as one section, cars with other vehicles as another, and tennis rackets with other sports equipment. By considering changes in the volume of exports of each item in a section (from past data) a selection of representative items was made.

For the IRP the situation is somewhat more complex. The current index covers the following main *groups* of items:

- (I) Food
- (II) Alcoholic Drink
- (III) Tobacco
- (IV) Housing
- (V) Fuel & Light
- (VI) Durable Household Goods
- (VII) Clothing & Footwear
- (VIII) Transport & Vehicles
- (IX) Miscellaneous Goods
- (X) Services
- (XI) Food bought and consumed outside the home

Each of the above groups consists of a number of sections, as for example, vegetables and fish within the food group. In the original construction of the index each section was considered separately and a number of items selected whose relative importance and representative nature would in combination show the same average price movement as the section as a whole. In all about 350 items were selected.

For all indices the sample of items has to be reviewed periodically – what is representative at one point in time may not be some time later. New products enter the market and old ones are withdrawn. Over a long period of time the pattern of demand can alter radically. Quality is another consideration – some goods change so much that they are barely recognisable as the same items a few years later and so direct comparisons are misleading.

Sources of Data. Each index will have its own problems as far as data collection is concerned. Convenience, accuracy and cost are the appropriate criteria for selection of data sources. In most cases it becomes necessary to use several sources for a given index, or even for a single section within an index. Data on the volume of exports (for the EVI) could be obtained from Customs and Excise, trade associations, shipping and airline companies etc. When dealing with cars, for example, the trade association may be the best data source, and for wines and spirits, the Customs and Excise department.

The following examples from the current IRP illustrate alternative methods used within a single group. In the case of Food (Group I) most of the information is collected on a personal visit to retailers by local officers of the Department of Employment, who record prices of food items at the time of their visits. Five shops in each of 200 areas are visited on a monthly basis, the same shops on each occasion. Additional information about price changes of some branded foods, such as chocolates and biscuits, is collected by means of a postal questionnaire to a few selected manufacturers.

Some groups present special problems. For Durable Household Goods (Group VI), changes in styles and quality of such items as floor coverings and furniture make the collection of *detailed* information necessary, and this is only practicable from a small sample of shops in large urban areas. Again, postal questionnaires are used for some items e.g. drapery and china.

The Selection of a Base. The problem is essentially that of choosing a base period of time (rather than a particular point in time) during which no very unusual events occurred as far as the measured quantity was concerned. So, for the EVI, we would choose a month showing an average volume of exports, rather than a peak export month, or a month in which much time was lost through strikes. If this presents a difficult problem an average of several consecutive periods may be taken. In either case it is important to update the base period from time to time so that the comparisons relate to the recent, rather than the distant, past and so that the index numbers do not differ *too much* from 100 (which would make percentage interpretations difficult). An alternative method is to use a "chain" base where the base changes each time the index is recalculated; but this method makes multi-time period comparisons more difficult. The technique is illustrated, and compared with the fixed base method, for a simple (one commodity) price index in Exhibit 5. It can be seen that the chain based price relatives are obtained by dividing by the price in the *previous year*.

EXHIBIT 5: The Chain Based Index Number

Year	*Price (p)*	*Chain Base*		*Fixed Base Simple Index (1970 = 100)*
		Price Relative	*Simple Index (Previous Year = 100)*	
1970	10		–	100
1971	12	12/10 = 1.200	120	120
1972	12	12/12 = 1.000	100	120
1973	14	14/12 = 1.167	116.7	140
1974	16	16/14 = 1.143	114.3	160
1975	18	18/16 = 1.125	112.5	180

The chain base and fixed base indices may be "reconciled" by multiplying up the relevant chain base price relatives, e.g.

$$(1.20 \times 1.00 \times 1.167 \times 1.143 \times 1.125) \times 100$$

$$= 180$$

$$= \text{P.I. } 1975\ (1970 = 100)$$

As with the fixed base index a chain based index may be formed from weighted price relatives, by using base period or current period weights. In particular, the Laspeyre base year index in chain form, for year *i*, would have the following formulation:

$$\text{P.I. Year } i \text{ (year } i - 1 = 100) = \frac{\sum P_i Q_0}{\sum P_{i-1} Q_0} \times 100$$

But for a chain based index we also have the option of *previous year* weighting – a technique which overcomes many of the objections which apply to base year or current year weighting. Again, for year *i*, we have the following formulation:

$$\text{P.I. Year } i \text{ (year } i - 1 = 100) = \frac{\sum P_i Q_{i-1}}{\sum P_{i-1} Q_{i-1}} \times 100$$

Such an index makes good sense for comparisons over a period of a few years, but over a longer time span the interpretation becomes increasingly doubtful, owing to the large number of changes in weights.

As mentioned earlier, a compromise solution between the fixed base and chain based methods is to revise the base once every few years. This is the method adopted with the Index of Retail Prices, the latest revision being for January 1974. On each occasion when a new index has been brought out, index numbers on both the old and the new base have been calculated together for some months, so that it is possible to compare two index numbers with different bases. Hence a research worker who wishes to deflate a value series (e.g. values of sales over the period 1950 to 1970) will find that he does not have one continuous price index over the whole period, but several index numbers which have to be "spliced" or linked into one (Exhibit 6).

EXHIBIT 6: Splicing Together Index Numbers

Year	*Value of Sales £000s*	*Price Index Numbers 1958–66*			*"Spliced" Price Index 1964 = 100*
		(1954 = 100)			
1958	20	110			77.7
1959	37	112			79.1
1960	62	114			80.5
1961	83	120(1961 = 100)			84.7
1962	104		108		91.5
1963	116		112		94.9
1964	133		118(1964 = 100)		100
1965	160			105	105
1966	166			110	110

The method used for splicing is similar to that of the chain based calculations. As the most recent data is usually the most important, the whole series is revalued on the latest base year (1964). The index for 1962 and 1963 may be recalculated on the 1964 base by multiplying by (100/118) since 1964 is the year for which the index numbers are available on both bases. Hence we have:

$$\text{P.I. } 1963 = 112 \times \frac{100}{118} = 94.9,$$

$$\text{P.I. } 1962 = 108 \times \frac{100}{118} = 91.5.$$

The years 1961 to 1958 have their index based on the two linking years 1961 and 1964. Hence we multiply by (100/118) × (100/120) = 0.7062:

$$\text{P.I. } 1961 = 120 \times 0.7062 = 84.7,$$

$$\text{P.I. } 1960 = 114 \times 0.7062 = 80.5,$$

$$\text{P.I. } 1959 = 112 \times 0.7062 = 79.1.$$

REFERENCES

1. *Studies in Official Statistics No. 6*, Method of Construction and Calculation of the Index of Retail Prices, HMSO.
2. An account of a volume index – the U.K. Index of Industrial Production – is to be found in K.A. Yeomans, *Introducing Statistics*, Statistics for the Social Scientist: Vol. I Penguin, Chapter 4.

EXERCISES

1(E) Why are weights used in the construction of index numbers? What is meant by:

(a) base weighting;
(b) current weighting;

and what are the advantages and disadvantages of each?

(ICMA December 1971)

2(U) *General Index of Retail Prices*
17th March 1970 (16th January 1962 = 100)

	Item	*Weight*	*Index*
I	Food	255	137.6
II	Alcoholic Drink	66	143.0
III	Tobacco	64	135.8
IV	Housing	119	152.2
V	Fuel and Light	61	145.6
VI	Durable Household Goods	60	122.7
VII	Clothing & Footwear	86	121.7
VIII	Transport & Vehicles	126	127.5
IX	Miscellaneous Goods	65	137.7
X	Services	55	149.5
XI	Meals bought outside the home	43	140.5
		1,000	

Source. Department of Employment and Productivity. March 1970.

(a) Show that "all Items" index is 137.3
(b) It has been stated that food prices may rise by 20%. Calculate the value of the "All Items" index if this rise takes place while other prices remain constant.
(c) To what value would the index for food have to rise in order

that the "All Items" index would rise to 145.0? Assume other prices remain constant.

(d) If taxes were reduced so that a 10% cut in price index occurred in groups II, III and VIII, what would be the reduction in the "All Items" index?

3(E) (a) From the information stated below construct a quantity index for the products made by Multiproducts Ltd., for the period 1966–69, weighted as to 1966 prices, and

(b) explain the purpose of preparing a quantity index, state what the indices calculated in (a) above indicate about the production of Multiproducts Ltd., for the years 1966–69, and discuss the influence of individual products on the index:

MULTIPRODUCTS LTD

Product	*1966 Average Price £*	*1966 Production (000's)*	*1967 Production (000's)*	*1968 Production (000's)*	*1969 Production (000's)*
Pliers	2.00	62	65	66	90
Wrenches	3.00	138	120	110	80
Bolts	0.50	500	540	580	800
Drills	4.50	10	10	10	10

(ACCA June 1972)

4(U) INDEX OF RETAIL PRICES

17th January 1956 = 100 Monthly Averages		*16th January 1962 = 100 (17th January 1956 base index = 119.3) Monthly Averages*	
1956	102.0	1962	101.6
1957	105.8	1963	103.6
1958	109.0	1964	107.0
1959	109.6	1965	112.1
1960	110.7	1966	116.5
1961	114.5	1967	119.4
		1968	125.0
		February 1969	129.8

Source: Employment and Productivity Gazette March 1969.

Obtain the series 1956 to February 1969 based on 17th January 1956 = 100.

5(E) (a) Describe the construction of the Official Retail Prices Index used in the United Kingdom, and
(b) explain how it could be used by

(i) Businessmen,
(ii) Consumers, and
(iii) Trade Unions.

(ACCA June 1971)

6(U) *Average Weekly Earnings Male Manual Workers 21 years +* — *Index of Retail Prices (January 1962 = 100)*

		Average Weekly Earnings Male Manual Workers 21 years +	Index of Retail Prices (January 1962 = 100)
April	1964	£17.60	106.1
October	1964	£18.10	107.9
April	1965	£18.90	112.0
October	1965	£19.60	113.1
April	1966	£20.25	116.0
October	1966	£20.30	117.4
April	1967	£20.60	119.5
October	1967	£21.40	119.7
April	1968	£22.25	124.8

Source: Employment & Productivity Gazette.

(a) Obtain an Index of "Real" Wages with April 1964 = 100
(b) Have weekly earnings risen faster over the period than the index of prices?
(c) On the basis of these figures why cannot we say that the standard of living of the higher grades of the Civil Service has risen. What extra information is required to answer this question?

7. Index numbers are widely used in applied statistics. Define three main types of index number and describe the properties of each type in the context of one specific example (other than the index of retail prices) with which you are familiar.

(IPM June 1974)

8(E) The table below shows the index of retail prices in the United Kingdom between 1966 and 1970 (January 1962 = 100).

Year	1966	1967	1968	1969	1970
General Index, All items	116.5	119.4	125.0	131.8	140.2

(a) On what basis is the index constructed?
(b) What purposes does the index serve?
(c) Explain briefly the meaning of the results shown in the table.

(d) Assuming that the calculation of the retail price index is continued indefinitely, what factors (if any) would indicate that a new base year should be chosen?

(IPM November 1973)

9. The "simple aggregate" method, the "simple average of relatives" method, the "weighted aggregate" method and the "weighted average of relatives" method are procedures for calculating an index number.

(a) Define the four methods and compare their advantages and disadvantages.
(b) What is meant by the "time reversal" and the "factor reversal" properties of index numbers?[1]
(c) Demonstrate that Fisher's Ideal Index has the properties mentioned in (b) above.

(IPM Summer 1973)

10. (a) What is meant by an "index number" in statistics and what purposes do index numbers serve?
(b) How would you calculate an index of average earnings (all employees) in manufacturing industry in a particular country?
(c) Given access to data concerning average earnings in an industrial company and to national statistical information concerning prices, how would you examine trends in the real purchasing power of earnings in that company?

(IPM November 1972)

SOLUTIONS TO EVEN NUMERED EXERCISES

2. (a) Forming the "weighted sum" of the indices of all items we obtain

$$(255 \times 137.6) + (66 \times 143.0) + (64 \times 135.8) + \ldots$$
$$+ (55 \times 149.5) + (43 \times 140.5) = 137{,}318.3$$

Hence "All Items" Index = $(137{,}318.3/1{,}000) \simeq 137.3$.

(b) The food index would *increase* by 20% of 137.6 = 27.52, thus leading to an increase in the weighted sum of 255 × 27.52 =

[1] See Reference 2.

7,017.6. The All Items index would increase by 7,017.6/1,000 $\simeq$ 7.0 giving a new value of 142.3.

(c) The weighted sum of indices *excluding* food = (137,318.3 – 35,088.0) = 102,230.3. Thus weighted sum of *new* food index must be (145,000 – 102,230.3) = 42,769.7, in order to give All Items index of 145.0. Consequently new food index must be 42,768.7/255 = 167.7.

(d) With a 10% cut in price index of groups II, III and VIII, the weighted sum of the indices of all items will be *reduced* by:

$$(255 \times 13.76) + (66 \times 14.30) + (126 \times 12.75) = 6{,}059.1$$

New All Items index will then be (137,318.3 – 6,059.1)/1,000 $\simeq$ 131.3.

4. Multiply 1962 – 9 data by (119.3/100) = 1.193

1962	121.2
1963	123.6
1964	127.7
1965	133.7
1966	139.0
1967	142.4
1968	149.1
Feb 1969	154.9

6. (a)

	Wages Index (April 64 = 100)	*I.R.P. (April 64 = 100)*	*Index of 'Real' Wages (April 64 = 100)*
April 64	100.0	100.0	100.0
Oct. 64	102.8	101.7	101.1
April 65	107.4	105.6	101.7
Oct. 65	111.4	106.6	104.5
April 66	115.1	109.3	105.3
Oct. 66	115.3	110.7	104.2
April 67	117.0	112.8	103.7
Oct. 67	121.6	112.8	107.8
April 68	126.7	117.6	107.1

(b) Yes, consistently.

(c) An appropriate earnings index for higher grades of the Civil Service is unlikely to be identical to the wages index calculated. Further the IRP does not take into account any increases in National Insurance contributions or Income Tax and so does not measure the "cost of living" for the "average" man, let alone for higher civil servants.

SEMINAR UNIT 3 – TIME SERIES ANALYSIS

3.1 INTRODUCTION

For the purpose of monitoring the performance of a company (or a national economy) over time, data is often collected and published on a regular basis – weekly, monthly, quarterly or yearly, depending upon the particular item. Typical examples are production figures, sales volumes, number of absentees and stock levels for a company; exports and unemployment for the economy.

Here we are concerned with the analysis of any such *time series* (i.e. a set of data in chronological order), the initial purpose of which is to gain an understanding of the changes that took place, as revealed by the data, over a protracted period of time.

Although we shall concentrate on the applications of time series analysis to business and economics it is important to point out that there are applications in the sciences, engineering and many other fields. The medical scientist may be studying the effect of drugs over time, the engineer may be working on machine tool wear, the geographer on seasonal climatic changes.

The businessman is perhaps most interested in using time series as an aid to business forecasting, particularly in the area of sales, so that at an appropriate time budget allocations may be made for capital investment in machines, materials, labour and advertising for the year ahead. Other forecasts are made, in less detail, for plans which take longer to materialise, such as the building of new factories and the development of new products. National governments have similar interests: productivity, balance of payments, unemployment rate and other economic indicators are carefully watched for short- and long-term developments; changes in age distribution and patterns of migration have to be monitored so that provision can be made well in advance for pensions, schools and teachers, and hospitals. The more complex the area of application, the greater the sophistication required from the time series analysis. On a national scale particulary, a number of separate time series may have to be studied simultaneously involving an analysis of some complexity. In this Seminar Unit, however, we shall concentrate on the simpler forms of analysis on which the more sophisticated techniques are based.

As in some other branches of applied statistics we start by postulating some underlying model. Our models identify various *components* which, in combination, "explain" the observed time series. In conditions where the mechanisms at work are well understood, the choice of model is often an easy task; otherwise, several models may be tried out and the one adopted which explains the time series most adequately. In simple time series analysis it is conventional to identify four components, algebraically denoted by T, C, S and R, which we now introduce.

The Trend Component *(T)*. Consider the time series plotted in Exhibit 1, which may be said to possess an upward or rising trend. The *trend curve*, superimposed on the time series graph, attempts to display this more forcibly by following the general pattern of the series but "smoothing out" the minor fluctuations. In Exhibit 1 we make the distinction between the **actual value** Y and the **trend value** T (taken from the trend curve) at the same point in time. It should also be noted that the actual values are joined up by straight lines in this graphical representation: the lines have no physical significance, and are merely used to guide the eye.

EXHIBIT 1: Sales 1962–1974

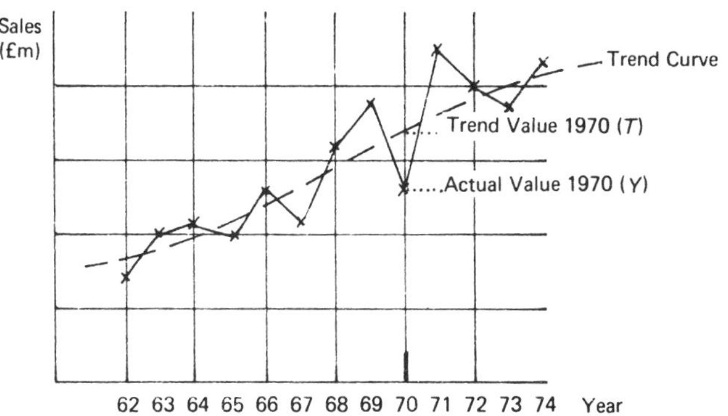

Cyclical Component *(C)*. This component is associated with the general up and down movements which occur in economic and business time series over a matter of a number of years. Traditionally such movements are attributed to "trade cycles", about which some controversy rages among economists. The cyclical component is often omitted in simple business models.

The Seasonal Component *(S)*. Many time series display rises and falls which occur at roughly predictable times in the year. The weather affects many time series in this way e.g. sales of ice cream, consumption of electricity, unemployment in the building trade. Seasonal effects may also be man-made and give rise, for example, to peaks in sales before Christmas, Easter and Mother's Day. The effect of any such seasonal influences is that in some weeks, months or quarters of the year the *actual values* of the series are regularly greater than the *trend values*, while in others they are regularly less.

Residual Component *(R).* Given an appropriate choice of model, much of the observed variation in the actual values may be explained in terms of the Trend, Cyclical and Seasonal components. But however good the model some "unexplained" variation, which we formally attribute to a Residual component, will remain due to such causes as:

(a) sampling errors in cases where the data is taken from sample surveys, e.g. sales and stock data from retail audit surveys.
(b) recording or measurement errors, e.g. in export statistics where some exports go unrecorded.
(c) "once only" effects, such as *un*seasonal spells of good weather, strikes, competitive advertising.

It is possible to incorporate large movements of type (c) into a particular model, but more frequently the analysis is carried out specifically to measure these effects. However, for our purposes, we shall assume that they are relatively small.

There are a number of ways in which Trend, Cyclical, Seasonal and Residual components are identified and calculated, depending on the particular model used. Here we are concerned with just two models – the Additive and Multiplicative models.

3.2 THE ADDITIVE MODEL

In this model it is assumed that the Actual value of the variable, Y (e.g. sales in £m) is made up of the *sum* of the four components T, C, S and R as represented by the following equation (for time period i):

$$Y_i = T_i + C_i + S_i + R_i$$

It is conventional to drop the suffix i, on the understanding that the equation applies separately to each time period:

$$Y = T + C + S + R \qquad (1)$$

The additive relationship is illustrated in Exhibits 2a and 2b.

EXHIBIT 2a: Actual Value Above the Trend Curve

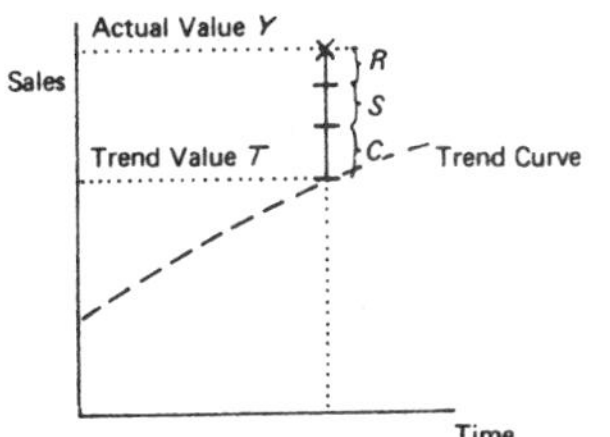

EXHIBIT 2b: Actual Value Below the Trend Curve

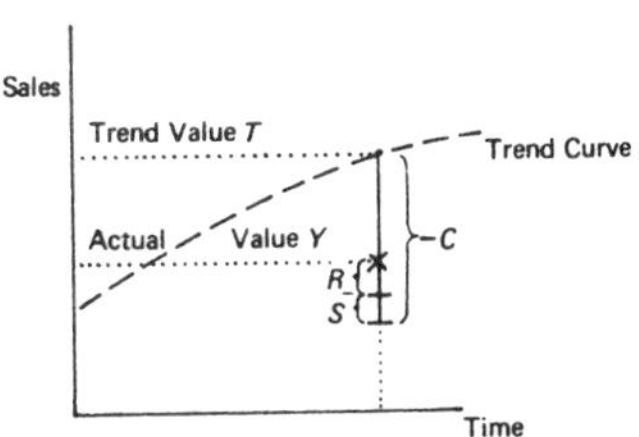

In Exhibit 2a the Actual value Y is above the trend curve: the components C, S and R are added to the trend value T until the Actual value is fully "explained". In Exhibit 2b the Actual value is below the trend curve; this is due to the value of C being negative at this point in time (as might occur during an economic recession), although S and R are both positive. The additive model allows for any or all of T, C, S and R to be positive or negative at any point in time and it assumes that their values are independent of each other – leading to 16 possible diagrams along the lines of Exhibits 2a and 2b.

The *analysis* of a time series using this model may be described in general as follows. First, the trend values are estimated (by one of the methods discussed in this Seminar Unit, Sections 3.4 or 3.8) and subtracted from the Actual values. Representing this symbolically, we subtract T from both sides of Equation (1) giving

$$Y - T = C + S + R. \quad (2)$$

We then estimate the cyclical component and subtract this:

$$Y - T - C = S + R. \quad (3)$$

Finally we estimate and subtract the seasonal component S,

$$Y - T - C - S = R \quad (4)$$

leaving a calculated value of the residual.

Taking a numerical example, for a particular time period where $Y = 163$, we may write Equation (1) as:

$$163 = T + C + S + R.$$

If T is estimated as 150, Equation (2) becomes

$$163 - 150 = 13 = C + S + R$$

If C is estimated as -10, Equation (3) becomes

$$13 - (-10) = 23 = S + R.$$

Finally if S is estimated as 20, Equation (4) becomes

$$23 - 20 = 3 = R,$$

our calculated value of the residual for this particular time period.

As a check on the validity of a model it is always useful to plot all the residuals on a time-scaled graph. If the values of the residual component are small and show no recognisable pattern of their own we may be satisfied that the model is a good fit. Alternatively, a systematic pattern shown by the residuals would indicate that we have chosen an inappropriate model: we might then search for a better one.

An important feature of the additive model is that the *units* in which T, C, S and R are measured are the same as those of the Actual values Y (£m., or whatever), and thus the components have a direct interpretation. In our numerical example $S = 20$ and so 20 units of sales, out of the total of 163, are attributable to the particular seasonal characteristics of that time of year.

3.3 THE MULTIPLICATIVE MODEL

For many time series the fluctuations in Actual values brought about by seasonal, cyclical and residual influences will depend on the going value of the trend. For instance, the sales boost due to Christmas is likely to be less for a firm with a £5m turnover than when that firm grows to a turnover of £10m. The additive model we have just discussed assumes, however, that the various components act *independently*. Provided we are dealing with a time series with a shallow trend curve (or over a limited period of time) the inaccuracies brought about by this assumption will not be important; but where the trend curve is steeply rising, or the analysis carried out over a large number of years, the additive model is often found to give a very poor fit to the data, as evidenced by a graphical plot of the residuals.[(1)] A more reasonable assumption in such conditions is that the seasonal, cyclical and residual components are *in proportion* to the trend value – the assumption embodied in the Multiplicative model:

$$Y = T \times C \times S \times R \tag{5}$$

[(1)]In such cases the residuals tend to be small over only a limited period of the data and/or they may display "seasonal" characteristics of their own (see Exercise 4).

The analysis of a series by this model involves carrying out successive *divisions* on Equation (5) as each component is estimated. Thus after estimating the trend values (T) we obtain

$$\frac{Y}{T} = C \times S \times R, \tag{6}$$

followed by

$$\frac{Y}{TC} = S \times R, \tag{7}$$

and finally

$$\frac{Y}{TCS} = R. \tag{8}$$

So if $Y = 163$ (as in the previous section) and T is estimated as 150, Equation (6) gives

$$\frac{163}{150} = 1.087 = C \times S \times R.$$

If C is then estimated as 0.90, Equation (7) gives

$$\frac{1.087}{1.30} = 0.929 = R.$$

Note that in contrast to the additive model, although the Trend component has the same units as the Actual values, the values of C, S and R are pure ratios. The value of $S = 1.30$, for example, means that, resulting from the seasonal characteristics of the time of year, the sales are 30% higher than average.

3.4 ESTIMATING THE TREND (T)

A number of methods are available for trend estimation, prior to an analysis by the additive or multiplicative (or other more complex) models. Those considered here involve no assumptions about the form of the trend curve: later we consider other methods for trend estimation which do require such assumptions.

Drawing a Freehand Curve. This is the simplest method and it is often carried out as a first step to help decide which of the more formal methods, shortly to be discussed, is the most appropriate. As a method in its own right it is open to objections on grounds of inaccuracy, too much variation from person to person and the fact that it cannot be incorporated within an analysis by computer.

The Method of Partial Averages. The data series is first divided up into two or more "blocks" (preferably with the same number of time periods for each) and the average value found for each block. These averages are then plotted on a time-scaled graph (each being located at the chronological centre of its block) and the points joined up by straight lines or a smooth curve. In particular, where the data series is simply divided into two halves and the two averages joined by a straight line, the method is called *semi-averages* – the line obtained in this manner is often regarded as an approximation to a least-squares regression line (CU14).

In any serious application these first two methods have little to commend them except as a preliminary to a more detailed analysis.

Moving Averages. Before discussing this method in detail let us work out one simple example – the calculation of a 3-year moving average for the sales data shown in Exhibit 4. If we take the 1969, 1970 and 1971 values and form their average we obtain $(23 + 27 + 25)/3 = 75/3 = 25.0$. This is a 3-year average for 1970, because 1970 is chronologically the centre of the three years involved. *Moving* one year forward, the 3-year average for 1971 is obtained as the average of the 1970, 1971 and 1972 values, $(27 + 25 + 30)/3 = 27.3$ – hence the term moving average. In general, we may calculate an n-period moving average for a set of data provided n is less (preferably, considerably less) than the total number of time periods, N, under analysis (in our example $n = 3$ and $N = 7$).

EXHIBIT 4: A 3-Year Moving Average

Year	*Sales*	*3 year moving total*	*3 year moving average*
1969	23	–	–
1970	27	75	25.0
1971	25	82	27.3
1972	30	89	29.7
1973	34	93	31.0
1974	29	99	33.0
1975	36	–	–

EXHIBIT 5: The Data and the Trend Curve

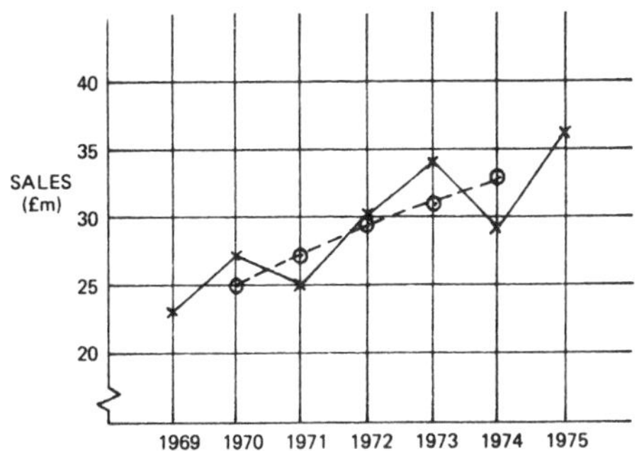

EXHIBIT 6: Calculating the Moving Totals

The total for the first 3 years (1969, 1970, 1971) = 23 + 27 + 25	=	75 (1970 Moving Total)
The first year (1969) is then dropped		−23
The fourth year (1972) is next added		+30
So the total for 1971	=	82 (1971 Moving Total)
The *new* first year (1970) is dropped		−27
and the *new* fourth year (1973) is added		+34
So the 1972 total	=	89 (1972 Moving Total)
etc.		

This simple example does serve to illustrate two important features of moving averages in general.

- From a computational point of view it is more convenient to calculate moving *totals* for the whole series in the manner shown in Exhibit 6. The moving average trend values are then obtained by dividing by n, giving the results shown in Exhibit 4.
- From Exhibit 4 we see that it is not possible to calculate a moving average for the first (1969) and last (1975) year. Generally, with an (odd) n-period moving average, the first and last $(n - 1)/2$ periods will not have a moving average (in our case $n = 3$, so $(n - 1)/2 = 2/2 = 1$). This is a serious limitation if one is hoping to extrapolate the trend values for the purpose of forecasting several periods ahead. In particular, where a large n is used, the trend curve terminates several periods before the end of the known data.

In our example we chose an odd number (3) for n, and this enabled us to position each moving total (and moving average) in the chronological centre of the data from which it was calculated. Thus the first moving average calculated (25.0) is positioned against its middle year (1970), both in the table (Exhibit 4) and on the graph (Exhibit 5). This would enable us to proceed immediately with an analysis by the additive or multiplicative models. For the additive model we would calculate:

for 1970, $Y-T = 27 - 25 = 2.0$;
for 1971, $Y-T = 25 - 27.3 = -2.3$;
for 1972, $Y-T = 30 - 29.7 = 0.3$, etc;

and if using the multiplicative model:

for 1970, $Y/T = 27/25 = 1.08$;
for 1971, $Y/T = 25/27.3 = 0.92$, etc.

Similarly, for a moving average based on any odd number of periods (e.g. n = 5,7,9,11 etc.), we would obtain a conveniently positioned set of moving averages. If, however, we chose an even number for n (and we shall see later why this is often desirable) the correct positioning of the moving average values is *between* the original data. (Exhibit 7 shows the correct positioning of 2-year moving totals between their relevant periods.) This is most inconvenient for the purpose of analysis by the additive or multiplicative models; and so the n-period moving totals are themselves next totalled in pairs to give what is called *2n–point* moving totals, positioned *in line* with the original data. (It is easier to carry out this procedure than to describe it, as an inspection of Exhibit 7 will show.) Finally, the moving averages are obtained by division by $2n$ (= 4, in Exhibit 7). This procedure is adopted whenever n is chosen as an even number – another example is given in Section 3.6, of an 8-point moving average.

We have seen that it is somewhat simpler to calculate a moving average when n is odd rather than even; but what value for n should one chose? The general rule is that n should correspond as nearly as possible to the number of periods between consecutive peaks (or troughs) in the data, as this will result in a trend curve which smooths out, as far as possible, the undulations in the original series. In Exhibit 8 the peaks occur at seven year intervals; therefore a seven year moving average is appropriate.

With seasonal data collected on a quarterly, monthly or weekly basis the appropriate values of n are 4, 12 and 52 respectively, as the peaks and troughs will occur at roughly yearly intervals. Unfortunately all three values of n are even numbers and so the moving averages will have

EXHIBIT 7: Trend Calculation[1] With $n = 2$

Year	*Sales (£m)*	*n = 2 Year Moving Total*	*2n = 4 Point Moving Total*	*4 Point Moving Average (Trend)*
1969	23			
		50		
1970	27		102	25.5
		52		
1971	25		107	26.75
		55		
1972	30		119	29.75
		64		
1973	34		127	31.75
		63		
1974	29		128	32.0
		65		
1975	36			

EXHIBIT 8: A Cyclical Time Series

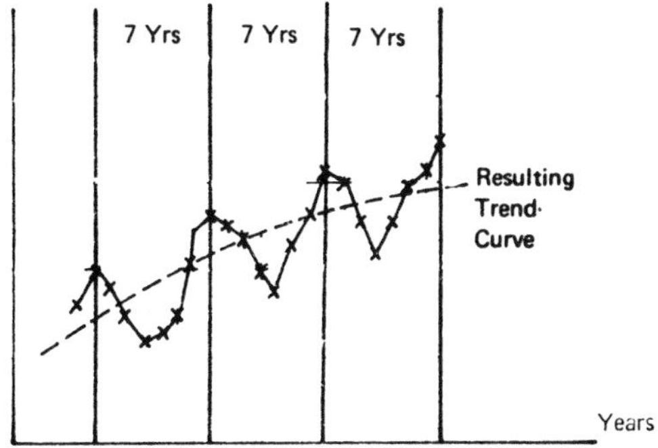

[1] It can be seen from Exhibit 7 that the 1970 4-point moving total, for example, is made up as follows:

$$\begin{aligned} 102 &= 50 + 52 \\ &= (23 + 27) + (27 + 25) \\ &= (23 \times 1) + (27 \times 2) + (25 \times 1); \end{aligned}$$

that is, the mid-value of 27 is included in both the totals 50 and 52 and so contributes twice to the overall total. (Similar arguments apply to the construction of a moving average using any even number for n.) This is just one example of a "weighting" system for time series data: A summary of early work in this field may be found in *An Introduction to the Theory of Statistics*, G.U. Yule and M.G. Kendall.

to be formed from 8-point (for quarterly data), 24-point (monthly data) and 104-point (weekly data) moving totals.

If an "incorrect" value for n is used the trend curve will display seasonal or cyclical characteristics of its own and this can give rise to misleading interpretations, especially when making forward projections of the trend curve. (See Exercise 2.)

Exponential Smoothing. This method, which is commonly used in computerised short-term sales analysis, really deserves a separate Seminar Unit of its own, but that would take us well outside the scope of the present course. Here are the briefest of details of "simple" exponential smoothing. For each period a "smoothed" value is calculated by updating the smoothed value obtained for the previous period in the following manner:

$$\text{New smoothed value} = \text{previous smoothed value} + \alpha\,(\text{Latest actual value} - \text{previous smoothed value}) \qquad (9)$$

The constant α is fixed at a value between 0 and 1 (often at 0.1 or 0.2) and determines the extent of the smoothing: the role played by α is rather similar to that of n in moving averages. Exhibit 9 gives an example, using some weekly sales figures, of the calculations involved when α is chosen as 0.2.

EXHIBIT 9: Simple Exponential Smoothing ($\alpha = 0.2$)

Week No.	*Actual Demand*	*Smoothed Demand*
1	15	16.0 (assumed)
2	20	16.0 + 0.2 (20 − 16.0) = 16.8
3	10	16.8 + 0.2 (10 − 16.8) = 15.4
4	30	15.4 + 0.2 (30 − 15.4) = 18.3
5	25	18.3 + 0.2 (25 − 18.3) = 19.6

From this example it can be seen that, as compared with moving averages, exponential smoothing does not require the storage of the last n-period's data – only the previous period's smoothed value need be stored. It is for this reason (and other theoretical considerations) that exponential smoothing has found favour in many industrial applications.

3.5 PREAMBLE TO THE ESTIMATION OF SEASONAL (*S*) AND RESIDUAL (*R*) COMPONENTS

We shall restrict our attention in the discussion on these matters to examples where:

- The cyclical component (C) is omitted from the model on which the analysis is based. We have already stated that this is often the case in practice unless a specialist analyst is employed.
- Moving averages are used for trend estimation. (With some other methods, e.g. exponential smoothing, a rather different approach is used for estimating the seasonal and residual components.)
- The data is collected and presented at three monthly intervals. Such quarterly data is often encountered since many official statistics are published in this way. The analysis of monthly or weekly data proceeds along very similar lines although the arithmetic is more lengthy.

Using the additive model, $Y = T + S + R$, the estimation of the seasonal (and residual) components from quarterly data is carried out by the *Method of Quarterly Deviations* which is discussed in detail in the next section. The corresponding treatment using the multiplicative model, $Y = T \times S \times R$, is called the *Method of Seasonal Indices* and is considered briefly in Section 3.7. As previously noted, the additive model is more appropriate where there is only a moderate trend in the series; the multiplicative model where the trend is relatively steep.

3.6 THE METHOD OF QUARTERLY DEVIATIONS

Because of the importance of this method we shall examine one example, the data for which is given in Exhibit 10, in some detail.

EXHIBIT 10: Quarterly Production (Tons) 1970–73

Year/Quarter	*I*	*II*	*III*	*IV*
1970	78	62	56	71
1971	84	64	61	82
1972	92	70	63	85
1973	100	81	72	96

The steps in the analysis are to list the data in column form, as shown in Exhibit 11, and then:

(a) Obtain the 4-quarter moving totals, listing the results in the "centre" of each block of four.

(b) Obtain the 8-point moving totals by summing the 4-quarter totals in pairs.

(c) Divide the 8-point moving totals by 8 to obtain the moving average trend values (note that there are no trend values for the first two and last two quarters).

(d) Subtract the trend values from the quarterly data to obtain the "quarterly deviations". Symbolically,

$$Y - T = S + R$$

e.g. for 1970 quarter 3, 56 – 67.500 = – 11.500.

(e) Form the *Table of Quarterly Deviations* (Exhibit 12) by listing the quarterly deviations ($Y - T$) for each year under their quarterly headings. The purpose of this table is to enable us to find the *average* amount by which the Actual values exceed or fall below the trend curve for each of the four quarters.

EXHIBIT 11: Method of Quarterly Deviations

Year	*Quarter*	*Production (tons) (Y)*	*4-Qtr Moving Total*	*8-Point Moving Total*	*Trend (T)*	*Deviation from Trend (Y – T)*
1970	1	78		–	–	–
	2	62	267	–	–	–
	3	56	273	540	67.500	–11.500
	4	71	275	548	68.500	+2.500
1971	1	84	280	555	69.375	+14.625
	2	64	291	571	71.375	–7.375
	3	61	299	590	73.750	–12.750
	4	82	305	604	75.500	+6.500
1972	1	92	307	612	76.500	+15.500
	2	70	310	617	77.125	–7.125
	3	63	318	628	78.500	–15.500
	4	85	329	647	80.875	+4.125
1973	1	100	338	667	83.375	+16.625
	2	81	349	687	85.875	–4.875
	3	72		–	–	–
	4	96		–	–	–

EXHIBIT 12: Table of Quarterly Deviations from the Trend

Year/Quarter	*1*	*2*	*3*	*4*
1970	–	–	–11.500	+2.500
1971	+14.625	–7.375	–12.750	+6.500
1972	+15.500	–7.125	–15.500	+4.125
1973	+16.625	–4.875	–	–
Total	+46.750	–19.375	–39.750	+13.125
Average	+15.583	–6.458	–13.250	+4.375
Correction	–0.063	–0.063	–0.063	–0.063
Corrected Average (= Seasonal Variation)	+15.520 S_1	–6.521 S_2	–13.313 S_3	+4.312 S_4

(f) Find the average quarterly deviation for *each* quarter. In our case there are three deviations for each quarter and so the totals for each quarter are divided by 3.

(g) Check to see if these average quarterly deviations sum to zero: if they do not we must then correct them so that they do.[1] In the example, +15.583 – 6.458 – 13.250 + 4.375 = +0.25. The total is 0.25 too *large* and hence we must *subtract* 0.25/4 = 0.063 from each of the four average deviations. The resulting corrected figures are our values for the seasonal component of the series, called **seasonal variations** (S_1, S_2, S_3 and S_4, where the suffix now refers to the quarter, irrespective of the year).

We may interpret the results obtained so far as showing that during the first quarter of the year production is on average 15.52 tons above the trend, while in the second quarter it is typically 6.52 tons below the trend, and so on. These quantities are often important in their own right, but they also play a major role in ensuring that we are not misled by the raw unadjusted data (Y) of a time series.

Seasonal Adjustments. In our example the production tends to be high in the first quarter and low in the second quarter. If the management were to look at the production figures in isolation they could well be needlessly subject to alternate bouts of optimism and pessimism as each set of quarterly figures is published. A better overall impression of the performance of the company is provided by examining the **seasonally adjusted** or **deseasonalised** production figures. These are obtained by subtracting the seasonal variations (S) from the actual production figures. In terms of our model,

Seasonally Adjusted Production = $Y - S = T + R$;

that is, the resulting figures show the trend, free from seasonal influences, but still subject to residual fluctuations. Exhibit 13 gives the results for

EXHIBIT 13: Seasonal Adjustments

Year	*Quarter*	*Production (Y)*	*Seasonal Variation S*	*Seasonally Adjusted Production Y – S*
1971	1	84	+15.52	68.48
	2	64	–6.52	70.52
	3	61	–13.31	74.31
	4	82	+4.31	77.69

(1) Just as the trend values should be free of any seasonal influences, the values of the seasonal component should be free of any trend influences. This is the reason for correcting their values so they sum to zero.

1971 from which it can be seen that although the production is low in the middle two quarters, the seasonally adjusted production shows a continuing rise, indicating that the long term trend is upwards.

Calculation of Residuals. The final step in the analysis is to check if our model, $Y = T + S + R$, is a good fit. To do this we need to check the size and pattern of the residuals in one of two formally identical ways. We may either subtract the seasonal variations (S) from the Quarterly Deviations ($Y - T$),

$$(Y - T) - S = R,$$

or alternatively, subtract the trend values (T) from the seasonally adjusted figures ($Y - S$),

$$(Y - S) - T = R,$$

if the seasonally adjusted figures have been calculated for other purposes: Exhibit 14 shows the latter method of calculation.

EXHIBIT 14: Calculation of Residuals

Year	*Quarter*	*Production* Y	*Seasonal Variation* S	*Seasonally Adjusted* $(Y - S)$	*Trend* T	*Unexplained or Residual Variation* $R = (Y - S) - T$
1970	1	78	–	–	–	–
	2	62	–	–	–	–
	3	56	−13.313	69.313	67.500	1.813
	4	71	+4.312	66.688	68.500	−1.812
1971	1	84	+15.520	68.480	69.375	−0.895
	2	64	−6.521	70.521	71.375	−0.854
	3	61	+13.313	74.313	73.750	+0.563
	4	82	+4.312	77.688	75.500	+2.188
1972	1	92	+15.520	76.480	76.500	−0.020
	2	70	−6.521	76.521	77.125	−0.504
	3	63	−13.313	76.313	78.500	−1.187
	4	85	+4.312	80.688	80.875	−0.187
1973	1	100	+15.520	84.480	83.375	+1.105
	2	81	−6.521	87.521	85.875	+1.646
	3	72	–	–	–	–
	4	96	–	–	–	–

The residuals should now be examined to determine if a pattern can be discerned (if it can we might consider fitting another model). A first step in this analysis would be to plot a graph of the residuals; however, the analysis now becomes too complicated for a first course and we leave it to the experienced analyst. It should be pointed out that the very process of taking a moving average will tend to generate its own patter of residuals with which the experienced analyst will be familiar.

3.7 THE METHOD OF SEASONAL INDICES

The additive model has been considered at some length. The mulitplicative model, $Y = T \times S \times R$, is considered in less detail since the pattern of the calculations is easily deduced. Firstly, the trend is determined by the method of moving averages giving the same results as for the additive model. Then, for each quarter, we *divide* Y by the trend value T (giving $Y/T = S \times R$) which, by convention, we express as a percentage as shown in Exhibit 15.

EXHIBIT 15: Initial Calculations for the Method of Seasonal Indices

Year	*Quarter*	*Production* Y	*Trend* T	*Percentage of Trend* $\frac{Y}{T} \times \frac{100}{1}$
1970	1	78	–	–
	2	62	–	–
	3	56	67.500	82.963
	4	71	68.500	103.650
1971	1	84	69.375	121.081
	2	64	71.375	89.667
	3	61	73.750	82.712
	4	82	75.500	108.609
1972	1	92	77.125	120.261
	2	70	77.125	90.762
	3	63	78.500	80.255
	4	85	80.875	105.100
1973	1	100	83.375	119.940
	2	81	85.875	94.323
	3	72	–	–
	4	96	–	–

The **seasonal indices** themselves may now be obtained by averaging the Percentage of Trend values over the four quarters of the year, as in the additive model. Finally the residual component may be calculated by dividing each Percentage of Trend by the appropriate Seasonal Index.

3.8 MATHEMATICAL TREND EQUATIONS

We now return to methods for trend estimation. Unlike those discussed earlier these methods make assumptions about the mathematical form of the trend curve. The simplest "curve" is a straight line which we may represent as:

$$\hat{Y} = a + bX,$$

where in the context of time series X is the time variable and $\widehat{Y}$ is identified with the trend value, T. The calculation of the best fitting line has already been studied under regression analysis in CU14. For time series the calculations are greatly eased by coding the time scale (on the horizontal x-axis) to simple values around zero as shown in Exhibit 16.

EXHIBIT 16: Coding of the Time Scale

Year (X)	*Coded value of X*
1968	−3
1969	−2
1970	−1
1971	0
1972	+1
1973	+2
1974	+3

This simplifies the arithmetic considerably, but the equation of the trend line then needs interpretation for use by the layman. (What we are doing is simply to update the calendar – where the calendar year 0 denotes the birth of Christ – for our own requirements.) A warning should be given about testing the significance of the regression coefficients obtained from time series data: it is rarely the case that the Y-errors (deviations from the trend) are independent and normally distributed as is required for the standard tests (CU18).

A straight line will not usually suffice to represent many actual time series met with in business. As a result we may wish to fit a "higher polynomial", e.g. a quadratic

$$\widehat{Y} = a + bX + cX^2$$

or a cubic

$$\widehat{Y} = a + bX + cX^2 + dX^3, \text{etc.}$$

In practice it is rarely worthwhile going beyond polynomials involving X^6. Although this can be done the simplicity of the trend curve is then lost and the higher power terms (e.g. eX^4, fX^5, gX^6) dominate the equation for even quite small values of X. Standard computer programs are available to fit the more complex equations and determine at which stage the adding of extra polynomial term should stop. If the equations are being fitted with the aid of a desk calculator it is easier to use "Orthogonal Polynomials". These can be found in sets of statistical tables and some advanced texts together with instructions for their use.

Exponential Growth Equations. A number of mathematical models have been put forward as being suitable for representing growth. This work originated in the field of forecasting population growth in the United States. Similar models were found to work in bacteria growth, and the methods have been tried, with varying degrees of success, in forecasting the growth of new industries (fertilisers, electronics) and the sales of new products (instant coffee and other grocery items) from the time of their initial launch until they achieve a position in the market. By comparing the growth curves of previously successful and unsuccessful products it is hoped to spot an unsuccessful product early enough to withdraw it before too much capital is invested in its promotion.

The general form of these growth curves is that of a stylised S. A slow introductory period, followed by a period of rapid growth and finally a tailing-off period is typical of many growth situations as in Exhibit 17.

EXHIBIT 17: A Typical Growth Curve

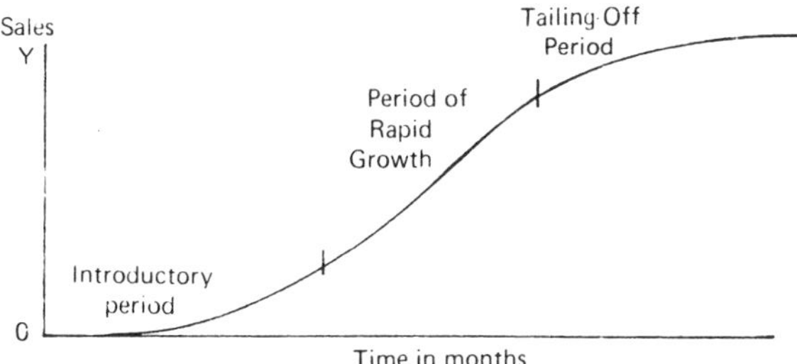

The simplest form of exponential trend is the logarithmic equation $Y = ab^x$ which may be converted to a linear form by taking logarithms of both sides, i.e. $\log Y = \log a + x \log b$. This is the semi-log equation which plots as a straight line on semi-log graph paper.

If the natural graph shows that the data appears to be approaching some upper limit then the "modified exponential" curve $Y = k + ab^x$ may be appropriate, where k is the upper limit or asymptote. The most well known growth curves, however, are:

(a) the *logistic* which covers a range of curves of the general form

$$Y = \frac{k}{1 + e^{f(x)}}$$

where $f(x)$ is a polynomial, $a + bX + cX^2 \ldots$

(b) The *Gompertz* curve which has the general form

$$Y = k \cdot a^{b^x}.$$

The above types of curves cannot be fitted easily to empirical data by hand, but computer programs are available for fitting purposes. A paper by F.R. Oliver "Methods of Estimating the Logistic Growth Function", Applied Statistics 1964, finds that the simple methods proposed by some textbooks are "highly unsatisfactory".

READING

A similar treatment, including evaluation of the Cyclical Component, may be found in:
K.A. Yeomans, *Introducing Statistics*, Statistics for the Social Scientist, Volume One. Penguin. Chapter 6.

REFERENCE

Further details of exponential smoothing are given in:
G.A. Coutie and others *Short-term Forecasting:* I.C.I. Monograph No. 2. Oliver & Boyd.

EXERCISES

1(E) Specify and describe the various movements which may be identified when analysing economic and business statistics over a long time span.

(ACCA June 1972)

2(U) *How many terms in a moving average trend estimate?*
Consider taking an idealised example of quarterly data built up by strict adherence to the additive model and, for simplicity, with no residual component,

$$Y = T + S.$$

Let T start at 1 in Year 1 Qtr. 1 and increase by one unit per quarter until it reaches 12 in Year 3 Qtr. 4 (i.e. a linear trend).
Let S for the four quarters be $S_1 = +1$, $S_2 = 0$, $S_3 = -1$, $S_4 = 0$.
Construct the actual series and graph it. *Estimate* the trend values

using a Moving Average with (a) $n = 3$, (b) $n = 4$ (i.e. 8-pt.), and (c) $n = 5$. Draw these three trend lines on your graph. What do you conclude?

3(E) Given a time series consisting of the monthly averages of time lost due to sickness in a particular industry over a long period of time, denoted t_{ij}, where the suffix i denotes the year and the suffix j denotes the month of the year, how would you examine the following characteristics:

(a) long-term trends
(b) cyclical variations
(c) seasonal variations?

On the basis of these analyses, how would you forecast the result for a given month in the year following the last year of the series?

(IPM June 1974)

4(U) *Additive or Multiplicative Model?*

Year	*1970*	*1971*	*1972*	*1973*	*1974*	*1975*	*1976*
Qtr							
1	0	8	16	24	32	40	48
2	1	5	9	13	17	21	25
3	1	3	5	7	9	11	N/A
4	3	7	11	15	19	23	N/A

Draw a graph of this artificial time series. Analyse it by the method of quarterly deviations (to save time, work to the nearest whole number) and plot the residuals on a separate graph.
What do you conclude about the suitability of the additive model for analysing this series. Would the multiplicative model be more appropriate? Try it and see.

5(E) The following table shows the numbers of days of incapacity (millions) of men and women in Great Britain during the period 1954–67.

Year	54	55	56	57	58	59	60	61	62	63	64	65	66	67
Men	187	188	180	203	199	191	200	205	213	212	218	235	229	251
Women	90	88	83	90	84	78	79	77	76	75	76	76	72	77

(a) Construct a four-year moving average of these two time series.

(b) Hence or otherwise calculate a forecast of the figure for 1968 for each series.
(c) If you are asked to interpret these data in terms of the sickness absence of individuals, what further information would you require and how would you analyse the combined data?
(d) Why do you think the two series show different trends?

(IPM June 1973)

6(E) *Unemployment in Great Britiain (thousands)*

	1969	*1970*	*1971*	*1972*	*1973*
February	591	523	568	572	718
May	624	578	606	602	591
August	721	755	859	926	571
November	925	861	885	770	N/A

Source: "Monthly Digest of Statistics.

Examine the above series for evidence of trends and seasonal movements using the method of quarterly deviations.
HND (Business Studies). Calculators were provided.

7(E) The table below shows the number of disputes taking place in a nationalised industry during the period 1966–71 in each successive three-month period.

Year	*Quarter*	*Number*	*Year*	*Quarter*	*Number*
1966	1	6	1969	1	9
	2	6		2	9
	3	9		3	17
	4	7		4	11
1967	1	5	1970	1	14
	2	6		2	13
	3	10		3	19
	4	6		4	15
1968	1	6	1971	1	15
	2	6		2	16
	3	12		3	21
	4	8		4	16

(a) Adjust this time series for seasonal variations.
(b) Apply the method of moving averages to estimate the long-term trend.
(c) Estimate the number of disputes for each of the four quarters of 1972.

(IPM November 1972)

SOLUTIONS TO EVEN NUMBERED EXERCISES

2.

Year	*Qtr*	*T*	*S*	*Y* (= *T* + *S*)	*3Q MT*	*3Q MA*	*4Q MT*	*8pt. MT*	*8pt. MA*	*5Q MT*	*5Q MA*
1	1	1	+1	2							
1	2	2	0	2	6	2.0					
							10				
1	3	3	−1	2	8	2.7		24	3.0	16	3.2
							14				
1	4	4	0	4	12	4.0		32	4.0	20	4.0
							18				
2	1	5	+1	6	16	5.3		40	5.0	24	4.8
							22				
2	2	6	0	6	18	6.0		48	6.0	30	6.0
							26				
2	3	7	−1	6	20	6.7		56	7.0	36	7.2
							30				
2	4	8	0	8	24	8.0		64	8.0	40	8.0
							34				
3	1	9	+1	10	28	9.3		72	9.0	44	8.8
							38				
3	2	10	0	10	30	10.0		80	10.0	50	10.0
							42				
3	3	11	−1	10	32	10.7					
3	4	12	0	12							

It can be seen that the 8-pt Moving Average accurately identifies the trend. By comparison the 3Q Moving Average *over-estimates* the trend value when the seasonal component is negative; under-estimates it when the seasonal component is positive. The 5Q Moving Average errs in the opposite sense.

4. The residuals for the additive model are shown below.

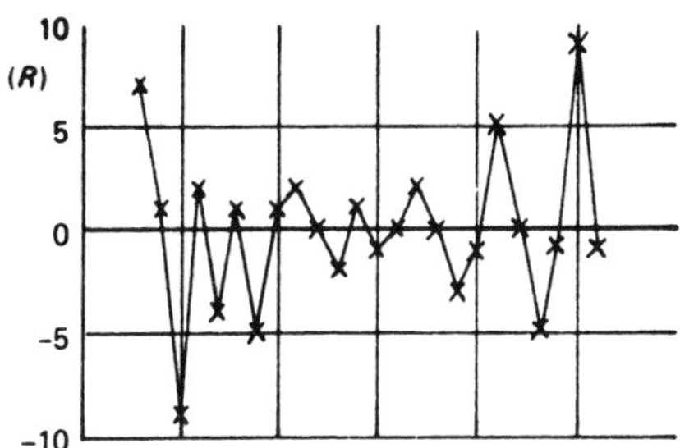

As can be seen the model is a poor fit at the beginning and end of the series.

6.

		Y	*4-Q MT*	*8-pt MT*	*8-pt MA (= T)*	*Deviations (Y − T)*	*Seasonal Variation (S)*	*Residual (Y − T) − S = R (to 1 d.p.)*
1969	F	591						
	M	624						
			2,861					
	A	721		5,654	706.750	14.250	99.484	−85.3
			2,783					
	N	925		5,540	692.500	232.500	140.046	92.5
			2,747					
1970	F	523		5,528	691.000	−168.000	−117.984	−50.0
			2,781					
	M	578		5,498	687.250	−109.250	−121.547	12.3
			2,717					
	A	755		5,479	684.875	70.125	99.484	−29.4
			2,762					
	N	861		5,552	694.000	167.000	140.046	27.0
			2,790					
1971	F	568		5,684	710.500	−142.500	−117.984	−24.5
			2,894					
	M	606		5,812	726.500	−120.500	−121.547	1.0
			2,918					
	N	859		5,840	730.000	129.000	99.484	29.5
			2,922					
	A	885		5,840	730.000	155.000	140.046	15.0
			2,918					
1972	F	572		5,903	737.875	−165.875	−117.984	−47.9
			2,985					
	M	602		5,855	731.875	−129.875	−121.547	−8.4
			2,870					
	A	926		5,886	735750	190.250	99.484	90.7
			3,016					
	N	770		6,021	757.625	12.375	140.046	−127.6
			3,005					
1973	F	718		5,655	706.875	11.125	−117.984	129.1
			2,650					
	M	591						
	A	571						
	N	−						

N.B. The residuals have been calculated in the form $(Y - T) - S$ rather than $(Y - S) - T$, as in the example in the text.

Table of Quarterly Deviations

	F	M	A	N
1969	−	−	14.250	232.500
1970	−168.000	−109.250	70.125	167.000
1971	−142.500	−120.500	129.000	155.000
1972	−165.875	−129.875	190.250	12.375
1973	11.125	−	−	−
Total	−465.250	−359.625	404.625	566.875
Average	−116.312	−119.875	101.156	141.718
Correction	−1.672	−1.672	−1.672	−1.672
Seasonal Var	−117.984	−121.547	99.484	140.046

N.B. Total of Averages = 6.687. Hence correction = 6.687/4 = 1.672.

SEMINAR UNIT 4 – STATISTICAL QUALITY CONTROL (SQC) CHARTS

4.1 INTRODUCTION

SQC charts are most commonly to be found associated with production machines in manufacturing industry, but they may, in fact be applied to any "process" where a monitoring of current performance is required. The essence of the technique is that a sample of "production" is taken from time to time and a measure of its quality plotted on a graph on which "control limits" have been drawn. (Glance at Exhibit 7.) A succession of sample results, plotted in this fashion, gives a visual impression of the current quality standard and enables sub-standard conditions to be identified and corrected. The control limits, which determine when action should be taken, are based upon some underlying probability distribution appropriate to the process. It goes without saying that such control charts have to be simple to operate as they are often used by unskilled persons under everyday working conditions.

The charts are either related to measured characteristics – height, weight, hardness, density etc. – or to attributes such as dents, scratches, wiring faults or missing parts. The main types of charts in operation are:

• Average and Range • Cusum (i.e. Cumulative Sum)	"Variables" Control Charts
• Number (or Percentage) of Defectives • Number of Defects (or Defects per Unit)	"Attributes" Control Charts

The traditional Average and Range chart is based on normal distribution theory, and is considered in some detail in this Seminar Unit as it illustrates most of the principles and practice of SQC charts in general. (Cusum charts are a more recent development and under some conditions are more sensitive to changes in a process than the older method, although the theory on which they are based is somewhat more complex.) The "attributes" control charts are based on the binomial and Poisson distributions.

The following technical jargon is often used in the context of SQC:

- *Chance Causes*. It is assumed that the quality of the process is following some probability distribution. When this is true it is said that only "chance causes" are operating and the process is "in control". This concept allows minor random fluctuations to occur (e.g. due to vibrations, temperature variations or slight changes in raw material quality) but under mass production conditions these minor fluctuations are not commercially important.

- *Assignable Cause*. If the quality characteristic departs from the assumed probability model then "something" has caused the change (e.g. a nut has loosened, a tube blocked). The function of the control chart is to detect when this happens. Typically the process would be switched off, the fault (assignable cause) corrected and production restarted.
- *Process Capability*. In order to determine the appropriate probability model (and hence the control limits) a study of the process is carried out under stable, routine working conditions i.e. when no assignable causes are present. This corresponds to a large random sample check of production, and from it we are able to estimate the population parameters of the probability distribution.

4.2 THE PROBABILITY BASIS FOR CONTROL LIMITS

In production decision making we have to decide, on sample evidence, either (a) to let the machine run on (so risking making defective products) or (b) to stop the machine for adjustment (so risking lost production if the process is in fact not faulty). As we have seen in the Course Units, sample results from the same population are liable to differ, and such variations can be described in probability terms.

The "measure" of quality employed in SQC is either a sample mean $\bar{X}$, the number of defectives r in a sample (or the percentage of defectives $P_{\%}$) or the number of defects m in a unit of production. Under the usual assumptions (discussed in CU's 11, 9 and 10) the sampling distribution of these statistics may be considered as normal, binomial and Poisson respectively.

EXHIBIT 1: Sampling Distributions

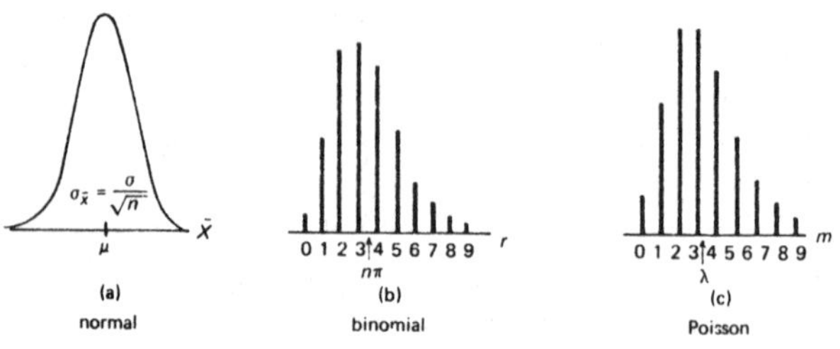

Referring to Exhibit 1, if the values of $n\pi$ and λ are greater than 10 then the binomial and Poisson distributions may be approximated by the normal distribution. Assuming that this criterion is satisfied we can make the following probability statements (where the constants 1.96 and 3.09 are taken from normal distribution tables):

(a) 95% of sample values will lie in the range given by:

$\mu \pm 1.96\frac{\sigma}{\sqrt{n}}$, for $\bar{X}$;

$n\pi \pm 1.96\sqrt{n\pi(1-\pi)}$, for r defectives in a sample of n;

(or $\pi_{\%} \pm 1.96\sqrt{\frac{\pi_{\%}(100-\pi_{\%})}{n}}$, for a percentage $P_{\%}$ of defectives)

$\lambda \pm 1.96\sqrt{\lambda}$, for the number of defects m.

(b) 99.8% of sample values will lie in the range given by:

$\mu \pm 3.09\frac{\sigma}{\sqrt{n}}$, for $\bar{X}$;

$n\pi \pm 3.09\sqrt{n\pi(1-\pi)}$, for r defectives in a sample of n;

(or $\pi_{\%} \pm 3.09\sqrt{\frac{\pi_{\%}(100-\pi_{\%})}{n}}$, for a percentage $P_{\%}$ of defectives)

$\lambda \pm 3.09\sqrt{\lambda}$, for the number of defects m.

In making these statements we have assumed that only chance causes are operating.

Because of the similarity of the three models under the stated conditions we will consider first the case of the sample mean $\bar{X}$, and return later to the other two.

4.3 THE AVERAGE AND RANGE CHART

In Exhibit 2 we show the two pairs of limits for the normal distribution.

These two sets of limits, based upon the 95% and 99.8% probabilities, form the basis for the Average Chart. It is not as useful to plot the sample averages in the vertical form of the above distribution: instead, the chart for sample averages is arranged horizontally, as in Exhibit 3, so that the sample results may be seen in their time sequence.

EXHIBIT 2: Control Limits Related to the Normal Distribution

EXHIBIT 3: The Control Chart for Sample Averages

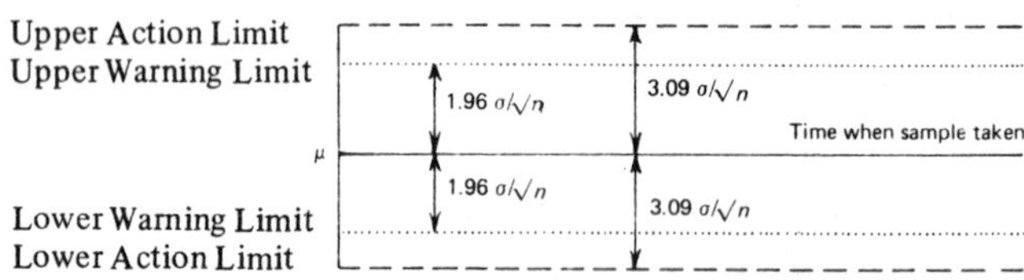

The assumption[(1)] being tested each time that a sample is taken from the process is that there has been no change in the population mean μ, i.e. only chance causes are operating. Under this assumption 95 out of every hundred sample means should, on average, lie *within* the Inner (Warning) Limits and 998 out of every 1,000 sample means should similarly lie *within* the Outer (Action) Limits. In Exhibit 4 this same chart is redrawn giving the probabilities of a sample mean falling *outside* these limits when only chance causes are operating.

So if the process mean has not changed there is a probability of 2.5% (1/40) that a sample mean will lie above the Upper Warning Limit, and the same probability that it will lie below the Lower Warning Limit (i.e. a probability of 5% or 1/20 that it will lie outside one or other of these limits). Now a process may be checked many times per day: if it is checked every 15 minutes of an 8 hour shift, then 96 sample checks will be made per three-shift day, and we would expect around five of these to fall outside a Warning Limit even when the process is "in control".

(1) Readers who have completed CU17 will recognise this statement as a *null hypothesis*. SQC can consequently be regarded as a sequence of significance tests over time.

EXHIBIT 4: Probability of Sample Means Falling Outside Control Limits

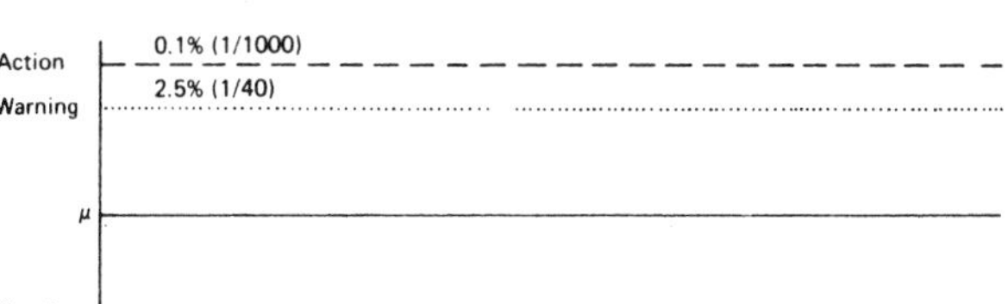

The term *Warning* Limit is used to indicate to the operator (or quality control staff) that *possibly* something has changed, i.e. an assignable cause of variation may have entered the system. Now suppose a particular sample mean plots outside a Warning Limit but within the Action Limit. The procedure is then to take a second sample immediately, and if this sample mean falls *within* the Warning Limits it can be assumed that the first sample was one of the expected 1 in 20 results that will fall outside the limits by chance. If, however, the second sample is also outside the Warning Limits then the probability of two successive random events of this type is

$$\frac{1}{20} \times \frac{1}{20} = \frac{1}{400}$$

(the multiplication rule for the probabilities of independent events is applicable). Thus if only chance causes are operating this is a very rare event indeed; and so such an event is taken as strong evidence that an assignable cause has entered the system. The process is then stopped or some other appropriate action taken to locate the cause of the trouble.

By similar reasoning, there is a 0.2% (1/500) probability that a sample average will fall outside either of the Action Limits. As the name implies immediate corrective action is taken without waiting for further evidence from another sample. A risk of 1 in 500 of making a wrong decision in correcting the process when there is no assignable cause is worth taking, since 499 times out of every 500 the decision to stop and correct the process will be the correct one to take.

In summary, the procedure for operating an Average Control Chart is as follows.

(a) Take a sample of n items; calculate and plot the average of the sample.
(b) If the sample average plots within the Inner (Warning) Limits allow the process to continue.
(c) If the sample average plots outside the Inner (Warning) Limits

take a second sample immediately, calculate and plot the new average, and:

- if the new average plots within the Inner Limits allow the process to continue
- if the new average also plots outside the Inner Limits take action to correct the process.

(d) If the sample average plots outside the Outer (Action) Limits take action immediately to correct the process.

An assumption that we have made in controlling changes in the population mean μ is that the population standard deviation has remained constant. Clearly some assignable causes could affect the variability of the process, while possibily not affecting the mean, such as a nut working loose, more variable raw materials or a less skilled operator. In order to control the variability of the process a second chart is kept. This second chart has exactly the same probability limits – Warning and Action, as for the averages chart – but they are based upon the sampling distribution of a statistic measuring variation. (In SQC it is more common to use the sample range W than the sample standard deviation s, for this purpose.) It is beyond the scope of this course to obtain the sampling distributions of the sample range and standard deviation – it is sufficient for our purposes to point out that for the sample sizes commonly employed in SQC these distributions are not symmetrical, but are positively skewed as shown in Exhibit 5.

EXHIBIT 5: Sampling Distribution and Probability Control Limits for the Sample Standard Deviation and Range

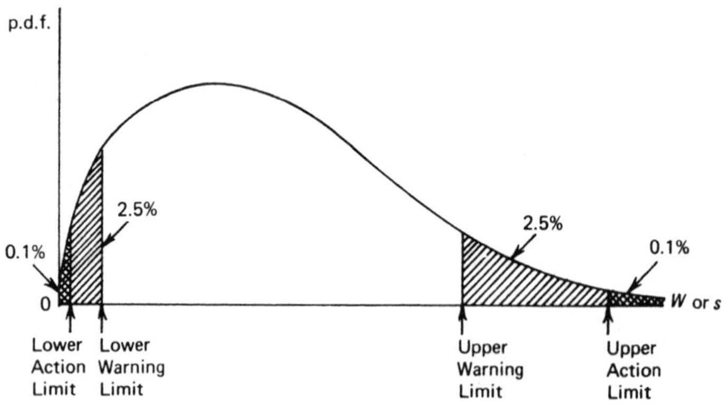

In practice the sample standard deviation s is only employed when the sample size, n, is large ($n > 12$), as the sample range is inefficient for such sample sizes. For smaller samples, the sample range is nearly always used since it is simpler to compute and is reasonably efficient. (For the concept of efficiency see CU1 and CU5.)

The essential point to grasp is that the probability of the sample statistic falling outside an Action or Warning limit (when the population parameter has not changed) is the same whether the statistic is the average, standard deviation or range. The computation of these limits has been simplified by the publication of tables for values $n \geqslant 2$ for both the standard deviation and the range. In the example below we consider in detail the construction of a specimen Average and Range chart, making use of values taken from published tables.

4.4 COMPUTING AN AVERAGE AND RANGE CHART

Example 1: The control procedure for the diameter of silver-copper wire is to take a sample of four measurements at fifteen minute intervals. The results of ten such samples, each of four measurements, are given below in Exhibit 6. We use these results to illustrate the construction of the charts.

EXHIBIT 6: The Diameter of Silver-Copper Wire

Sample No.	*Measurement No.* 1	2	3	4	*Sample Average ($\bar{X}$)*	*Sample Range (W)*
1	.020"	.020"	.022"	.018"	.020"	.004"
2	.021"	.022"	.020"	.021"	.021"	.002"
3	.020"	.021"	.019"	.020"	.020"	.002"
4	.018"	.018"	.020"	.016"	.018"	.004"
5	.019"	.020"	.018"	.019"	.019"	.002"
6	.019"	.019"	.022"	.019"	.020"	.003"
7	.018"	.019"	.020"	.023"	.020"	.005"
8	.022"	.022"	.022"	.022"	.022"	.000"
9	.020"	.024"	.022"	.022"	.022"	.004"
10	.018"	.016"	.018"	.020"	.018"	.002"

Step 1. For each sample compute the average ($\bar{X}$) and range (W) as shown.

Step 2. Compute the grand average,

$$\hat{\mu} = \frac{\Sigma \bar{X}}{10} = \frac{0.200}{10} = 0.020''$$

Step 3. Compute the average range,

$$\bar{W} = \frac{\sum W}{10} = \frac{0.028}{10} = 0.0028''$$

Step 4. From SQC tables (with sample size $n = 4$) we obtain the following factors:[1]

for the Average Chart

Warning Limits factor	$A'_{.025} = 0.476$
Action Limits factor	$A'_{.001} = 0.750$

for the Range Chart

Lower Action Limit	$D'_{.999} = 0.10$
Lower Warning Limit	$D'_{.975} = 0.29$
Upper Warning Limit	$D'_{.025} = 1.93$
Upper Action Limit	$D'_{.001} = 2.57$

Step 5. The limits for the Averages Charts are:

Warning Limits at $\hat{\mu} \pm A'_{.025}\ \bar{W}$

$= .020'' \pm 0.476 \times .0028$
$= .020'' \pm .0013''$
$= .0187''$ and $.0213''$

Action Limits at $\hat{\mu} \pm A'_{.001}\ \bar{W}$

$= .020'' \pm 0.750 \times .0028$
$= .020'' \pm .0021''$
$= .0179''$ and $.0221''$

Step 6. The Limits for the Range Chart are:

Lower Action	$D'_{.999}\ \bar{W} = 0.10 \times .0028 = .0003''$
Lower Warning	$D'_{.975}\ \bar{W} = 0.29 \times .0028 = .0008''$
Upper Warning	$D'_{.025}\ \bar{W} = 1.93 \times .0028 = .0054''$
Upper Action	$D'_{.001}\ \bar{W} = 2.57 \times .0028 = .0072''$

The control chart, complete with Action and Warning Limits and the sample results plotted, is shown in Exhibit 7, to which the following points apply:

---

[1] The factors for the charts are published in two forms: those for use with the standard deviation σ (denoted by A and D), and those for use with the average range $\bar{W}$ (denoted by A' and D'). Both forms will be found in *Statistical Tables*, J. Murdoch and J.A. Barnes (Macmillan).

- The Average Chart is plotted above, rather than alongside, the Range Chart. Plotting in this manner aids the interpretation of the results.
- The lower limits on the Range Chart have been omitted. This is common in practice since it would be unusual to adjust the process because it was working better (less variable) than is expected. These limits are retained when it is important to detect improvements, such as when carrying out machine development trials.
- Whilst the sample ranges are "in control" throughout, successive averages (samples 8, 9, 10) are outside the Warning Limits and some form of action should have been taken.(1)
- In Exhibit 7 the points have not been joined. Practice varies: some users regard the samples as being independent random samples and do not join the points; others like to emphasise the "time series" aspect of the charts and prefer to join them.

4.5 ATTRIBUTES CONTROL CHARTS

In Section 4.2 we saw how limits may be obtained when the normal distribution is a good approximation to the binomial and Poisson distributions. Below we give an example of each.

Example 2. A sample of 200 cans is taken from every two hours production and the number of cans with minor faults (small dents, scratches) recorded. The number of defective cans (in 200) recorded for 10 successive samples is as follows:

5, 18, 14, 4, 9, 10, 8, 12, 13, 17.

The average *number* of defectives per sample of 200 is

$$\frac{110}{10} = 11.0 \ (= n\hat{\pi}).$$

Hence the average *percentage* defective

$$\hat{\pi}_{\%} = \frac{11}{200} \times \frac{100}{1} = 5.5\%.$$

---

(1) Here we have used the same samples to calculate the control limits and to illustrate an "out of control" situation. This is strictly incorrect – the control limits, to have any meaning at all, must be obtained from a (large) sample where the process is "in control" throughout. Only then can they be used to detect (statistically) situations where assignable causes are present. In the example we have condensed two processes into one in the interest of brevity.

EXHIBIT 7: Average and Range Chart for Silver-Copper Wire

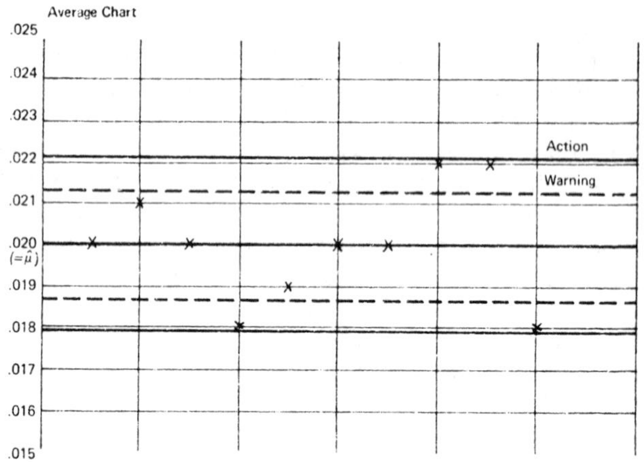

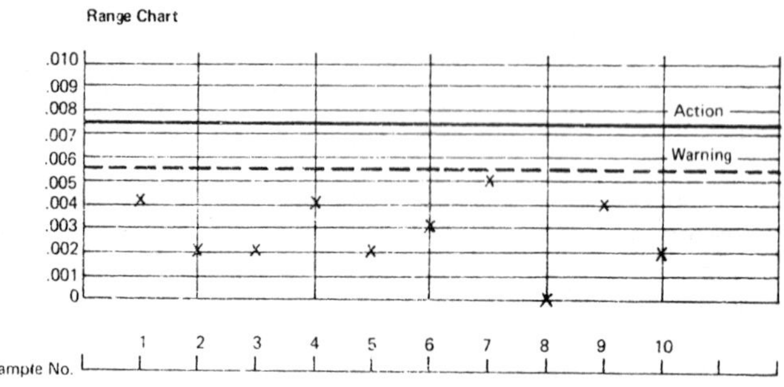

(The "hats" over π and $\pi_{\%}$ indicate that their values are estimated from the data.)

We may obtain a chart in one of two ways depending on whether we wish to plot the number of defective cans found in each sample or the percentage ($P_{\%}$). To obtain the limits we use SD's of $\sqrt{n\hat{\pi}(1-\hat{\pi})}$ and

$$\sqrt{\frac{\hat{\pi}_{\%}(100-\hat{\pi}_{\%})}{n}},$$

respectively for the two types of chart.

For the *number of defectives* d-chart we obtain the limits as

Warning Limits $11 \pm 1.96\sqrt{200(.055)(.945)}$,

using $n\hat{\pi} \pm 1.96\sqrt{n\hat{\pi}(1-\hat{\pi})}$.

$= 11 \pm 6.32$ (i.e. 4.68 and 17.32)

Action Limits $11 \pm 3.09\sqrt{200(.055)(.945)}$,

using $n\hat{\pi} \pm 3.09\sqrt{n\hat{\pi}(1-\hat{\pi})}$.

$= 11 \pm 9.92$ (i.e. 1.08 and 20.92)

For the *percentage of defectives* p-chart we obtain the limits as

Warning Limits $5.5 \pm 1.96\sqrt{\dfrac{5.5 \times 945}{200}}$

using $\hat{\pi}_{\%} \pm 1.96\sqrt{\dfrac{\hat{\pi}_{\%}(100-\hat{\pi}_{\%})}{n}}$

$= 5.5 \pm 3.16$ (i.e. 2.34 and 8.66)%

Action Limits $5.5 \pm 3.09\sqrt{\dfrac{5.5 \times 94.5}{200}}$

using $\hat{\pi}_{\%} \pm 3.09\sqrt{\dfrac{\hat{\pi}_{\%}(100-\hat{\pi}_{\%})}{n}}$

$= 5.5 \pm 4.96$ (i.e. 0.54 and 1 0.46)%

If the p-chart were being used then the sample results would be converted into percentage form, i.e. for the above 10 samples:

2.5%, 9%, 7%, 2%, 4.5%, 5%, 4%, 6%, 6.5%, 8.5%.

The d- and p-charts are shown in Exhibit 8. This illustrates, perhaps more clearly than the data, that the two charts are identical apart from a scale factor.

Example 3. In a mass-produced car assembly plant each car is given a final inspection for visual defects – blemishes, paintwork, tears. The number of defects on 12 successive cars is: 20, 22, 13, 14, 19, 23, 16, 10, 11, 22, 12, 10.

The average number of defects = 192/12 = 16.0. For this type of chart we use the Poisson probability distribution which has mean λ and standard deviation $\sqrt{\lambda}$.

Hence the limits for the *number of defects* c-charts are obtained as

Warning Limits $16 \pm 1.96\sqrt{16}$, using $\lambda \pm 1.96\sqrt{\lambda}$

$= 16 \pm 7.84$ (i.e. 8.16 and 23.84)

EXHIBIT 8: d- and p-Charts

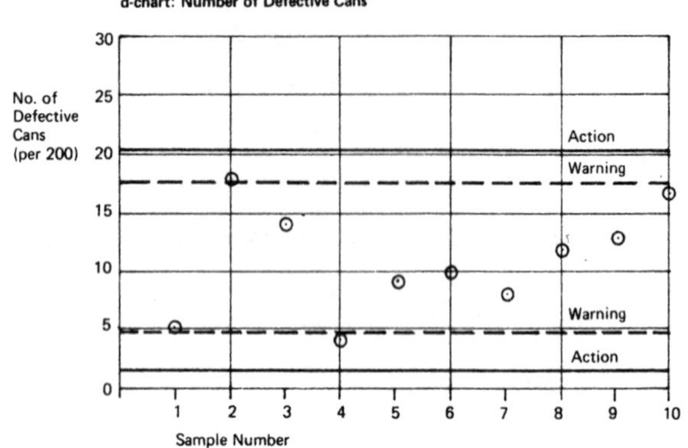

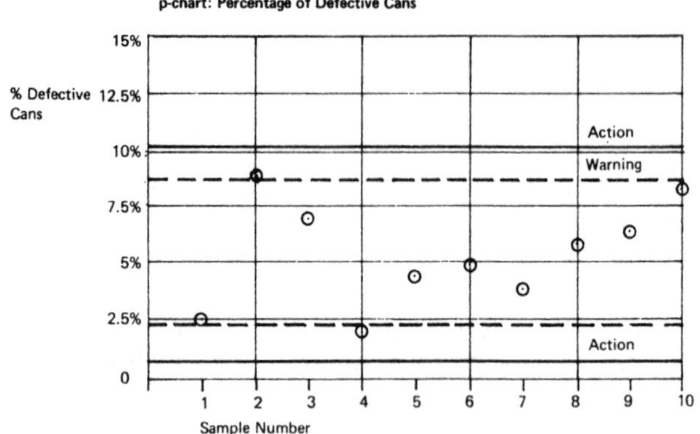

Action Limits $16 \pm 3.09\sqrt{16}$, using $\lambda \pm 3.09\sqrt{\lambda}$

$= 16 \pm 12.36$ (i.e. 3.64 and 28.36)

The chart is plotted in Exhibit 9.

A number of points arise out of Exhibits 8 and 9

- Management are accepting that minor faults do occur; they are not directly trying to eliminate them by means of the charts, but only to ensure that they are controlled to acceptable minimum – acceptable, that is, for the commercial price being paid. Unacceptable faults, e.g. missing wheelnuts, would be checked and controlled by a more rigorous procedure.

- As in the case of a Range chart the lower limits are often omitted for routine production charts.
- Since the Poisson distribution may be used to approximate to the binomial distribution we could have obtained the limits for the d-chart (number of defectives) by using the Poisson distribution – as in the c-charts.

EXHIBIT 9: c-Chart. Number of Visual Defects

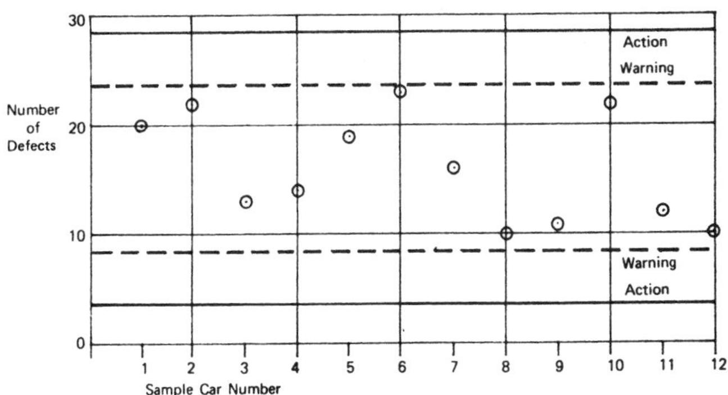

EXHIBIT 10: c-Chart. Number of Errors

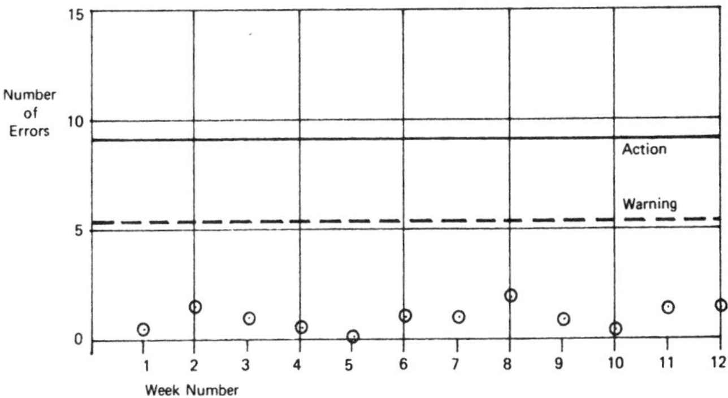

In the three attribute charts above we have been able to use the normal distribution factors – 1.96 and 3.09 – because the sample size and number of defects have been large enough for the approximation to be sufficiently accurate. What if this is not so?

If the sample size for the d-chart or the number of defects for the c-chart are so small that we are concerned about the validity of using the normal approximation for obtaining our limits then we may use instead tables of the cumulative distribution to obtain our limits. The method is illustrated for the c-charts using tables of the cumulative Poisson distribution.

Example 4. A Sales Department, concerned at the trouble caused by an incorrectly completed invoice, carries out a weekly quality control check of its sales invoices. It records the number of invoices found containing a clerical error. Over a period of 12 weeks the number of errors found were:

Week No.	1	2	3	4	5	6	7	8	9	10	11	12
Number of Errors (C)	1	3	2	1	0	2	2	4	2	1	3	3

Over a longer period it has been established that the number of errors found conformed to a Poisson distribution

The average number of errors $\bar{C} = \frac{\sum C}{12} = \frac{24}{12} = 2.0$

From tables of the cumulative Poisson distribution with $\lambda = 2.0$ we obtain the following:

Number of Mistakes (r)	*Probability of r or more Mistakes*	
0	1.0000	
1	.8647	
2	.5940	
3	.3233	
4	.1429	
5	.0527	
←		← .0250
6	.0166	
7	.0045	
8	.0011	
←		← .0010
9	.0002	

We may obtain the Upper Warning Limit by finding its associated probability 1/40 = .0250 in the cumulative probability column. As the Poisson distribution is a discrete distribution we will rarely find exactly the value .0250, but we note that the nearest values above and below .0250 are .0527 and .0166. These two values are associated with 5 or more and 6 or more mistakes respectively (as arrowed in the table). Hence the Upper Warning Limit is drawn on the control chart – Exhibit 10 – such that a value of 5 mistakes is "in control" while 6 mistakes is outside the Warning Limit. By similar reasoning the Upper Action Limit (1/1,000 or .001) is placed between 8 and 9 mistakes. In Exhibit 10 to all the points are "in control"; however, had a point been "out of control" then Management would have good reason to try to determine why the quality of the clerical work had deteriorated. Note that in this example the two lower control limits would have been set at zero.

4.6* SQC AND SPECIFICATION TOLERANCES

In our discussion of Average and Range Charts (Section 4.3) no mention was made of specification tolerances, the range of values (or length, weight, strength etc.) within which the product is deemed to function satisfactorily. The assumption made in Section 4.3 was that provided the process was kept in *statistical* control by means of the Average and Range charts the products would remain within tolerance. But clearly if a precision product (one with a narrow tolerance) is being manufactured by a process with high variability (even when statistically "in control") there would be no guarantee of this: and the effort spent on SQC would be largely wasted. Conversely if a product with a wide tolerance is produced by a process with low variability it would not matter if the process mean wandered out of control to some extent: to apply our previously discussed forms of control to such a process would involve needless interruptions of production.

The moral of this is that before applying SQC to a manufacturing process one needs to assess the process variability *relative to* the specification tolerance. To do this the Relative Precision Index is often used, given by:

$$\text{RPI} = \frac{\text{Specification Tolerance}}{\text{Average Range}},$$

where the specification tolerance is the *range of acceptable values*, e.g. for a specification of 1.00 ± 0.01 cm, the specification tolerance is 0.02 cm. By reference to standard tables[1] the RPI tells us whether the process is high, medium or low relative precision. For instance, for sample sizes of 4, a process has low relative precision if the RPI is less than 3, medium if the RPI lies between 3 and 4 and high if it exceeds 4.

---

[1]BS 2564: 1955, reproduced in *Statistical Tables*, J. Murdoch and J.A. Barnes (Macmillan).

Example 5. Referring to the control of silver-copper wire (Example 1, Section 4.4), suppose the specification is (0.020 ± 0.005)″. Then as the average range, from samples of 4, was found to be 0.0028″;

$$\text{RPI} = \frac{0.01}{0.0028} \simeq 3.6,$$

indicating *medium* relative precision.

For a process with *low* relative precision there is little value in employing SQC as rejects are inevitable. Normally the alternatives are:

- review the specification (i.e. increase the tolerance)
- change the manufacturing process to one with lower variability
- employ 100% inspection.

With *medium* relative precision SQC is effective and our discussion in Section 3 refers to this category. With *high* relative precision SQC is also effective but a different form of Average chart, with "modified" control limits, is employed; the Range chart is abandoned for this category (see Reference).

REFERENCES

1. E.L. Grant *Statistical Quality Control 3rd Ed.* McGraw-Hill.
2. A.G. Hopper *Basic Statistical Quality Control* McGraw-Hill.

EXERCISES

1(D) (a) Discuss the statistical theory underlying the "Average and Range" quality control chart.

(b) Describe how in principle you would proceed if you were asked to install such a chart on a process on which no data was available prior to the request.

(c) What procedures might you adopt if a process dimension appeared to be "in control" and yet a substantial proportion of production was outside the manufacturing tolerances specified for that dimension?

BA (Business Studies)

2(E) Random samples are taken each hour from the mass production of compression springs and the free length of these measured giving the following results. The engineering tolerances for the springs are 1.50 ± .03 cms.

Sample No.	*1*	*2*	*3*	*4*	*5*	*6*	*7*	*8*
	1.51	1.49	1.53	1.49	1.54	1.56	1.54	1.56
Length	1.53	1.50	1.51	1.50	1.51	1.52	1.54	1.50
in cms.	1.49	1.48	1.51	1.52	1.52	1.52	1.53	1.52
	1.51	1.49	1.49	1.53	1.51	1.52	1.51	1.50

(a) Construct an average and range chart, enter the control lines, and plot the data, and comment on the results obtained.

(b) Using the average range, estimate the standard deviation of the process, and the proportion of springs outside the tolerances.

HND (Business Studies)

SEMINAR UNIT 5 – SAMPLING METHODS FOR SOCIAL AND MARKET RESEARCH

5.1 INTRODUCTION

The fundamental method of sampling, the simple random sample, has already been illustrated and discussed in CU15. We now consider some additional methods of sampling that are used in marketing and social surveys. They are usually either more accurate or are more practicable for some other reason – principally that of cost or time. It will be seen that there are often a number of ways of sampling a particular population. One may distinguish a well-designed sample from a poorly-designed sample by the use of two basic criteria: the avoidance of bias in the selection procedure, and the smallest sampling error for a given cost (or some other measure of outlay of resources).

5.2 BIAS AND SAMPLING ERROR

The term *bias* is used loosely to refer to any influence which tends to make the results calculated from samples differ *consistently* from the "true" result; but it is also used technically to refer to the *amount* by which the *average* value of a sample statistic (from all possible samples selected using a given procedure) differs from the corresponding population parameter (Exhibit 1).

EXHIBIT 1

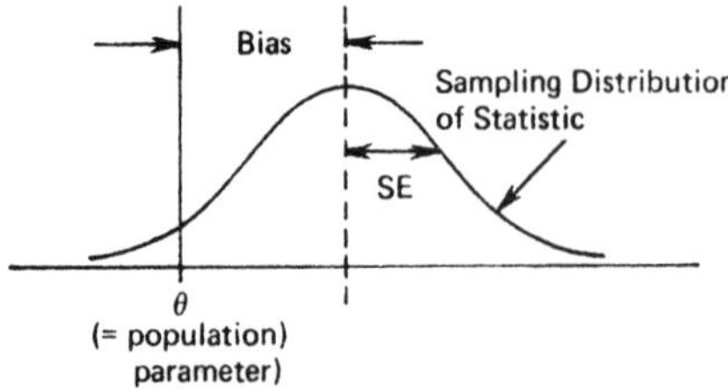

Here we consider only biases which result from the actual process of selecting the sample: other forms of bias, such as those resulting from the poor wording of questionnaires, have previously been covered in SU1a and SU1b. Such bias can arise in several ways:

- if the selection is carried out by a non-random process, such as in the quota sampling of respondents, discussed later. Generally the most serious faults in both marketing and industrial applications arise from selecting the "most convenient" sample – housewives in the shopping centre or the top box of components in a delivery.

- if the list from which the sample members are selected (Sampling Frame) does not include all the members of the population or if it contains duplicates. If a company, for example, is not listed in an industrial survey that is used as a sampling frame it obviously cannot be selected for the sample. If, on the other hand, the company happens to be listed twice then it has twice the chance of being chosen.
- if some of the selected members of the sample refuse to co-operate, or are difficult to contact or interview. This is the "non-response" problem, and a number of techniques are available (see Reference for a summary) which attempt to compensate for the missing members. In many surveys it is often the very rich and the very poor that fall into this category: similarly one finds young adults, working housewives and some other sections of the population under-represented in many surveys.

It is frequently overlooked that a very large sample will not, of itself, compensate for the effect of such bias. We may therefore compare a very large sample with a precise thermometer. The thermometer may give a precision of 0.01 °C, but if it has been miscalibrated it may read consistently 0.5 °C too high. Similarly a very large sample may nominally have a standard error of 1%, but if a substantial section of the sample refuses to co-operate then it may well be 5% in error.

It should be apparent from this illustration that the standard error of a sample statistic (CU15), the measure of the sampling error, is only a useful indicator of accuracy in conditions where the sampling method is free from bias. This comment should be borne in mind when the various sampling methods are compared because although it is often possible to evaluate the standard error of a given method, it is much more difficult to assess the extent of bias.

5.3 BASIC METHODS OF SAMPLING

The two basic methods of sampling, *random* and *systematic* sampling, require the use of a sampling frame. (Other methods which do not – *quota, random-route* and *cluster* sampling – are dealt with later.)

The Sampling Frame. In market research in the United Kingdom the most convenient sampling frame of the general public is the Electoral Register and we have already seen how it may be used for the selection of the sample (CU15). But a sampling frame does not have to take the form of an actual *list* of members of the population provided that the members are separately identifiable. (A map showing the location of public houses would suffice for a survey of draught beer sales.) Whatever the form of the sampling frame, it should be as comprehensive, complete and up-to-data as possible, in order to keep bias to a minimum; and it must enable the researcher to calculate the probability of selection (given a particular sampling method) of each member of the population. Such probabilities are required to "weight" the individual observations in the calculation of sample statistics and for the evaluation of standard errors. The techniques used to correct for unequal probability of selection are discussed later in Section 5.9.

Simple Random Sampling. this has previously been discussed and we may summarise the position by saying that each possible sample (i.e. combination of a given number of members) has the *same* probability of being chosen. This can be achieved by giving each member a number and using random number tables to identify the sample members. A simple random sample is rarely used for large scale surveys carried out in marketing operations because the difficulties of interviewing widely scattered respondents may mean that the fieldwork is too expensive. The method *could* be used in test town operations (selecting a sample of shops or households in a small area) or for sampling office records (sales invoices etc.), but a *stratified* sample (Section 5.4) would usually be more efficient for both. More appropriate uses of a simple random sample are in quality control work, such as when selecting individual items for testing from a consignment. Despite such narrow areas of direct application, simple random sampling may be used as a component part of an overall sample design as, for instance, in stratified random sampling.

Systematic Sampling. This is a method whereby every *k*th member of a sampling frame is chosen for inclusion in the sample. Strictly, the selection process should start from some randomly chosen member. The method is often more convenient than simple random sampling (consider selecting every 10th card from a card index, for example), and provided that the sampling frame itself can be considered to be randomly ordered, the method approximates to simple random sampling. Where the sampling frame is deliberately ordered (e.g. companies arranged by turnover, or by number of employees) a *more* representative sample may well be drawn, resulting in smaller standard errors. But in contrast, if there is any periodicity in the sampling frame (as, for example, where every fifth house is a corner house) the sample could be very unrepresentative. As with simple random sampling, the method is rarely used as the sole means by which the sample is drawn in large scale surveys, but it is frequently used *at some stage* of a total sampling process. An application of this type is "sampling with probability proportional to size", Section 5.6.

5.4 STRATIFIED RANDOM SAMPLING

Before discussing this method in general let us take one example of its use. In a marketing survey, concerned with the sales of razor blades, the retail grocers in a test town were subdivided into "multiples", co-operatives, and large, medium and small independent shops. A simple random sample of shops was then selected from *each of these categories* (strata), the mean sales of razor blades was calculated for each sample, and the results finally combined.

The essential difference between this method of sampling and a simple random sample covering all shops is that we are free to choose the individual sample sizes within each stratum. The strata themselves may be constructed on the basis of any convenient classifying variable (in our case, shop type), and provided that this variable is related to the item under study (sales of razor blades) the procedure outlined will lead to more accurate results than a simple random sample of the same size. It is thus not surprising that all well designed surveys carried out for marketing purposes involve some element of stratification by geographic area, town size, social class of respondent or other relevant criteria.

To put it formally, stratified random sampling consists of the following steps:

(a) The entire population of sampling units is divided into mutually exclusive sub-populations called strata.
(b) Within each stratum a separate random sample is selected from all members of the population in that stratum.
(c) From the sample obtained in each stratum, a separate stratum mean (or other statistic) is calculated.
(d) The stratum statistics from (c) are properly weighted to form a combined estimate for the entire population.
(e) The variance of the statistics (i.e. the squares of their standard errors) are also calculated separately within each stratum and then properly weighted and added to form a combined estimate for the population.

General Theory. Assume that the population is made up of N sampling units which are divided into h strata:

$$N = N_1 + N_2 + N_3 + \ldots + N_i + \ldots + N_h,$$

where N_i represents the number of sampling units in the ith stratum. Similarly let us *choose* to divide up a sample of total size n among the strata so that,

$$n = n_1 + n_2 + n_3 + \ldots + n_i + \ldots + n_h.$$

The *sampling fractions* for each stratum are then

$$\frac{n_1}{N_1}, \frac{n_2}{N_2}, \frac{n_3}{N_3}, \ldots, \frac{n_i}{N_i}, \ldots, \frac{n_h}{N_h}.$$

We can obtain the mean and variance of the ith stratum as:

$$\bar{Y}_i = \frac{1}{n_i} \sum^{n_i} Y \tag{1}$$

(for simplification we omit the within-stratum suffix on Y) and[1]

$$S_i^2 = \frac{1}{n_i - 1} \sum^{n_i} (Y - \bar{Y}_i)^2. \tag{2}$$

The *overall* sample mean and the variance of this overall mean are obtained as weighted averages of the individual stratum results, the weights reflecting the relative size of each stratum in the population. The following results assume simple random sampling *within* each stratum for unbiased estimates. The overall sample mean,

$$\bar{\bar{Y}} = \frac{\sum^{h} N_i \bar{Y}_i}{\sum N_i} = \frac{\sum^{h} N_i \bar{Y}_i}{N} = \sum^{h} \left(\frac{N_i}{N} \right) \bar{Y}_i \tag{3}$$

and its variance

$$S_{\bar{\bar{Y}}}^2 = \sum^{h} \left(1 - \frac{n_i}{N_i}\right) \left(\frac{N_i}{N}\right)^2 \frac{S_i^2}{n_i} \tag{4}$$

This variance formula can be seen to consist of three parts:

---

[1] It is conventional in this work to denote unbiased estimates of variance by upper case S's. So whereas for sample size m

$$s^2 = \frac{1}{m} \sum (X_i - \bar{X})^2, \; S^2 = \hat{\sigma}^2 = \frac{1}{m-1} \sum (X_i - \bar{X})^2.$$

(i) $\left(1 - \frac{n_i}{N_i}\right)$, the finite population correction (CU15). This term may be omitted when $n_i < N_i/20$ without seriously affecting the final value of $S^2_{\bar{\bar{Y}}}$.

(ii) $\left(\frac{N_i}{N}\right)^2$, which arises from the general result (CU12) that, for a constant k and random variable X, if $Y = kX$ then $\text{Var}(Y) = k^2 \text{Var}(X)$.

(iii) $\frac{S_i^2}{n_i}$, the variance (standard error squared) of the sample mean within stratum i.

Representative Sampling. An important special case of stratified random sampling occurs when we choose to make the sampling fractions the *same* for all strata, i.e.

$$\frac{n_1}{N_1} = \frac{n_2}{N_2} \ldots = \frac{n_i}{N_i} \ldots = \frac{n_h}{N_h} \left(= \frac{n}{N}\right), \tag{5}$$

which may be achieved by choosing n_i as $n(N_i/N)$.

This is referred to as *proportional* or (more commonly) *representative* stratified sampling. So if 15% of all grocers shops are co-operatives, using a representative stratified sample 15% of our sample of shops will also be co-operatives. Representative samples have the advantage of somewhat simpler formulae than those for the general case given in (3) and (4), as we now show.

For a representative sample the overall mean can be expressed as

$$\bar{\bar{Y}} = \frac{1}{n} \sum^{n} Y; \tag{6}$$

that is, a simple mean of the *whole* sample. Thus a representative sample is said to be "self-weighting" – we may estimate the population mean (by $\bar{\bar{Y}}$) from the whole sample without having to calculate the individual strata means if these are not required for other purposes. The variance of $\bar{\bar{Y}}$ simplifies to

$$S^2_{\bar{\bar{Y}}} = \frac{1}{Nn} \sum^{h} (N_i - n_i) S_i^2. \tag{7}$$

As in the general case a further simplification can be effected if each n_i is small in comparison with N_i by approximating the $(N_i - n_i)$ term by N_i. (Proofs of results (6) and (7) from the general results are left to the reader as an exercise.)

Representative samples are commonly used in surveys of the human population (where they are highly stratified) because of the ease of subsequent calculations. However, under some conditions disproportionate samples are deliberately employed when either (a) worthwhile gains in precision result from doing so, or (b) it is desired to subject a particular stratum to special analysis.

Disproportionate Sampling–Optimal Allocation. It can be shown that for a given cost of sampling the maximum precision in the overall sample mean is obtained when the strata samples sizes are chosen according to the following formula:

$$n_i = n \cdot \frac{N_i S_i / \sqrt{c_i}}{\sum (N_i S_i / \sqrt{c_i})}, \qquad (8)$$

where for the ith stratum N_i is the stratum size, S_i the stratum standard deviation and c_i the unit cost of sampling (i.e. collecting the required information from the sample member). The following observations apply to this important result:

- If the unit cost of sampling (c_i) is roughly equal for all strata we may approximate (8) by

  $$n_i = n \cdot \frac{N_i S_i}{\sum N_i S_i}.$$

  But whatever the costs, this formula ensures maximum precision for a given overall sample size n.
- It is not worth using this optimum allocation method unless $S_i/\sqrt{c_i}$ varies substantially among the strata, otherwise the increase in accuracy is offset by the extra calculation involved in analysing weighted samples (in comparison with a representative sample).
- It can be shown that small departures from the optimum strata sample sizes have little effect on the overall accuracy; consequently rough estimates of S_i and c_i are acceptable.
- Most of the maximum achievable gain in precision may be obtained with only a small number of strata, provided that the strata vary considerably with respect to $S_i/\sqrt{c_i}$.

An Example of Stratified Sampling. In an industrial survey a company classified its customers into 4 strata according to the previous years' sales to each. The number of firms and the standard deviation of sales (in units of £100) in each stratum are given in Exhibit 2.

EXHIBIT 2: Some Details of Sales to a Company's Customers

Size of Account	*Number of Firms* N_i	*Standard Deviation of Sales* S_i
Small	2,000	10
Medium	1,200	20
Large	600	80
Very Large	200	400
	N = 4,000	

The company now wishes to interview buyers at 100 of its customers to assist in a sales forecasting exercise. Assuming the cost of interviewing buyers does not vary significantly between the strata how should the sample of 100 be allocated among the 4 strata? Exhibit 3 shows the calculations involved for both representative and optimum sampling.

EXHIBIT 3: Calculation of Strata Sample Sizes

(a) *Representative (Proportional) Sampling*

Size	N_i	(N_i/N)	$n_i = n(N_i/N)$
Small	2,000	.50	50
Medium	1,200	.30	30
Large	600	.15	15
Very Large	200	.05	5
	N = 4,000	1.00	100

(b) *Optimum Sampling*

Size	N_i	S_i	N_iS_i	$\frac{N_iS_i}{\Sigma N_iS_i}$	$n_i = n \cdot \frac{N_iS_i}{\Sigma N_iS_i}$
Small	2.000	10	20,000	.116	12
Medium	1,200	20	24,000	.140	14
Large	600	80	48,000	.279	28
Very Large	200	400	80,000	.465	46
	4,000		172,000 (= ΣN_iS_i)	1.000	100

It can be seen that the optimum sample selects many more customers with large and very large accounts (where the variation in sales is higher) than does the representative sample. It is this emphasis on using a higher sampling fraction for the more variable strata that leads to substantial increases in accuracy of the optimum allocation method for institutional surveys of industrial firms, retail shops, wholesalers, hospitals, schools, etc. Indeed it is often the case in such work that the most important stratum is sampled 100% (i.e. $n_i = N_i$) and the other strata less intensively.

By way of comparison, the standard error of the overall sample mean is 9.4 using the representative sample and 4.0 with the optimum sample. (A simple random sample of the same size would have a standard error of about 30.) It is clear that considerable gains in precision are to be had by using stratified samples, and in particular by the use of the optimum allocation method.

5.5 MULTI-STAGE SAMPLING

Nearly all surveys of the human population carried out in areas larger than a test town are by means of a multi-stage sample. It works like this: initially a number of *first-stage* sampling units are selected (e.g. towns); then from each, a number of smaller localities (e.g. wards), *second-stage* units, are chosen; and from these a number of *final-stage* units (e.g. households, shops or individuals) are drawn. The advantages of this method are obvious. The researcher only requires, for example, a list of households in selected wards from selected towns. The fieldwork costs are also reduced (in comparison with other methods) as the final-stage units are grouped together and this results in a saving of time and expense in travelling between interviews.

The multi-stage survey will tend to result in higher sampling errors than a simple random or stratified random sample of the same size, as people who live in a particular area are often of a similar social class and their answers to some questions tend to be the same (and rather different from those who live in other areas which may not have been sampled). In order to reduce the standard error of the overall statistics from the survey it is common practice to employ stratification techniques in the sample design. Thus to counteract the similarity in response of groups of final-stage units, the first-stage units are chosen to be as different as possible. Often first-stage units are stratified both by geographical region and by such variables as political disposition or industrialisation (Section 5.8 gives some specific examples). Indeed, the decision on which *type* of first-stage unit to use is often dictated by the availability of relevant stratifying information. Now although stratification tends to reduce sampling errors it rarely overcomes the increase in sampling error due to multi-stage selection, with the result that the standard error is of the order of $1\frac{1}{2}$ times as large as a simple random sample of the same size – but this ratio may vary considerably according to the topic of the survey.

In some American texts reference is made to *Area Sampling*, in which the sampling units are land areas. The technique is essentially that of a multi-stage sample and the difference arises from the information available to the researcher in the construction of a sampling frame. Commonly the final-stage units are chosen by *cluster* sampling (Section 5.7).

In the next two sections we consider some special sampling techniques which are often employed *within* the context of a multi-stage sample.

5.6 SAMPLING WITH PROBABILITY PROPORTIONAL TO SIZE

This is a method of selecting first- or second-stage units in a multi-stage sample in such a way that the probability of selection is proportional to the size of the human population (or more usually, the number of the electorate). Thus a large town with a population of 120,000 would stand ten times the chance of being selected as a smaller town of 12,000 people. Correspondingly an individual person within the large town stands only one tenth the chance of being selected, so that in fact individuals in both towns actually stand an equal chance of being selected at the beginning of the sampling operations. The selection is carried out using the cumulative total of electorate as in the example shown (Exhibit 4).

Example. To select four constituencies using a systematic sample, with probability proportional to size.

EXHIBIT 4: Sampling Frame for Selection of 4 Constituencies

Constituency	*Electorate*	*Cumulative Electorate*
Bournemouth East	63,750	63,750
Bournemouth West	70,238	133,988
Gosport & Fareham	82,053	216,041
Poole	66,000	282,041
Portsmouth Langstone	91,587	373,628
Portsmouth South	53,915	427,543
Oxford	67,014	494,557
Reading	59,371	553,928
Portsmouth West	49,517	603,445
Southampton Test	66,572	670,017
Eton & Slough	56,725	726,742
Southampton Itchen	72,170	798,912

To obtain a systematic sample of 4 constituencies the grand total of electorate is divided by four: 798,912/4 = 199,728. A random number is then selected between 1 and 199,728. Suppose this turns out to be 068,234; when this is compared with the cumulative scale it selects Bournemouth West as the first member of the sample. The number 199,728 is added to 68,234 three times to give the numbers 267,962 – 467,690 – 667,418. These identify the other three constituencies as Poole, Oxford and Southampton Test respectively.

The advantage of this method of sampling is that self-weighted samples are drawn, so that corrections for unequal probabilities of selection are not required. In the particular example chosen the constituencies have been listed in ascending order of percentage Labour vote at a General Election. Thus this systematic sample is in fact a stratified sample by this measure of the social class of the community.

5.7 FINAL STAGE SAMPLING METHODS

The final stage units of a multi-stage sample of individuals or households may be selected by the use of a simple random or systematic sample, using the Electoral Register as the sampling frame. Other methods though are sometimes used in order to reduce costs, or where a sampling frame is unreliable (e.g. in new towns) or does not exist (e.g. for pet shops).

Quota Sampling. This is a commonly used sampling method in market research, as it leads to a saving in fieldwork costs compared with random sampling methods. Instead of being given the name and addresses of the required members of the sample, the interviewer is told to interview a certain number of people with specific characteristics (in an area which has been randomly chosen) such as those shown in Exhibit 5:

EXHIBIT 5: Requirements List for a Quota Sample

	Men				*Women*			
Social Class	*AB*	*C1*	*C2*	*DE*	*AB*	*C1*	*C2*	*DE*
Age Under 25	0	1	1	1	0	1	1	1
25–45	1	1	1	0	0	1	1	1
45–64	1	1	1	2	1	1	1	0

The quotas are worked out so that the overall sample will reflect accurately the known population characteristics, i.e. a quota sample is a non-random but representative stratified sample. However the method is biased as every member of the population does not stand a known chance of being selected; and unless an extra control is used, in addition to those specified in Exhibit 5, too many men in the distributive trades tend to be interviewed and too few working housewives. As the laws of chance are not allowed to operate, valid estimates of the sampling error cannot be made, nor tests of significance carried out. But in fairness to this method, it is only reasonable to point out that even simple random samples are biased owing to the previously mentioned non-response problem.

Random Route Sampling. This is a "quasi" random method used by many practitioners for sampling households or retail stores in urban areas. Firstly an address is selected at random from the Electoral Register. This address is used as a starting point by the interviewer who is then given a set of instructions to follow to identify further addresses without needing a list from the Electoral Register. These instructions take the form of telling the interviewer to proceed down the street taking alternately left and right hand turns at road junctions and calling at every kth address *en route*. Special instructions are needed for blocks of flats and other multiple dwellings, or where areas of parkland or industrial estates are encountered.

The justification for this method is that it is thought that much of the bias in the quota survey method is eliminated when the interviewer is not allowed to choose the respondent and the instructions themselves determine the address at which the interviewer must call. In terms of cost per interview the random route method tends to fall between the quota sampling and the fully random methods. While the office routine is simplified as compared with fully random sampling, the method of random route is more open to abuse by the interviewer in the field and it is difficult to check that the instructions have always been fully adhered to.

Cluster Sampling. This is a method where, for example, *all* housewives in a block of flats are interviewed or *all* chemists shops in a shopping street visited. Its rationale is essentially economic: where the major component of cost involved in sampling concerns the time involved in travelling between locations, it is sensible to make the best use of time when the interviewer has arrived. As with other methods discussed here the starting location is chosen randomly.

Cluster sampling is little used in the U.K. for marketing surveys (although it *is* used in Statistical Quality Control, where the *first* ten items, say, from a production line are subjected to tests). More use of the technique is made in the U.S. where the cluster is a "block" in a town.

Its main disadvantage is the high sampling errors which result from its use, although these may be reduced (with a corresponding increase in cost) by using a large number of small clusters rather than a small number of large clusters.

5.8 DESIGNING A MULTI-STAGE NATIONAL SAMPLE OF GREAT BRITAIN

The major task in designing a national sample of individuals or households is to compile a sampling frame of *first-stage* sampling units (of roughly town-size), stratified in an appropriate manner for the survey and from which the sampling process can start. The subsequent processes of selecting the first-stage, second-stage and final-stage units make use of the techniques we have already discussed in Sections 5.6 and 5.7.

The whole country is first divided up into geographical areas which, for market research purposes, are either

Company sales divisions, or
Registrar General's Standard Regions, or
I.T.V. regions.

Frequently, because of the high cost of interviewing in sparsely populated areas, the north of Scotland and central Wales are omitted. This stratification divides the country up into 6 to 10 more or less homogeneous regions.

The next step is to divide up each of these into the *first-stage* units, which are normally either Local Authority Areas (L.A.A.'s) or Parliamentary Constituencies. The former have the advantage that more stratifying information is available since government statistics (e.g. Censuses of Distribution and Population) are published in this way. Constituencies are sometimes preferred since they are more nearly equal in population (electorate). But the choice between these two sampling units will often depend on the subject of the enquiry. Thus an enquiry into soaps may use L.A.A.'s since data is available to stratify these by water hardness, while for an election opinion poll the constituency is more relevant.

If L.A.A.'s are used these are usually stratified within region by such features as industrialisation index, the percentage of houses over £200 rateable value and/or some other factor relevant to a particular survey such as water hardness, mentioned earlier.

If constituencies are employed then the strata often used are:

(a) conurbations,
(b) borough (mainly urban),
(c) county (mainly rural).

There are fewer factors available for further stratification in this case – usually one resorts to the percentage Labour vote at the last general election.

Whether constituencies or L.A.A.'s are used the first two stages of sampling are often carried out with probability proportional to size as illustrated in Section 5.6. The second-stage selection is into wards or polling districts depending on the degree of clustering required. Individuals or households may then be selected by simple random or systematic sampling, or by using one of the special techniques discussed in Section 5.7.

Typically, a national sample of 2,000 individuals would have about 100 first-stage units and 2 second-stage units with 10 interviews in each. From an accuracy (sampling error) point of view a large number of first-stage units is to be preferred. From a field cost viewpoint few first-stage units are desirable, and in practice these two requirements are compromised.

5.9 WEIGHTING FOR UNEQUAL PROBABILITY OF SELECTION

Problems sometimes arise when the sampling frame does not meet exactly the sampling requirements. For instance, the Electoral Register lists individual electors (those who will be 18 years or older by the following October), but the sample may require that only the head of the household, or the housewife, be interviewed. If an elector were selected and then the housewife interviewed, the resulting sample would over-represent large (adult) family households – a household of 6 adults would thus have three times the probability of being selected as a 2 adult household. This over representation may be corrected by either rejecting 2 out of every 3 six elector households (e.g by accepting for the sample only those households where the elector selected is the first elector listed for that household) or by applying weighting factors of 1/6 and 1/2 respectively to the interviews at the analysis stage. (See also Exercise 2(ii) in CU15.)

The reverse problem may apply when the unit selected is the household (e.g. when using the random route method of Section 7) but the sampling unit is the individual. Once an address has been identified a list of all the members of the household qualifying for interview could be made, and one of these persons selected at random. But it is easier to apply the "birthday" rule, i.e. the person whose birthday is nearest the day of the interview being chosen. It is not usual to make more than one interview per household and so if one person is selected from a household of five adults in this manner, a weight of 5 is given to the interview in the analysis.

REFERENCE

C.A. Moser and G. Kalton, *Survey Methods in Social Investigation*, Heinemann.

EXERCISES

1(E) The following terms are commonly used in sample surveys:

(a) simple random sampling
(b) stratified sampling
(c) systematic sampling
(d) quota sampling.

For each term, define the corresponding sampling method, comment on its advantages from both a practical and a theoretical viewpoint and give an example of a real-life situation in which the method might be used.

(IPM Nov 1972)

2(E) (a) Describe the following methods of sampling

(i) simple random sampling
(ii) stratified sampling
(iii) cluster sampling.

(b) If you were asked to estimate the cigarette consumption of a group of 1,000 adults living in a particular area by sampling, how would you proceed to select your sample.

(ACCA Dec 1972)

3(E) A large retail organisation Head Office receives several thousands of invoices every week. The invoices vary in value from a few £'s to several hundreds of £'s each. The computer invoicing system provides monthly totals. However, management wish to install a sampling scheme whereby weekly estimates may be obtained on random samples of say 250 invoices.

There is ample previous data to derive any estimates of numbers of invoices, measures of average value and variability etc.
Discuss the various types of sampling method that could be used. Give the advantages and disadvantages of using each method indicating how the invoices could be selected.

BA (Accounting and Finance)

4(E) Define and evaluate the sampling methods you would use to select a sample in the following surveys:

(a) ascertain urgently the likely readership of a proposed weekly magazine devoted mainly to television and radio programme details, with articles on the programmes and artists;
(b) to provide information concerning the present health of former employees of a particular industry who have been incapacitated by a common disease. It has been established that 5,000 persons were disabled in the period being considered and that a sample of 250 persons is sufficient;
(c) to assist a town council of a popular holiday resort, in connection with a proposed advertising campaign, with information of accommodation to be available during the coming season for visitors and tourists.

(ICMA 1972)

APPENDIX 1

Standard Data Sheets

SERIES A.

55	82	83	109	78	87	95	94	85	67
80	109	83	89	91	104	90	103	67	52
107	78	86	19	72	66	92	99	60	75
88	112	97	88	49	62	70	66	88	62
72	85	81	78	77	41	105	92	94	74
78	75	87	83	71	99	56	69	78	60
119	39	104	86	67	79	98	102	82	91
46	120	73	62	68	86	70	55	112	83
62	74	99	100	86	67	61	97	77	59
65	51	99	53	105	95	107	46	90	71

SERIES B.

169	110	92	142	204	95	93	79	115	102
261	130	144	443	141	129	130	93	65	67
127	103	120	144	27	111	114	106	49	133
47	34	37	149	120	103	239	69	68	161
216	49	125	189	46	102	79	160	85	139
55	136	114	142	113	116	191	193	23	69
101	48	277	129	108	75	122	252	77	134
86	48	57	186	135	180	89	190	161	124
217	34	83	71	128	158	54	85	102	118
43	124	78	35	63	46	80	105	119	77

APPENDIX 1 (continued)

SERIES C.

07 51 34 87 92 47 31 48 36 60 68 90 70 53 36 82 57 99 15 82
86 59 36 85 01 56 63 89 98 00 82 83 93 51 48 56 54 10 72 32
83 73 52 25 99 97 97 78 12 48 36 83 89 95 60 32 41 06 76 14
08 59 52 18 26 54 65 50 82 74 87 99 01 70 33 56 25 80 53 84
41 27 32 71 49 44 29 36 94 58 16 82 86 39 62 15 86 43 54 31
89 22 10 23 62 65 78 77 47 33 51 27 23 02 13 92 44 13 96 51
04 00 59 98 18 63 91 82 90 32 94 01 24 23 63 01 26 11 06 50
48 54 63 80 66 50 85 67 50 45 40 64 52 28 41 53 25 44 41 25
51 71 98 44 01 59 22 60 13 14 54 58 14 03 98 49 98 86 55 79
28 73 37 24 89 00 78 52 58 43 24 61 34 97 97 85 56 78 44 71

00 47 37 59 08 56 23 81 22 42 72 63 17 63 14 47 25 20 63 47
86 13 15 37 89 81 38 30 78 68 89 13 29 61 82 07 00 98 64 32
33 84 97 83 59 04 40 20 35 86 03 17 68 86 63 08 01 82 25 46
61 87 04 16 57 07 46 80 86 12 98 08 39 73 49 20 77 54 50 91
43 89 86 59 23 25 07 88 61 29 78 49 19 76 53 91 50 08 07 86
29 93 93 91 23 04 54 84 59 85 60 95 20 66 41 28 72 64 64 73
38 50 58 55 55 14 38 85 50 77 18 65 79 48 87 67 83 17 08 19
31 82 43 84 31 67 12 52 55 11 72 04 41 15 62 53 27 98 22 68
91 43 00 37 67 13 56 11 55 97 06 75 09 25 52 02 39 13 87 53
38 63 56 89 76 25 49 89 75 26 96 45 80 38 05 04 11 66 35 14

65 21 38 39 27 77 76 20 30 86 80 74 22 43 95 68 47 68 37 92
65 55 31 26 78 90 90 69 04 66 43 67 02 62 17 69 90 03 12 05
05 66 86 90 80 73 02 98 57 46 58 33 27 82 31 45 98 69 29 98
39 30 29 97 18 49 75 77 95 19 27 38 77 63 73 47 26 29 16 12
64 59 23 22 54 45 87 92 94 31 38 32 00 59 81 18 06 78 71 37
02 49 05 41 22 27 94 43 93 64 04 23 07 20 74 11 67 95 40 82
11 96 73 64 69 60 62 78 37 01 09 25 33 02 08 01 38 53 74 82
48 25 68 34 65 49 69 92 40 79 05 40 33 51 54 39 61 30 31 36
27 24 67 30 80 21 48 12 35 36 04 88 18 99 77 49 48 49 30 71
32 53 27 72 65 72 43 07 07 22 86 52 91 84 57 92 65 71 60 11

APPENDIX 2

Individual Binomial Probabilities

		π									
n	*r*	*.05*	*.10*	*.15*	*.20*	*.25*	*.30*	*.35*	*.40*	*.45*	*.50*
1	0	.9500	.9000	.8500	.8000	.7500	.7000	.6500	.6000	.5500	.5000
	1	.0500	.1000	.1500	.2000	.2500	.3000	.3500	.4000	.4500	.5000
2	0	.9025	.8100	.7225	.6400	.5625	.4900	.4225	.3600	.3025	.2500
	1	.0950	.1800	.2550	.3200	.3750	.4200	.4550	.4800	.4950	.5000
	2	.0025	.0100	.0225	.0400	.0625	.0900	.1225	.1600	.2025	.2500
3	0	.8574	.7290	.6141	.5120	.4219	.3430	.2746	.2160	.1664	.1250
	1	.1354	.2430	.3251	.3840	.4219	.4410	.4436	.4320	.4084	.3750
	2	.0071	.0270	.0574	.0960	.1406	.1890	.2389	.2880	.3341	.3750
	3	.0001	.0010	.0034	.0080	.0156	.0270	.0429	.0640	.0911	.1250
4	0	.8145	.6561	.5220	.4096	.3164	.2401	.1785	.1296	.0915	.0625
	1	.1715	.2916	.3685	.4096	.4219	.4116	.3845	.3456	.2995	.2500
	2	.0135	.0486	.0975	.1536	.2109	.2646	.3105	.3456	.3675	.3750
	3	.0005	.0036	.0115	.0256	.0469	.0756	.1115	.1536	.2005	.2500
	4	.0000	.0001	.0005	.0016	.0039	.0081	.0150	.0256	.0410	.0625
5	0	.7738	.5905	.4437	.3277	.2373	.1681	.1160	.0778	.0503	.0312
	1	.2036	.3280	.3915	.4096	.3955	.3602	.3124	.2592	.2059	.1562
	2	.0214	.0729	.1382	.2048	.2637	.3087	.3364	.3456	.3369	.3125
	3	.0011	.0081	.0244	.0512	.0879	.1323	.1811	.2304	.2757	.3125
	4	.0000	.0004	.0022	.0064	.0146	.0284	.0488	.0768	.1128	.1562
	5	.0000	.0000	.0001	.0003	.0010	.0024	.0053	.0102	.0185	.0312
6	0	.7351	.5314	.3771	.2621	.1780	.1176	.0754	.0467	.0277	.0156
	1	.2321	.3543	.3993	.3932	.3560	.3025	.2437	.1866	.1359	.0938
	2	.0305	.0984	.1762	.2458	.2966	.3241	.3280	.3110	.2780	.2344
	3	.0021	.0146	.0415	.0819	.1318	.1852	.2355	.2765	.3032	.3125
	4	.0001	.0012	.0055	.0154	.0330	.0595	.0951	.1382	.1861	.2344
	5	.0000	.0001	.0004	.0015	.0044	.0102	.0205	.0369	.0609	.0938
	6	.0000	.0000	.0000	.0001	.0002	.0007	.0018	.0041	.0083	.0516
7	0	.6983	.4783	.3206	.2097	.1335	.0824	.0490	.0280	.0152	.0078
	1	.2573	.3720	.3960	.3670	.3115	.2471	.1848	.1306	.0872	.0547
	2	.0406	.1240	.2097	.2753	.3115	.3177	.2985	.2613	.2140	.1641
	3	.0036	.0230	.0617	.1147	.1730	.2269	.2679	.2903	.2918	.2734
	4	.0002	.0026	.0109	.0287	.0577	.0972	.1442	.1935	.2388	.2734
	5	.0009	.0002	.0012	.0043	.0115	.0250	.0466	.0774	.1172	.1641
	6	.0000	.0000	.0001	.0004	.0013	.0036	.0084	.0172	.0320	.0547
	7	.0000	.0000	.0000	.0000	.0001	.0002	.0006	.0016	.0037	.0078

Appendix 2 (continued)

		π									
n	*r*	*.05*	*.10*	*.15*	*.20*	*.25*	*.30*	*.35*	*.40*	*.45*	*.50*
8	0	.6634	.4305	.2725	.1678	.1001	.0576	.0319	.0168	.0084	.0039
	1	.2793	.3826	.3847	.3355	.2670	.1977	.1373	.0896	.0548	.0312
	2	.0515	.1488	.2376	.2936	.3115	.2965	.2587	.2090	.1569	.1094
	3	.0054	.0331	.0839	.1468	.2076	.2541	.2786	.2787	.2568	.2188
	4	.0004	.0046	.0815	.0459	.0865	.1361	.1875	.2322	.2627	.2734
	5	.0000	.0004	.0026	.0092	.0231	.0467	.0808	.1239	.1719	.2188
	6	.0000	.0000	.0002	.0011	.0038	.0100	.0217	.0413	.0703	.1094
	7	.0000	.0000	.0000	.0001	.0004	.0012	.0033	.0079	.0164	.0312
	8	.0000	.0000	.0000	.0000	.0000	.0001	.0002	.0007	.0017	.0039
9	0	.6302	.3874	.2316	.1342	.0751	.0404	.0207	.0101	.0046	.0020
	1	.2985	.3874	.3679	.3020	.2253	.1556	.1004	.0605	.0339	.0176
	2	.0629	.1722	.2597	.3020	.3003	.2668	.2162	.1612	.1110	.0703
	3	.0077	.0446	.1069	.1762	.2336	.2668	.2716	.2508	.2119	.1641
	4	.0006	.0074	.0283	.0661	.1168	.1715	.2194	.2508	.2600	.2461
	5	.0000	.0008	.0050	.0165	.0389	.0735	.1181	.1672	.2128	.2461
	6	.0000	.0001	.0006	.0028	.0087	.0210	.0424	.0743	.1160	.1641
	7	.0000	.0000	.0000	.0003	.0012	.0039	.0098	.0212	.0407	.0703
	8	.0000	.0000	.0000	.0000	.0001	.0004	.0013	.0035	.0083	.0716
	9	.0000	.0000	.0000	.0000	.0000	.0000	.0001	.0003	.0008	.0020
10	0	.5987	.3487	.1969	.1074	.0563	.0282	.0135	.0060	.0025	.0010
	1	.3151	.3874	.3474	.2684	.1877	.1211	.0725	.0403	.0207	.0098
	2	.0746	.1937	.2759	.3020	.2816	.2335	.1757	.1209	.0763	.0439
	3	.0105	.0574	.1298	.2013	.2503	.2668	.2522	.2150	.1665	.1172
	4	.0010	.0112	.0401	.0881	.1460	.2001	.2377	.2508	.2384	.2051
	5	.0001	.0015	.0085	.0246	.0584	.1029	.1536	.2007	.2340	.2461
	6	.0000	.0001	.0012	.0055	.0162	.0368	.0689	.1115	.1596	.2051
	7	.0000	.0000	.0001	.0008	.0031	.0090	.0212	.0425	.0746	.1172
	8	.0000	.0000	.0000	.0001	.0004	.0014	.0043	.0106	.0229	.0439
	9	.0000	.0000	.0000	.0000	.0000	.0001	.0005	.0016	.0042	.0098
	10	.0000	.0000	.0000	.0000	.0000	.0000	.0000	.0001	.0003	.0010
20	0	.3585	.1216	.0388	.0115	.0032	.0008	.0002	.0000	.0000	.0000
	1	.3774	.2702	.1368	.0576	.0211	.0068	.0020	.0005	.0001	.0000
	2	.1887	.2852	.2293	.1369	.0669	.0278	.0100	.0031	.0008	.0002
	3	.0596	.1901	.2428	.2054	.1339	.0716	.0323	.0123	.0040	.0011
	4	.0133	.0898	.1821	.2182	.1897	.1304	.0738	.0350	.0139	.0046
	5	.0022	.0319	.1028	.1746	.2023	.1789	.1272	.0746	.0365	.0148
	6	.0003	.0089	.0454	.1091	.1686	.1916	.1712	.1244	.0746	.0370
	7	.0000	.0020	.0160	.0545	.1124	.1643	.1844	.1659	.1221	.0739

Appendix 2 (continued)

n	r	.05	.10	.15	.20	.25	.30	.35	.40	.45	.50
		π									
20	8	.0000	.0004	.0046	.0222	.0609	.1144	.1614	.1797	.1623	.1201
	9	.0000	.0001	.0011	.0074	.0271	.0654	.1158	.1597	.1771	.1602
	10	.0000	.0000	.0002	.0020	.0099	.0308	.0686	.1171	.1593	.1762
	11	.0000	.0000	.0000	.0005	.0030	.0120	.0336	.0710	.1185	.1602
	12	.0000	.0000	.0000	.0001	.0008	.0039	.0136	.0355	.0727	.1201
	13	.0000	.0000	.0000	.0000	.0002	.0010	.0045	.0146	.0366	.0739
	14	.0000	.0000	.0000	.0000	.0000	.0002	.0012	.0049	.0150	.0370
	15	.0000	.0000	.0000	.0000	.0000	.0000	.0003	.0013	.0049	.0148
	16	.0000	.0000	.0000	.0000	.0000	.0000	.0000	.0003	.0013	.0046
	17	.0000	.0000	.0000	.0000	.0000	.0000	.0000	.0000	.0002	.0011
	18	.0000	.0000	.0000	.0000	.0000	.0000	.0000	.0000	.0000	.0002
	19	.0000	.0000	.0000	.0000	.0000	.0000	.0000	.0000	.0000	.0000
	20	.0000	.0000	.0000	.0000	.0000	.0000	.0000	.0000	.0000	.0000

APPENDIX 3

Individual Poisson Probabilities

r	λ = .005	.01	.02	.03	.04	.05	.06	.07	.08	.09
0	.9950	.9900	.9802	.9704	.9608	.9512	.9418	.9324	.9231	.9139
1	.0050	.0099	.0192	.0291	.0384	.0476	.0565	.0653	.0738	.0823
2	.0000	.0000	.0002	.0004	.0008	.0012	.0017	.0023	.0030	.0037
3	.0000	.0000	.0000	.0000	.0000	.0000	.0000	.0001	.0001	.0001

r	λ = 0.1	0.2	0.3	0.4	0.5	0.6	0.7	0.8	0.9	1.0
0	.9048	.8187	.7408	.6703	.6065	.5488	.4966	.4493	.4066	.3679
1	.0905	.1637	.2222	.2681	.3033	.3293	.3476	.3595	.3659	.3679
2	.0045	.0164	.0333	.0536	.0758	.0988	.1217	.1438	.1647	.1839
3	.0002	.0011	.0033	.0072	.0126	.0198	.0284	.0383	.0494	.0613
4	.0000	.0001	.0002	.0007	.0016	.0030	.0050	.0077	.0111	.0153
5	.0000	.0000	.0000	.0001	.0002	.0004	.0007	.0012	.0020	.0031
6	.0000	.0000	.0000	.0000	.0000	.0000	.0001	.0002	.0003	.0005
7	.0000	.0000	.0000	.0000	.0000	.0000	.0000	.0000	.0000	.0001

r	λ = 1.1	1.2	1.3	1.4	1.5	1.6	1.7	1.8	1.9	2.0
0	.3329	.3012	.2725	.2466	.2231	.2019	.1827	.1653	.1496	.1353
1	.3662	.3614	.3543	.3452	.3347	.3230	.3106	.2975	.2842	.2707
2	.2014	.2169	.2303	.2417	.2510	.2584	.2640	.2678	.2700	.2707
3	.0738	.0867	.0998	.1128	.1255	.1378	.1496	.1607	.1710	.1804
4	.0203	.0260	.0324	.0395	.0471	.0551	.0636	.0723	.0812	.0902
5	.0045	.0062	.0084	.0111	.0141	.0176	.0216	.0260	.0309	.0361
6	.0008	.0012	.0018	.0026	.0035	.0047	.0061	.0078	.0098	.0120
7	.0001	.0002	.0003	.0005	.0008	.0011	.0015	.0020	.0027	.0034
8	.0000	.0000	.0001	.0001	.0001	.0002	.0003	.0005	.0006	.0009
9	.0000	.0000	.0000	.0000	.0000	.0000	.0001	.0001	.0001	.0002

r	λ = 2.1	2.2	2.3	2.4	2.5	2.6	2.7	2.8	2.9	3.0
0	.1225	.1108	.1003	.0907	.0821	.0743	.0672	.0608	.0550	.0498
1	.2572	.2438	.2306	.2177	.2052	.1931	.1815	.1703	.1596	.1494
2	.2700	.2681	.2652	.2613	.2565	.2510	.2450	.2384	.2314	.2240
3	.1890	.1966	.2033	.2090	.2138	.2176	.2205	.2225	.2237	.2240
4	.0992	.1082	.1169	.1254	.1336	.1414	.1488	.1557	.1622	.1680

Appendix 3 (continued)

r	λ 2.1	2.2	2.3	2.4	2.5	2.6	2.7	2.8	2.9	3.0
5	.0417	.0476	.0538	.0602	.0668	.0735	.0804	.0872	.0940	.1008
6	.0146	.0174	.0206	.0241	.0278	.0319	.0362	.0407	.0455	.0504
7	.0044	.0055	.0068	.0083	.0099	.0118	.0139	.0163	.0188	.0216
8	.0011	.0015	.0019	.0025	.0031	.0038	.0047	.0057	.0068	.0081
9	.0003	.0004	.0005	.0007	.0009	.0011	.0014	.0018	.0022	.0027
10	.0001	.0001	.0001	.0002	.0002	.0003	.0004	.0005	.0006	.0008
11	.0000	.0000	.0000	.0000	.0000	.0001	.0001	.0001	.0002	.0002
12	.0000	.0000	.0000	.0000	.0000	.0000	.0000	.0000	.0000	.0001

r	λ 3.1	3.2	3.3	3.4	3.5	3.6	3.7	3.8	3.9	4.0
0	.0450	.0408	.0369	.0334	.0302	.0273	.0247	.0224	.0202	.0183
1	.1397	.1304	.1217	.1135	.1057	.0984	.0915	.0850	.0789	.0733
2	.2165	.2087	.2008	.1929	.1850	.1771	.1692	.1615	.1539	.1465
3	.2237	.2226	.2209	.2186	.2158	.2125	.2087	.2046	.2001	.1954
4	.1734	.1781	.1823	.1858	.1888	.1912	.1931	.1944	.1951	.1954
5	.1075	.1140	.1203	.1264	.1322	.1377	.1429	.1477	.1522	.1563
6	.0555	.0608	.0662	.0716	.0771	.0826	.0881	.0936	.0989	.1042
7	.0246	.0278	.0312	.0348	.0385	.0425	.0466	.0508	.0551	.0595
8	.0095	.0111	.0129	.0148	.0169	.0191	.0215	.0241	.0269	.0298
9	.0033	.0040	.0047	.0056	.0066	.0076	.0089	.0102	.0116	.0132
10	.0010	.0013	.0016	.0019	.0023	.0028	.0033	.0039	.0045	.0053
11	.0003	.0004	.0005	.0006	.0007	.0009	.0011	.0013	.0016	.0019
12	.0001	.0001	.0001	.0002	.0002	.0003	.0003	.0004	.0005	.0006
13	.0000	.0000	.0000	.0000	.0001	.0001	.0001	.0001	.0002	.0002
14	.0000	.0000	.0000	.0000	.0000	.0000	.0000	.0000	.0000	.0001

	λ 4.1	4.2	4.3	4.4	4.5	4.6	4.7	4.8	4.9	5.0
0	.0166	.0150	.0136	.0123	.0111	.0101	.0091	.0082	.0074	.0067
1	.0679	.0630	.0583	.0540	.0500	.0462	.0427	.0395	.0365	.0337
2	.1393	.1323	.1254	.1188	.1125	.1063	.1005	.0948	.0894	.0842
3	.1904	.1852	.1798	.1743	.1687	.1631	.1574	.1517	.1460	.1404
4	.1951	.1944	.1933	.1917	.1898	.1875	.1849	.1820	.1789	.1755
5	.1600	.1633	.1662	.1687	.1708	.1725	.1738	.1747	.1753	.1755
6	.1093	.1143	.1191	.1237	.1281	.1323	.1362	.1398	.1432	.1462
7	.0640	.0686	.0732	.0778	.0824	.0869	.0914	.0959	.1002	.1044
8	.0328	.0360	.0393	.0428	.0463	.0500	.0537	.0575	.0614	.0653
9	.0150	.0168	.0188	.0209	.0232	.0255	.0280	.0307	.0334	.0363

Appendix 3 (continued)

r	λ									
	4.1	*4.2*	*4.3*	*4.4*	*4.5*	*4.6*	*4.7*	*4.8*	*4.9*	*5.0*
10	.0061	.0071	.0081	.0092	.0104	.0118	.0132	.0147	.0164	.0181
11	.0023	.0027	.0032	.0037	.0043	.0049	.0056	.0064	.0073	.0082
12	.0008	.0009	.0011	.0014	.0016	.0019	.0022	.0026	.0030	.0034
13	.0002	.0003	.0004	.0005	.0006	.0007	.0008	.0009	.0011	.0013
14	.0001	.0001	.0001	.0001	.0002	.0002	.0003	.0003	.0004	.0005
15	.0000	.0000	.0000	.0000	.0001	.0001	.0001	.0001	.00001	.0002

APPENDIX 4

Areas in the Tail of the Normal Distribution

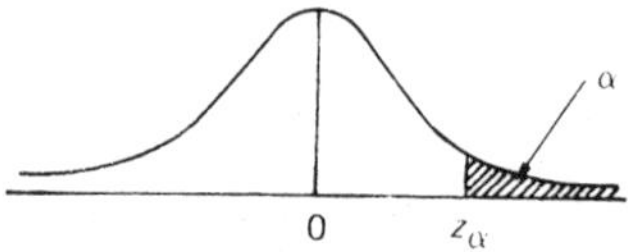

z	.00	.01	.02	.03	.04	.05	.06	.07	.08	.09
0.0	.5000	.4960	.4920	.4880	.4840	.4801	.4761	.4721	.4681	.4641
0.1	.4602	.4562	.4522	.4483	.4443	.4404	.4364	.4325	.4286	.4247
0.2	.4207	.4186	.4129	.4090	.4052	.4013	.3974	.3936	.3897	.3859
0.3	.3821	.3783	.3745	.3707	.3669	.3632	.3594	.3557	.3520	.3483
0.4	.3446	.3409	.3372	.3336	.3300	.3264	.3228	.3192	.3156	.3121
0.5	.3085	.3050	.3015	.2981	.2946	.2912	.2877	.2843	.2810	.2776
0.6	.2743	.2709	.2676	.2643	.2611	.2578	.2546	.2514	.2483	.2451
0.7	.2420	.2389	.2358	.2327	.2296	.2266	.2236	.2206	.2177	.2148
0.8	.2119	.2090	.2061	.2033	.2005	.1977	.1949	.1922	.1894	.1867
0.9	.1841	.1814	.1788	.1762	.1736	.1711	.1685	.1660	.1635	.1611
1.0	.1587	.1562	.1539	.1515	.1492	.1469	.1446	.1423	.1401	.1379
1.1	.1357	.1335	.1314	.1292	.1271	.1251	.1230	.1210	.1190	.1170
1.2	.1151	.1131	.1112	.1093	.1075	.1056	.1038	.1020	.1003	.0985
1.3	.0968	.0951	.0934	.0918	.0901	.0885	.0869	.0853	.0838	.0823
1.4	.0808	.0793	.0778	.0764	.0749	.0735	.0721	.0708	.0694	.0681
1.5	.0668	.0655	.0643	.0630	.0618	.0606	.0594	.0582	.0571	.0559
1.6	.0548	.0537	.0526	.0516	.0505	.0495	.0485	.0475	.0465	.0455
1.7	.0446	.0436	.0427	.0418	.0409	.0401	.0392	.0384	.0375	.0367
1.8	.0359	.0351	.0344	.0336	.0329	.0322	.0314	.0307	.0301	.0294
1.9	.0287	.0281	.0274	.0268	.0262	.0256	.0250	.0244	.0239	.0233
2.0	.0228	.0222	.0217	.0212	.0207	.0202	.0197	.0192	.0188	.0183
2.1	.0179	.0174	.0170	.0166	.0162	.0158	.0154	.0150	.0146	.0143
2.2	.0139	.0136	.0132	.0129	.0125	.0122	.0119	.0116	.0113	.0110
2.3	.0107	.0104	.0102	.0099	.0096	.0094	.0091	.0089	.0087	.0084
2.4	.0082	.0080	.0078	.0075	.0073	.0071	.0069	.0068	.0066	.0064
2.5	.0062	.0060	.0059	.0057	.0055	.0054	.0052	.0051	.0049	.0048
2.6	.0047	.0045	.0044	.0043	.0041	.0040	.0039	.0038	.0037	.0036
2.7	.0035	.0034	.0033	.0032	.0031	.0030	.0029	.0028	.0027	.0026
2.8	.0026	.0025	.0024	.0023	.0023	.0022	.0021	.0021	.0020	.0019
2.9	.0019	.0018	.0018	.0017	.0016	.0016	.0015	.0015	.0014	.0014
3.0	.0013	.0013	.0013	.0012	.0012	.0011	.0011	.0011	.0010	.0010
3.1	.0010									
3.2	.0007									
3.3	.0005									
3.4	.0003									
3.5	.0002									

APPENDIX 5

Critical Values of the χ^2 Distribution

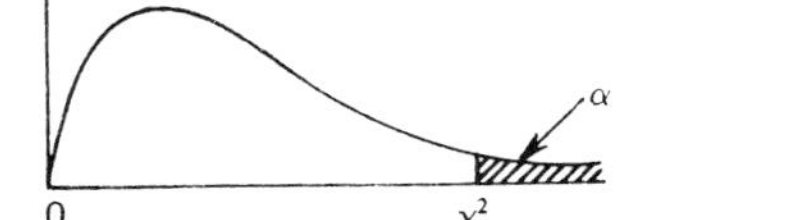

ν \ α	.995	.99	.975	.95	.90	.75	.50	.25	.10	..05	.025	.01	.005	α / ν
1	0.0000	0.0002	0.0010	0.004	0.016	0.102	0.455	1.323	2.71	3.84	5.02	6.63	7.88	1
2	0.0100	0.0201	0.0506	0.103	0.211	0.575	1.386	2.77	4.61	5.99	7.38	9.21	10.60	2
3	0.0717	0.115	0.216	0.352	0.584	1.213	2.37	4.11	6.25	7.81	9.35	11.34	12.84	3
4	0.207	0.297	0.484	0.711	1.064	1.923	3.36	5.39	7.78	9.49	11.14	13.28	14.86	4
5	0.412	0.554	0.831	1.145	1.610	2.67	4.35	6.63	9.24	11.07	12.83	15.09	16.75	5
6	0.676	0.872	1.237	1.635	2.20	3.45	5.35	7.84	10.64	12.59	14.45	16.81	18.55	6
7	0.989	1.239	1.690	2.17	2.83	4.25	6.35	9.04	12.02	14.07	16.01	18.48	20.3	7
8	1.344	1.646	2.18	2.73	3.49	5.07	7.34	10.22	13.36	15.51	17.53	20.1	22.0	8
9	1.735	2.09	2.70	3.33	4.17	5.90	8.34	11.39	14.68	16.92	19.02	21.7	23.6	9
10	2.16	2.56	3.25	3.94	4.87	6.74	9.34	12.55	15.99	18.31	20.5	23.2	25.2	10
11	2.60	3.05	3.82	4.57	5.58	7.58	10.34	13.70	17.28	19.68	21.9	24.7	26.8	11
12	3.07	3.57	4.40	5.23	6.30	8.44	11.34	14.85	18.55	21.0	23.3	26.2	28.3	12
13	3.57	4.11	5.01	5.89	7.04	9.30	12.34	15.98	19.81	22.4	24.7	27.7	29.8	13
14	4.07	4.66	5.63	6.57	7.79	10.17	13.34	17.12	21.1	23.7	26.1	29.1	31.3	14
15	4.60	5.23	6.26	7.26	8.55	11.04	14.34	18.25	22.3	25.0	27.5	30.6	32.8	15

Appendix 5 (continued)

ν \ α	.995	.99	.975	.95	.90	.75	.50	.25	.10	.05	.025	.01	.005	α / ν
16	5.14	5.81	6.91	7.96	9.31	11.91	15.34	19.37	23.5	26.3	28.8	32.0	34.3	16
17	5.70	6.41	7.56	8.67	10.09	12.79	16.34	20.5	24.8	27.6	30.2	33.4	35.7	17
18	6.26	7.01	8.23	9.39	10.86	13.68	17.34	21.6	26.0	28.9	31.5	34.8	37.2	18
19	6.84	7.63	8.91	10.12	11.65	14.56	18.34	22.7	27.2	30.1	32.9	36.2	38.6	19
20	7.43	8.26	9.59	10.85	12.44	15.45	19.34	23.8	28.4	31.4	34.2	37.6	40.0	20
21	8.03	8.90	10.28	11.59	13.24	16.34	20.3	24.9	29.6	32.7	35.5	38.9	41.4	21
22	8.64	9.54	10.98	12.34	14.04	17.24	21.3	26.0	30.8	33.9	36.8	40.3	42.8	22
23	9.26	10.20	11.69	13.09	14.85	18.14	22.3	27.1	32.0	35.2	38.1	41.6	44.2	23
24	9.89	10.86	12.40	13.85	15.66	19.04	23.3	28.2	33.2	36.4	39.4	43.0	45.6	24
25	10.52	11.52	13.12	14.61	16.47	19.94	24.3	29.3	34.4	37.7	40.6	44.3	46.9	25
26	11.16	12.20	13.84	15.38	17.29	20.8	25.3	30.4	35.6	38.9	41.9	45.6	48.3	26
27	11.81	12.88	14.57	16.15	18.11	21.7	26.3	31.5	36.7	40.1	43.2	47.0	49.6	27
28	12.46	13.56	15.31	16.93	18.94	22.7	27.3	32.6	37.9	41.3	44.5	48.3	51.0	28
29	13.12	14.26	16.05	17.71	19.77	23.6	28.3	33.7	39.1	42.6	45.7	49.6	52.3	29
30	13.79	14.95	16.79	18.49	20.6	24.5	29.3	34.8	40.3	43.8	47.0	50.9	53.7	30
40	20.7	22.2	24.4	26.5	29.1	33.7	39.3	45.6	51.8	55.8	59.3	63.7	66.8	40
50	28.0	29.7	32.4	34.8	37.7	42.9	49.3	56.3	63.2	67.5	71.4	76.2	79.5	50
60	35.5	37.5	40.5	43.2	46.5	52.3	59.3	67.0	74.4	79.1	83.3	88.4	92.0	60
70	43.3	45.4	48.8	51.7	55.3	61.7	69.3	77.6	85.5	90.5	95.0	100.4	104.2	70
80	51.2	53.5	57.2	60.4	64.3	71.1	79.3	88.1	96.6	101.9	106.6	112.3	116.3	80
90	59.2	61.8	65.6	69.1	73.3	80.6	89.3	98.6	107.6	113.1	118.1	124.1	128.3	90
100	67.3	70.1	74.2	77.9	82.4	90.1	99.3	109.1	118.5	124.3	129.6	135.8	140.2	100

APPENDIX 6

Critical Values of the t-Distribution

One-Sided Test: α, 0, t_α

Two-Sided Test: $\alpha/2$, $-t_{\alpha/2}$, 0, $t_{\alpha/2}$, $\alpha/2$

	α (one-sided) or α/2 (two-sided)	*.25*	*.10*	*.05*	*.025*	*.01*	*.005*
	$\nu = 1$	1.000	3.078	6.314	12.706	31.821	63.657
	2	.816	1.886	2.920	4.303	6.965	9.925
	3	.765	1.638	2.353	3.182	4.541	5.841
	4	.741	1.533	2.132	2.776	3.747	4.604
	5	.727	1.476	2.015	2.571	3.365	4.032
	6	.718	1.440	1.943	2.447	3.143	3.707
	7	.711	1.415	1.895	2.365	2.998	3.499
	8	.706	1.397	1.860	2.306	2.896	3.355
	9	.703	1.383	1.833	2.262	2.821	3.250
	10	.700	1.372	1.812	2.228	2.764	3.169
	11	.697	1.363	1.796	2.201	2.718	3.106
	12	.695	1.356	1.782	2.179	2.681	3.055
	13	.694	1.350	1.771	2.160	2.650	3.102
	14	.692	1.345	1.761	2.145	2.624	2.977
Degrees of Freedom	15	.691	1.341	1.753	2.131	2.602	2.947
	16	.690	1.337	1.746	2.120	2.583	2.921
	17	.689	1.333	1.740	2.110	2.567	2.898
	18	.688	1.330	1.734	2.101	2.552	2.878
	19	.688	1.328	1.729	2.093	2.539	2.861
	20	.687	1.325	1.725	2.086	2.528	2.845
	21	.686	1.323	1.721	2.080	2.518	2.831
	22	.686	1.321	1.717	2.074	2.508	2.819
	23	.685	1.319	1.714	2.069	2.500	2.807
	24	.685	1.318	1.711	2.064	2.492	2.397
	25	.684	1.316	1.708	2.060	2.485	2.787
	26	.684	1.315	1.706	2.056	2.479	2.779
	27	.684	1.314	1.703	2.052	2.473	2.771
	28	.683	1.313	1.701	2.048	2.467	2.763
	29	.683	1.311	1.699	2.045	2.462	2.756
	30	.683	1.310	1.697	2.042	2.457	2.750
	40	.681	1.303	1.684	2.021	2.423	2.704
	60	.679	1.296	1.671	2.000	2.390	2.660
	120	.677	1.289	1.658	1.980	2.358	2.617
	∞	.674	1.282	1.645	1.960	2.326	2.576

APPENDIX 7

Critical Values of the *F*-Distribution

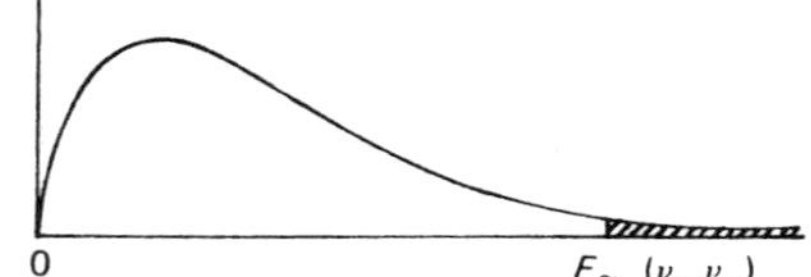

Upper figures give $F_{0.05}$ (5% points) and lower bold face figures, $F_{0.01}$ (1% points).

	ν_2	*ν_1 degrees of freedom (for larger variance estimate)*												ν_2
		1	*2*	*3*	*4*	*5*	*6*	*7*	*8*	*10*	*12*	*24*	∞	
Degrees of Freedom of Smaller Variance Estimate	1	161 **4,052**	200 **4,999**	216 **5,403**	225 **5,625**	230 **5,764**	234 **5,859**	237 **5,928**	239 **5,981**	242 **6,056**	244 **6,106**	249 **6,234**	254 **6,366**	1
	2	18.51 **98.49**	19.00 **99.00**	19.16 **99.17**	19.25 **99.25**	19.30 **99.30**	19.33 **99.33**	19.36 **99.34**	19.37 **99.36**	19.39 **99.40**	19.41 **99.42**	19.45 **99.46**	19.50 **99.50**	2
	3	10.13 **34.12**	9.55 **30.82**	9.28 **29.46**	9.12 **28.71**	9.01 **28.24**	8.94 **27.91**	8.88 **27.67**	8.84 **29.49**	8.78 **27.23**	8.74 **27.05**	8.64 **26.60**	8.53 **26.12**	3
	4	7.71 **21.20**	6.94 **18.00**	6.59 **16.69**	6.39 **15.98**	6.26 **15.52**	6.16 **15.21**	6.09 **14.98**	6.04 **14.80**	5.96 **14.54**	5.91 **14.37**	5.77 **13.93**	5.63 **13.46**	4

	5	6.61 **16.26**	5.79 **13.27**	5.41 **12.06**	5.19 **11.39**	5.05 **10.97**	4.95 **10.67**	4.88 **10.45**	4.82 **10.27**	4.74 **10.05**	4.68 **9.89**	4.53 **9.47**	4.36 **9.02**	5
	6	5.99 **13.74**	5.14 **10.92**	4.76 **9.78**	4.53 **9.15**	4.39 **8.75**	4.28 **8.47**	4.21 **8.26**	4.15 **8.10**	4.06 **7.87**	4.00 **7.72**	3.84 **7.31**	3.67 **6.88**	6
	7	5.59 **12.25**	4.74 **9.55**	4.35 **8.45**	4.12 **7.85**	3.97 **7.46**	3.87 **7.19**	3.79 **7.00**	3.73 **6.84**	3.63 **6.62**	3.57 **6.47**	3.41 **6.07**	3.23 **5.65**	7
	8	5.32 **11.26**	4.46 **8.65**	4.07 **7.59**	3.84 **7.01**	3.69 **6.63**	3.58 **6.37**	3.50 **6.19**	3.44 **6.03**	3.34 **5.82**	3.28 **5.67**	3.12 **5.28**	2.93 **4.86**	8
Degrees of Freedom of Smaller Variance Estimate	9	5.12 **10.56**	4.26 **8.02**	3.86 **6.99**	3.63 **6.42**	3.48 **6.06**	3.37 **5.80**	3.29 **5.62**	3.23 **5.47**	3.13 **5.26**	3.07 **5.11**	2.90 **4.73**	2.71 **4.31**	9
	10	4.96 **10.04**	4.10 **7.56**	3.71 **6.55**	3.48 **5.99**	3.33 **5.64**	3.22 **5.39**	3.14 **5.21**	3.07 **5.06**	2.97 **4.85**	2.91 **4.71**	2.74 **4.33**	2.54 **3.91**	10
	11	4.84 **9.65**	3.98 **7.20**	3.59 **6.22**	3.36 **5.67**	3.20 **5.32**	3.09 **5.07**	3.01 **4.88**	2.95 **4.74**	2.86 **4.54**	2.79 **4.40**	2.61 **4.02**	2.40 **3.60**	11
	12	4.75 **9.33**	3.88 **6.93**	3.49 **5.95**	3.26 **5.41**	3.11 **5.06**	3.00 **4.82**	2.92 **4.65**	2.85 **4.50**	2.76 **4.30**	2.69 **4.16**	2.50 **3.78**	2.30 **3.36**	12
	13	4.67 **9.07**	3.80 **6.70**	3.41 **5.74**	3.18 **5.20**	3.02 **4.86**	2.92 **4.62**	2.84 **4.44**	2.77 **4.30**	2.67 **4.10**	2.60 **3.96**	2.42 **3.59**	2.21 **3.16**	13
	14	4.60 **8.86**	3.74 **6.51**	3.24 **5.56**	3.11 **5.03**	2.96 **4.69**	2.85 **4.46**	2.77 **4.28**	2.70 **4.14**	2.60 **3.94**	2.53 **3.80**	2.35 **3.43**	2.13 **3.00**	14

Appendix 7 (continued)

	ν_2	ν_1 degrees of freedom (for larger variance estimate)												ν_2
		1	*2*	*3*	*4*	*5*	*6*	*7*	*8*	*10*	*12*	*24*	∞	
	15	4.54 **8.68**	3.68 **6.36**	3.29 **5.42**	3.06 **4.89**	2.90 **4.56**	2.79 **4.32**	2.70 **4.14**	2.64 **4.00**	2.55 **3.80**	2.48 **3.67**	2.29 **3.29**	2.07 **2.87**	15
	16	4.49 **8.53**	3.63 **6.23**	3.24 **5.29**	3.01 **4.77**	2.85 **4.44**	2.74 **4.20**	2.66 **4.03**	2.59 **3.89**	2.49 **3.69**	2.42 **3.55**	2.24 **3.18**	2.01 **2.75**	16
	17	4.45 **8.40**	3.59 **6.11**	3.20 **5.18**	2.96 **4.67**	2.81 **4.34**	2.70 **4.10**	2.62 **3.93**	2.55 **3.79**	2.45 **3.59**	2.38 **3.45**	2.19 **3.08**	1.96 **2.65**	17
Degrees of Freedom of Smaller Variance Estimate	18	4.41 **8.28**	3.55 **6.01**	3.16 **5.09**	2.93 **4.58**	2.77 **4.25**	2.66 **4.01**	2.58 **3.85**	2.51 **3.71**	2.41 **3.51**	2.34 **3.37**	2.15 **3.00**	1.92 **2.57**	18
	19	4.38 **8.18**	3.52 **5.93**	3.13 **5.01**	2.90 **4.50**	2.74 **4.17**	2.63 **3.94**	2.55 **3.77**	2.48 **3.63**	2.38 **3.43**	2.31 **3.30**	2.11 **2.92**	2.88 **2.49**	19
	20	4.35 **8.10**	3.49 **5.85**	3.10 **4.94**	2.87 **4.43**	2.71 **4.10**	2.60 **3.87**	2.52 **3.71**	2.45 **3.56**	2.35 **3.37**	2.28 **3.23**	2.08 **2.86**	1.84 **2.42**	20
	21	4.32 **8.02**	3.47 **5.78**	3.07 **4.87**	2.84 **4.37**	2.68 **4.04**	2.57 **3.81**	2.49 **3.65**	2.42 **3.51**	2.32 **3.31**	2.25 **3.17**	2.05 **2.80**	1.81 **2.36**	21
	22	4.30 **7.94**	3.44 **5.72**	3.05 **4.82**	2.82 **4.31**	2.66 **3.99**	2.55 **3.76**	2.47 **3.59**	2.40 **3.45**	2.30 **3.26**	3.23 **3.12**	2.03 **2.75**	1.78 **2.31**	22
	23	4.28 **7.88**	3.42 **5.66**	3.03 **4.76**	2.80 **4.26**	2.64 **3.94**	2.53 **3.71**	2.45 **3.54**	2.38 **3.41**	2.28 **3.21**	2.20 **3.07**	2.00 **2.70**	1.76 **2.26**	23

24	4.26 **7.82**	3.40 **5.61**	3.01 **4.72**	2.78 **4.22**	2.62 **3.90**	2.51 **3.67**	2.43 **3.50**	2.36 **3.36**	2.26 **3.17**	2.18 **3.03**	1.98 **2.66**	1.73 **2.21**	24
25	4.24 **7.77**	3.38 **5.57**	2.99 **4.68**	2.76 **4.18**	2.60 **3.86**	2.49 **3.63**	2.41 **3.46**	2.34 **3.32**	2.24 **3.13**	2.16 **2.99**	1.96 **2.62**	1.71 **2.17**	25
26	4.22 **7.72**	3.37 **5.53**	2.98 **4.64**	2.74 **4.14**	2.59 **3.82**	2.47 **3.59**	2.39 **3.42**	2.32 **3.29**	2.22 **3.09**	2.15 **2.96**	1.95 **2.58**	1.69 **2.13**	26

APPENDIX 8

Greek and Other Mathematical Notation

Greek letter	English name	Pronunciation
α	alpha	alpha
β	beta	beeta
γ	gamma	gamma
θ	theta	theeta
λ	lambda	lambda
μ	mu	mew
ν	nu	new
π	pi	pie
ρ	rho	roe
σ, Σ (capital)	sigma	sigma
χ	chi	ki
ω, Ω (capital)	omega	omega

Symbol	Meaning
$X = a$	X equals a
$X \neq a$	X is not equal to a
$X \simeq a$	X is approximately equal to a
$X < a$	X is less than a
$X > a$	X is greater than a
$X \leqslant a$	X is less than or equal to a
$X \geqslant a$	X is greater than or equal to a
$b < X < a$	X lies between a and b
$b \leqslant X \leqslant a$	X lies between a and b (inclusive)
$n!$	n-factorial $= n \times (n-1) \times (n-2) \times \ldots \times 2 \times 1$

INDEX

Roman and Greek letters, used as mathematical symbols, are included in this index. Greek letters will be found indexed at or near the beginning of the alphabetic section appropriate to their English names (see Appendix 8 on facing page). E.g. μ is indexed under M (for mu).

An "n" following a page number indicates that the information is contained within a footnote on that page.

E

P

S

T